Inorganic Infrared and Raman Spectra

Consulting Editor

P Sykes, MSc, PhD

Fellow of Christ's College
University of Cambridge

Other titles in the European Chemistry Series

Avery: The Quantum Theory of Atoms, Molecules, and Photons
Banwell: Fundamentals of Molecular Spectroscopy
Barnard: Theoretical Basis of Inorganic Chemistry
Barnard and Chayen: Modern Methods of Chemical Analysis
Barnard and Mansell: Fundamentals of Physical Chemistry
Benson: Mechanisms of Inorganic Reactions in Solution
Brennan and Tipper: A Laboratory Manual of Experiments in Physical Chemistry
Bu'lock: The Biosynthesis of Natural Products
Chapman: Introduction to Lipids
Cox and Kemp: Introductory Photochemistry
Garratt: Aromaticity. An Introduction to Aromatic Compounds
Gimblett: Introduction to the Kinetics of Chemical Chain Reactions
Hallas: Organic Steriochemistry
Kemp: Practical Organic Chemistry
Kemp: Qualitative Organic Analysis
Krebs: Fundamentals of Inorganic Crystal Chemistry
Sutton: Electronic Spectra of Transition Metal Complexes
Williams: Introduction to the Chemistry of Enzyme Action
Williams and Fleming: Spectroscopic Methods in Organic Chemistry
Williams and Fleming: Spectroscopic Problems in Organic Chemistry

Inorganic Infrared and Raman Spectra

S D Ross
Lecturer in Physical Chemistry,
Chelsea College of Science and Technology,
London University.

London · New York · St Louis · San Francisco · Düsseldorf
Johannesburg · Kuala Lumpur · Mexico · Montreal · New Delhi
Panama · Paris · Rio de Janeiro · Singapore · Sydney · Toronto

Published by
McGRAW-HILL Book Company (UK) Limited
MAIDENHEAD · BERKSHIRE · ENGLAND

07 094179 3

PRINTED AND BOUND IN GREAT BRITAIN

Contents

Preface

The aim of this book is twofold. Firstly, it is intended to provide research workers who employ vibrational spectroscopy with the necessary theory to enable them to predict vibrational selection rules and to calculate vibrational frequencies both for internal (or molecular) modes and for external (or lattice) modes. Secondly, it collects together infrared and Raman spectroscopic data on inorganic compounds. The original literature search covered the period 1945–68; during the production of the book, the search was extended to the end of 1970, and the text has been rewritten in some places to take account of the more important new information. Studies of the vibrational spectra of inorganic compounds have been mainly directed to one of four ends. Many of the earlier papers, of which that by Miller and Wilkins [957] is an example, were analytical in bias, in that they were concerned with the identification of compounds rather than the interpretation of their spectra. All such studies fall outside the scope of this work, except that the appropriate references to the compounds studied may be found in the formula index and bibliography. A second field, which is largely excluded from consideration, is the study of bonding in metal complexes. A considerable part of the book by Nakamoto [13] deals with such studies, as does Adams' more recent book, *Metal-Ligand and Related Vibrations* [1]. Such studies of metal complexes as are mentioned in this book are considered mainly in relation to the skeletal vibrations of the system, as, for instance, the vibrations of the system $Co^{III}N_4O_2$ in a complex such as $Co^{III}(NH_3)_4CO_3Cl$. The material which is collected, therefore, deals with the other two fields of study, namely, the deductions about molecular symmetry which can be made from spectroscopic studies on the free molecule, and the way in which violations of the free-molecule selection rules in the solid (or occasionally the liquid) state can give information about the structure of these condensed phases. This bias, which is that of my own research interests, necessitates an exposition of the theory which differs in many ways from that found in existing books. The theory of normal vibrations set out by Herzberg [9] remained the accepted norm of treatment for some years. If one compares Herzberg's approach with that of (say) Nakamoto, nearly twenty years later, two particularly striking differences become apparent. The first is in the determination of the number of vibrations of a given symmetry species. Herzberg's treatment, which is followed in some recent books, such as that by Jaffe and Orchin [10], is couched in terms of the degrees of freedom contributed by a nucleus to a vibration of given symmetry type. Most of the books which were published subsequent to *Molecular Vibrations* by Wilson, Decius and Cross [18] use the methods of group theory to solve this problem. The second difference is in the treatment of the solution of the secular determinant. Herzberg deals with the solution in terms of symmetry coordinates, so that the secular determinant is factored into blocks corresponding to the symmetry species, but these blocks have the latent root $\lambda = 4\pi^2\nu^2c$ in every term instead of in the diagonal terms only. For instance, in his treatment of nonlinear XY_2, the potential energy $2V$ and the kinetic energy $2T$ are given by

$$\begin{aligned} 2V &= c_{11}S_1^2 + 2c_{12}S_1S_2 + c_{22}S_2^2 + c_{33}S_3^2, \\ 2T &= d_{11}\dot{S}_1^2 + 2d_{12}\dot{S}_1\dot{S}_2 + d_{22}\dot{S}_2^2 + d_{33}\dot{S}_3^2, \end{aligned}$$

where S_1, S_2 are symmetry coordinates of species A_1 and S_3 is of species B_1. The solution for the A_1 block comes out as

$$\begin{vmatrix} c_{11} - \lambda d_{11} & c_{12} - \lambda d_{12} \\ c_{21} - \lambda d_{21} & c_{22} - \lambda d_{22} \end{vmatrix} = 0.$$

On the other hand, the almost universally used Wilson *FG*-matrix method [1516, 1517] would give the corresponding determinant as

$$\begin{vmatrix} F_{11}G_{11} + F_{21}G_{12} - \lambda & G_{11}F_{12} + G_{12}F_{22} \\ G_{21}F_{11} + G_{22}F_{21} & G_{21}F_{12} + F_{22}G_{22} - \lambda \end{vmatrix} = 0,$$

and a solution is much easier to arrive at with the determinant in this form. Further, the potential fields used in Herzberg are not very realistic; only the central-force and simple valence-force fields are dealt with in detail, though the Urey–Bradley field in its simplest form is mentioned. The first book in which more general force fields are discussed is that by Wilson, Decius and Cross referred to above; they discussed the Urey–Bradley field and the orbital valence force field due to Heath and Linnett [626] but in their sample vibrational analysis they used the most general valence-force field. The vibrational analysis of $CHCl_3$ given by Colthup, Daly and Wiberley in their book *Introduction to Infra-Red and Raman Spectroscopy* [5] also used this field; their treatment of this normal-coordinate analysis, based on the papers by Meister and Cleveland [938, 939], is very clear. Nakamoto [13] has carried out a normal-coordinate treatment of H_2O using a general valence-force field, and in Appendix III to his book has set out a normal-coordinate treatment of a copper complex with acetylacetone, in which a Urey–Bradley force field is employed. A recent book by Ferraro and Ziomek [7] deals with the group-theoretical approach to selection rules in isolated molecules and gives normal-coordinate treatments of H_2O, NH_3, CH_4 and UF_6, using the general valence-force field in each case. The secular equations for these and some other simple systems are given, without derivation, in the recent book by Siebert [15]. More recently, Gans [7a] has dealt admirably and in considerable detail with force constant calculations. The feature common to all these treatments is that they are confined to the vibrations of discrete free molecules. However, for effective studies of the solid state, a rather more extensive body of theory is required. The differences between the spectra of ions in solution and in the crystal lattice have been observed and attributed to interactions in the lattice for many years, and the appropriate theories for predicting the effect on vibrational selection rules were developed in 1939 by Bhagavantam and Venkatarayudu [144] and in 1946 by Halford [606]. These two treatments, which are known respectively as factor-group analysis and site-group analysis, have been dealt with in only one previous book [2], though some review articles exist [973, 975]. In addition, the methods of factor-group analysis make it possible to predict the number and activity of the external modes of a crystal, and the accessibility of the region 400–20 cm^{-1}, in which so many of these modes occur, has stimulated interest in that field. Such investigations have become increasingly important in solid-state physics as well as in inorganic and structural chemistry. These external modes, as stated in chapter 3, are of two types, translational and rotational. In general, in solids, the rotational modes are restricted rotations rather than free rotations, and are often referred to as liberations. Whereas the methods for calculating internal vibrational frequencies are well known and extensively dealt with in the literature and in the books referred to above, the methods for calculating external-mode frequencies are scattered about the journals, and are brought together for the first time in this book. Particular mention should be made of the contribution of Blackman [150–154] in the late 1930s in using the Born–von Kármán theory of specific heats as a basis for calculating lattice translational frequencies, and of the adaptation of Wilson's *FG*-matrix method to the same problem by Shimanouchi, Tsuboi and Miyazawa in 1961 [1291]. The theory of librational frequencies is rather better known owing to the interest in systems such as ethane, and the calculations of such frequencies employ Mathieu's equation, which arises from the solution of the Schrödinger equation with a potential energy of the form $2V = V_0(1 - \cos n\phi)$. Most of the recent work on inorganic systems has been carried out by Lippincott and co-workers; the series of articles by Miller and co-workers on restricted rotors attached to a framework has obvious applications to the study of inorganic systems such as $Y_3X—Y—XY_3$ where Y is oxygen and X may be (for instance) V, P, S, Se, Cr or Si, through the examples he considered are largely organic (454–7, 955).

There is one further area of interest not treated in existing books, and again the interest has arisen because of the accessibility of the low-frequency end of the spectrum. In many crystals it is meaningless to discuss the vibrational spectra either in terms of the stoichiometric formula unit or, indeed, in terms of any free molecule at all. A good example of the first situation is afforded by the perovskites ABX_3 such as $KNiF_3$ and $BaTiO_3$. Some early studies treated these as assemblies of A^{n+} and $BX_3{}^{n-}$ ions, but consideration of the structure and bonding of these materials shows that the vibrations are more properly treated in terms of the BX_6-coordination polyhedron and the isolated A cation. Here, the A—X bonds are weak compared with the B—X bonds. The second case is exemplified by the spinels, A_2BX_4. Again one can formally represent such compounds in terms of isolated A cations and BX_4 anions, and, in systems such as Na_2WO_4 where the WO_4 tetrahedra are much more strongly bound than the NaO_6 octahedra, this is a good approximation, but in Cr_2CuO_4 it is a poor approximation and discussion in terms of CuO_4 and CrO_6 vibrating units would be meaningless. The same situation arises in Al_2O_3, La_2O_3, SiO_2 and many other crystals. In consequence, a chapter has been included dealing with the vibrational spectra of compounds with no discrete molecular vibrating unit, based on factor-group analysis of the unit cell. Again, the interest in this field is by no means confined to the chemist; many of these compounds are studied by physicists, mineralogists and geochemists.

Finally, it is necessary to discuss the level of the mathematical treatment in this book. I have tried to concentrate on setting out the calculations which lead to the required numerical results, without paying very much attention to rigorous proofs concerning the basis of the calculations. For instance, there is no formal proof of the fact that an n-atomic nonlinear molecule has $3n - 6$ modes of vibration, nor is there any proof of the important *great orthogonality theorem*. In my own opinion, these matters may be taken for granted by the practising spectroscopist. I would therefore expressly disclaim any intention of making this book into a textbook —its function is to provide the research worker, in whatever discipline, who is interested in the spectra of inorganic compounds, with the available experimental observations and the means of assessing their significance.

Acknowledgments

I wish to express my sincere thanks to Dr G R Wilkinson, Reader in Physics at King's College, London, and Mr R F Branch, Lecturer in Pharmaceutical Chemistry in the Department of Pharmacy, Chelsea College, London, for reading the manuscript and offering many helpful comments.

I am also most grateful to Mrs J M Barrow for typing the manuscript, to my wife for her help with the indexing and proof-reading, and to the library staffs at Chelsea College, the Patent Office, and the Science Museum for their help during the course of the literature search.

Finally I wish to thank the staff of the Physico-Chemical Measurements Unit, Harwell, for the Perkin-Elmer 225 spectra used in illustrating Part II.

S D Ross

Part One

1. Symmetry effects and the number and activity of absorption bands

1.1 Point-group symmetry

The number of bands observed in the spectrum of a molecule depends on its symmetry. Much attention has been paid to the effect of the point group symmetry, which is dependent on the geometry of the molecule, but more recently it has been observed that space group symmetry, particularly the symmetry of the site occupied by the molecule in the crystal lattice, may play a part in determining the nature of the vibrational spectrum. This modification due to site symmetry is particularly important in the study of the spectra of inorganic molecules, since in very many cases the site symmetry is lower than the point symmetry and causes additional bands to appear.

Point-group symmetry is classified according to the possession of the following symmetry elements:

(a) A centre of symmetry, designated i. This exists in a molecule where each atom with coordinates (x, y, z) referred to the centre of symmetry as origin is identical with another atom whose coordinates are $(-x, -y, -z)$. Of the 32 crystal point groups, 11 are centro-symmetric.
(b) An n-fold rotation axis, designated C_n. This exists where rotation of the molecule by $2\pi/n$ gives a configuration indistinguishable from the original. The integer n is referred to as the order of the axis.
(c) Planes of symmetry, designated σ_v, σ_h and σ_d, depending on whether the plane is oriented vertically, horizontally or diagonally with respect to the rotation axis of highest order. Such planes divide the molecule into two portions which are mirror images.
(d) Rotation–reflection axes, designated S_n. Rotation by $2\pi/n$ about such an axis, followed by reflection in a plane perpendicular to the axis, produces a configuration indistinguishable from the original.

(e) The identity, I. This is formally required in order that the symmetry operations of the molecule shall constitute a group.

Molecules may belong to any point group, though the majority of them belong to one of the 32 crystal point groups. In the crystal point groups, the orders of rotation and rotation–reflection axes may only be 1, 2, 3, 4 or 6. Linear molecules form the largest class of those not belonging to the crystal point groups: unsymmetrical molecules such as HCN belong to the group $C_{\infty v}$ and symmetrical molecules such as CO_2 belong to the group $D_{\infty h}$. Ruthenocene and ferrocene belong to the groups D_{5h} and D_{5d} respectively, while the S_8 molecule belongs to the group D_{4d}.

The symbols used above for designating point groups are the Schönflies symbols, which are generally used by spectroscopists. There is another system, the Hermann–Mauguin system, which is used by crystallographers. The Schönflies symbols indicate the independent elements required to generate all the symmetry operations of the group. The simplest type of group is that designated C_n, in which the only element is the n-fold rotation axis, generating a total of n elements, one of which (C_n^n, that is, C_n repeated n times) is the identity element I. The elements generated may be written in the form C_n^k (where $k = 1, 2, \ldots, n$) and are usually reduced to their lowest terms. For instance, an axis C_6 would generate the six elements $C_6^1, C_6^2, C_6^3, C_6^4, C_6^5, C_6^6$. Of these, C_6^2, C_6^3, C_6^4 and C_6^6 would be written C_3^1, C_2, C_3^2 and C_1 ($=I$) respectively. The only other type of group whose operations are all generated by one symmetry element is the type S_n. Where n is even the operations generated are S_n^k ($k = 1, \ldots, n$). The operation S_n^k when k is even is equivalent to $C_{n/k}$, and is so designated in the character table. For instance, the element S_4 gives rise to the four operations S_4^1, $S_4^2 = C_2$, S_4^3, $S_4^4 = I$.

Where n is odd, the operations generated are the same as those generated by a combination of C_n and σ_h, and such groups, which contain $2n$ elements, are designated C_{nh}. As an example, consider the group S_3 ($=C_{3h}$). We have

$$
\begin{aligned}
S_3^1 &= C_3 \times \sigma_h \\
S_3^2 &= C_3^2 \\
S_3^3 &= \sigma_h \\
S_3^4 &= C_3 \\
S_3^5 &= C_3^2 \times \sigma_h \\
S_3^6 &= I.
\end{aligned}
$$

When a group has n twofold axes perpendicular to the C_n-axis of highest order, it is designated D_n. It contains a total of $2n$ operations, namely the n twofold axes and the n operations generated by the C_n-axis. If planes of symmetry (σ_d) bisect the twofold axes, groups D_{nd} are formed; if there is a plane of symmetry (σ_h) perpendicular to the principal C_n-axis, groups D_{nh} are formed. Each of these contains $4n$ symmetry elements.

The only other type of axial point group is that in which vertical planes of symmetry contain the principal C_n-axis; there must necessarily be n such planes, giving rise to the group C_{nv} with $2n$ operations. When n is odd, these planes are indistinguishable; when n is even, half the planes contain molecular axes and the other half bisect interbond angles. These are conventionally designated σ_v and σ_d respectively. The same distinction is made for the groups D_{nh} where n is even. (In the groups C_{2v}, D_{2h}, the planes are labelled according to their orientation with respect to a set of Cartesian coordinate axes, of which the principal axis is the z-axis.)

The point groups generated above include 27 of the 32 crystal point groups; these are the so-called axial point groups. The remaining five point groups are known as the cubic point groups; three of these (T, T_h, T_d) have symmetry operations based on the elements of a tetrahedron, while the other two (O, O_h) have operations based on the elements of an octahedron.

The group T consists of the twelve rotation operations of a tetrahedron; these are easily visualized if we consider the tetrahedron as being inscribed in a cube. There are three C_2 parallel to the cube faces and four C_3 which are the body diagonals of the cube. Combining these operations with the inversion element i gives rise to the group T_h, the additional elements being four S_6 coincident with the C_3, and three σ_h arising from the combination of the three C_2 with the inversion element. Tetrahedral molecules have the point group T_d, which, like T_h, has 24 operations. In addition to the rotation operations there are six σ_d which are normal to a cube face and pass through an apex of the tetrahedron; the three C_2 are also S_4.

The group O consists of the 24 rotation operations of a regular octahedron. Again considering the octahedron inscribed in a cube, with the apices at the centre of the faces, the symmetry elements are the four body diagonals (C_3), three axes parallel to the cube faces (C_4) and the six axes joining the centre of the octahedron to the centres of the octahedral faces (C_2); these six C_2 are parallel to the face diagonals of the cube. Combination of the operations generated by these elements with the inversion element leads to the 48 operations of the group O_h; the three C_4 are also S_4 and the four C_3 are also S_6. There are nine planes of symmetry, three σ_h and six σ_d.

The crystal point groups with their symmetry elements are given in Table 1.1; Table 1.2 gives the symmetry elements of some molecular point groups which are not crystal point groups.

When the point symmetry of a molecule is lowered, due to distortion or to coordination, the symmetry elements that it retains will be some of

Table 1.1 Symmetry elements of crystal point groups.

System	Point group symmetry: S	Point group symmetry: H/M	Symmetry elements	Number of operations
Triclinic	C_1	1	$I = C_1$	1
	$C_i = S_2$	$\bar{1}$	$I, i(=S_2)$	2
Monoclinic	C_2	2	I, C_2	2
	$C_S = C_{1h} = C_{1v}$	m	I, σ	2
	C_{2h}	$2/m$	I, C_2, σ_h, i	4
Orthorhombic	C_{2v}	$mm2$	$I, C_2, 2\sigma$	4
	D_2	222	$I, 3C_2$	4
	D_{2h}	mmm	$I, 3C_2, 3\sigma, i$	8
Tetragonal	C_4	4	I, C_4	4
	S_4	$\bar{4}$	$I, S_4\ (=C_2)$	4
	D_4	422	$I, C_4\ (=C_2), 2C_2', 2C_2''$	8
	C_{4v}	$4mm$	$I, C_4, 2\sigma_v, 2\sigma_d$	8
	C_{4h}	$4/m$	$I, C_4\ (=S_4), \sigma_h, i$	8
	D_{2d}	$\bar{4}2m$	$I, S_4\ (=C_2), 2C_2', 2\sigma_d$	8
	D_{4h}	$4/mmm$	$I, C_4\ (=S_4), 2C_2', 2C_2'', 2\sigma_v, 2\sigma_d, i$	16
Trigonal	C_3	3	I, C_3	3
	$C_{3i} = S_6$	$\bar{3}$	$I, S_6\ (=C_3), i$	6
	D_3	32	$I, C_3, 3C_2$	6
	C_{3v}	$3m$	$I, C_3, 3\sigma_v$	6
	D_{3d}	$\bar{3}m$	$I, S_6\ (=C_3), 3C_2, 3\sigma_d, i$	12
Hexagonal	C_6	6	I, C_6	6
	C_{3h}	$\bar{6}$	$I, S_3\ (=C_3), \sigma_h$	6
	D_6	622	$I, C_6, 3C_2', 3C_2''$	12
	D_{3h}	$\bar{6}m2$	$I, C_3\ (=S_3), 3C_2, 3\sigma_v, \sigma_h$	12
	C_{6h}	$6/m$	$I, C_6\ (=S_6), \sigma_h, i$	12
	C_{6v}	$6mm$	$I, C_6, 3\sigma_v, 3\sigma_d$	12
	D_{6h}	$6/mmm$	$I, C_6\ (=S_6), 3C_2', 3C_2'', 3\sigma_v, 3\sigma_d, \sigma_h, i$	24
Cubic	T	23	$I, 3C_2, 4C_3$	12
	T_h	$m3$	$I, 3C_2, 4C_3\ (=S_6), 3\sigma_h, i$	24
	T_d	$\bar{4}3m$	$I, 3C_2\ (=S_4), 4C_3, 6\sigma_d$	24
	O	432	$I, 3C_2, 4C_3, 3C_4$	24
	O_h	$m3m$	$I, 3C_2, 4C_3\ (=S_6), 3C_4\ (=S_4), 3\sigma_h, 6\sigma_d, i$	48

those of the original point group; the new point group is said to be a subgroup of the old group and it is possible to construct so-called correlation tables which list the changes of vibrational species and activities due to this lowering of symmetry.

Table 1.2 Some important molecular point groups and their symmetry elements.

Molecule	Point group symmetry		Symmetry elements	Number of operations
	S	H/M		
HCN	$C_{\infty v}$	∞m	I, $1C_{\infty}$, $\infty\sigma_v$	∞
CO_2	$D_{\infty h}$	∞/mm	I, $1C_{\infty}$, $\infty\sigma_v$, $1\sigma_h$, i	∞
S_8	D_{4d}	$\bar{8}4m$	I, C_4 ($=S_8$), $4C_2'$, $4\sigma_d$	16
Ferrocene	D_{5d}	$\bar{5}m$	I, C_5 ($=S_{10}$), $5C_2$, $5\sigma_d$, i	20
Ruthenocene	D_{5h}	$\overline{10}m2$	I, C_5, $5C_2$, σ_h, $5\sigma_v$	20

1.2 Application of group theory to the prediction of vibration spectra: general principles

The symmetry operations of a point group are a group in the mathematical sense because they obey the following rules:

(a) They include an identity element I such that $AI = IA = A$, where A is any element of the group.
(b) They obey the combinative law whereby the product C of the multiplication of any two elements of the group A, B is an element of the group. It is important to realize that multiplication is not necessarily commutative; in general $AB \neq BA$, though there are some groups, such as C_{2v}, illustrated below, in which $AB = BA$. Noncommutative groups are always those representing molecules which have at least one C_n ($n > 2$) and which display degenerate vibrations.
(c) They obey the associative law $(AB)C = A(BC)$. Note that, in representing the performance of successive operations by multiplication of group elements, the operation performed first is written to the right.
(d) Every element A in the group has an inverse A^{-1} in the group such that $AA^{-1} = A^{-1}A = I$.

Table 1.3 Multiplication table for the symmetry operations of the group C_{2v}.

	I	C_2	σ_v'	σ_v''
I	I	C_2	σ_v'	σ_v''
C_2	C_2	I	σ_v''	σ_v'
σ_v'	σ_v'	σ_v''	I	C_2
σ_v''	σ_v''	σ_v'	C_2	I

Such a multiplication table as the one above may be constructed for any point group. The operation written along the top is taken to be carried out first. Consider the result $C_2 \times \sigma'_v = \sigma''_v$. A bent triatomic molecule like H_2O belongs to the point group C_{2v}. Label the H atoms 1, 2 and the O atom 3.

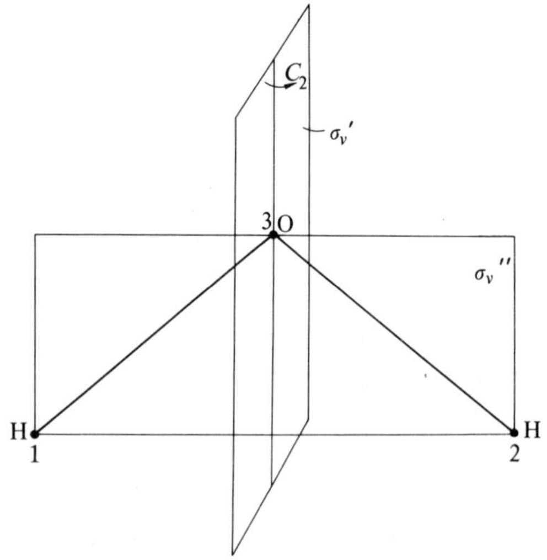

Fig. 1.1 H_2O, a triatomic molecule belonging to the point group C_{2v}.

When we say $C_2 \times \sigma'_v = \sigma''_v$, we mean, therefore, that the configuration adopted by the molecule after performance of the operation σ'_v followed by the operation C_2 is identical to that adopted after performance of the single operation σ''_v. Note that, because in this case multiplication is commutative, the order in which we perform the operations σ'_v, C_2 does not affect the final configuration. A simple example of the more general situation, in which multiplication is not commutative, is afforded by the multiplication table of the group C_{3v} (Table 1.4).

Table 1.4 Multiplication table for the symmetry operations of the group C_{3v}.

	I	C_3^1	C_3^2	σ_{v1}	σ_{v2}	σ_{v3}
I	I	C_3^1	C_3^2	σ_{v1}	σ_{v2}	σ_{v3}
C_3^1	C_3^1	C_3^2	I	σ_{v3}	σ_{v1}	σ_{v2}
C_3^2	C_3^2	I	C_3^1	σ_{v2}	σ_{v3}	σ_{v1}
σ_{v1}	σ_{v1}	σ_{v2}	σ_{v3}	I	C_3^1	C_3^2
σ_{v2}	σ_{v2}	σ_{v3}	σ_{v1}	C_3^2	I	C_3^1
σ_{v3}	σ_{v3}	σ_{v1}	σ_{v2}	C_3^1	C_3^2	I

This table is not symmetrical about the leading diagonal, and the elements on that diagonal are not all I. As an example, consider the NH_3

molecule; let the three H atoms be 1, 2, 3. We have from Table 1.4 the two results $\sigma_{v2}\sigma_{v1} = C_3^2$; $\sigma_{v1}\sigma_{v2} = C_3^1$. We designate the planes through atoms 1, 2, 3 in the original configuration as σ_{v1}, σ_{v2}, σ_{v3}, respectively.

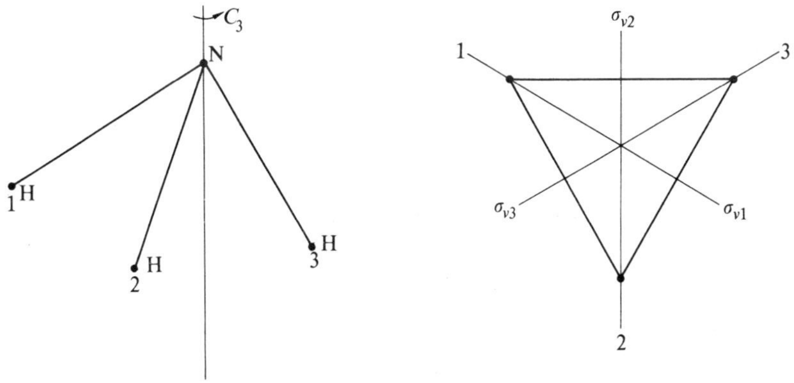

Fig. 1.2 NH_3, a tetratomic molecule belonging to the point group C_{3v}.

Under the symmetry operations, the atoms change positions but the elements do not; the designation always refers to the original configuration.

An important concept illustrated by considering group multiplication tables is that of class.

Consider two elements of a group, and let these be A, X. Then by the combinative law the product $X^{-1}AX$ is equal to some other element (say B) of the group, that is,

$$B = X^{-1}AX.$$

The elements A and B are said to be *conjugate*. The complete set of elements which are conjugate to each other form a class. The number of elements in the set is called the *order* of the class and it must be an integral divisor of the number of elements in the group. From the multiplication table of the group C_{3v} we can determine the distribution of the elements among the classes.

It is apparent that the identity element I always forms a class by itself; if we consider any element X, we have, since $XI = IX$,

$$X^{-1}IX = X^{-1}XI = I.$$

Members of the same class will always be operations of the same kind, though the converse is not necessarily true; in the group C_{2v}, the operations σ_v', σ_v'' are in different classes, since there is no element X of the group such that $X^{-1}\sigma_v'X = \sigma_v''$.

We can show that, in the group C_{3v}, the operations C_3^1, C_3^2 belong to the same class. Let $C_3^1 = A$, and let X be σ_{v1}. Then X^{-1} is also σ_{v1}, since $\sigma_{v1}\sigma_{v1} = I$.

$$\begin{aligned} \sigma_{v1}C_3^1\sigma_{v1} &= \sigma_{v1}(C_3^1\sigma_{v1}) && \text{by the associative law} \\ &= \sigma_{v1}\sigma_{v3} && \text{from Table 1.4} \\ &= C_3^2 && \text{from Table 1.4} \end{aligned}$$

therefore C_3^1, C_3^2 are in the same class. By working out all the transformations of this kind, it is possible to show that C_3^1, C_3^2 form a class (of order 2) and σ_{v1}, σ_{v2}, σ_{v3} form a class (of order 3). Together with the identity element I this makes a total of 3 classes.

It is useful at this stage to introduce the idea of class multiplication. Consider the three classes

$$\begin{aligned} \mathbf{C}_1 &= I \\ \mathbf{C}_2 &= C_3^1, C_3^2 \\ \mathbf{C}_3 &= \sigma_{v1}, \sigma_{v2}, \sigma_{v3}. \end{aligned}$$

If $\mathbf{C}$ is any complete class of the group and X is an element of the group, then

$$X^{-1}\mathbf{C}X = \mathbf{C}. \tag{1.1}$$

The converse of this statement is that any collection of elements obeying (1.1) must be composed of complete classes. If we multiply two classes $\mathbf{C}_j$, $\mathbf{C}_k$, then by (1.1)

$$\begin{aligned} \mathbf{C}_j\mathbf{C}_k &= X^{-1}\mathbf{C}_jXX^{-1}\mathbf{C}_kX \\ &= X^{-1}(\mathbf{C}_j\mathbf{C}_k)X \end{aligned}$$

for any element X of the group. The converse of this is that the product $\mathbf{C}_j\mathbf{C}_k$ must be composed of complete classes, and it can be written in the form

$$\mathbf{C}_j\mathbf{C}_k = \sum_m C_{jkm}\mathbf{C}_m \tag{1.2}$$

where C_{jkm} denotes the number of times the class $\mathbf{C}_m$ appears in the product $\mathbf{C}_j\mathbf{C}_k$.

We need the products $\mathbf{C}_j\mathbf{C}_k$ ($j, k = 1, 2, 3$) and we can get these from Table 1.4.

Class multiplication is commutative, since the product of two classes will contain the same elements regardless of the order of multiplication. We see that the products contain the elements of various classes and we can tabulate the coefficients C_{jkm} ($=C_{kjm}$). This is done in Table 1.6.

These coefficients can be used in the construction of the character table

Table 1.5 Class multiplication tables for the group C_{3v}.

$$\text{Class: } \mathbf{C}_1 \times \mathbf{C}_1 = I \times I = I = \mathbf{C}_1$$

$$\mathbf{C}_1 \times \mathbf{C}_2 = I \times C_3^1 + I \times C_3^2 = C_3^1 + C_3^2 = \mathbf{C}_2$$

$$\mathbf{C}_1 \times \mathbf{C}_3 = I \times \sigma_{v1} + I \times \sigma_{v2} + I \times \sigma_{v3} = \sigma_{v1} + \sigma_{v2} + \sigma_{v3} = \mathbf{C}_3$$

$$\mathbf{C}_2 \times \mathbf{C}_2 = \begin{array}{c|cc} & C_3^1 & C_3^2 \\ \hline C_3^1 & C_3^2 & I \\ C_3^2 & I & C_3^1 \end{array} = 2\mathbf{C}_1 + \mathbf{C}_2$$

$$\mathbf{C}_3 \times \mathbf{C}_2 = \begin{array}{c|cc} & C_3^1 & C_3^2 \\ \hline \sigma_{v1} & \sigma_{v2} & \sigma_{v3} \\ \sigma_{v2} & \sigma_{v3} & \sigma_{v1} \\ \sigma_{v3} & \sigma_{v1} & \sigma_{v2} \end{array} = 2\mathbf{C}_3$$

$$\mathbf{C}_3 \times \mathbf{C}_3 = \begin{array}{c|ccc} & \sigma_{v1} & \sigma_{v2} & \sigma_{v3} \\ \hline \sigma_{v1} & I & C_3^1 & C_3^2 \\ \sigma_{v2} & C_3^2 & I & C_3^1 \\ \sigma_{v3} & C_3^1 & C_3^2 & I \end{array} = 3\mathbf{C}_1 + 3\mathbf{C}_2$$

Table 1.6 Coefficients C_{jkm} for classes of the group C_{3v}.

$jk = kj$ \ m	1	2	3
11	1	0	0
22	2	1	0
33	3	3	0
12	0	1	0
23	0	0	2
31	0	0	1

of the group. Character tables are a more convenient way of expressing the information contained in group multiplication tables such as Tables 1.3, 1.4. Such information can be expressed by replacing the elements of the group with suitable functions that multiply in the same way. These functions can be written as matrices; if we have the result $\sigma_{v1}C_3^1 = \sigma_{v2}$, we can write down matrices which *represent* these symmetry operations, in the sense that the product of the matrices representing σ_{v1}, C_3^1, multiplied in the correct order, is equal to the matrix representing σ_{v2}.

Matrix multiplication

A matrix is simply an array of numbers arranged in rows and columns. If there are r rows and s columns, the matrix is called an $r \times s$ matrix.

Matrices can be multiplied only if they are conformable, i.e., if the number of columns in one matrix is the same as the number of rows in the other. If two matrices are square ($r = s$), which is the case for all matrices that can represent symmetry operations, they can be multiplied if they have the same number of rows (or columns) and are said to be conformable both ways; that is, if A, B are both (say) 2×2 matrices, both the products AB and BA can be formed. Matrix multiplication, like group element multiplication, is not, in general, commutative.

The elements P_{ij} of the product P of two matrices Q, R, which are both $n \times n$ matrices are given by

$$P_{ij} = \sum_{k=1}^{n} q_{ik} r_{kj}. \tag{1.3}$$

Consider two matrices which are 2×2, so that

$$Q = \begin{bmatrix} q_{11} & q_{12} \\ q_{21} & q_{22} \end{bmatrix}, \qquad R = \begin{bmatrix} r_{11} & r_{12} \\ r_{21} & r_{22} \end{bmatrix}.$$

Then, multiplying them in this order gives

$$P = \begin{bmatrix} q_{11}r_{11} + q_{12}r_{21} & q_{11}r_{12} + q_{12}r_{22} \\ q_{21}r_{11} + q_{22}r_{21} & q_{21}r_{12} + q_{22}r_{22} \end{bmatrix}. \tag{1.4a}$$

Whereas if they are multiplied in the reverse order, giving

$$P'_{ij} = \sum_{k=1}^{n} r_{ik} q_{kj},$$

the product P' is given by

$$P' = \begin{bmatrix} r_{11}q_{11} + r_{12}q_{21} & r_{11}q_{12} + r_{12}q_{22} \\ r_{21}q_{11} + r_{22}q_{21} & r_{21}q_{12} + r_{22}q_{22} \end{bmatrix}. \tag{1.4b}$$

Suppose that the two matrices

$$Q = \begin{bmatrix} 1 & 0 \\ 0 & -1 \end{bmatrix}, \qquad R = \begin{bmatrix} -\frac{1}{2} & \sqrt{3}/2 \\ \sqrt{3}/2 & \frac{1}{2} \end{bmatrix}$$

are chosen to represent the symmetry operations σ_{v1}, σ_{v2} respectively. Then if we multiply them in the order QR, as in (1.4a), the product is

$$\begin{bmatrix} -\frac{1}{2} & \sqrt{3}/2 \\ -\sqrt{3}/2 & -\frac{1}{2} \end{bmatrix}, \tag{1.5a}$$

Whereas if we multiply them in the order RQ, as in (1.4b), the product is

$$\begin{bmatrix} -\frac{1}{2} & -\sqrt{3}/2 \\ \sqrt{3}/2 & -\frac{1}{2} \end{bmatrix}. \tag{1.5b}$$

According to Table 1.4, we have

$$\sigma_{v1}\sigma_{v2} = C_3^1, \qquad \sigma_{v2}\sigma_{v1} = C_3^2.$$

Since C_3^1 denotes a clockwise rotation of $1 \times 2\pi/3$, the matrix representing this is written in the form

$$\begin{bmatrix} \cos\theta & +\sin\theta \\ -\sin\theta & \cos\theta \end{bmatrix} = \begin{bmatrix} \cos(2\pi/3) & +\sin(2\pi/3) \\ -\sin(2\pi/3) & \cos(2\pi/3) \end{bmatrix}$$

and since $\cos(2\pi/3) = -\frac{1}{2}$, $\sin(2\pi/3) = \sqrt{3}/2$, we see that the matrix (1.5a) does represent C_3^1; similarly for C_3^2 the quantities required are $\cos(4\pi/3) = -\frac{1}{2}$, $\sin(4\pi/3) = -\sqrt{3}/2$, and the matrix required to represent C_3^2 is (1.5b).

In representing the results obtained from group multiplication tables, we do not need to use the whole matrix, but merely its *character* (sometimes called the trace) χ since this is invariant under a similarity transformation. The character is the sum of the elements of the leading diagonal

$$\chi(a) = \sum_i a_{ii}$$

The matrices (1.5a), (1.5b), which are the products QR, RQ, both have the character -1; they both represent rotations about the C_3-axis and therefore belong to the same class. All matrices representing operations in the same class have the same character (such matrices are said to be conjugate). Although we can construct an infinite number of matrix representations of a group we are interested mainly in a set of representations called the irreducible representations of the group.

Suppose we have a set of matrices $A_1, A_2, \ldots, A_h$, where h is the order of the group, and let them all be square $n \times n$ matrices. It may be possible to find a matrix S such that every matrix $A_1', A_2', \ldots, A_h'$ produced by the transformation

$$A' = S^{-1}AS$$

is blocked out (partitioned) in the same way; this process is the reduction of the representation. When no further transformation exists by which the reduction can be repeated, the matrices $A_1', A_2', \ldots, A_h'$ are irreducible representations of the group, and can be written as a partitioned matrix in which all the elements are on, or distributed symmetrically about, the leading diagonal, as below.

The matrices $A_1, A_2, \ldots, A_h$ do not necessarily constitute the complete set of irreducible representations of the group. If, however, the reducible representation chosen is the so-called regular representation of the group, the matrices will contain all the irreducible representations of the group. Further, each irreducible representation, of dimensionality l, will appear

l times. We can construct the regular representation of a group from its group multiplication table. Table 1.4 above relates to the point group C_{3v}. The rows of the table show the effect of applying any given operation successively to all the operations of the group. We have

$$\begin{aligned}
I\,(I, C_3^1, C_3^2, \sigma_{v1}, \sigma_{v2}, \sigma_{v3}) &= (I, C_3^1, C_3^2, \sigma_{v1}, \sigma_{v2}, \sigma_{v3})\\
C_3^1\,(I, C_3^1, C_3^2, \sigma_{v1}, \sigma_{v2}, \sigma_{v3}) &= (C_3^1, C_3^2, I, \sigma_{v3}, \sigma_{v1}, \sigma_{v2})\\
C_3^2\,(I, C_3^1, C_3^2, \sigma_{v1}, \sigma_{v2}, \sigma_{v3}) &= (C_3^2, I, C_3^1, \sigma_{v2}, \sigma_{v3}, \sigma_{v1})\\
\sigma_{v1}\,(I, C_3^1, C_3^2, \sigma_{v1}, \sigma_{v2}, \sigma_{v3}) &= (\sigma_{v1}, \sigma_{v2}, \sigma_{v3}, I, C_3^1, C_3^2)\\
\sigma_{v2}\,(I, C_3^1, C_3^2, \sigma_{v1}, \sigma_{v2}, \sigma_{v3}) &= (\sigma_{v2}, \sigma_{v3}, \sigma_{v1}, C_3^2, I, C_3^1)\\
\sigma_{v3}\,(I, C_3^1, C_3^2, \sigma_{v1}, \sigma_{v2}, \sigma_{v3}) &= (\sigma_{v3}, \sigma_{v1}, \sigma_{v2}, C_3^1, C_3^2, I).
\end{aligned}$$

From the above transformations, the matrices $A_1, A_2, \ldots, A_6$ representing the six symmetry operations are

$$A_1(I) = \begin{bmatrix} 1&0&0&0&0&0\\ 0&1&0&0&0&0\\ 0&0&1&0&0&0\\ 0&0&0&1&0&0\\ 0&0&0&0&1&0\\ 0&0&0&0&0&1 \end{bmatrix} \qquad A_2(C_3^1) = \begin{bmatrix} 0&1&0&0&0&0\\ 0&0&1&0&0&0\\ 1&0&0&0&0&0\\ 0&0&0&0&0&1\\ 0&0&0&1&0&0\\ 0&0&0&0&1&0 \end{bmatrix}$$

$$A_3(C_3^2) = \begin{bmatrix} 0&0&1&0&0&0\\ 1&0&0&0&0&0\\ 0&1&0&0&0&0\\ 0&0&0&0&1&0\\ 0&0&0&0&0&1\\ 0&0&0&1&0&0 \end{bmatrix} \qquad A_4(\sigma_{v1}) = \begin{bmatrix} 0&0&0&1&0&0\\ 0&0&0&0&1&0\\ 0&0&0&0&0&1\\ 1&0&0&0&0&0\\ 0&1&0&0&0&0\\ 0&0&1&0&0&0 \end{bmatrix}$$

$$A_5(\sigma_{v2}) = \begin{bmatrix} 0&0&0&0&1&0\\ 0&0&0&0&0&1\\ 0&0&0&1&0&0\\ 0&0&1&0&0&0\\ 1&0&0&0&0&0\\ 0&1&0&0&0&0 \end{bmatrix} \qquad A_6(\sigma_{v3}) = \begin{bmatrix} 0&0&0&0&0&1\\ 0&0&0&1&0&0\\ 0&0&0&0&1&0\\ 0&1&0&0&0&0\\ 0&0&1&0&0&0\\ 1&0&0&0&0&0 \end{bmatrix}.$$

Now there is a matrix S such that for every matrix A of the regular representation

$$A' = S^{-1}AS$$

and every A' is blocked out in the same way. The matrix is

$$S = \begin{bmatrix} B & B & B & B & B & B \\ B & B & B & -B & -B & -B \\ C & -C/2 & -C/2 & -C & C/2 & C/2 \\ 0 & \frac{1}{2} & -\frac{1}{2} & 0 & \frac{1}{2} & -\frac{1}{2} \\ 0 & -\frac{1}{2} & \frac{1}{2} & 0 & \frac{1}{2} & -\frac{1}{2} \\ C & -C/2 & -C/2 & C & -C/2 & -C/2 \end{bmatrix},$$

where $B = 6^{-1/2}$, $C = 3^{-1/2}$. The matrices A' formed from $A' = S^{-1}AS$ are

$$A_1'(I) = \begin{bmatrix} 1 & 0 & 0 & 0 & 0 & 0 \\ 0 & 1 & 0 & 0 & 0 & 0 \\ 0 & 0 & 1 & 0 & 0 & 0 \\ 0 & 0 & 0 & 1 & 0 & 0 \\ 0 & 0 & 0 & 0 & 1 & 0 \\ 0 & 0 & 0 & 0 & 0 & 1 \end{bmatrix}$$

$$A_2'(C_3^1) = \begin{bmatrix} 1 & 0 & 0 & 0 & 0 & 0 \\ 0 & 1 & 0 & 0 & 0 & 0 \\ 0 & 0 & -\frac{1}{2} & \sqrt{3}/2 & 0 & 0 \\ 0 & 0 & -\sqrt{3}/2 & -\frac{1}{2} & 0 & 0 \\ 0 & 0 & 0 & 0 & -\frac{1}{2} & \sqrt{3}/2 \\ 0 & 0 & 0 & 0 & -\sqrt{3}/2 & -\frac{1}{2} \end{bmatrix}$$

$$A_3'(C_3^2) = \begin{bmatrix} 1 & 0 & 0 & 0 & 0 & 0 \\ 0 & 1 & 0 & 0 & 0 & 0 \\ 0 & 0 & -\frac{1}{2} & -\sqrt{3}/2 & 0 & 0 \\ 0 & 0 & \sqrt{3}/2 & -\frac{1}{2} & 0 & 0 \\ 0 & 0 & 0 & 0 & -\frac{1}{2} & -\sqrt{3}/2 \\ 0 & 0 & 0 & 0 & \sqrt{3}/2 & -\frac{1}{2} \end{bmatrix}$$

$$A'_4(\sigma_{v1}) = \begin{bmatrix} 1 & 0 & 0 & 0 & 0 & 0 \\ 0 & -1 & 0 & 0 & 0 & 0 \\ 0 & 0 & -1 & 0 & 0 & 0 \\ 0 & 0 & 0 & 1 & 0 & 0 \\ 0 & 0 & 0 & 0 & -1 & 0 \\ 0 & 0 & 0 & 0 & 0 & 1 \end{bmatrix}$$

$$A'_5(\sigma_{v2}) = \begin{bmatrix} 1 & 0 & 0 & 0 & 0 & 0 \\ 0 & -1 & 0 & 0 & 0 & 0 \\ 0 & 0 & \frac{1}{2} & \sqrt{3}/2 & 0 & 0 \\ 0 & 0 & \sqrt{3}/2 & -\frac{1}{2} & 0 & 0 \\ 0 & 0 & 0 & 0 & \frac{1}{2} & \sqrt{3}/2 \\ 0 & 0 & 0 & 0 & \sqrt{3}/2 & -\frac{1}{2} \end{bmatrix}$$

$$A'_6(\sigma_{v3}) = \begin{bmatrix} 1 & 0 & 0 & 0 & 0 & 0 \\ 0 & -1 & 0 & 0 & 0 & 0 \\ 0 & 0 & \frac{1}{2} & -\sqrt{3}/2 & 0 & 0 \\ 0 & 0 & -\sqrt{3}/2 & -\frac{1}{2} & 0 & 0 \\ 0 & 0 & 0 & 0 & \frac{1}{2} & -\sqrt{3}/2 \\ 0 & 0 & 0 & 0 & -\sqrt{3}/2 & -\frac{1}{2} \end{bmatrix}$$

The matrices $A'_1, A'_2, \ldots, A'_6$ are all blocked out in the same way, since the elements of the matrices which are, in general, nonzero, are $A'_{(11)}$, $A'_{(22)}$, $A'_{(33)}$, $A'_{(34)}$, $A'_{(43)}$, $A'_{(44)}$, $A'_{(55)}$, $A'_{(56)}$, $A'_{(65)}$ and $A'_{(66)}$. Two important points should be noted. Firstly, the characters of corresponding matrices $A_1, A'_1; A_2, A'_2, \ldots, A_h, A'_h$ are unchanged by the similarity transformation $A' = S^{-1}AS$. Secondly, the characters of operations belonging to the same class are identical; $\chi(C_3^1) = \chi(C_3^2)$; $\chi(\sigma_{v1}) = \chi(\sigma_{v2}) = \chi(\sigma_{v3})$. If we denote the regular representation as Γ_{reg}, then we have

	I	$2C_3$	$3\sigma_v$
Γ_{reg}	6	0	0

The character table of the group C_{3v} is shown in Table 1.7 and either by inspection or by application of eq. (1.18) following, we see that $\Gamma_{\text{reg}} = 1A_1 + 1A_2 + 2E$. The first two are one-dimensional and the last is a two-

Table 1.7 Character table of the group C_{3v}.

	I	$2C_3$	$3\sigma_v$
A_1	1	1	1
A_2	1	1	-1
E	2	1	0

dimensional representation; they occur once and twice respectively. Note also that the one-dimensional representations correspond to the 1×1 'blocks' in the matrices A', while the two-dimensional representations correspond to 2×2 blocks. Thus the irreducible representations of the point group C_{3v} are as shown in Table 1.8.

Table 1.8 Irreducible representations of the point group C_{3v}.

	Γ_1	Γ_2	Γ_3, Γ_4
I	1	1	$\begin{bmatrix} 1 & 0 \\ 0 & 1 \end{bmatrix}$
C_3^1	1	1	$\begin{bmatrix} -\frac{1}{2} & \sqrt{3}/2 \\ -\sqrt{3}/2 & -\frac{1}{2} \end{bmatrix}$
C_3^2	1	1	$\begin{bmatrix} -\frac{1}{2} & -\sqrt{3}/2 \\ \sqrt{3}/2 & -\frac{1}{2} \end{bmatrix}$
σ_{v1}	1	-1	$\begin{bmatrix} -1 & 0 \\ 0 & 1 \end{bmatrix}$
σ_{v2}	1	-1	$\begin{bmatrix} \frac{1}{2} & \sqrt{3}/2 \\ \sqrt{3}/2 & -\frac{1}{2} \end{bmatrix}$
σ_{v3}	1	-1	$\begin{bmatrix} \frac{1}{2} & -\sqrt{3}/2 \\ -\sqrt{3}/2 & -\frac{1}{2} \end{bmatrix}$

Since the characters χ are invariant under similarity transformations, we need only these and not the matrices themselves. The advantage of using the characters is that the character table can be worked out by the use of the rules given below, without constructing the matrices constituting the irreducible representations.

Consider a group of order h, and let the dimensionality of the ith irreducible representation be l_i. Let R signify any symmetry operation of the group, and let an element in the mth row and nth column of the matrix representing R in the ith irreducible representation be denoted $\Gamma^{(i)}(R)_{mn}$. The rules for the construction of the character table depend on the fact that the collection of matrices constituting any given irreducible representation contain sets of elements (each matrix contributing one element to the set) which behave like the components of a vector in h-dimensional space. This statement is expressed symbolically as

$$\sum_R \Gamma^{(i)}(R)_{mn}\Gamma^{(j)}(R)_{pq} = \frac{h}{\sqrt{l_i l_j}}\delta_{ij}\,\delta_{mp}\,\delta_{nq}, \tag{1.6}$$

where

$$\delta_{ij} = \begin{cases} 0 & \text{when } i \neq j \\ 1 & \text{when } i = j \end{cases}$$

and is known as the *great orthogonality theorem.*

The physical interpretation of (1.6) is easier to visualize if it is reduced by imposing certain conditions:

(i) If $i \neq j$,

$$\sum_R \Gamma^{(i)}(R)_{mn}\Gamma^{(j)}(R)_{pq} = 0. \tag{1.7}$$

This implies that vectors chosen from matrices which are different representations are orthogonal.

(ii) If $i = j$, but either or both the conditions $m \neq p$, $n \neq q$ hold,

$$\sum_R \Gamma^{(i)}(R)_{mn}\Gamma^{(i)}(R)_{pq} = 0, \tag{1.8}$$

which implies that vectors chosen from different sets of elements in matrices of the same representation are orthogonal.

(iii) If $i = j$, $m = p$, $n = q$, then

$$\sum_R \Gamma^{(i)}(R)_{mn}\Gamma^{(i)}(R)_{mn} = h/l_i, \tag{1.9}$$

which implies that the square of the length of any vector is h/l_i.

From these equations, combined with the definition of character, which we can write in the form

$$\chi^{(i)} = \sum_m \Gamma^{(i)}(R)_{mm} \tag{1.10}$$

we can formulate the rules to be used in the construction of character tables.

Rule 1. The sum of the squares of the dimensionalities of the irreducible representations of a group is equal to the order of a group.

$$\sum_i l_i^2 = h. \tag{1.11}$$

(There will always be at least one one-dimensional representation; this is the totally symmetric representation in which the character for every symmetry operation is $+1$.)

Rule 2. In any representation the characters of operations in the same class are identical. (Operations in the same class are represented by conjugate matrices.)

Rule 3. The sum of the squares of the characters of any irreducible representation is equal to the order of the group.

$$\sum_R [\chi^{(i)}(R)]^2 = h. \tag{1.12}$$

Rule 4. Vectors whose components are the characters of two different irreducible representations are orthogonal; it follows from this that the sum of the squares of the characters for any operation is equal to h/N, where N is the number of operations in the class. The character of the identity in the ith representation is always equal to the dimensionality l_i.

Rule 5. The elements within the ith row of a character table (i.e., those of the ith irreducible representation) are related by the expression

$$N_j\chi^{(i)}(\mathbf{C}_j)N_k\chi^{(i)}(\mathbf{C}_k) = l_i \sum_i C_{jk,m}N_m\chi^{(i)}(\mathbf{C}_m), \tag{1.13}$$

where N_j, N_k, N_m are the numbers of operations in the jth, kth, mth class and the $C_{jk,m}$ are the class multiplication coefficients defined by eq. (1.2).

Rule 6. The number of irreducible representations of a group is equal to the number of classes in the group.

We can combine Rules 3 and 4 to give another useful relation. Rule 3 is expressed by eq. (1.12); Rule 4 can be written in the form

$$\sum_R \chi^{(i)}(R)\chi^{(j)}(R) = 0 \qquad \text{if } i \neq j. \tag{1.14}$$

Since $[\chi^{(i)}(R)]^2 = h$, we can put

$$\sum_R \chi^{(i)}(R)\chi^{(j)}(R) = h\,\delta_{ij} \qquad (0 \text{ when } i \neq j,\ h \text{ when } i = j). \tag{1.15}$$

Equation (1.15) helps us in the calculation of the number of times each irreducible representation appears in any given reducible representation. Knowledge of this enables us to determine how many vibrations of each type a molecule will have, and whether they are active in either or both of the infrared and Raman effects.

We have already stated that for any reducible representation, we can find a similarity transformation which will reduce it to a partitioned matrix in which the matrix corresponding to the jth irreducible representation may appear a_j times.

Since the character $\chi(R)$ of the matrix corresponding to the operation R is invariant under a similarity transformation, we can write

$$\chi(R) = \sum_j a_j\chi^{(j)}(R). \tag{1.16}$$

Then

$$\begin{aligned}\chi(R)\chi^{(i)}(R) &= \sum_R \sum_j a_j\chi^{(j)}(R)\chi^{(i)}(R)\\ &= \sum_j \sum_R a_j\chi^{(j)}(R)\chi^{(i)}(R).\end{aligned}$$

For each term in the sum over j we have

$$\begin{aligned}\sum_R a_j\chi^{(j)}(R)\chi^{(i)}(R) &= a_j \sum_R \chi^{(j)}(R)\chi^{(i)}(R)\\ &= a_j h\, \delta_{ij} \qquad \text{(from 1.15).}\end{aligned} \tag{1.17}$$

Since all terms are zero unless $i = j$, we can write (1.17) as

$$\sum_R \chi(R)\chi^{(j)}(R) = ha_j;$$

therefore,

$$a_j = \frac{1}{h}\sum_R \chi(R)\chi^{(j)}(R). \tag{1.18}$$

1.3 Application of the general principles to the construction of character tables

This section is devoted to the illustration of the general principles by constructing the character table of the point group T_d. The procedure discussed here is suitable for the construction of character tables all of whose elements are real numbers.

We first refer to Table 1.1 and we find that the symmetry operations of which the group is made up are I, $3C_2$, $8C_3$, $6S_4$, $6\sigma_d$. This gives a total of 24 operations, divided among 5 classes.

By rule 6, we know that there must be 5 irreducible representations, and for the moment we may label these $\Gamma^{(1)}$, $\Gamma^{(2)}, \ldots, \Gamma^{(5)}$. Rule 1 (eq. 1.11) tells us that the sum of the squares of the dimensionalities of these 5 representations must be 24, and we also know that at least one of these is one-dimensional. (This totally symmetric representation is always labelled $\Gamma^{(1)}$ and the character of every operation is $+1$.) We therefore have

$$l_2^2 + l_3^2 + l_4^2 + l_5^2 = 23,$$

giving by inspection

$$l_2 = 1, \qquad l_3 = 2, \qquad l_4 = l_5 = 3.$$

(The remaining representations are conventionally labelled in order of increasing dimensionality.)

Since we now know the dimensionalities of the 5 irreducible representations, we know the character of the identity operation for each representation: by Rule 4 this is equal to the dimensionality.

We can tabulate the information so far obtained as in Table 1.9(a).

Table 1.9(a) Part of the character table for the point group T_d.

	I	$8C_3$	$3C_2$	$6S_4$	$6\sigma_d$
$\Gamma^{(1)}$	1	1	1	1	1
$\Gamma^{(2)}$	1				
$\Gamma^{(3)}$	2				
$\Gamma^{(4)}$	3				
$\Gamma^{(5)}$	3				

We now apply Rule 4; it states that vectors whose components are the characters of two different irreducible representations are orthogonal; for the representations $\Gamma^{(1)}$, $\Gamma^{(2)}$ this means

$$\sum_R \chi^{(1)}(R)\chi^{(2)}(R) = 0.$$

Therefore

$$1 \times 1 \times 1 + 8 \times 1 \times \chi^{(2)}(C_3) + 3 \times 1 \times \chi^{(2)}(C_2) + 6 \times 1 \times \chi^{(2)}(S_4) + 6 \times 1 \times \chi^{(2)}(\sigma_d) = 0. \tag{1.19}$$

Each term is of the form [Number of operations in the class × character of the operation in $\Gamma^{(1)}$ × character of the operation in $\Gamma^{(2)}$]. Again, by inspection we must have

$$\chi^{(2)}(C_3) = \chi^{(2)}(C_2) = +1,$$
$$\chi^{(2)}(S_4) = \chi^{(2)}(\sigma_d) = -1.$$

We can therefore complete the row for $\Gamma^{(2)}$, which will read in full

$$1, 1, 1, -1, -1.$$

We still need to determine twelve elements of the character table. Using Rule 3 for rows 3, 4, 5, we have three relations of the form

$$\sum_R (\chi^{(i)}(R))^2 = 24.$$

Using Rule 4 for columns 2, 3, 4, 5, we have four relations of the type

$$\sum_{i=1}^{5} (\chi^{(i)}(R))^2 = h/N.$$

Since columns 2, 3, 4, 5 refer to the classes C_3, C_2, S_4, σ_d, which contain 8, 3, 6, 6 operations respectively, h/N will be successively 3, 8, 4, 4. This makes a total of seven equations, and we can obtain the remaining five by applying Rule 4 in the sense of the orthogonality relation between rows (as used in (1.19)). Take the relations between rows 1 and 3 and rows 2 and 3.

$$1 \times 1 \times 2 + 8 \times 1 \times \chi^{(3)}(C_3) + 3 \times 1 \times \chi^{(3)}(C_2) + 6 \times 1 \times \chi^{(3)}(S_4) + 6 \times 1 \times \chi^{(3)}(\sigma_d) = 0. \qquad (1.20a)$$

$$1 \times 1 \times 2 + 8 \times 1 \times \chi^{(3)}(C_3) + 3 \times 1 \times \chi^{(3)}(C_2) + 6 \times -1 \times \chi^{(3)}(S_4) + 6 \times -1 \times \chi^{(3)}(\sigma_d) = 0. \qquad (1.20b)$$

From these,

$$\chi^{(3)}(\sigma_d) = \chi^{(3)}(S_4) = 0.$$

Then since

$$\sum_R (\chi^{(3)}(R))^2 = 24,$$

we must have

$$\chi^{(3)}(C_3) = \pm 1, \qquad \chi^{(3)}(C_2) = \pm 2. \qquad (1.21)$$

From either (1.20a) or (1.20b), in conjunction with (1.21),

$$\chi^{(3)}(C_3) = -1, \qquad \chi^{(3)}(C_2) = +2.$$

The row for $\Gamma^{(3)}$ now reads 2, -1, 2, 0, 0. The orthogonality rule for column 2, which gives

$$\sum_{i=1}^{5} (\chi^{(i)}(C_3))^2 = 3,$$

now immediately gives

$$(\chi^{(4)}(C_3))^2 = (\chi^{(5)}(C_3))^2 = 0,$$

since

$$(\chi^{(1)}(C_3))^2 + (\chi^{(2)}(C_3))^2 + (\chi^{(3)}(C_3))^2 = 3. \qquad (1.22)$$

The information obtained so far is now tabulated as Table 1.9(b). The orthogonality relations between rows 3 and 4 and rows 3 and 5 show that $\chi^{(4)}(C_2) = \chi^{(5)}(C_2) = -1$; we have

$$1 \times 2 \times 3 + 8 \times -1 \times 0 + 3 \times 2 \times \chi^{(4)}(C_2) + 6 \times 0 \times \chi^{(4)}(S_4) + 6 \times 0 \times \chi^{(4)}\sigma_d = 0. \qquad (1.23)$$

Table 1.9(b) Part of the character table for the point group T_d.

	I	$8C_3$	$3C_2$	$6S_4$	$6\sigma_d$	
$\Gamma^{(1)}$	1	1	1	1	1	
$\Gamma^{(2)}$	1	1	1	-1	-1	(from 1.19)
$\Gamma^{(3)}$	2	-1	2	0	0	(from 1.20, 1.21)
$\Gamma^{(4)}$	3	0				
$\Gamma^{(5)}$	3	0				(from 1.22)

with a similar relation for $\chi^{(5)}$. The relations for the individual rows 4 and 5 are each of the form

$$-1 \times 3^2 + 8 \times 0^2 + 3 \times (-1)^2 + 6(\chi(S_4))^2 + 6(\chi(\sigma_d))^2 = 24,$$

giving

$$\chi(S_4) = \pm 1, \qquad \chi(\sigma_d) = \pm 1. \tag{1.24}$$

By the orthogonality relation between rows 4 and 5, if $\chi^{(4)}(S_4) = +1$, then $\chi^{(5)}(S_4) = -1$. Also, by the relation between either rows 1 and 4 or 2 and 4, if $\chi^{(4)}(S_4) = +1$, $\chi^{(4)}(\sigma_d) = -1$. Going back to the relation between rows 4 and 5, $\chi^{(5)}(\sigma_d)$ under these conditions is $+1$. The complete table therefore reads as shown in Table 1.10.

Table 1.10 Completed character table for the point group T_d.

	I	$8C_3$	$3C_2$	$6S_4$	$6\sigma_d$
$\Gamma^{(1)}$	1	1	1	1	1
$\Gamma^{(2)}$	1	1	1	-1	-1
$\Gamma^{(3)}$	2	-1	2	0	0
$\Gamma^{(4)}$	3	0	-1	1	-1
$\Gamma^{(5)}$	3	0	-1	-1	1

The labelling of the irreducible representations as $\Gamma^{(i)}$ is not very informative and the normal labelling employs the Mulliken symbols *A*, *B*, *E*, *F* (or *T*) with appropriate subscripts or superscripts.

The symbols *A*, *B* denote one-dimensional representations. Those which are symmetric with respect to rotation about the proper axis of highest order are designated *A*; those which are antisymmetric are *B*. In Table 1.10, the two one-dimensional representations $\Gamma^{(1)}$, $\Gamma^{(2)}$ are both symmetric with respect to the operation $C_3(\chi^{(1)}(C_3) = \chi^{(2)}(C_3) = +1)$ and are therefore of type *A*.

The symbols *E*, *F* (or *T*) denote two- and three-dimensional representations respectively. In Table 1.10, $\Gamma^{(3)}$ is of type *E* and $\Gamma^{(4)}$, $\Gamma^{(5)}$ are of type *F*.

Subscripts *g*, *u*, denote symmetry or antisymmetry with respect to the inversion operation. Superscripts ′, ″ denote symmetry or antisymmetry with respect to reflection at a horizontal plane of symmetry, σ_h. Since the point group T_d possesses neither i nor σ_h, these designations will not appear.

For one-dimensional representations, numerical subscripts 1, 2 denote symmetry or antisymmetry with respect to reflection at a vertical or diagonal plane. Since $\chi^{(1)}(\sigma_d) = +1$, $\Gamma^{(1)}$ is an A_1 representation, while $\chi^{(2)}(\sigma_d) = -1$, so $\Gamma^{(2)}$ is an A_2 representation. The subscripts do not have this significance in representations of higher dimensionality.

A special system of numerical subscripts exists for the one-dimensional representations of the groups D_2 and D_{2h}. In these groups there are three mutually perpendicular two-fold axes; the B representations are symmetrical with respect to rotation about only one such axis. By convention, the B_1, B_2 and B_3 representations are symmetric with respect to rotation about the z-, y- and x-axes respectively.

Although we shall show (in the next section) how to determine the infrared or Raman activity of vibrations, this is usually read directly from the character table. In addition to the characters, the tables contain two additional columns. One of these has the symbols x, y, z (or T_x, T_y, T_z), R_x, R_y, R_z opposite the irreducible representations; this means that translations in, or rotations about, these directions, transform in the same way as the irreducible representations. Vibrations belonging to irreducible representations which transform in the same way as x, y or z are infrared active. The other additional column contains various binary combinations of coordinates; if any such combination transforms in the same way as an irreducible representation, vibrations belonging to that representation are Raman active. The complete character table, with these additional columns, is given in Table 1.11. This shows that only vibrations of the

Table 1.11 Complete character table of the point group T_d.

I	II					III	IV
	I	$8C_3$	$3C_2$	$6S_4$	$6\sigma_d$		
A_1	1	1	1	1	1		$x^2+y^2+z^2$
A_2	1	1	1	-1	-1		
E	2	-1	2	0	0		$2x^2-y^2-z^2$, x^2-y^2
F_1	3	0	-1	1	-1	R_x, R_y, R_z	
F_2	3	0	-1	-1	1	x, y, z	xy, yz, zx

type F_2 are infrared active; such vibrations are also Raman active, as are those of types A_1 and E. A_2 and F_1 vibrations are inactive in both effects.

Although we can tell from inspection of the character table what activity the vibrations will display, we cannot tell how many belong to each species. However, by using the relation

$$a_j = \frac{1}{h} \sum_R \chi(R)\chi^{(j)}(R) \qquad (1.18)$$

we can determine this, and we can also determine the infrared and Raman activity without knowing the coordinate transformations of columns III and IV of Table 1.11.

1.4 Calculation of the number and activity of vibrations of a molecule XY_4 belonging to the point group T_d

(a) Determination of the infrared activity

We must begin by constructing the character $\chi(\mu)$ of the reducible representation of the dipole moment, $\Gamma(\mu)$, and then use (1.18) to determine how many times each of the irreducible representations occurs in it.

Consider rotation of a line joining the origin to a point (x, y, z) through an angle θ, in a clockwise sense, in the xy-plane. The new coordinates (x', y', z') are related to the old by the equations

$$\begin{aligned} x' &= x \cos\theta + y \sin\theta, \\ y' &= -x \sin\theta + y \cos\theta, \\ z' &= z. \end{aligned} \qquad (1.25)$$

In matrix form we have

$$\begin{bmatrix} x' \\ y' \\ z' \end{bmatrix} = \begin{bmatrix} \cos\theta & \sin\theta & 0 \\ -\sin\theta & \cos\theta & 0 \\ 0 & 0 & 1 \end{bmatrix} \begin{bmatrix} x \\ y \\ z \end{bmatrix}. \qquad (1.26)$$

The character of this matrix is $1 + 2\cos\theta$, and therefore the character in $\Gamma(\mu)$ for all such proper rotations through an angle θ is $1 + 2\cos\theta$.

If the rotation is combined with a reflection at the plane, we have

$$\begin{aligned} x'' &= +x \cos\theta + y \sin\theta, \\ y'' &= -x \sin\theta + y \cos\theta, \\ z'' &= -z, \end{aligned} \qquad (1.27)$$

and constructing the matrix corresponding to (1.26), we see that the character for such operations (called improper rotations) is $(-1 + 2\cos\theta)$. The character of the inversion operation is formally obtained by putting

$\theta = 180°$ in the expression $\chi(i) = (-1 + 2\cos\theta)$; the character of the matrix expressing the transformation

$$x''' = -x, \qquad y''' = -y, \qquad z''' = -z$$

is clearly -3 which is $(-1 + 2\cos 180°)$.

For the reflection operation the character is again $\chi(\sigma) = (-1 + 2\cos\theta)$ with $\theta = 0°$; the character of the matrix expressing the transformation

$$x^+ = -x, \qquad y^+ = y, \qquad z^+ = z$$

is -1, which is equal to $(-1 + 2\cos\theta°)$. These characters for i and σ correspond to the equivalence of these operations with S_2 and S_1 respectively. To obtain $\Gamma(\mu)$ we therefore construct Table 1.12. In this and all

Table 1.12 Calculation of $\chi(\mu)$ for the point group T_d.

	I	$8C_3$	$3C_2$	$6S_4$	$6\sigma_d$
Type of rotation		Proper		Improper	
$\theta°$	0	120	180	90	0
$\chi(\mu) = (\pm 1 + 2\cos\theta)$	3	0	-1	-1	$+1$

subsequent tables of this nature, + signs in the expression for the character refer to the proper rotations C_n^k ($I = C_1^1$); minus signs refer to the improper rotations S_n^k, $\sigma(=S_1^1)$ and $i\,(=S_2^1)$.

By comparison with Table 1.11, we can see that $\chi(\mu) = \chi(F_2)$, i.e., F_2 occurs once in $\Gamma(\mu)$ and no other irreducible representation occurs at all; therefore only vibrations of type F_2 are infrared active. We can check that this is so by applying (1.18).

Since there are 24 operations, $h = 24$.

$$a_{(A_1)} = \tfrac{1}{24}[1.1.3 + 8.1.0 + 3.1.-1 + 6.1.-1 + 6.1.1] = 0 \qquad (1.28a)$$

$$a_{(A_2)} = \tfrac{1}{24}[1.1.3 + 8.1.0 + 3.1.-1 + 6.1.-1 + 6.1.-1] = 0 \qquad (1.28b)$$

$$a_{(E)} = \tfrac{1}{24}[1.2.3 + 8.-1.0 + 3.2.-1 + 6.-1.0 + 6.1.0] = 0 \qquad (1.28c)$$

$$a_{(F_1)} = \tfrac{1}{24}[1.3.3 + 8.0.0 + 3.-1.-1 + 6.1.-1 + 6.1.-1] = 0 \qquad (1.28d)$$

$$a_{(F_2)} = \tfrac{1}{24}[1.3.3 + 8.0.0 + 3.-1.-1 + 6.-1.-1 + 6.1.1] = 1 \qquad (1.28e)$$

In (1.28a–e), every term has been written out in full in the form (Number of operations in the class × Character of the operation in the irreducible representation × Character of the operation in the reducible representation).

If vibrations of an irreducible representation are infrared active, a_j will be equal to 1; if they are inactive, a_j will be zero. Consequently, this tells us only the activity and not the number of active bands. To determine this, we have to find out how many times each irreducible representation appears in the reducible representation for vibration. This will be discussed in (c) below.

(b) Determination of the Raman activity

We must construct the character $\chi(\alpha)$ of the reducible representation for polarizability, $\Gamma(\alpha)$. Consider the polarizability referred to two sets of axes $Ox, Oy, Oz; Ox', Oy', Oz'$. If a rotation by an angle θ about the z-axis causes the components of the polarizability to undergo the changes

$$\alpha_{xx} \rightarrow \alpha_{x'x'}, \qquad \alpha_{yy} \rightarrow \alpha_{y'y'}, \text{ etc.,}$$

we have

$$\begin{bmatrix} \alpha_{x'x'} \\ \alpha_{y'y'} \\ \alpha_{z'z'} \\ \alpha_{x'y'} \\ \alpha_{y'z'} \\ \alpha_{z'x'} \end{bmatrix} = \begin{bmatrix} \cos^2\theta & \sin^2\theta & 0 & 2\sin\theta\cos\theta & 0 & 0 \\ \sin^2\theta & \cos^2\theta & 0 & -2\sin\theta\cos\theta & 0 & 0 \\ 0 & 0 & 1 & 0 & 0 & 0 \\ -\sin\theta\cos\theta & \sin\theta\cos\theta & 0 & 2\cos^2\theta - 1 & 0 & 0 \\ 0 & 0 & 0 & 0 & \pm\cos\theta & \pm\sin\theta \\ 0 & 0 & 0 & 0 & \pm\sin\theta & \pm\cos\theta \end{bmatrix} \begin{bmatrix} \alpha_{xx} \\ \alpha_{yy} \\ \alpha_{zz} \\ \alpha_{xy} \\ \alpha_{yz} \\ \alpha_{zx} \end{bmatrix}. \tag{1.29}$$

The character of the transformation matrix is

$$4\cos^2\theta \pm 2\cos\theta = 2\cos\theta(\pm 1 + 2\cos\theta),$$

for proper and improper rotations respectively. Table 1.13 gives $\chi(\alpha)$ for the operations of the group T_d.

Table 1.13 Calculation of $\chi(\alpha)$ for the point group T_d.

	I	$8C_3$	$3C_2$	$6S_4$	$6\sigma_d$
$\theta°$	0	120	180	90	0
$\chi(\alpha) = 2\cos\theta\,(\pm 1 + 2\cos\theta)$	6	0	2	0	2

Following the procedure used in the previous section, we have

$$a_{(A_1)} = \tfrac{1}{24}[1.1.6 + 8.1.0 + 3.1.2 + 6.1.0 + 6.1.2] = 1 \tag{1.30a}$$

$$a_{(A_2)} = \tfrac{1}{24}[1.1.6 + 8.1.0 + 3.1.2 + 6.-1.0 + 6.-1.2] = 0 \tag{1.30b}$$

$$a_{(E)} = \tfrac{1}{24}[1.2.6 + 8.-1.0 + 3.2.2 + 6.0.0 + 6.0.2] = 1 \tag{1.30c}$$

$$a_{(F_1)} = \tfrac{1}{24}[1.3.6 + 8.0.0 + 3.-1.2 + 6.1.0 + 6.-1.2] = 0 \tag{1.30d}$$

$$a_{(F_2)} = \tfrac{1}{24}[1.3.6 + 8.0.0 + 3.-1.2 + 6.-1.0 + 6.1.2] = 1. \tag{1.30e}$$

Vibrations of the irreducible representations A_1, E and F_2 are therefore Raman active.

(c) Determination of the number of bands belonging to each irreducible representation

We need to construct the character $\chi(v)$ of the reducible representation for vibration, $\Gamma(v)$, and this is done by calculating $\chi(m)$, the character of the reducible representation for all motions of the molecule, and subtracting from it the characters for translation ($\chi(t)$) and rotation ($\chi(r)$). ($\chi(t)$ is in fact equal to $\chi(\mu)$, since the dipole moment components transform like the coordinates.) The character for rotation, $\chi(r)$, is equal to $\pm\chi(t)$, while $\chi(m)$ is equal to $N(\pm 1 + 2\cos\theta)$, where N is the number of atoms left unshifted by each operation, and the signs +, −, refer, as in $\chi(\mu)$, to proper and improper rotations respectively.

Again, using the procedure of (1.29) and (1.30), we have

$$a_{(A_1)} = \tfrac{1}{24}[1.1.9 + 8.1.0 + 3.1.1 + 6.1.-1 + 6.1.3] = 1 \tag{1.31a}$$

$$a_{(A_2)} = \tfrac{1}{24}[1.1.9 + 8.1.0 + 3.1.1 + 6.-1.-1 + 6.-1.3] = 0 \tag{1.31b}$$

$$a_{(E)} = \tfrac{1}{24}[1.2.9 + 8.-1.0 + 3.2.1 + 6.0.-1 + 6.0.3] = 1 \tag{1.31c}$$

$$a_{(F_1)} = \tfrac{1}{24}[1.3.9 + 8.0.0 + 3.-1.1 + 6.1.-1 + 6.-1.3] = 0 \quad (1.31d)$$

$$a_{(F_2)} = \tfrac{1}{24}[1.3.9 + 8.0.0 + 3.-1.1 + 6.-1.-1 + 6.1.3] = 2. \quad (1.31e)$$

Table 1.14 Calculation of $\chi(v)$ for a molecule XY_4 of the point group T_d.

Operation	I	$8C_3$	$3C_2$	$6S_4$	$6\sigma_d$
Rotation type		Proper		Improper	
$\theta°$	0	120	180	90	0
$\cos\theta$	1	$-\frac{1}{2}$	-1	0	1
$\pm 1 + 2\cos\theta$	3	0	-1	-1	1
N	5	3	1	1	3
$\chi(m) = N(\pm 1 + 2\cos\theta)$	15	0	-1	-1	3
$\chi(t) = \chi(\mu) = (\pm 1 + 2\cos\theta)$	3	0	-1	-1	1
$\chi(r) = (+1 \pm 2\cos\theta) = \pm\chi(t)$	3	0	-1	1	-1
$\chi(v) = \chi(m) - \chi(t) - \chi(r)$	9	0	1	-1	3

The XY_4 molecule of T_d symmetry therefore has one A_1 vibration which is Raman active, one E vibration which is Raman active, and two F_2 vibrations which are both infrared and Raman active. No vibrations of the types A_2 and F_1 exist.

We can check the correctness of the calculations (1.29), (1.30) and (1.31) by comparison with Table 1.11. We should have

$$\chi(\mu) = \chi(F_2) \quad (1.32a)$$

$$\chi(\alpha) = \chi(A_1) + \chi(E) + \chi(F_2) \quad (1.32b)$$

$$\chi(v) = \chi(A_1) + \chi(E) + 2\chi(F_2). \quad (1.32c)$$

Table 1.15 Check of 1.32(a–c).

	I	$8C_3$	$3C_2$	$6S_4$	$6\sigma_d$	
$\chi(A_1)$	1	1	1	1	1	
$\chi(E)$	2	-1	2	0	0	
$\chi(F_2)$	3	0	-1	-1	1	
	6	0	2	0	2	$= \chi(\alpha)$ (Table 1.13)
$\chi(F_2)$	3	0	-1	-1	1	$= \chi(\mu)$ (Table 1.12)
	9	0	1	-1	3	$= \chi(v)$ (Table 1.14)

The calculations carried out in detail in this section can very frequently be made by inspection, as in the obvious case $\chi(\mu) = \chi(F_2)$.

2. Principles of normal coordinate analysis

2.1 The secular equation; normal and symmetry coordinates

When a molecule vibrates, the displacements of the N atoms from their equilibrium positions may be represented in terms of coordinates $q_1, q_2, \ldots, q_{3N}$.

Both the potential energy, V, and the kinetic energy, T, are expressible in terms of these coordinates; we have

$$2V = \sum_{i,j=1}^{3N} b_{ij} q_i q_j, \tag{2.1}$$

$$2T = \sum_{i,j=1}^{3N} a_{ij} \dot{q}_i \dot{q}_j, \tag{2.2}$$

where

$$b_{ij} = \left(\frac{\partial^2 V}{\partial q_i \, \partial q_j} \right)_0,$$

and the a_{ij} are functions of the atomic masses.

Since

$$\frac{\mathrm{d}}{\mathrm{d}t}\left(\frac{\partial L}{\partial \dot{q}_k} \right) - \frac{\partial L}{\partial q_k} = 0, \tag{2.3}$$

where $L = T - V$, and since T, V are functions of $\dot{q}$, q only, respectively, we may rewrite (2.3) as

$$\frac{\mathrm{d}}{\mathrm{d}t}\left(\frac{\partial T}{\partial \dot{q}_k} \right) + \frac{\partial V}{\partial q_k} = 0, \qquad k = 1, 2, \ldots, 3N - 6. \tag{2.4}$$

Differentiating (2.1), (2.2) with respect to q_k, $\dot{q}_k$, and substituting in (2.4), we obtain $3N - 6$ equations of the type

$$\sum_j (a_{jk}\ddot{q}_j + b_{jk}q_j) = 0, \tag{2.5}$$

the general solution of which is

$$q_j = A_j \sin (\lambda^{1/2}t + \varepsilon), \tag{2.6}$$

where $\lambda = 4\pi^2\nu^2$, and A, ε are the amplitude and phase, respectively, of the wave motion.

Then from (2.5), (2.6),

$$\sum_j (b_{jk} - a_{jk}\lambda)A_j = 0. \tag{2.7}$$

Apart from the trivial solution $A_j = 0$, the solutions are given by the condition

$$\det |b_{jk} - a_{jk}\lambda| = 0. \tag{2.8}$$

This determinant is the secular determinant, and will give the eigenvalues of λ. The evaluation of the determinant is simplified by transformation to a new set of coordinates Q_k, which are such that the cross-product terms in (2.1), (2.2) disappear; we employ the orthogonal transformation

$$q_k = \sum_{l=1}^{3N} C_{kl}Q_l, \tag{2.9}$$

upon which (2.1), (2.2) become

$$2V = \sum_l \lambda_l' Q_l^2 \tag{2.10}$$

$$2T = \sum_l \dot{Q}_l^2. \tag{2.11}$$

This normal coordinate transformation gives, analogously to (2.6), the expression

$$Q_l = B_l \sin (\lambda_l'^{1/2}t + \varepsilon') \tag{2.12}$$

but, since from (2.9) the q_k are linear combinations of the Q_l, $\lambda = \lambda'$. To each value of λ which is a root of the secular equation, there corresponds a so-called normal vibration of the molecule in which all coordinates Q_l vibrate with the same frequency and phase; thus the atoms attain their positions of maximum displacement simultaneously.

The solution of the secular determinant is greatly simplified if the initial coordinates are chosen such that their symmetries correspond to the irreducible representations of the molecule. If this is done, the determinant may be factored into blocks, each of which corresponds to an irreducible representation for which vibrations exist.

The required coordinates are called the symmetry coordinates, S, and these are constructed from the internal coordinates of the molecule, which are the changes in bond lengths and bond angles. The types of internal coordinate employed are bond stretching, r, bond angle deformation, ϕ, ϕ', out-of-plane bending, δ, and torsion, τ. The following rules due to Decius [332] tell us how many internal coordinates of each type are required in order to describe the normal vibrations of the molecule.

Let the molecule contain b bonds, irrespective of their nature, and a atoms; let there be a_1 atoms which participate in one bond only. If the numbers of coordinates of the various classes required are n_r, n_ϕ, n'_ϕ, n_δ, n_τ, then in the general case we have

$$\begin{aligned} n_r &= b \\ n_\phi &= 4b - 3a + a_1 \\ n_\tau &= b - a_1. \end{aligned} \qquad (2.13)$$

Consider the tetrahedral molecule XY_4. We have $b = 4$, $a = 5$, $a_1 = 4$. Then

$$\begin{aligned} n_r &= b && = 4 \\ n_\phi &= 4b - 3a + a_1 = 16 - 15 + 4 && = 5 \\ n_\tau &= b - a_1 = 4 - 4 && = 0. \end{aligned}$$

This gives nine coordinates, which is the correct number.

There are some special cases to which (2.13) is not directly applicable.

(i) The molecule contains an atom at which more than two coplanar bonds meet. The number of bonds which meet at an atom is called the multiplicity, m, of that atom. In this case, we need to replace $m - 2$ of the ϕ's by δ's. For example, the carbonate ion has a central atom of multiplicity 3. From (2.13), we would have

$$\begin{aligned} n_r &= b && = 3 \\ n_\phi &= 4b - 3a + a_1 = 12 - 12 + 3 && = 3 \\ n_\tau &= b - a_1 = 3 - 3 && = 0. \end{aligned}$$

Since $m = 3$, we must replace one of the in-plane bending coordinates ϕ by an out-of-plane coordinate, δ; we therefore have

$$n_r = 3, \qquad n_\phi = 2, \qquad n_\delta = 1, \qquad n_\tau = 0.$$

(ii) The molecule contains a linear sub-molecule of l bonds. In this case, $l - 1$ of the torsional coordinates are replaced by bending coordinates ϕ' which are in a plane perpendicular to the bending coordinates ϕ, these planes intersecting along the line of the atoms of the linear sub-molecule. Consider ketene

$$H_2C{=}C{=}O.$$

We have $b = 4$, $a = 5$, $a_1 = 3$, $l = 2$ (the C=C=O system of 2 bonds constituting the linear sub-molecule). Then (2.13) gives

$$\begin{aligned} n_r &= b && && = 4 \\ n_\phi &= 4b - 3a + a_1 &&= 16 - 15 + 3 && = 4 \\ n_\tau &= b - a_1 &&= 4 - 3 && = 1. \end{aligned}$$

But $l - 1$ of the τ must be replaced by ϕ'. Since $l - 1 = 1$, we have

$$n_r = 4, \qquad n_\phi = 4, \qquad n_{\phi'} = 1, \qquad n_\tau = 0.$$

For linear molecules we have simply

$$\begin{aligned} n_r &= b = a - 1, \\ n_\phi &= n_{\phi'} = a - 2, \end{aligned}$$

giving, correctly, $3a - 5$ coordinates; all the bending modes are doubly degenerate.

Having chosen the internal coordinates I_j according to these rules, we can construct the symmetry coordinates S_i, which will be given by

$$S_i = \sum_j U_{ij} I_j, \tag{2.14}$$

where I_j is the jth internal coordinate and the coefficients U_{ij} are chosen such that the S_i form an orthonormal set. This requires that

$$\sum_j U_{ij}^2 = 1 \qquad \text{(for normalization)} \tag{2.15}$$

and

$$\sum_k U_{ik} U_{jk} = 0 \qquad \text{(for orthogonality).} \tag{2.16}$$

There is no unique set of symmetry coordinates, but an orthonormal set can be constructed using the relation due to Nielson and Berryman [1075a].

$$S_i^{(\gamma)} = \sum_R \chi^{(\gamma)}(R) R I_k. \tag{2.17}$$

This means that, to construct a coordinate of the representation γ, we take the kth internal coordinate I_k and perform on it the symmetry operation R. This forms a new internal coordinate RI_k, which appears, multiplied by the character of R in the representation γ, in the symmetry coordinate $S_i^{(\gamma)}$.

In order to determine the coordinates formed from the generating coordinate (which may be either a single internal coordinate or a linear combination of internal coordinates), we construct a table showing the internal coordinates chosen and those to which they transform under the various symmetry operations of the molecule.

For the tetrahedral molecule XY_4 we see from (2.13) that we require four stretching and five bending coordinates. However, as Fig. 2.1 shows, we have a total of ten internal coordinates, four stretching and six bending; we therefore have a redundant coordinate. This redundancy arises because the sum of the angle displacements around the central atom is necessarily zero. Many other molecules will display one or more redundancies, and it is useful to have a rule for determining the irreducible representation to which the redundant coordinates belong. We may construct the character χ_{int} for the reducible representation in internal coordinates; it is simply the number of internal coordinates left unshifted by the symmetry operation. If we subtract from this $\chi(v)$, the character for vibration, which we constructed in chapter 1, Table 1.14, we shall be left with the character of the reducible representation, χ_{red}, for redundant coordinates. From (1.18), or by inspection, we can determine the irreducible representation to which the redundant coordinate belongs; we can often (as in the case of angle displacements around the central atom) eliminate it directly; alternatively it may be easier to leave it in, when a zero root in the secular determinant, corresponding to the redundant condition, should be obtained.

Table 2.1 Redundancy condition in a molecule XY_4 of point group T_d.

	I	$8C_3$	$3C_2$	$6S_4$	$6\sigma_d$	
$\chi(r)$ (stretching)	4	1	0	0	2	
$\chi(\phi)$ (bending)	6	0	2	0	2	
χ_{int}	10	1	2	0	4	
$\chi(v)$	9	0	1	−1	3	(Table 1.14)
χ_{red}	1	1	1	1	1	(= $\chi(A_1)$ from Table 1.11)

Since $\chi(r) = \chi(A_1) + \chi(F_2)$, and $\chi(\phi) = \chi(A_1) + \chi(E) + \chi(F_2)$, and we know that four stretching and five bending coordinates are required, we can see that the redundant coordinate is the A_1 bending coordinate.

We therefore need to construct one A_1 and one F_2 stretching coordinate, and one E and one F_2 bending coordinate. We begin by defining the internal coordinates and constructing the table of transformations.

In the transformation table (Table 2.2), the operations $C_{3(1)}$–$C_{3(4)}$ are rotations, about the C_3 axes coincident with r_1–r_4, in a clockwise sense; $C_{3(5)}$–$C_{3(8)}$ represent the corresponding anticlockwise rotations. The

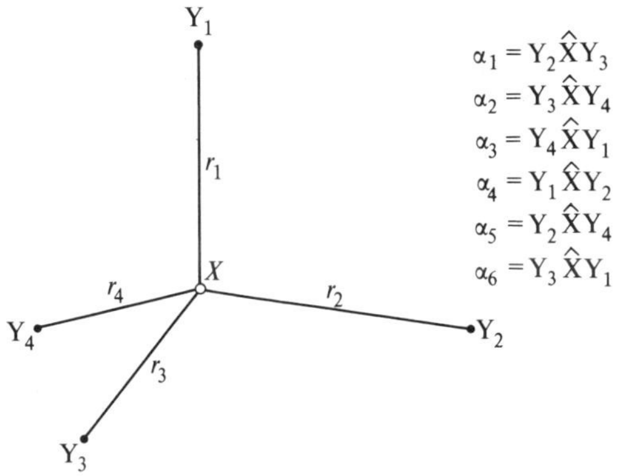

Fig. 2.1 Internal coordinates of the molecule XY_4 with T_d symmetry.

planes $\sigma_{d(1)}$–$\sigma_{d(6)}$ contain the angles α_1–α_6; the C_2-axes all pass through X and the centres of the opposite edges Y_1Y_2–Y_3Y_4, Y_2Y_3–Y_4Y_1, Y_2Y_4–Y_3Y_1. The S_4 coincide with the C_2, $S_{4(1)}$–$S_{4(3)}$ and $S_{4(4)}$–$S_{4(6)}$ representing clockwise and anticlockwise rotation respectively.

We now have all the information required for the construction of the symmetry coordinates.

(a) The A_1 stretching coordinate

Using r_1 as the generating coordinate, we have

$$S_1(A_1) = \sum_R \chi(A_1)(R).Rr_1.$$

Every operation has the character $+1$ in the A_1 representation, and from this and Table 2.2 we have

$$S_1(A_1) = \frac{1}{N}(6r_1 + 6r_2 + 6r_3 + 6r_4).$$

For normalization, we require that

$$\frac{1}{N}(6^2 + 6^2 + 6^2 + 6^2)^{1/2} = 1.$$

Hence, $N = 12$, and so

$$S_1(A_1) = \tfrac{1}{12}(6r_1 + 6r_2 + 6r_3 + 6r_4)$$
$$= \underline{\tfrac{1}{2}(r_1 + r_2 + r_3 + r_4).} \qquad (2.18)$$

We can check that this coordinate belongs to the A_1 species by performing each operation R, and confirming from Table 2.2 that

$$RS_1(A_1) = \chi(A_1)(R).S_1.$$

Table 2.2 Transformations of the internal coordinates under the symmetry operations of the point group T_d.

Internal coordinate	Operation I	$C_{3(1)}$	(2)	(3)	(4)	(5)	(6)	(7)	(8)	$C_{2(1)}$	(2)	(3)	$S_{4(1)}$	(2)	(3)	(4)	(5)	(6)	$\sigma_{d(1)}$	(2)	(3)	(4)	(5)	(6)
r_1	r_1	r_1	r_3	r_4	r_2	r_1	r_4	r_2	r_3	r_2	r_4	r_3	r_3	r_2	r_4	r_4	r_3	r_2	r_4	r_2	r_1	r_1	r_3	r_1
r_2	r_2	r_4	r_2	r_1	r_3	r_3	r_2	r_4	r_1	r_1	r_3	r_4	r_4	r_4	r_1	r_3	r_1	r_3	r_2	r_1	r_3	r_2	r_2	r_4
r_3	r_3	r_2	r_4	r_3	r_1	r_4	r_1	r_3	r_2	r_4	r_2	r_1	r_2	r_1	r_2	r_1	r_4	r_4	r_3	r_3	r_2	r_4	r_1	r_3
r_4	r_4	r_3	r_1	r_2	r_4	r_2	r_3	r_1	r_4	r_3	r_1	r_2	r_1	r_3	r_3	r_2	r_2	r_1	r_1	r_4	r_4	r_3	r_4	r_2
α_1	α_1	α_5	α_5	α_6	α_6	α_2	α_4	α_2	α_4	α_3	α_1	α_3	α_5	α_3	α_4	α_6	α_3	α_2	α_1	α_6	α_1	α_5	α_4	α_2
α_2	α_2	α_1	α_3	α_1	α_3	α_5	α_6	α_6	α_5	α_2	α_4	α_4	α_4	α_6	α_1	α_4	α_5	α_3	α_6	α_2	α_5	α_2	α_3	α_1
α_3	α_3	α_6	α_6	α_5	α_5	α_4	α_2	α_4	α_2	α_1	α_3	α_1	α_6	α_1	α_2	α_5	α_1	α_4	α_3	α_5	α_3	α_6	α_2	α_4
α_4	α_4	α_3	α_4	α_3	α_1	α_6	α_5	α_5	α_6	α_4	α_2	α_2	α_2	α_5	α_3	α_2	α_6	α_1	α_5	α_4	α_6	α_4	α_1	α_3
α_5	α_5	α_2	α_1	α_4	α_2	α_1	α_1	α_3	α_3	α_6	α_6	α_5	α_3	α_2	α_6	α_1	α_4	α_6	α_4	α_3	α_2	α_1	α_5	α_5
α_6	α_6	α_4	α_2	α_2	α_4	α_3	α_3	α_1	α_1	α_5	α_5	α_6	α_1	α_4	α_5	α_3	α_2	α_5	α_2	α_1	α_4	α_3	α_6	α_6

Since $\chi(A_1)(R)$ is $+1$ for every operation, each operation should give $RS_1 = S_1$. This check is valid for coordinates belonging to any one-dimensional representation.

(b) The *E* bending coordinate

Using α_1 as a generating coordinate, and $\chi(E)$ from Table 1.11,

	I	$8C_3$	$3C_2$	$6S_4$	$6\sigma_d$
$\chi(E)$	2	-1	2	0	0

we have

$$\begin{aligned} S_{2a}(E) &= 2\alpha_1 - (2\alpha_5 + 2\alpha_6 + 2\alpha_4 + 2\alpha_2) + 2(\alpha_1 + 2\alpha_3) \\ &= 2(2\alpha_1 - \alpha_2 + 2\alpha_3 - \alpha_4 - \alpha_5 - \alpha_6) \end{aligned}$$

and on normalizing,

$$S_{2a}(E) = 12^{-1/2}(2\alpha_1 - \alpha_2 + 2\alpha_3 - \alpha_4 - \alpha_5 - \alpha_6). \qquad (2.19a)$$

In order to check the correctness of a degenerate pair of coordinates, we need both members of the pair. They must be orthogonal, and we find that no single coordinate will give an orthogonal coordinate to S_{2a}; however, the linear combination $(\alpha_2 - \alpha_5)$ on applying (2.17) is found to give

$$S_{2b}(E) = \tfrac{1}{2}(\alpha_2 + \alpha_4 - \alpha_5 - \alpha_6). \qquad (2.19b)$$

Using (2.16), we have

$$\begin{aligned} \sum U_{S_{2a} \cdot S_{2b}} &= (2 \times 12^{-1/2} \times 0) + (-12^{-1/2} \times \tfrac{1}{2}) + (2 \times 12^{-1/2} \times 0) \\ &\quad + (-12^{-1/2} \times \tfrac{1}{2}) + (-12^{-1/2} \times -\tfrac{1}{2}) + (-12^{-1/2} \times -\tfrac{1}{2}) = 0. \end{aligned}$$

When degenerate coordinates undergo a symmetry operation, they transform to a linear combination of themselves. We have

$$\begin{aligned} RS_{2a} &= xS_{2a} + yS_{2b} \\ RS_{2b} &= x'S_{2a} + y'S_{2b}. \end{aligned}$$

If we form the matrix

$$\begin{bmatrix} x & y \\ x' & y' \end{bmatrix}$$

the required check is that the character of the matrix, $x + y'$, is equal to the character of R in the representation to which the coordinates belong.

$$\begin{aligned} I(S_{2a}) &= 1.S_{2a} + 0.S_{2b} \\ I(S_{2b}) &= 0.S_{2a} + 1.S_{2b} \\ &= 2 = \chi(E)(I). \end{aligned}$$

For the operation $C_{3(1)}$ we have, from Table 2.2,

$$C_{3(1)}(S_{2a}) = 12^{-1/2}(2\alpha_5 - \alpha_1 + 2\alpha_6 - \alpha_3 - \alpha_2 - \alpha_4) \quad (2.20a)$$
$$C_{3(1)}(S_{2b}) = \tfrac{1}{2}(\alpha_3 + \alpha_1 - \alpha_4 - \alpha_2) \quad (2.20b)$$
$$C_{3(1)}(S_{2a}) = xS_{2a} + yS_{2b}$$
$$C_{3(1)}(S_{2b}) = x'S_{2a} + y'S_{2b}.$$

Equating the coefficients of α_1, α_2 in (2.19a), (2.19b), (2.20a), we have

$$-12^{-1/2} = 2 \times 12^{-1/2}x + 0y$$
$$-12^{-1/2} = -12^{-1/2}x + \tfrac{1}{2}y,$$

from which $x = -\frac{1}{2}$, $y = -12^{-1/2}$, and equating the same coefficients in (2.19a), (2.19b), (2.20b),

$$\tfrac{1}{2} = 2 \times 12^{-1/2}x' + 0y'$$
$$-\tfrac{1}{2} = -12^{-1/2}x' + \tfrac{1}{2}y',$$

from which $x' = 12^{1/2} \times 4^{-1} = 3^{1/2} \times 2^{-1}$, $y' = -\frac{1}{2}$. The matrix is therefore

$$\begin{bmatrix} -\frac{1}{2} & -12^{-1/2} \\ 3^{1/2} \times 2^{-1} & -\frac{1}{2} \end{bmatrix}$$

and its character is equal to -1, which is equal to $\chi(E)(C_3)$.

The reader may check that the matrices constructed in the same way for the operations C_2, S_4 and σ_d are, in that order,

$$\begin{bmatrix} 1 & 0 \\ 0 & 1 \end{bmatrix}; \qquad \chi = 2 = \chi(E)(C_2),$$

$$\begin{bmatrix} -\frac{1}{2} & -1 \times 12^{-1/2} \\ -3^{1/2} \times 2^{-1} & \frac{1}{2} \end{bmatrix}; \qquad \chi = 0 = \chi(E)(S_4),$$

$$\begin{bmatrix} 1 & 0 \\ 0 & -1 \end{bmatrix}; \qquad \chi = 0 = \chi(E)(\sigma_d).$$

(c) The F_2 stretching coordinate

It is not always necessary to choose a single internal coordinate as generating coordinate; in the case of the F_2 stretching coordinate, a set of orthogonal symmetry coordinates is most easily constructed from a linear combination.

The internal coordinate r_1 would generate a symmetry coordinate

$$3r_1 - (r_2 + r_3 + r_4) - (2r_2 + 2r_3 + 2r_4) + (3r_1 + r_2 + r_3 + r_4) = 6r_1 - 2r_2 - 2r_3 - 2r_4.$$

Similarly, r_2 would generate $6r_2 - 2r_3 - 2r_4 - 2r_1$. Consequently the linear combination $(r_1 + r_2)$ would give, when normalized,

$$S_{3a}(F_2) = \tfrac{1}{2}(r_1 + r_2 - r_3 - r_4). \tag{2.21a}$$

Orthogonal coordinates $S_{3b}(F_2)$, $S_{3c}(F_2)$ may be derived from $(r_1 + r_3)$ and $(r_1 + r_4)$.

$$S_{3b}(F_2) = \tfrac{1}{2}(r_1 - r_2 + r_3 - r_4), \tag{2.21b}$$
$$S_{3c}(F_2) = \tfrac{1}{2}(r_1 - r_2 - r_3 + r_4). \tag{2.21c}$$

The transformation properties are checked in the same way as those of the E bending coordinate; we shall now have

$$\begin{aligned} R(S_{3a}) &= x(S_{3a}) + y(S_{3b}) + z(S_{3c}) \\ R(S_{3b}) &= x'(S_{3a}) + y'(S_{3b}) + z'(S_{3c}) \\ R(S_{3c}) &= x''(S_{3a}) + y''(S_{3b}) + z''(S_{3c}) \end{aligned}$$

and it will be necessary for $x + y' + z''$ to be equal to $\chi(F_2)(R)$.

(d) The F_2 bending coordinate

Having chosen the set of F_2 stretching coordinates generated by $(r_1 + r_2)$, $(r_1 + r_3)$, $(r_1 + r_4)$, we must choose, as generators of the bending coordinate, internal coordinates which are invariant to the same symmetry operations. One possible set of choices is the angles enclosed by the bonds, namely $\alpha_4(Y_1XY_2)$, $\alpha_6(Y_3XY_1)$ and $\alpha_3(Y_1XY_4)$.

From Table 2.2, we can check that $r_1 + r_2$ is invariant under I, $C_{2(1)}$, $\sigma_{d(2)}$ and $\sigma_{d(4)}$, and that α_4 is invariant under the same set of operations. Using α_4 as the generator, we have

$$\begin{aligned} S_{4a}(F_2) &= 3\alpha_4 - (\alpha_4 + 2\alpha_2) - (2\alpha_2 + \alpha_3 + \alpha_4 + \alpha_5 + \alpha_6) \\ &\quad + (2\alpha_4 + \alpha_3 + \alpha_4 + \alpha_5 + \alpha_6) \\ &= 4(\alpha_4 - \alpha_2) \\ &= 2^{-1/2}(\alpha_4 - \alpha_2) \text{ on normalization.} \end{aligned} \tag{2.22a}$$

Similarly, we have

$$S_{4b}(F_2), \text{ generated from } \alpha_6, = 2^{-1/2}(\alpha_6 - \alpha_5) \tag{2.22b}$$

and

$$S_{4c}(F_2), \text{ generated from } \alpha_3, = 2^{-1/2}(\alpha_3 - \alpha_1). \tag{2.22c}$$

We need to use only one of the pairs S_{3i}, S_{4i} ($i = a, b, c$) in the actual normal coordinate analysis; however, we need to construct all the members of the degenerate set in order to verify the transformation properties and the orthogonality.

2.2 Potential and kinetic energy

(a) The construction of the *F* matrix

If, instead of using the displacement coordinates q_i, q_j of eq. (2.1), we use internal coordinates I_i, I_j, we may write

$$2V = \sum f_{ij} I_i I_j, \tag{2.23a}$$

where the f_{ij} are force constants and $f_{ij} = f_{ji}$. On transforming the symmetry coordinates, we can rewrite (2.23a) in the form

$$2V = \sum F_{ij} S_i S_j, \tag{2.23b}$$

where the F_{ij} $(=F_{ji})$ are linear combinations of the f_{ij}. From the above, we have

$$\sum f_{ij} I_i I_j = \sum F_{ij} S_i S_j. \tag{2.24}$$

Each side of (2.24) is expressible in matrix form. If transposes of the matrices I, S, are $I^\dagger$, $S^\dagger$, then

$$I^\dagger f I = S^\dagger F S \tag{2.25}$$

where I, S are the column matrices consisting of the internal and symmetry coordinates respectively. But S is obtained from I by a linear transformation (2.14) which is also expressible as a matrix U. That is,

$$S = UI \tag{2.26}$$

$$\therefore\ U^{-1}S = U^{-1}(UI) = EI = I. \tag{2.27}$$

(E is the unit matrix, and the inverse U^{-1} of U is defined by the relation $U^{-1}U = E$.)

Since the symmetry coordinates are orthonormal, the inverse U^{-1} and the transpose $U^\dagger$ of U are equal, so (2.27) becomes

$$I = U^\dagger S. \tag{2.28}$$

$I^\dagger$ is the transposed product of the transposes of $U^\dagger$ and S:

$$I^\dagger = (U^\dagger S)^\dagger = S^\dagger U, \tag{2.29}$$

so (2.25) becomes

$$S^\dagger U f U^\dagger S = S^\dagger F S$$

$$\therefore\ \underline{U f U^\dagger = F.} \tag{2.30}$$

Now we have already obtained the symmetry coordinates belonging to each irreducible representation; if there are m internal coordinates and n symmetry coordinates, the U matrix will have n rows and m columns.

The U matrices for the irreducible representations of the vibrations of tetrahedral XY_4 are as shown in Tables 2.3–2.5, from (2.18), (2.19), (2.22) and (2.23). Having obtained the U matrices, we now need to write down

Table 2.3 The matrix $U(A_1)$.

	r_1	r_2	r_3	r_4	α_1	α_2	α_3	α_4	α_5	α_6
S_1	$\frac{1}{2}$	$\frac{1}{2}$	$\frac{1}{2}$	$\frac{1}{2}$	0	0	0	0	0	0

Table 2.4 The matrix $U(E)$.

	r_1	r_2	r_3	r_4	α_1	α_2	α_3	α_4	α_5	α_6
S_{2a}	0	0	0	0	$2\times 12^{-1/2}$	$-12^{-1/2}$	$2\times 12^{-1/2}$	$-12^{-1/2}$	$-12^{-1/2}$	$-12^{-1/2}$
or										
S_{2b}	0	0	0	0	0	$\frac{1}{2}$	0	$\frac{1}{2}$	$-\frac{1}{2}$	$-\frac{1}{2}$

Table 2.5 The matrix $U(F_2)$.

	r_1	r_2	r_3	r_4	α_1	α_2	α_3	α_4	α_5	α_6
S_{3a}	$\frac{1}{2}$	$\frac{1}{2}$	$-\frac{1}{2}$	$-\frac{1}{2}$	0	0	0	0	0	0
S_{4a}	0	0	0	0	0	$-2^{-1/2}$	0	$2^{-1/2}$	0	0
or										
S_{3b}	$\frac{1}{2}$	$-\frac{1}{2}$	$\frac{1}{2}$	$-\frac{1}{2}$	0	0	0	0	0	0
S_{4b}	0	0	0	0	0	0	0	0	$-2^{-1/2}$	$2^{-1/2}$
or										
S_{3c}	$\frac{1}{2}$	$-\frac{1}{2}$	$-\frac{1}{2}$	$\frac{1}{2}$	0	0	0	0	0	0
S_{4c}	0	0	0	0	$-2^{-1/2}$	0	$2^{-1/2}$	0	0	0

the f matrix. This is simply a convenient way of expressing the potential-energy function (2.23a), which, if we include all interactions, takes the form

$$2V = f_r(\sum r_i^2) + d^2 f_\alpha(\sum \alpha_i^2) + 2f_{rr}(\sum_{i<j} r_{ij}) + 2d^2 f_{\alpha\alpha}(\sum_{i<j} \alpha_{ij}) + 2d^2 f_{\alpha\alpha'}(\sum_{i<j} \alpha'_{ij}) + 2df_{r\alpha}(\sum r_i\alpha_j) + 2df_{r\alpha'}(\sum r_i\alpha'_j). \quad (2.31)$$

In this expression, the force constants $f_{\alpha\alpha}, f_{\alpha\alpha'}$ refer to bending interactions between angle coordinates with and without a common side respectively; similarly $f_{r\alpha}, f_{r\alpha'}$ refer respectively to stretch–bend interactions where the bond specified by the stretching coordinate is, or is not, a side of the angle specified by the bending coordinate. Multiplication by d or d^2

causes the units of the FG product to have the same dimensions, T^{-2}; the force constant unit normally employed is the millidyne per angstrom (mdyn $Å^{-1}$).

Expressing (2.31) as a matrix (the f matrix) we obtain Table 2.6. We now carry out the matrix multiplication $F = UfU^{\dagger}$ for each irreducible representation.

Table 2.6 The f matrix.

	r_1	r_2	r_3	r_4	α_1	α_2	α_3	α_4	α_5	α_6
r_1	f_r	f_{rr}	f_{rr}	f_{rr}	$df_{r\alpha'}$	$df_{r\alpha'}$	$df_{r\alpha}$	$df_{r\alpha}$	$df_{r\alpha'}$	$df_{r\alpha}$
r_2		f_r	f_{rr}	f_{rr}	$df_{r\alpha}$	$df_{r\alpha'}$	$df_{r\alpha'}$	$df_{r\alpha}$	$df_{r\alpha}$	$df_{r\alpha'}$
r_3			f_r	f_{rr}	$df_{r\alpha}$	$df_{r\alpha}$	$df_{r\alpha'}$	$df_{r\alpha'}$	$df_{r\alpha'}$	$df_{r\alpha}$
r_4				f_r	$df_{r\alpha'}$	$df_{r\alpha}$	$df_{r\alpha}$	$df_{r\alpha'}$	$df_{r\alpha}$	$df_{r\alpha'}$
α_1					d^2f_{α}	$d^2f_{\alpha\alpha}$	$d^2f_{\alpha\alpha'}$	$d^2f_{\alpha\alpha}$	$d^2f_{\alpha\alpha}$	$d^2f_{\alpha\alpha}$
α_2						d^2f_{α}	$d^2f_{\alpha\alpha}$	$d^2f_{\alpha\alpha'}$	$d^2f_{\alpha\alpha}$	$d^2f_{\alpha\alpha}$
α_3							d^2f_{α}	$d^2f_{\alpha\alpha}$	$d^2f_{\alpha\alpha}$	$d^2f_{\alpha\alpha}$
α_4								d^2f_{α}	$d^2f_{\alpha\alpha}$	$d^2f_{\alpha\alpha}$
α_5									d^2f_{α}	$d^2f_{\alpha\alpha'}$
α_6										d^2f_{α}

We first form the product $fU^{\dagger} = Z$ and then $F = UZ$. The rule for matrix multiplication is

$$c_{rs} = \sum_t a_{rt} b_{ts},$$

that is,

$$Z_{rs} = \sum_t f_{rt} U^{\dagger}{}_{ts} = \sum_t f_{rt} U_{st}.$$

Using the $U(A_1)$ matrix (Table 2.3), we have

$$Z(A_1) = \begin{bmatrix} \frac{1}{2}f_r + \frac{3}{2}f_{rr} \\ \frac{1}{2}f_r + \frac{3}{2}f_{rr} \\ \frac{1}{2}f_r + \frac{3}{2}f_{rr} \\ \frac{1}{2}f_r + \frac{3}{2}f_{rr} \\ 0 \\ \vdots \\ 0 \end{bmatrix}$$

and $F(A_1) = \sum_t U_{rt}(A_1) Z_{ts}(A_1)$.

Since $U(A_1)$ has one row and ten columns, and $Z(A_1)$ has ten rows and one column, the product UZ will have one row and one column and this is the single element

$$F(A_1) = F_{11} = f_r + 3f_{rr}. \tag{2.32}$$

In forming $F(E)$ we may use either version of Table 2.4; less work is involved if we use the matrix based on S_{2b}. The $Z(E)$ matrix is

$$Z(E) = \begin{bmatrix} 0 \\ \vdots \\ 0 \\ \frac{1}{2}A \\ 0 \\ \frac{1}{2}A \\ -\frac{1}{2}A \\ -\frac{1}{2}A \end{bmatrix},$$

where

$$A = d^2(f_\alpha + f_{\alpha\alpha'} - 2f_{\alpha\alpha}), \quad \text{and} \quad F(E) = F_{22} = d^2(f_\alpha + f_{\alpha\alpha'} - 2f_{\alpha\alpha}). \tag{2.33}$$

Finally $F(F_2)$ may be formed using any one of three versions of Table 2.5; if we use the U matrix based on S_{3a}, we obtain

$$Z(F_2) = \begin{bmatrix} \frac{1}{2}(f_r - f_{rr}) & 2^{-1/2}d(f_{r\alpha} - f_{r\alpha'}) \\ \frac{1}{2}(f_r - f_{rr}) & 2^{-1/2}d(f_{r\alpha} - f_{r\alpha'}) \\ -\frac{1}{2}(f_r - f_{rr}) & -2^{-1/2}d(f_{r\alpha} - f_{r\alpha'}) \\ -\frac{1}{2}(f_r - f_{rr}) & -2^{-1/2}d(f_{r\alpha} - f_{r\alpha'}) \\ 0 & 0 \\ -d(f_{r\alpha} - f'_{r\alpha}) & -2^{-1/2}d^2(f_\alpha - f_{\alpha\alpha'}) \\ 0 & 0 \\ d(f_{r\alpha} - f_{r\alpha'}) & -2^{-1/2}d^2(f_\alpha - f_{\alpha\alpha'}) \\ 0 & 0 \\ 0 & 0 \end{bmatrix},$$

and the 2×2 matrix $F(F_2)$ is

$$\begin{bmatrix} f_r - f_{rr} & \sqrt{2}\, d(f_{r\alpha} - f_{r\alpha'}) \\ \sqrt{2}\, d(f_{r\alpha} - f_{r\alpha'}) & d^2(f_\alpha - f_{\alpha\alpha'}) \end{bmatrix}. \tag{2.34}$$

(These elements are labelled

$$\begin{bmatrix} F_{33} & F_{34} \\ F_{43} & F_{44} \end{bmatrix},$$

corresponding to the numbers of the symmetry coordinates. It is not necessary to work out both F_{34} and F_{43}, since they are equal, but it is a useful check on the correctness of the multiplication.)

In the expressions (2.32), (2.33), we have used the general valence force field (GVFF), in which all quadratic terms are included. If we had a tetrahedral molecule XABCD, the f matrix would have contained $\frac{1}{2}n(n + 1) = 55$ different force constants. A molecule XY_4 requires only 7 force constants in the f matrix; this is still 3 more than the number of vibrations. In the factored F matrices we see that $f_{\alpha\alpha}$ does not appear on its own, and $f_{r\alpha}$, $f_{r\alpha'}$ appear only as the linear combination $f_{r\alpha} - f_{r\alpha'}$. This reduces the number of independent force constants to five. Even in this favourable case, we have more force constants than vibrations; it is nearly always necessary to modify the GVFF function, in order that the number of independent force constants should not exceed the number of vibrations, though additional information about force constants may be obtained from isotopic substitution, mean amplitudes of vibration and Coriolis coupling constants; this may enable more constants to be included.

(b) Choice of potential field

Two simple types of field, originally due to Bjerrum, are the central force field and the simple valence force field. In the central force field, forces are assumed to act only along the lines joining atoms. Since Dennison showed in 1926 that the central force field predicted a zero bending frequency for linear XY_2 molecules, this field has been little used in the calculation of molecular vibration frequencies; it is still sometimes used in simple calculations of ionic lattice frequencies.

The potential function (2.31), under the assumption of central forces, would become

$$2V = f_r(\sum r_i^2) + f_{r'}(\sum r_i) + f_q(\sum q_i^2) + f_q(\sum q_i), \qquad (2.35)$$

where the q_i are the changes in distances between non-bonded atoms. The linear terms arise because, under the central force field, all forces do not vanish at equilibrium. However, consideration of these forces gives the relation (for a tetrahedral XY_4 molecule):

$$f_{r'} = -\sqrt{6}\, f_{q'},$$

reducing the number of independent constants to three.

The simple valence force field (SVFF) has proved more useful; the forces considered are those directly resisting changes in bond length and bond angle. No forces between non-bonded atoms are included, and the potential function is

$$2V = f_r(\sum r_i^2) + f_\alpha(\sum \alpha^2). \qquad (2.36)$$

The stretching force constants f_r for various bonds were found to be approximately transferable from one molecule to another, leading to the idea of characteristic frequencies for various functional groups.

The assumptions underlying the two fields discussed above were combined by Urey and Bradley [1449a] in the force field named after them (UBFF). This field includes both valence forces and central forces between non-bonded atoms. Like the central force field, the UBFF includes linear and quadratic terms, and for a tetrahedral XY_4 molecule the potential function can be written

$$2V = f_r(\sum r_i^2) + f_q(\sum q_i^2) + f_\alpha(\sum f_\alpha^2) + f_{r'}(\sum r_i) + f_{q'}(\sum q_i) + f_{\alpha'}(\sum \alpha_i). \qquad (2.37a)$$

Although this expression contains six force constants, they are not all required, since the central-force relation $f_{r'} = -\sqrt{6}\, f_{q'}$ is found to apply. Further, the form of the potential field imposes a relation between f_q and $f_{q'}$; if we assume $V = a/r^m - b/r^n$, we find the condition $f_{q'}/f_q = -1/(n+1)$. Finally, the number of independent constants may be reduced to three, by neglecting $f_{\alpha'}$. In most UBFF calculations f_r, f_q, f_α and $f_{q'}$ are replaced by the symbols K, F, H and F' ($F' = -\frac{1}{10}F$).

This simple Urey–Bradley force field has been studied by many workers, and modifications have been made, e.g., by Torkington [1435], Janz and Mikawa [698] and Shimanouchi and co-workers [1284]. Because of the more exact nature of the potential function, UBFF force constants are more accurately transferable than SVFF force constants. The advantage of the UBFF is that the interaction constants explicitly required in the GVFF are lumped together, thus reducing their number; even if some interaction constants are included in a modified UBFF, there is usually a saving. For instance, the modified UBFF applied to octahedral complexes by Shimanouchi and Nakagawa requires seven quadratic constants, whereas the GVFF for the same molecule (assuming O_h symmetry) requires fourteen. However, this saving can equally well be achieved if the GVFF function is initially written out in full and relations derived between the GVFF constants on purely chemical grounds. Such an approach has been attempted by Jones [710] in his study of $Ni(CO)_4$. Of the 21 GVFF constants required, 8 occur as linear combinations, leaving 17 independent constants. Using the hypothesis that the interaction constants arise mainly from π-electron interactions, he developed relationships whereby the 17 independent constants were reduced to 10, of which two were negligible.

The general valence force field is usually taken to be described by a potential function of the type given in eq. (2.31), in which all quadratic terms have been included. Such a force field has also been referred to as the most general harmonic force field (MGHFF). Duncan [396] has shown that this general force field is poorly approximated to by the UBFF for molecules in which the atomic repulsions are non-coplanar. In effect the UBFF neglects angle–angle and (non-bonded) bond–angle interactions in, for instance, pyramidal XY_3 molecules such as the group V halides and hydrides. If additional constants are included to represent these, there is no effective saving of force constants by using a MUBFF rather than a GVFF. The same considerations apply to tetrahedral molecules. Curtis [305] has considered a 'most general' Urey–Bradley force field, the potential function being

$$2V = \sum K_i \Delta r_i^2 + \sum H_i \Delta\alpha_i^2 + \sum F_i \Delta q_i^2 + 2\sum q_i F_i' \Delta q_i + 2\sum k_{ij} \Delta r_i \Delta r_j + 2\sum h_{ij} \Delta\alpha_i \Delta\alpha_j + 2\sum' h_{ij} \Delta\alpha_i \Delta\alpha_j + 2\sum f_{ij} \Delta r_i \Delta\alpha_j + 2\sum' f_{ij} \Delta r_i \Delta\alpha_j \quad (2.37b)$$

where the primed summations refer to coordinates with no common bond, the unprimed to those with a common bond. In his analysis of the applicability of such a field to $CO_3^=$, NO_3^-, NO_2^- and NH_4^+, he found that the vibrational frequencies of the three planar anions could be computed using K, H, F, F', k and f only, though in all cases the relation between F' and F of $F' = -0{\cdot}1F$ was not even approximately adhered to. For NH_4^+ an additional constant $\rho = -\frac{27}{32}r^2F'$, expressing the intermolecular tension, was required in the calculation; further, the values of H for the E and F_2 blocks differ. The general impression obtained from these two studies is that even a substantially modified UBFF is satisfactory only for planar molecules. The application of the UBFF to octahedral molecules has been criticized by Kim *et al.* [758]. These workers based their criticisms on the suggestion by Aldous and Mills [46, 47] that force fields expressed in terms of coordinates which include redundant angular coordinates round a central atom should not be employed. Neither the UBFF nor the GVFF is exempt from this criticism. Kim therefore resurrected the orbital valency force field (OVFF) in which this difficulty does not arise. Heath and Linnett [626] were the first to conceive the OVFF. They first applied it to the problem of the out-of-plane bending mode (A_2'') of planar XY_3 molecules. In both the UBFF and GVFF the force constant characterizing this mode is not related to any of the in-plane constants. The basic assumption of the OVFF for such a molecule is that the bond-forming orbitals of X are set at definite angles to one another, and the most stable bond is produced when one of these orbitals overlaps the bond-forming orbital of Y to the maximum extent. In planar XY_3 the restoring forces for an angular displacement out of the plane and in the plane are equal. Thus we may write a contribution to the total potential energy due to an angular displacement θ of a bond from the axis of the orbital (ij) as $k_\theta \Delta\theta_{ij}^2$. Heath and Linnett gave the total potential energy for a planar XY_3 molecule as

$$2V = 2B' \sum_j \Delta r_{ij} + k_1 \sum_j \Delta r_{ij}^2 + k_\theta \sum_j \Delta\theta_{ij}^2 - 2B \sum_{jk} \Delta R_{jk} + 2A \sum_{jk} \Delta R_{jk}^2,$$

where i denotes the central atom, j, k the outer atoms, Δr_{ij} and ΔR_{jk} are increases in bond length and non-bonded distance respectively. The second and third terms are the OVFF terms; the last two represent repulsions between non-bonded atoms expressed in the form

$$V = A\,\Delta R^2 - B\,\Delta R$$

where

$$2A = \frac{d^2V}{dR^2} \quad \text{and} \quad B = -\frac{dV}{dR}.$$

If a functional relationship between V and R is assumed, a ratio of A to B/R is automatically fixed.

The first term is included to balance the fourth such that V is a minimum when $\Delta r = \Delta R = \Delta\theta = 0$. Heath and Linnett then extended their treatment to tetrahedral halides [627], tetrahedral anions [628] and octahedral halides [629]; the treatment passed the not very rigorous test of proving superior to the simple valence force field. Krebs and Mueller [779, 783] in two fairly recent papers, studied first the $CS_3^=$ ion, and then tetrahedral oxo-anions. They also reviewed previous use of the OVFF and UBFF to give very similar results. For tetrahedral oxo-anions they found that negative values of B/R, which are meaningless, were obtained if a Lennard–Jones repulsive potential $V = a/R^{12}$ was assumed.

This gives $A = 6{\cdot}5B/R$, and not only leads to meaningless B/R values but fails to give the correct values for $\nu_2(E)$ and $\nu_4(F_2)$. The situation is highly reminiscent of that obtained with the UBFF where satisfactory results cannot be obtained for many non-planar molecules unless the idea of a purely van der Waals potential between non-bonded atoms, leading to $E' = -0{\cdot}1F$, is abandoned. Krebs and Mueller suggested the inclusion of Coulomb forces in the non-bonded potential. This leads to the abandonment of the fixed ratio of A to B/R, and physically meaningful values of B/R are then obtainable, as are the correct bending frequencies. In the context of these findings, Kim's statement that

> it is difficult to understand why such an appealingly simple idea (as the OVFF) has not been more thoroughly tested on various types of molecules

finds a ready reply. However, Kim did succeed in getting much closer agreement with a four-constant OVFF function, for the fifteen hexafluorides studied, than could be obtained with a four-constant UBFF function, particularly with respect to the bending modes. Kim also agreed with the conclusion of Krebs and Mueller that Coulomb forces between non-bonded atoms must be taken into consideration.

Finally, it must be pointed out that force constants have significance only with respect to the potential field in which they appear. In the majority of cases, it would seem to be satisfactory to determine the form of the F matrix elements, and to calculate these without splitting them up into their component GVFF force constants. This will be discussed further in the next section.

(c) Construction of the G matrix

It has been shown [18] that the kinetic energy can be expressed as

$$2T = \sum_{ij} (g^{-1})_{ij}\dot{I}_i\dot{I}_j \tag{2.38a}$$

in internal coordinates, or as

$$2T = \sum_{ij} (G^{-1})_{ij}\dot{S}_i\dot{S}_j \tag{2.38b}$$

in symmetry coordinates.

Following the steps (2.24–2.30) for the construction of the F matrix, we find that the relation

$$\underline{G = UgU^{\dagger}} \tag{2.39}$$

is to be used in the construction of the G matrix.

The g elements may be derived from the expression

$$g_{tu} = \frac{1}{m_i} B_{ti}B_{ui}, \qquad t, u = 1, 2, \ldots, 3N - 6, \tag{2.40}$$

where the coefficients B_{ti} form the matrix whereby the Cartesian displacement coordinates $(x_1, \ldots, x_{3N-6})$ are transformed to the internal coordinates

$$I_t = \sum B_{ti}x_i. \tag{2.41}$$

In matrix form $g = BM^{-1}B^{\dagger}$, where M^{-1} is the diagonal matrix consisting of the reciprocal masses of the atoms. Instead of using these Cartesian displacement coordinates, we may use a vector $\mathbf{V}_a$ for each atom, whose components in the directions of the axes are the Cartesian displacement coordinates. Then (2.41) becomes

$$I_t = \sum_a \mathbf{S}_{ta}.\mathbf{V}_a \tag{2.42}$$

where $\mathbf{S}_{ta}$ is a vector associated with the internal coordinate I_t and the atom a. Let all atoms but a be in their equilibrium positions. Then the direction of $\mathbf{S}_{ta}$ will be one such that a displacement of a will produce the maximum change in I_t, while the magnitude $|S_{ta}|$ will be equal to the increase in I_t produced by unit displacement of a in that direction.

The construction of vectors $\mathbf{S}_{ta}$ for various types of displacement is dealt with by Wilson, Decius and Cross [18]; once the $\mathbf{S}_{ta}$ are known, the g elements can be found from

$$g_{tu} = \sum_{a=1}^{N} \mu_a \mathbf{S}_{ta} . \mathbf{S}_{ua} \tag{2.43}$$

and in this case we have $g = SM^{-1}S^{+}$, where $\mu_a = 1/m_a$.

The g matrix for a bent triatomic molecule is now constructed from (2.41). Consider a bent triatomic molecule, XY_2, lying in the xz-plane with the z-axis corresponding to the C_2-axis of the molecule. We can define internal coordinates, and by eq. (2.13) we know these to be 2 stretching and 1 in-plane bending: an obvious choice is Δr_1, Δr_2, $\Delta\alpha$.

Figure 2.2 illustrates this.

To obtain the g matrix we must first construct the B matrix. This relates the displacements of the internal coordinates (I) to the displacements of the Cartesian coordinates (X) and we express this relation as $I = BX$. Figures 2.3a, b and c show the relations between the displacements of the Cartesian coordinates and of r_1, r_2 and α respectively.

In matrix form, $I = BX$, where I, X are the column vectors of internal and Cartesian coordinates respectively:

$$\begin{bmatrix} \Delta r_1 \\ \Delta r_2 \\ \Delta\alpha \end{bmatrix} = \begin{bmatrix} -S & 0 & -C & 0 & 0 & 0 & S & 0 & C \\ 0 & 0 & 0 & S & 0 & -C & -S & 0 & C \\ -\frac{C}{r} & 0 & \frac{S}{r} & \frac{C}{r} & 0 & \frac{S}{r} & 0 & 0 & -\frac{2S}{r} \end{bmatrix} \begin{bmatrix} \Delta x_1 \\ \Delta y_1 \\ \Delta z_1 \\ \Delta x_2 \\ \Delta y_2 \\ \Delta z_2 \\ \Delta x_3 \\ \Delta y_3 \\ \Delta z_3 \end{bmatrix}$$

where $S = \sin(\alpha/2)$, $C = \cos(\alpha/2)$, r is the equilibrium X–Y bond length.

The matrix M^{-1} in the equation $g = BM^{-1}B^{\dagger}$ is a 9×9 diagonal matrix: if its elements are m_{ii} then $m_{ii} = \mu_1$ ($i = 1, 2, 3$); $m_{ii} = \mu_2 = \mu_1$ ($i = 4, 5, 6$); $m_{ii} = \mu_3$ ($i = 7, 8, 9$).

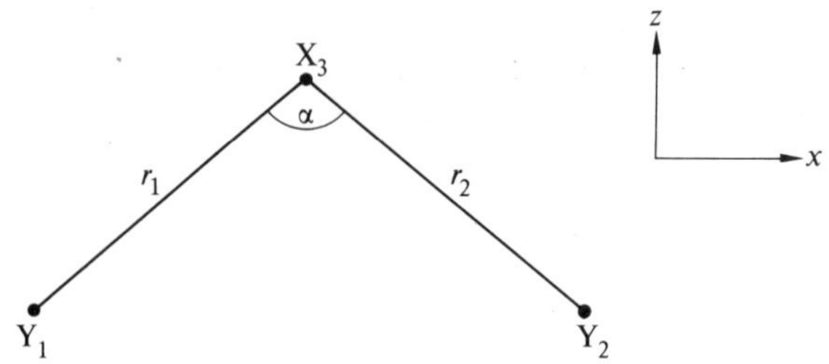

Fig. 2.2 Internal coordinates of a triatomic molecule XY_2, of C_{2v} symmetry.

(a)

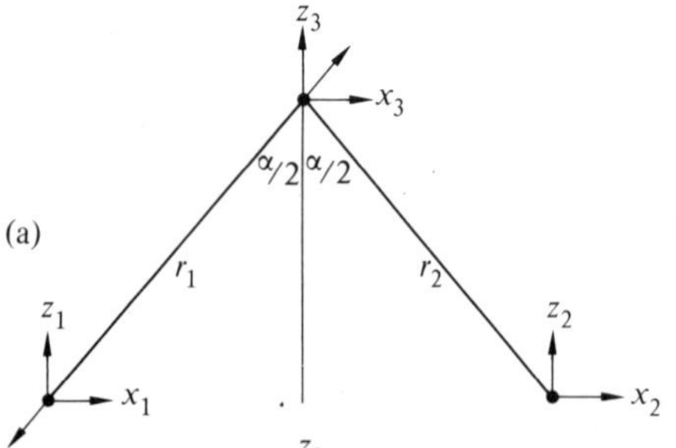

$$\Delta r_1 = -\Delta x_1 \sin \alpha/2 - \Delta z_1 \cos \alpha/2 + \Delta x_3 \sin \alpha/2 + \Delta z_3 \cos \alpha/2$$

(b)

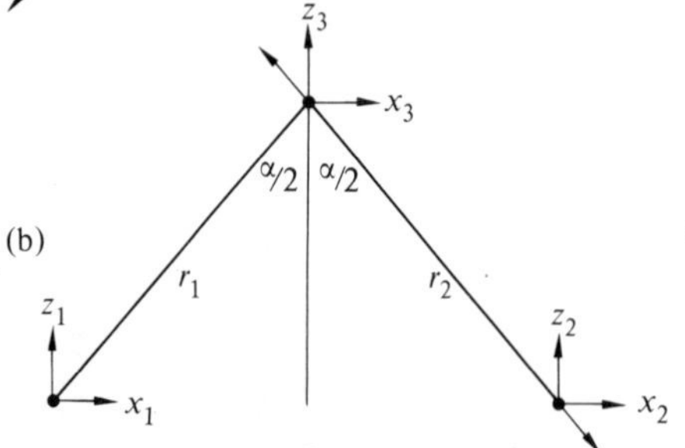

$$\Delta r_2 = \Delta x_2 \sin \alpha/2 - \Delta z_2 \cos \alpha/2 - \Delta x_3 \sin \alpha/2 + \Delta z_3 \cos \alpha/2$$

(c)

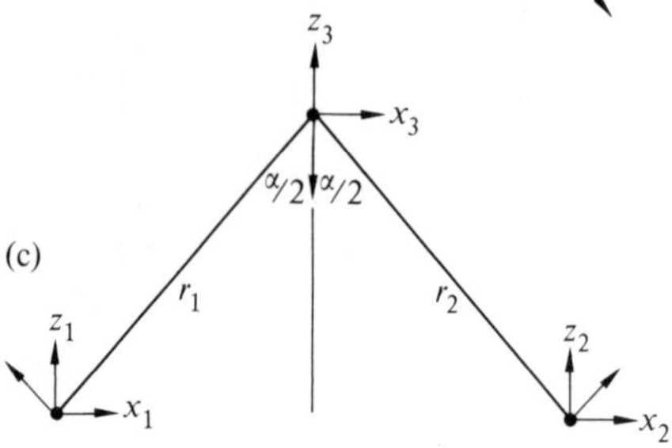

$$\Delta\alpha = 1/r_1 (-\Delta x_1 \cos \alpha/2 + \Delta z_1 \sin \alpha/2 + \Delta x_3 \cos \alpha/2 - \Delta z_3 \sin \alpha/2) + 1/r_2 (\Delta x_2 \cos \alpha/2 + \Delta z_2 \sin \alpha/2 - \Delta x_3 \cos \alpha/2 - \Delta z_3 \sin \alpha/2)$$

and since $r_1 = r_2 = r$

$$\Delta\alpha = 1/r\ (-\Delta x_1 \cos \alpha/2 + \Delta z_1 \sin \alpha/2 + \Delta x_2 \cos \alpha/2 + \Delta z_2 \sin \alpha/2 - 2\Delta z_3 \sin \alpha/2)$$

Fig. 2.3 The relations, between the displacements of the internal and Cartesian coordinates, which make up the B matrix. Each diagram corresponds to one row of the B matrix.

The matrix multiplication $M^{-1}B^{\dagger}$ gives a matrix of nine rows and three columns

$$M^{-1}B^{\dagger} = \begin{bmatrix} -\mu_1 S & 0 & -\mu_1 C/r \\ 0 & 0 & 0 \\ -\mu_1 C & 0 & \mu_1 S/r \\ 0 & \mu_1 S & \mu_1 C/r \\ 0 & 0 & 0 \\ 0 & -\mu_1 C & \mu_1 S/r \\ \mu_3 S & -\mu_3 S & 0 \\ 0 & 0 & 0 \\ \mu_3 C & \mu_3 C & -2\mu_3 S/r \end{bmatrix}.$$

Then $BM^{-1}B^{\dagger}$ will be a 3×3 matrix:

$$\begin{bmatrix} \mu_1(S^2 + C^2) + \mu_3(S^2 + C^2) & \mu_3(C^2 - S^2) & -\dfrac{2\mu_3}{r} CS \\ & (\mu_1 + \mu_3)(S^2 + C^2) & -\dfrac{2\mu_3}{r} CS \\ & & \dfrac{2}{r^2}(\mu_1 + 2\mu_3 S^2) \end{bmatrix}$$

Only the upper half is given since $g_{ij} = g_{ji}$. On making use of the standard trigonometric identities, the above matrix becomes

$$\begin{bmatrix} \mu_1 + \mu_3 & \mu_3 \cos\alpha & -\dfrac{\mu_3}{r}\sin\alpha \\ & \mu_1 + \mu_3 & -\dfrac{\mu_3}{r}\sin\alpha \\ & & \dfrac{2}{r^2}(\mu_1 + \mu_3(1 - \cos\alpha)) \end{bmatrix}.$$

The labour involved in evaluating the g matrix is much reduced if displacement vectors rather than Cartesian coordinate displacements are considered.

Expanding (2.41), we have

$$\begin{aligned} I_t = {} & (B_{tx_1}x_1 + B_{ty_1}y_1 + B_{tz_1}z_1) \\ & + (B_{tx_2}x_2 + B_{ty_2}y_2 + B_{tz_2}z_2) + \cdots \\ & + (B_{tx_n}x_n + B_{ty_n}y_n + B_{tz_n}z_n). \end{aligned}$$

If we consider the vectors

$$\mathbf{V}_a \text{ with components } x_a, y_a, z_a$$
$$\mathbf{S}_{ta} \text{ with components } B_{tx_a}, B_{ty_a}, B_{tz_a},$$

where a runs over the n atoms of the molecule, then we have

$$I_t = \sum_{a=1}^{n} \mathbf{S}_{ta} \cdot \mathbf{V}_a. \tag{2.42}$$

Consider

$$g_{tu} = \sum_{a=1}^{n} \mu_a \mathbf{S}_{ta} \cdot \mathbf{S}_{ua}$$
$$= \mu_a (B_{tx_a} B_{ux_a} + B_{ty_a} B_{uy_a} + B_{tz_a} B_{uz_a}), \tag{2.43}$$

from (2.42).

The utility of the expression (2.43) depends on the possibility of finding a way of expressing the vectors $\mathbf{S}_{ta}$; this is done by constructing unit vectors $\boldsymbol{\varepsilon}_{pq}$ along the bonds in the molecule; with this nomenclature the direction of the vector is from atom p to atom q. The $\mathbf{S}_{ta}$ are expressible in terms of the $\boldsymbol{\varepsilon}_{pq}$.

Consider I_t involving the stretching of a bond between two atoms p, q. Then

$$\mathbf{S}_{t1} = -\boldsymbol{\varepsilon}_{pq}, = \boldsymbol{\varepsilon}_{qp}, \qquad \mathbf{S}_{t2} = \boldsymbol{\varepsilon}_{pq}.$$

For I_t as an angle bending, consider the system of Fig. 2.4. The directions of $\mathbf{S}_{t1}$, $\mathbf{S}_{t2}$ are those in which displacement of the atoms 1, 2 will have the greatest effect in increasing the bond angle ϕ, that is, perpendicular to and coplanar with the bond directions.

Since unit displacement of atom 1 along the bond direction 31 will cause ϕ to increase by $1/r_{31}$, the length of $\mathbf{S}_{t1}$ is $1/r_{31}$.

Similarly the length of $\mathbf{S}_{t2}$ is $1/r_{32}$, while that of $\mathbf{S}_{t3}$ is equal to $-1/r_{31} - 1/r_{32}$; this follows from the fact that we may displace the atom 3 by some amount and then perform a rigid translation of equal and

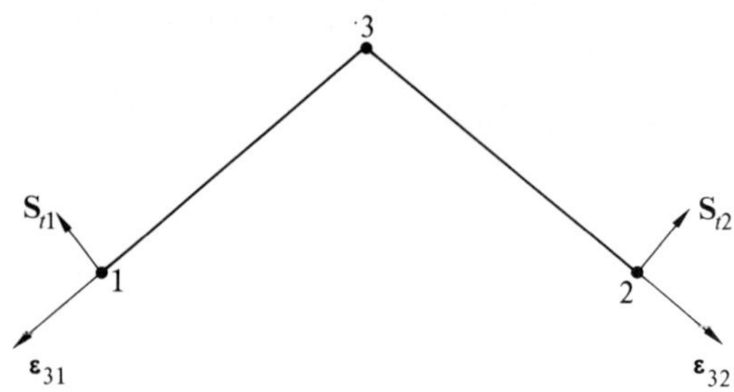

Fig. 2.4 Displacement vectors for a triatomic molecule of C_{2v} symmetry.

opposite magnitude, bringing atom 3 back to its original position; atoms 1 and 2 are then each displaced by amounts equal and opposite to the displacement of atom 3.

A convenient way of constructing the $\mathbf{S}_{ta}$ so that the g matrix can be set up in the form

$$g = SM^{-1}S^{\dagger}$$

is given by Wilson, Decius and Cross [18] and illustrated for the bent XY_2 molecule by Nakamoto [13].

In practice, we do not often need to construct the g matrix elements in this way, since tabulations of the g elements have been made for all common cases; e.g., by Ferigle and Meister [460] for linear molecules, and Shimanouchi [1285] for planar molecules; but the most comprehensive treatment is that of Decius [331]. His tabulation is the most commonly used, and the expressions given have been specialized to the important cases where the valence angles are 90°, 109° 28′ and 120°. He has also devised a scheme of nomenclature for the g elements, so that the task of constructing the G elements simply reduces to selection of the appropriate g elements and performance of the transformation $G = UgU^{\dagger}$.

If we write down a g matrix with the same suffixes as the f matrix, the G elements will be the same linear combinations of the g elements as the F elements were of the f elements. In the case of the tetrahedral XY_4 molecule we have, analogous to (2.32)–(2.34),

$$\begin{aligned}
G_{11}(A_1) &= g_r + 3g_{rr} \\
G_{22}(E) &= g_\alpha + g_{\alpha\alpha'} - 2g_{\alpha\alpha} \\
G_{33}(F_2) &= g_r - g_{rr} \\
G_{34}(F_2) &= G_{43}(F_2) = 2^{1/2}(g_{r\alpha} - g_{r\alpha'}) \\
G_{44}(F_2) &= g_\alpha - g_{\alpha\alpha'}.
\end{aligned} \tag{2.44}$$

In the Decius notation (see Appendix 3) the g elements in (2.44) take the following form:

$$\begin{aligned}
g_r &= g_{rr}^2 = \mu_1 + \mu_2 = \mu_x + \mu_y \\
g_{rr} &= g_{rr}^1 = -\frac{\mu_1}{3} = -\frac{\mu_x}{3} \\
g_\alpha &= g_{\phi\phi}^3 = 2\rho^2\mu_1 + \tfrac{8}{3}\rho^2\mu_2 = 2\rho^2\mu_y + \tfrac{8}{3}\rho^2\mu_x \\
g_{\alpha\alpha} &= g_{\phi\phi}^2\binom{1}{1} = -\tfrac{1}{2}\rho^2\mu_1 = -\tfrac{1}{2}\rho^2\mu_y \\
g_{\alpha\alpha'} &= g_{\phi\phi}^1\binom{1}{1} = -\tfrac{8}{3}\rho^2\mu_1 = -\tfrac{8}{3}\rho^2\mu_x \\
g_{r\alpha} &= g_{r\phi}^2 = -2.2^{1/2}.3^{-1}\rho\mu_2 = -2.2^{1/2}.3^{-1}\rho\mu_x \\
g_{r\alpha'} &= g_{r\phi}^1\binom{1}{1} = 2.2^{1/2}.3^{-1}\rho\mu_1 = 2.2^{1/2}.3^{-1}\rho\mu_x,
\end{aligned} \tag{2.45}$$

where ρ is the reciprocal of the X–Y distance.

It is important to be careful in the labelling, since the atom with a given designation (1, 2) in the Decius notation is not always the same atom in the molecule. In (2.45), atom 1 is the central atom (X) in g^1_{rr}, $g^1_{\phi\phi}\binom{1}{1}$ and $g^1_{r\phi}\binom{1}{1}$; whereas in $g^3_{\phi\phi}$, $g^2_{\phi\phi}\binom{1}{1}$ and $g^2_{r\phi}$, atom 1 is the terminal atom (Y).

Substitution of (2.45) in (2.44) gives

$$\begin{aligned} G_{11}(A) &= \mu_y \\ G_{22}(E) &= 3\rho^2\mu_y \\ G_{33}(F_2) &= \mu_y + \tfrac{4}{3}\mu_x, \\ G_{34}(F_2) &= -\tfrac{8}{3}\rho\mu_x, \\ G_{44}(F_2) &= \rho^2(2\mu_y + \tfrac{16}{3}\mu_x). \end{aligned} \tag{2.46}$$

Note that any product $F_{ij}G_{ij}$ will have zero dimension in length if the F_{ij} are defined as in (2.32)–(2.34).

2.3 The secular equation

(a) Calculation of the vibrational frequencies

We have, from (2.23b) and (2.38b), the expressions

$$2V = \sum_{ij} F_{ij}S_iS_j,$$

$$2T = \sum_{ij} (G^{-1})_{ij}\dot{S}_i\dot{S}_j.$$

The frequencies of the normal vibrations, in terms of λ ($=4\pi^2\nu^2$) are given by solving the equation

$$\det |F - G^{-1}\lambda| = 0, \tag{2.47a}$$

or, in full

$$\begin{vmatrix} F_{11} - (G^{-1}_{11})\lambda & F_{12} - (G^{-1}_{12})\lambda & \dots & F_{1n} - (G^{-1}_{1n})\lambda \\ F_{21} - (G^{-1}_{21})\lambda & F_{22} - (G^{-1}_{22})\lambda & \dots & F_{2n} - (G^{-1}_{2n})\lambda \\ \vdots & & & \vdots \\ F_{n1} - (G^{-1}_{n1})\lambda & F_{n2} - (G^{-1}_{n2})\lambda & \dots & F_{nn} - (G^{-1}_{nn})\lambda \end{vmatrix} = 0. \tag{2.47b}$$

The secular determinant is not usually solved in this form, owing to the inconvenience of λ appearing in every term and the labour required in inverting the G matrix to get the (G^{-1}_{ij}). Both these difficulties are avoided if (2.47) is multiplied by the determinant of the G matrix, giving

$$\det |GF - E\lambda| = 0. \tag{2.48a}$$

Since both the F and G matrices are symmetrical, the alternative form

$$\det |FG - E\lambda| = 0$$

is equivalent. We may rewrite (2.48a) in full as

$$\begin{vmatrix} (\sum G_{1t}F_{t1}) - \lambda & \sum G_{1t}F_{t2} & \cdots & \sum G_{1t}F_{tn} \\ \sum G_{2t}F_{t1} & (\sum G_{2t}F_{t2}) - \lambda & \cdots & \sum G_{2t}F_{tn} \\ \vdots & \vdots & & \vdots \\ \sum G_{nt}F_{t1} & \sum G_{nt}F_{t2} & \cdots & (\sum G_{nt}F_{tn}) - \lambda \end{vmatrix} = 0, \qquad (2.48b)$$

where the summation over t is from 1 to n.

This, the most commonly used procedure, is generally referred to as the Wilson FG matrix method (Wilson [1516–17]). Other forms of the secular determinant are discussed in Wilson, Decius and Cross [18].

An equation such as (2.48b) has to be solved for the vibrations of each irreducible representation. For the tetrahedral XY_4 molecule, we have two linear equations, corresponding to the single vibrations of the A_1 and E representations, and one quadratic, corresponding to the two vibrations of the F_2 representations.

We define λ as $4\pi^2\nu^2$ where ν is in $\sec^{-1}$, or $4\pi^2\nu^2c^2$ where ν is in cm^{-1}. If the G elements are expressed in atomic mass units (amu), then

$$\lambda = \frac{4\pi^2c^2\nu^2}{N_0}, \qquad \text{where } \nu \text{ is in cm}^{-1}$$

and

$$\nu\,(\text{cm}^{-1}) = \frac{N_0^{1/2}}{4\pi^2c^2}\sqrt{\lambda}$$

$$= 1303{\cdot}16\sqrt{\lambda}.$$

The dimensions of λ must be the same as those of the product FG. If λ is to have the unit $\sec^{-2}$, the dimensions of F and G elements of various kinds could be as in Table 2.7. These dimensions of F can be achieved by

Table 2.7 Dimensions of F and G.

Type of element	Dimensions of G	Dimensions of F
Stretching (principal and interaction)	M^{-1}	MT^{-2}
Stretch/bend interaction	$M^{-1}L^{-1}$	MLT^{-2}
Bending (principal and interaction)	$M^{-1}L^{-2}$	ML^2T^{-2}

expressing stretching, stretch/bend and bending elements in mdyn Å^{-1}, mdyn and mdyn Å respectively, since the dimensions of the mdyn are MLT^{-2}. This, in turn, can be ensured by expressing the force constant f in the form f_r (stretching), $df_{r\alpha}$ (stretch/bend) and d^2f_α (bending).

If all force constants are expressed in mdyn Å^{-1}, and all masses as atomic masses, then we have

$$\nu\,(\text{cm}^{-1}) = 1303{\cdot}16\sqrt{\lambda}, \tag{2.49}$$

where the units of λ are sec^{-2} (as can also be seen from $FG = \lambda$, since the dimensions of F and G are MLT^{-2} × L^{-1} and M^{-1} respectively), or

$$\lambda = 0{\cdot}58885\left(\frac{\nu}{1000}\right)^2.$$

We shall now carry out this calculation for the $SO_4^{=}$ ion, using the force constants, atomic masses and interatomic distances given in Table 2.8.

Table 2.8 Data required for the calculation of the *F* and *G* elements.

$f_r = 6{\cdot}53$ mdyn Å^{-1}	$\mu_x = \mu_S = 0{\cdot}03125$ amu^{-1}
$f_{rr} = 0{\cdot}87$ mdyn Å^{-1}	$\mu_y = \mu_O = 0{\cdot}06250$ amu^{-1}
$d^2 f_\alpha = 1{\cdot}50$ mdyn Å	$\rho = \dfrac{1}{r_{S-O}} = 0{\cdot}6667$ Å^{-1}
$d^2 f_{\alpha\alpha} = 0{\cdot}29$ mdyn Å	
$d^2 f_{\alpha\alpha'} = -0.29$ mdyn Å	
$d(f_{r\alpha} - f_{r\alpha'}) = 0{\cdot}0$ mdyn	

Table 2.9 Calculated *F* and *G* matrix elements.

$F_{11} = 9{\cdot}13$ mdyn Å^{-1}	$G_{11} = 0{\cdot}06250$ amu^{-1}
$F_{22} = 1{\cdot}44$ mdyn Å	$G_{22} = 0{\cdot}08333$ amu^{-1} Å^{-2}
$F_{33} = 5{\cdot}66$ mdyn Å^{-1}	$G_{33} = 0{\cdot}10417$ amu^{-1}
$F_{34} = 0{\cdot}00$ mdyn	$G_{34} = -0{\cdot}05555$ amu^{-1} Å^{-1}
$F_{44} = 2{\cdot}73$ mdyn Å	$G_{44} = 0{\cdot}12784$ amu^{-1} Å^{-2}

(a) The A_1 stretching frequency:

$$F_{11}G_{11} - \lambda_1(A_1) = 0$$
$$9{\cdot}13 \times 0{\cdot}06250 = \lambda_1(A_1) = 0{\cdot}561.$$

From (2.49)

$$\underline{\nu_1 = 976\ \text{cm}^{-1}}.$$

(b) The E bending frequency:

$$F_{22}G_{22} - \lambda_2(E) = 0,$$
$$1{\cdot}44 \times 0{\cdot}08333 = \lambda_2(E) = 0{\cdot}12,$$
$$\underline{\nu_2 = 451\ \text{cm}^{-1}}.$$

(c) The F_2 stretching and bending frequencies:
Expanding the secular determinant, we have

$$\lambda^2 - (F_{33}G_{33} + 2F_{34}G_{34} + F_{44}G_{44})\lambda + \begin{vmatrix} F_{33} & F_{34} \\ F_{43} & F_{44} \end{vmatrix} \begin{vmatrix} G_{33} & G_{34} \\ G_{43} & G_{44} \end{vmatrix} = 0. \quad (2.50)$$

From Table 2.9,

$$\begin{aligned} F_{33}G_{33} &= 0{\cdot}5908 \\ F_{44}G_{44} &= 0{\cdot}3498 \\ F_{34}G_{34} &= 0{\cdot}0 \end{aligned}$$

$$\begin{aligned} F_{33}F_{44} &= 15{\cdot}48 \\ G_{33}G_{44} &= 0{\cdot}01331 \\ G_{34}^2 &= 0{\cdot}00308 \end{aligned}$$

(2.50) becomes

$$\lambda^2 - 0{\cdot}9406\lambda + 0{\cdot}1581 = 0$$

giving (to 2 figures) the two roots $\lambda_3 = 0{\cdot}72$, $\lambda_4 = 0{\cdot}22$, to which correspond

$$\underline{\nu_3 = 1106\ \text{cm}^{-1}}, \qquad \underline{\nu_4 = 611\ \text{cm}^{-1}}.$$

Table 2.10 Comparison of calculated and observed frequencies (cm^{-1}).

	Calculated	Observed [1216]
ν_1	976	983
ν_2	451	450
ν_3	1106	1105
ν_4	611	611

It was stated in section 2.1 above that the inclusion of a redundant coordinate in the set of symmetry coordinates leads to a zero root in the secular determinant. This arises because there will be some linear combinations of f elements which are indeterminate and have no physical significance, since they represent the resistance to an 'impossible displacement'. The corresponding linear combinations of g elements are identically zero. We may illustrate this point by considering the effect of leaving in the A_1 set of coordinates for the tetrahedral molecule the coordinate $6^{-1/2}(\alpha_1 + \alpha_2 + \alpha_3 + \alpha_4 + \alpha_5 + \alpha_6)$. This is a redundant bending coordinate, whose existence was demonstrated by Table 2.1. Let this coordinate be labelled S_1'; then the U matrix for the A_1 representation acquires a second line, becoming

	r_1	r_2	r_3	r_4	α_1	α_2	α_3	α_4	α_5	α_6
S_1	$\frac{1}{2}$	$\frac{1}{2}$	$\frac{1}{2}$	$\frac{1}{2}$	0	0	0	0	0	0
S_1'	0	0	0	0	$6^{-1/2}$	$6^{-1/2}$	$6^{-1/2}$	$6^{-1/2}$	$6^{-1/2}$	$6^{-1/2}$

Following the procedure given earlier, and using the GVFF, we obtain for the $F(A_1)$ matrix

$$F(A_1) = \begin{vmatrix} f_r + f_{rr} & \sqrt{2}\, d(f_{r\alpha} + f_{r\alpha'}) \\ \sqrt{2}\, d(f_{r\alpha} + f_{r\alpha'}) & d^2(f_\alpha + 4f_{\alpha\alpha} + f_{\alpha\alpha'}) \end{vmatrix}.$$

From eq. (2.47), we see that $g_{r\alpha} = -g_{r\alpha'}$; thus the off-diagonal terms in $G(A_1)$ are zero; similarly, the term $g_\alpha + 4g_{\alpha\alpha} + g_{\alpha\alpha'}$ is $2\rho^2\mu_y + \frac{8}{3}\rho^2\mu_x + 4(-\frac{1}{2}\rho^2\mu_y) + (-\frac{8}{3}\rho^2\mu_x) = 0$.

Thus, in the $F(A_1)$ matrix, only the combination $F_{11} = f_r + f_{rr}$ is meaningful, and, in the $G(A_1)$ matrix, only the term $G_{11} = \mu_y$ is nonzero. In this particular case we have one meaningful F element and one observable frequency, so that this element, F_{11}, is determinable. In general, there will be more meaningful F elements than frequencies, so that additional information must be obtained, as discussed at the end of section 2.2(a) above.

(b) Band assignments and the potential-energy distribution

When the solutions of the secular determinant have been obtained, it is necessary to assign the calculated frequencies to the various modes of motion. In many cases this can be done by inspection, but where frequency values are close together, or where there is a possibility of vibrational coupling, it may be necessary to carry out the assignment on a more quantitative basis.

The vibrations of a molecule may be described in terms of normal coordinates Q_l which are related to the symmetry coordinates S_k by the linear transformation

$$S_k = \sum_l L_{kl} Q_l \tag{2.51}$$

where the L_k are the normalized amplitudes of vibration, which are obtained as follows:

Consider a vibration, of frequency λ_k, satisfying the determinantal equation

$$\det |GF - E\lambda| = 0.$$

Then det $|GF - E\lambda_k|\,|L_k| = 0$, and if we consider all values of λ which are solutions of the secular determinant,

$$GFL = L\Lambda \tag{2.52}$$

where Λ is the diagonal matrix of such values $(\lambda_a, \lambda_b, \ldots, \lambda_n)$. For each such value λ_k, we can obtain a set of homogeneous equations

$$\begin{aligned}
&(\textstyle\sum G_{1t}F_{t1} - \lambda_k)L_{1k} + (\sum G_{1t}F_{t2})L_{2k} + \ldots + (\sum G_{1t}F_{tn})L_{nk} = 0,\\
&(\textstyle\sum G_{2t}F_{t1})L_{1k} + (\sum G_{2t}F_{t2} - \lambda_k)L_{2k} + \ldots + (\sum G_{2t}F_{tn})L_{nk} = 0,\\
&\quad\vdots\\
&(\textstyle\sum G_{nt}F_{t1})L_{1k} + (\sum G_{nt}F_{t2})L_{2k} + \ldots + (\sum G_{nt}F_{tn} - \lambda_k)L_{nk} = 0.
\end{aligned} \tag{2.53}$$

This set of homogeneous equations will give the ratios of the amplitudes $L_{1k}:L_{2k}:\cdots:L_{nk}$ for the solution λ_k; a similar set can be constructed for every other eigenvalue.

Morino and Kuchitsu [997a] have shown that the potential energy distribution rather than the normalized amplitudes is a more satisfactory quantity to use in band assignments; for a vibration of frequency ν_a, associated with a normal coordinate Q_a, the potential energy of the molecule is given by

$$2V = Q_a^2 \sum_{ij} F_{ij}L_{ia}L_{ja}. \tag{2.54}$$

Such terms are large only when $i = j$, since the diagonal force constants F_{ii} are much greater than the off-diagonal constants F_{ij}. Consequently we need calculate only the terms of the form

$$F_{ii}L_{ia}^2.$$

Since the normalization condition $L^{\dagger}FL = \Lambda$ gives relations of the form

$$\sum_{ij} F_{ij}L_{ia}L_{ja} = \lambda_a, \tag{2.55a}$$

neglect of cross terms gives

$$\sum_{i} F_{ii}L_{ia}^2 = \lambda_a. \tag{2.55b}$$

We can therefore get a normalized potential-energy distribution, whose terms are

$$V_{ia} = \frac{F_{ii}L_{ia}^2}{\lambda_a},$$

where V_{ia} is the contribution of the ith symmetry coordinate to the potential energy of the vibration whose frequency is ν_a.

If one term in the potential-energy distribution, V_{ia}, is much greater than any of the other terms V_{ja}, V_{ka} . . ., we are justified in regarding the mode ν_a as a pure vibration of the ith coordinate; if two or more terms are roughly equal, we must regard the vibration as coupled.

We will now calculate the potential-energy distribution for the F_2 vibrations of the $SO_4^=$ ion.

The homogeneous equation (2.53) will be

$$\begin{aligned}(G_{33}F_{33} + G_{34}F_{43} - \lambda_k)L_{3k} + (G_{33}F_{34} + G_{34}F_{44})L_{4k} &= 0,\\ (G_{43}F_{33} + G_{44}F_{43})L_{3k} + (G_{43}F_{34} + G_{44}F_{44} - \lambda_k)L_{4k} &= 0.\end{aligned} \tag{2.56}$$

Using the data in Table 2.9, we have

$$\begin{aligned}G_{33}F_{33} &= 0{\cdot}5908, & G_{43}F_{33} &= -0{\cdot}3144,\\ G_{34}F_{44} &= -0{\cdot}1517, & G_{44}F_{44} &= 0{\cdot}3498.\end{aligned}$$

All other GF products are zero since $F_{34} = F_{43} = 0$.

The two values of λ obtained were 0·72, 0·22; let these be λ_a and λ_b respectively. Then, rounding off the GF products to two figures, we have for λ_a,

$$\begin{aligned}-0{\cdot}13L_{3a} - 0{\cdot}15L_{4a} &= 0,\\ -0{\cdot}31L_{3a} - 0{\cdot}37L_{4a} &= 0.\end{aligned} \tag{2.57a}$$

From these equations, $L_{3a}/L_{4a} \cong -1{\cdot}18$,
and for λ_b,

$$\begin{aligned}0{\cdot}35L_{3b} - 0{\cdot}15L_{4b} &= 0,\\ -0{\cdot}31L_{3b} + 0{\cdot}13L_{4b} &= 0,\end{aligned} \tag{2.57b}$$

from which $L_{3b}/L_{4b} = +0{\cdot}42$.

If we apply the normalization condition $L^{\dagger}FL = \Lambda$, we have

$$\begin{vmatrix}L_{3a} & L_{4a}\\ L_{3b} & L_{4b}\end{vmatrix}\begin{vmatrix}F_{33} & F_{34}\\ F_{43} & F_{44}\end{vmatrix}\begin{vmatrix}L_{3a} & L_{3b}\\ L_{4a} & L_{4b}\end{vmatrix} = \begin{vmatrix}\lambda_a & 0\\ 0 & \lambda_b\end{vmatrix}.$$

After multiplication of the left-hand side and equating terms in the corresponding positions, we have (remembering that $F_{34} = 0$)

$$F_{33}L_{3a}^2 + F_{44}L_{4a}^2 = \lambda_a, \tag{2.58a}$$

$$F_{33}L_{3b}^2 + F_{44}L_{4b}^2 = \lambda_b, \tag{2.58b}$$

$$F_{33}L_{3a}L_{3b} + F_{44}L_{4a}L_{4b} = 0. \tag{2.58c}$$

Substituting the results of (2.57a), (2.57b) in (2.58a), (2.58b) respectively, and using the values of F_{33}, F_{44} from Table 2.9, we obtain

$$5{\cdot}66 \times (-1{\cdot}18L_{4a})^2 + 2{\cdot}73L_{4a}^2 = 0{\cdot}72, \tag{2.59a}$$

$$5{\cdot}66 \times (0{\cdot}42L_{4b})^2 + 2{\cdot}73L_{4b}^2 = 0{\cdot}22, \tag{2.59b}$$

that is,

$$10{\cdot}61L_{4a}^2 = 0{\cdot}72, \qquad V_{4a} = \frac{F_{44}L_{4a}^2}{\lambda_a} = 0{\cdot}26.$$

From (2.58a),

$$V_{3a} = \frac{F_{33}L_{3a}^2}{\lambda_a} = 0{\cdot}74$$

and

$$3{\cdot}72L_{4b}^2 = 0{\cdot}22, \qquad V_{4b} = \frac{F_{44}L_{4b}^2}{\lambda_b} = 0{\cdot}74,$$

and from (2.58b),

$$V_{3b} = \frac{F_{33}L_{3b}^2}{\lambda_b} = 0{\cdot}26.$$

The potential energy distribution is therefore

	ν_3	ν_4
S_3	0·74	0·26
S_4	0·26	0·74

that is, ν_3 is primarily a vibration of the stretching coordinate S_3, while ν_4 is primarily a vibration of the bending coordinate S_4.

An alternative description of the potential energy depends on the evaluation of the Jacobian matrix elements J_{ai}.

Consider a number of roots λ_a, λ_b, $\lambda_c, \ldots$ of the determinantal equation $|GF - E\lambda| = 0$, and designate the force constants F_i, F_j, $F_k \ldots$. Then the Jacobian matrix element J_{ai} will be defined as

$$\left(\frac{\partial \lambda_a}{\partial F_i}\right)_{F_j, F_k \cdot \ \ldots},$$

and the contribution of the force constant F_i to the potential energy of the vibration corresponding to the root λ_a is then $(F_i/\lambda_a)J_{ai}$.

The J_{ai} may therefore be obtained numerically, by solving the determinantal equation for successive sets of F values, only one force constant being allowed to vary in each set. Let F_i be varied, and plot λ_a, λ_b, $\lambda_c, \ldots$ against F_i. Then J_{ai} for any value of λ_a is the slope of the graph of λ_a against F_i for that value of λ_a. A recent example of the use of this method is the analysis of the lattice vibrations of NiF_2 and $KNiF_3$ [1445].

Useful information about band assignments is sometimes obtainable from measurements on isotopically substituted molecules. Consider a molecule for which the roots of the vibrational secular equation are $\lambda_1, \lambda_2, \ldots, \lambda_n$; then $\Pi(\lambda_n) = |F|\,|G|$. If, on isotopic substitution, the roots become $\lambda'_1, \lambda'_2, \ldots, \lambda'_n$, and $|F|$ is unchanged, then $\Pi(\lambda'_n) = |F|\,|G'|$. Since $\lambda = 4\pi^2c^2\nu$,

$$\frac{\Pi(\nu)}{\Pi(\nu')} = \left(\frac{G}{G'}\right)^{1/2}. \tag{2.60}$$

The method is easiest to apply when isotopic substitution does not alter the molecular symmetry, as in $^{10}BF_3$, $^{11}BF_3$, since then eq. (2.60) is true for the vibrations of each symmetry species. Equation (2.60) represents the Teller–Redlich product rule in its simplest form. Wilson, Decius and Cross [18] deal with the more complex situation where the molecular symmetry is altered by substitution, as in CH_4, CH_3D, and also with the case in which the symmetry species contains translational and rotational modes.

3. Crystal spectra

3.1 Introduction

The spectra of molecules fixed in a lattice differ from those of free molecules in two ways. In the first place, the selection rules which operate are affected by the crystal symmetry as well as the molecular symmetry; secondly, frequencies appear which are due to modes of motion of the lattice, either translational or rotational. Such frequencies are often referred to as external frequencies, to distinguish them from the so-called internal vibrations of the individual molecules. A discussion of crystal spectra should therefore deal with two problems; one, the determination of selection rules, for both the internal and external vibrations; the other, the prediction of frequencies of the external modes, and, ideally, a statement of the criteria whereby translational and rotational modes can be distinguished from each other.

3.2 The determination of selection rules

A convenient starting point for discussion is the nature of the vibrational potential energy of the crystal. Hornig [678] has shown that this is expressible in the form

$$V = \sum_{n} V_n + \sum_{\substack{n \\ (n \neq k)}} \sum_{k} V_{nk} + V_l + V_{ln}, \tag{3.1}$$

where these four terms represent, respectively, the potential energies due to internal vibrations, interaction between internal vibrations, lattice modes, and interaction between lattice and internal modes. The two most frequently used approaches to the problem of determining selection rules are the site-group approximation, which effectively uses only the first term of the potential-energy expression, and the factor-group approximation, which takes the first and second terms into account.

(a) The site-group approximation

The theory of the site-group approximation was developed by Halford [606]. He considered the molecule as moving in a potential field whose symmetry is that of the crystal. The crystal belongs to one of the 230 space groups, and the group, S, of operations of any space group can be decomposed into a subgroup T of translational operations, and a subgroup U of rotational operations [14, 17]. The subgroup U will be isomorphic with one of the 32 crystal point groups, and is referred to as the unit cell group, since it describes the symmetry of the unit cell. Since $S = U \times T$, the subgroup U is a factor group of S, and irreducible representations of U are also irreducible representations of S. The symmetry of any point in the lattice can be specified in terms of a site group, which is a subgroup of the space group. Now the centre of mass of a molecule is invariant under the operations of the point group to which the molecule belongs, and its equilibrium position is generally on what Halford terms a site. Halford's definition of a site appears to be equivalent to what crystallographers call a special position; points of symmetry C_1, i.e., general positions, are not called sites. Two useful rules follow from the requirement that the equilibrium position of the centre of mass should be a site; first, the site group must be a subgroup of the molecular point group, and, second, the positions of the centres of mass of the molecules must constitute a complete set of sites.

These rules will also be true of an atom, not at the centre of mass, which is invariant to all the symmetry operations of the group; for instance the X atom in pyramidal XY_3.

Having determined the site group, we must then find the irreducible representations of the site group corresponding to those of the molecular point group. Tables for this purpose (correlation tables) are given in the books by Adams [1], Herzberg [9] (incomplete), Cotton [6] (incomplete) and Wilson, Decius and Cross [18] as well as in Appendix 2 of this work. The selection rules for the site group are then found by looking up the character table of the point group corresponding to the site group. For instance, consider zircon, $ZrSiO_4$. The silicate ions occupy D_{2d} sites, and the molecular point group is T_d. We therefore require a correlation between T_d and D_{2d}. This correlation is shown in Table 3.1. Thus neither ν_1 nor

Table 3.1 Correlation between T_d and D_{2d}.

	T_d	D_{2d}
ν_1	A_1(R)	A_1(R)
ν_2	E(R)	A_1(R) + B_1(R)
ν_3, ν_4	F_2(ir, R)	B_2(ir) + E(ir, R)

ν_2 of the free ion would appear in the infrared spectrum of the crystal; ν_3 and ν_4 would both appear as doublets in the infrared and as singlets in the Raman. This treatment can be very useful in predicting the general nature of crystal spectra, though some cases are known where the selection rules predicted by the site symmetry treatment are not adhered to. A good example of a system obeying the rules is YPO_4, which crystallizes in the zircon structure and has the selection rules given above. In the example shown in Table 3.2, the allocation of a site to the PO_4 group is unambiguous

Table 3.2 Frequencies (cm^{-1}) for YPO_4.

	Raman spectrum of the free $PO_4^{\equiv}$ ion [15]	Infrared spectrum of $PO_4^{\equiv}$ in crystalline YPO_4 [653]
ν_1	980	—
ν_2	363	—
ν_3	1082	1058, 990
ν_4	515	641, 521

once the space group, D_{4h}^{19}, and Z, the unit cell content, $=4$, are determined, since the only sites available are two fourfold sets of D_{2d} sites which must be occupied by the Y and P atoms, and a set of sixteen sites occupied by the sixteen oxygen atoms.

A case where alternatives exist is $AgClO_3$, crystallizing in the space group C_{4h}^5 with $Z = 8$. The eight chlorine atoms, if crystallographically equivalent, must occupy a set of eightfold sites. In C_{4h}^5 there are three such sets available, of symmetry C_s, C_2 and C_i. Since the ClO_3^- ion has the molecular point group C_{3v}, and the Cl atom is invariant to all the symmetry operations of that group, it must occupy a site whose group is a subgroup of the molecular point group C_{3v}. The only possibility among the above three is C_s; the correlation is ν_1, ν_2 $(A_1, (\text{ir, R})) \rightarrow A'(\text{ir, R})$; ν_3, ν_4 $(E(\text{ir, R})) \rightarrow A' + A''(\text{ir, R})$. Consequently all six vibrations of the ClO_3^- group in crystalline $AgClO_3$ should be observed. Again the agreement with expectation is good.

Table 3.3 Frequencies (cm^{-1}) for $AgClO_3$.

	Free ClO_3^- (R, aqueous solution) [13]	Solid $AgClO_3$ (ir) [1204]
ν_1	932	935
ν_2	613	612
ν_3	982	987, 967
ν_4	479	493, 470

We may conclude that the site group provides an adequate basis for determining the selection rules associated with internal vibrations in crystals. It can be used when only the space group and unit cell content of the crystal is known. Its main disadvantage is its lack of applicability to the determination of selection rules for external vibrations.

(b) The factor-group approximation

Bhagavantam and Venkatarayudu first developed this treatment, which attempts to determine the number and activity of the vibrations of the complete unit cell. For a crystal composed of N unit cells, each containing n atoms, there are $3nN$ modes of motion altogether. However, only those modes in which equivalent atoms are in phase are active as fundamentals in the infrared or Raman: the remaining modes may be seen as combinations of very low intensity. This being so, it is necessary to determine only the modes of motion of one unit cell, and this can be done by considering the factor group, which is, in this sense, the subgroup U of the space group consisting of its rotation operations. The factor group is isomorphic with one of the 32 crystallographic point groups, and is designated by the Schönflies space-group symbol without its numerical superscript; e.g., the factor group of $ZrSiO_4$, whose space group is D_{4h}^{19}, is D_{4h}. In addition to the factor group we need to know the symmetry operations to which each atom is invariant, that is, the site group for each atom position, and also the unit cell content. Since this information is frequently unavailable, it is often impossible to carry out a factor-group analysis.

The determination of the selection rules for vibrational activity is carried out exactly as for the free molecule, using the character table of the point group with which the factor group is isomorphic. For infrared activity we construct the reducible representation for the dipole moment; its character $\chi(\mu)$ is given by

$$\chi(\mu) = \pm 1 + 2\cos\theta. \tag{3.2a}$$

For Raman activity, the character $\chi(\alpha)$ of the reducible representation for polarizability is

$$\chi(\alpha) = 2\cos\theta(\pm 1 + 2\cos\theta). \tag{3.2b}$$

If a_j is the number of times the jth irreducible representation is contained in the reducible representation, and the character of the jth irreducible representation is $\chi^{(j)}(R)$, then

$$a_j = \frac{1}{h}\sum_R \chi^{(j)}(R)\chi(R), \tag{3.3}$$

where h is the order of the group, and $\chi(R)$ is $\chi(\mu)$ or $\chi(\alpha)$, for determining infrared and Raman activity respectively. If $a_j = 0$, vibrations of the jth irreducible representation are inactive; if $a_j \neq 0$, they are active. This information can in any case be obtained from the character table of the point group, just as for internal molecular vibrations.

The next step in the factor-group analysis is the classification of the vibrations of the unit cell into various types. Of the $3n$ modes associated with an n-atomic unit cell, 3 represent translations of the crystal as a whole. These so-called acoustic modes are not infrared or Raman active as fundamentals, but may be observed as combinations of low intensity with active fundamentals; they can also be studied by neutron scattering, Brillouin scattering and elastic constant measurements. The remaining $3n - 3$ modes are divided into internal and external modes, and the external modes may be translational or rotational. Consider a crystal made up entirely of nonlinear polyatomic molecules with p atoms in the molecule, and let the unit cell contain Z such molecules. Then there will be $3Zp$ modes, made up of 3 acoustic, $3Z$ rotational, $3(Z - 1)$ translational and $Z(3p - 6)$ internal. For instance, XeF_4 crystallizes with 2 molecules in the unit cell; the 30 modes therefore comprise 3 acoustic, 6 rotational, 3 translational and 18 internal. The majority of inorganic compounds, however, do not have the formula unit as a structural unit; e.g., $MgSO_4$ does not crystallize as a single $MgSO_4$ molecule. Instead, we have some monatomic and some polyatomic groups. Suppose the unit cell contains Z molecules, as before, and let there be a total of S structural groups, of which m are monatomic and q polyatomic with p atoms in the molecule. Then the total number of modes will consist of 3 acoustic, $3(S - 1)$ translational, $3q$ rotational and $(3p - 6)q$ internal. Consider a molecule MXO_4, with $Z = 4$. Then there are 8 structural groups, of which 4 are monatomic and 4 polyatomic with 5 atoms in the molecule. Since the unit cell contains 24 atoms, there are 72 modes; these are 3 acoustic, $3(8 - 1) = 21$ translational, $3 \times 4 = 12$ rotational and $(3 \times 5 - 6)4 = 36$ internal. Working out the number of modes of each type in this way provides a useful partial check on the correctness of a factor-group analysis.

Recently, Adams and Newton [1a] have compiled a set of tables for factor group analysis, illustrating their use for $CuCl_2$, TiO_2 and K_2PtCl_4.

A very clear exposition of the technique of factor-group analysis is given in the papers of Mitra [973] on brucite, $Mg(OH)_2$, and Mitra and Gielisse [975] who dealt with cyclopropane and with $CaSO_4, 2H_2O$; in this latter example they followed the treatment of Hass and Sutherland [618]. A useful point arising from their analysis of gypsum is the validity of treating the unit cell first as consisting of $CaSO_4$ molecules only, and then as consisting of H_2O molecules only, thus separating the internal modes of the polyatomic groups. As an illustration of the method we now consider $LaPO_4$ in the well-established monazite form. In this case we have $Z = 4$, $s = 8$, $m = 4$, $p = 5$, $q = 4$, as in the hypothetical example above. We should therefore expect the same distribution of modes (3 acoustic, 21 translational, 12 rotational, 36 internal).

Monazite crystallizes in the space group C_{2h}^5, and the factor group is therefore C_{2h}. All atoms are on positions C_1 and are therefore invariant only to the identity operation of the factor group. In order to determine the distribution of modes between types, we need to know the characters of the reducible representations for the total number of modes, and also for acoustic, translational and rotational modes. The number of internal modes is obtained by difference. The requisite characters are

Total: $$\chi(n) = N_R(\pm 1 + 2\cos\theta) = N_R\chi(T), \tag{3.4a}$$

Acoustic: $$\chi(T) = \pm 1 + 2\cos\theta, \tag{3.4b}$$

Translational: $$\chi(T') = (N_R(s) - 1)(\pm 1 + 2\cos\theta) = (N_R(s) - 1)\chi(T), \tag{3.4c}$$

Rotational: $$\chi(R) = (N_R(s - m))\chi(P). \tag{3.4d}$$

In this notation, which is similar to that of Mitra and Gielisse, N_R, $N_R(s)$ and $N_R(s - m)$ are the numbers, respectively, of atoms, structural groups, and polyatomic groups remaining invariant under the operation R. The value of $\chi(P)$ is $1 \pm 2\cos\theta$ for nonlinear polyatomic groups, and $\pm 2\cos\theta$ for linear polyatomic groups, except for the operations C_2 and σ_v, where it is zero for linear polyatomic groups. The + and − signs refer to proper and improper rotations respectively.

As for molecular vibrations, we now set up a calculation table (Table 3.4) giving the characters of the reducible representations required.

Table 3.4 Calculation table for the factor-group analysis of monazite (factor group C_{2h}).

Mode type	Character	Proper		Improper	
		I	C_2	$\sigma_h \equiv S_1$	$i \equiv S_2$
	θ	0	180	0	180
	$\cos\theta$	1	−1	1	−1
Acoustic	$\pm 1 + 2\cos\theta = \chi(T)$	3	−1	1	−3
	$1 \pm 2\cos\theta = \chi(P)$	3	−1	−1	3
	N_R	24	0	0	0
	$N_R(s)$	8	0	0	0
	$N_R(s) - 1$	7	−1	−1	−1
	$N_R(s - m)$	4	0	0	0
Translational	$(N_R(s) - 1)\chi(T) = \chi(T')$	21	1	−1	3
Rotational	$(N_R(s - m))\chi(P) = \chi(R')$	12	0	0	0
Total	$N_R\chi(T) = \chi(n)$	72	0	0	0
Internal	$\chi(n) - [\chi(T) + \chi(T') + \chi(R')]$	36	0	0	0

Table 3.5 Character table for the group C_{2h}.

	I	C_2	σ_h	i		
A_g	1	1	1	1	$\alpha_{xx}, \alpha_{yy}, \alpha_{zz}, \alpha_{xy};$	R_z
A_u	1	1	−1	−1	T_z	
B_g	1	−1	−1	1	$\alpha_{yz}, \alpha_{zx};$	R_x, R_y
B_u	1	−1	1	−1	$T_x, T_y.$	

We need to use (3.3) to find the number of modes of each type belonging to each irreducible representation. The procedure is exactly the same as that detailed in section 1.4(c) and leads to the result shown in Table 3.6.

Table 3.6 Distribution of modes by type and symmetry class.

Symmetry class	A_g	A_u	B_g	B_u	Total
Total	18	18	18	18	72
Acoustic	0	1	0	2	3
Translational	6	5	6	4	21
Rotational	3	3	3	3	12
Internal	9	9	9	9	36

Although in principle one could work out the selection rules from (3.3) and (3.2a, b) the character table shows that in this centrosymmetric group the g modes are Raman active and the u modes infrared active.

The above analysis gives the division of the internal modes into symmetry classes but does not relate these 36 modes to the modes ν_1, ν_2, ν_3, ν_4 of the free $PO_4^{\equiv}$ ion. It is not possible to carry out a direct correlation because the factor group, C_{2h}, is not a subgroup of the molecular point group, T_d; however, in accordance with the rules laid down by Halford, the site group, C_1, of the $PO_4^{\equiv}$ ion, is a subgroup of both C_{2h} and T_d. Incidentally, the requirement that the site group be a subgroup of the molecular point group eliminates the possibility of the $PO_4^{\equiv}$ ions occupying the C_i sites of the group C_{2h}^5, since C_i is not a subgroup of T_d. However, inspection suggests that each component of an internal mode of the free ion gives rise to 4 components, one of each symmetry class, in the unit cell. We would represent this on a correlation diagram, arranged as shown in Fig. 3.1.

The predicted effect of the interaction within the unit cell is to split each component of each internal fundamental into two infrared active (A_u + B_u) and two Raman active (A_g + B_g) components giving rise to 2, 4, 6 and 6 lines for ν_1, ν_2, ν_3 and ν_4 respectively, in each effect. In practice, as Halford has pointed out [606], such splitting (correlation splitting) is rarely observed, particularly under the normal condition of examining a polycrystalline sample at room temperature. The interactions between molecules are, of course, very weak compared to those within molecules. Hass and Sutherland did observe splitting of the sulphate ion frequencies in $CaSO_4, 2H_2O$, but in this case the site-group and factor-group approximations predict the same number of frequencies. Since the unit cell of gypsum contains two sulphate ions, they give rise to a total of 18 frequencies, all infrared active under the C_2 site symmetry. Polyatomic cations can also undergo vibrational modes and these will be independent of anion mass.

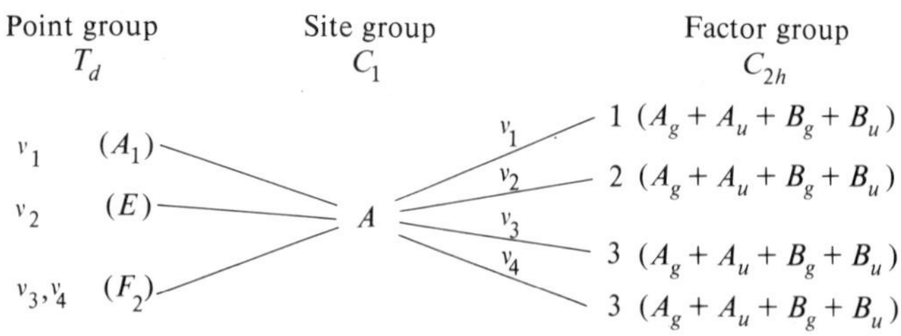

Fig. 3.1 Correlation diagram for monazite.

They also note that the very large magnitude of the splitting (ν_4, at about 650 cm^{-1}, is split by 70 cm^{-1}) may be connected with anion distortion due to hydrogen bonding. Characteristically one might expect, having regard to the relative strengths of the inter- and intramolecular forces, that correlation splitting might lead to a separation of components equivalent to about 1–2 per cent of the mode frequency.

3.3 The calculation of external mode frequencies

(a) Introduction

Many studies of internal molecular vibrations have revealed structure in the fundamental bands which is not attributable to crystal effects causing a breakdown of free-molecule selection rules. The bands making up this fine structure have often been assigned (in many cases arbitrarily) to combinations of the internal fundamentals with either translational or rotational lattice modes. Examples include the splitting of bands in NO_3^- and $CO_3^=$ observed by Buijs and Schutte [206–8, 1255] and by Lippincott and co-workers [1248, 1499], and the structure of bands in the ammonium halides observed by Wagner and Hornig [1476–7]. During recent years, the development of far infrared spectroscopy has led to the direct observation of many lattice frequencies, but assignments as between translational and rotational modes are still arbitrary in many cases, and no coherent body of theory on either the identification of modes or the calculation of their frequencies has been assembled. The first requirement in discussing this subject is to find criteria for distinguishing between translational and rotational modes: incidentally it should be noted that it is not free rotational but restricted rotational modes which are mainly observed, and it is these so-called librational modes which will be discussed. Criteria for distinguishing between translational and librational modes have been suggested by Schroeder, Weir and Lippincott [1248], Bloor [160] and Hezel and Ross [655], and are discussed below.

Translational modes of a cation against an anion are most readily identified by their strong dependence on the cation mass; in an ionic lattice AB their frequencies depend on the reduced mass of the 'molecule' AB. Approximate values of translational frequencies may be calculated from specific heat or elastic constant data, and the force constants arrived at in such calculations are usually in the region 0·1–1 mdyn $Å^{-1}$. They are often strong in infrared absorption and weak in Raman scattering, and may be observed in reflection as a broad band which includes both the longitudinal and transverse modes. Anionic librational mode frequencies, in contrast, are independent of cation mass. In the carbonates of the aragonite structure, the translational mode frequencies are 276, 214 ($CaCO_3$), 220, 176 ($SrCO_3$), 214, 156 ($BaCO_3$), while Schroeder *et al.* [1248] have suggested that the librational frequencies of these compounds all lie near 30 cm^{-1}.

These bands are usually weak in the infrared unless enhanced by anharmonicity, and their intensity is temperature dependent, since the temperature strongly affects the population of the librational levels. Since the energy quanta involved are often small, a series of bands of roughly constant separation, sometimes ending in a continuum marking the onset of free rotation, may be found. This cut-off frequency corresponds to the barrier height restricting rotation, and this is usually of the order of 1 kcal $mole^{-1}$.

(b) The calculation of translational mode frequencies

The three approaches largely used in this connection are those based on the Debye theory, those based on the Born–von Kármán theory and those which employ the Wilson *FG* matrix method.

In the Debye theory, the crystal is treated as a homogeneous, elastic, isotropic medium in which vibrations of any frequency up to some limit ν_{max} can travel. The value of ν_{max} depends on the crystal structure, and is related to the Debye characteristic temperature θ_D by

$$h\nu_{max} = k\theta_D \tag{3.5}$$

where h and k are Planck's and Boltzmann's constants respectively. The problem then becomes that of finding an experimentally observable parameter related in a known way to θ_D, and calculating ν_{max} from θ_D by (3.5).

Observable parameters whose values are related to θ_D include melting point, thermal expansion, temperature dependence of electrical resistivity and X-ray and neutron scattering intensities, but the two most frequently studied are specific heats and elastic constants. The theory underlying the relationships between these parameters and θ_D is treated in numerous standard works on thermodynamics or statistical mechanics, e.g., [1198], and will not be given here. The relations themselves are

$$C_v = 464{\cdot}4\left(\frac{T}{\theta_D}\right)^3 \text{ cal deg}^{-1}\text{ mole}^{-1}, \tag{3.6}$$

where T is the temperature in °K, and

$$(\theta_D^{-3}) = \frac{4\pi V}{9N}\left(\frac{h}{k}\right)^3\left[2\left(\frac{2(1+\sigma)}{3(1-2\sigma)}\right)^{3/2} + \left(\frac{1+\sigma}{3(1-\sigma)}\right)^{3/2}\right](\beta\rho)^{3/2}, \tag{3.7}$$

where V = volume
σ = Poisson's ratio (bulk modulus/rigidity modulus)
β = compressibility
ρ = density.

This relation holds for isotropic solids. Over the range $0 < T < \theta_D/50$, in which the Debye cube law (3.6) is obeyed, the θ_D values derived from specific heat and elastic constant measurements should be identical. Data for a few alkali halides, derived from elastic constant measurements by Alers and Neighbours [49] and from specific heat data, are given: the ν_{max} so obtained are compared with observed lattice frequencies (Table 3.7). The predicted agreement between the θ_D values obtained by the two

Table 3.7 Debye characteristic temperatures and observed lattice frequencies (cm^{-1}) for some alkali halides.

	Elastic constant		Specific heat		(infrared)
	θ_D	ν_{max}	θ_D	ν_{max}	ν_{obs}
LiF	734·1	510	740·0	514	307
NaCl	321·9	224	320·0	222·4	164
KCl	237·1	164·8	235·1	163·4	142
KBr	172·8	120·0	174·0	120·9	98
KI	131·5	91·4	130·1	90·4	62

methods is attained, but the ν_{max} calculated is not equal to ν_{obs}. Maradudin *et al.* [11] have pointed out that this is to be expected, the ν_{max} being merely a normalizing factor and not a frequency of the lattice. Even if some relation between ν_{max} and ν_{obs} could be established, there are still two reasons why estimates of lattice frequencies based on the Debye theory are unsatisfactory.

In the first place, the frequency distribution of the lattice is not usually a Debye distribution. Secondly, all methods of obtaining θ_D involve some average of the frequency distribution. If P is some parameter from which θ_D is to be calculated, and $g(\nu)$ the frequency distribution, then

$$P = \int W g(\nu)\, d\nu \tag{3.8}$$

where W is a weighting factor. Since W is, in general, temperature dependent, whereas the essence of the Debye theory is that θ_D is temperature independent, a variation in lattice frequency with temperature will be predicted by (3.8) in contradiction to Debye theory. We must conclude that the Debye theory cannot be successfully used in the calculation of lattice translational frequencies.

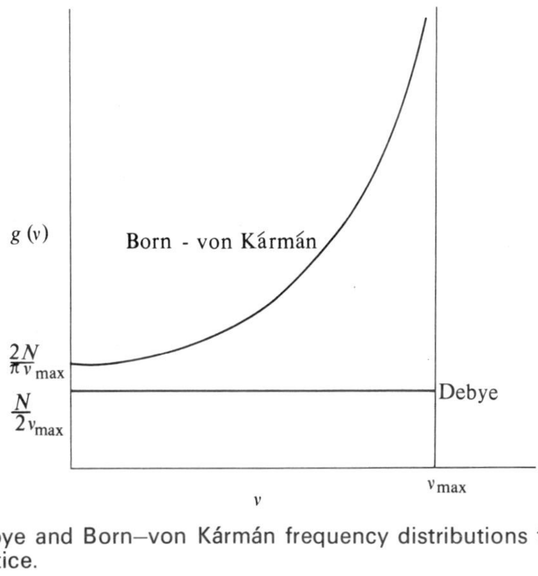

Fig. 3.2 Debye and Born–von Kármán frequency distributions for a linear monatomic lattice.

The contrast between the Debye and Born–von Kármán theories is well illustrated by the frequency-distribution curves (Fig. 3.2) which show $g(\nu)$ as a function of ν. For the linear lattice containing only one type of particle, the Debye distribution function is given by

$$g(\nu) = \frac{N}{2\nu_{max}} \qquad 0 < \nu < \nu_{max}.$$

On the other hand, the Born–von Kármán theory gives

$$g(\nu) = \frac{2N}{\pi\nu_{max}}(\nu_{max}^2 - \nu^2)^{-1/2}.$$

The more familiar three-dimensional Debye distribution is given by

$$g(\nu) = \frac{9N\nu^2}{\nu_{max}^3} \qquad 0 < \nu < \nu_{max},$$

and Fig. 3.3 shows that this is very unlike the frequency-distribution curve obtained by numerical analysis of observed dispersion curves for real monatomic solids.

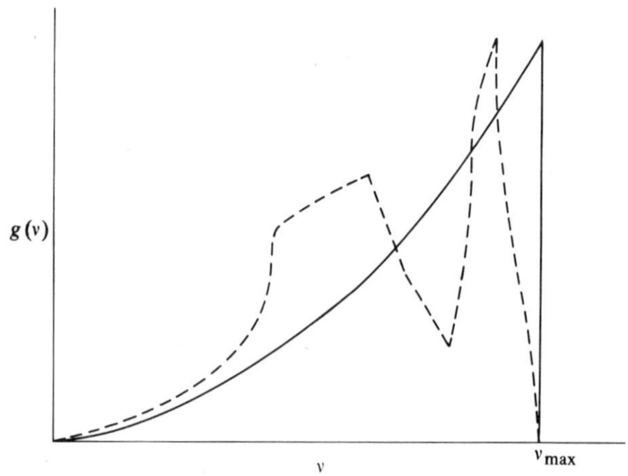

Fig. 3.3 Debye distribution (full line) and frequency distribution obtained by numerical analysis of dispersion curves for a monatomic metal lattice (dotted line). The areas under the two curves are equal.

The theory of Born and von Kármán has, until recently, received much less attention than the Debye theory. Since it is atomistic, and thus does not regard the lattice as a continuous elastic medium, it gives a more realistic picture of lattice modes. The theory has been well treated in the review articles by Mitra [973], and Mitra and Gielisse [975], where the linear diatomic chain has been dealt with. More detailed treatments are given in the monographs by Brillouin [4], Born and Huang [3], Hadni [8], Maradudin *et al.* [11], Smith [16] and Zbinden [19]. As long ago as 1935, Blackman had used the Born–von Kármán theory, and in a series of articles [150–4] he dealt with the frequency distributions of the monatomic and diatomic one-, two- and three-dimensional lattices. Subsequent calculations such as those by Kellermann on KCl [739, 740], Smith on diamond [1343] and Karo and Hardy on NaI [722] showed that the method could give quite accurate results for the observed infrared or Raman active lattice fundamentals. The two major difficulties in the way of using the Born–von Kármán theory are

(i) that there is very little theoretical basis for choosing suitable force constants,
(ii) that none of the numerous force fields which have been suggested have a very wide range of applicability.

Consider a crystal composed of N unit cells, with n particles in each unit cell: such a crystal has $3nN$ degrees of freedom. Three of these correspond to the $3N$ modes obtained by taking translations of each unit cell as a whole with different phase relations (the so-called acoustic modes) while the remaining $3nN - 3$ constitute the optical modes and include those frequencies active in the infrared or Raman effect. Since, in a macroscopic system, N will be of the order of Avogadro's number, the frequency spectrum needs to be described in terms of a frequency-distribution function. This function, previously referred to as $g(\nu)$, tells us the number of modes whose frequencies lie between the values ν and $\nu + d\nu$. As with molecular vibrations, we proceed by setting up the equations of motion and solving the resulting secular determinant. In this case we get a relation, called the dispersion relation, between the frequency ν and the wave vector $\mathbf{k}$ $(=2\pi/\lambda)$. Two types of solution are obtained; those in which $\nu \to 0$ as $\mathbf{k} \to 0$ correspond to the acoustic branch, while those for which ν tends to a finite limit as $\mathbf{k} \to 0$ represent the optical branch (see Fig. 3.4).

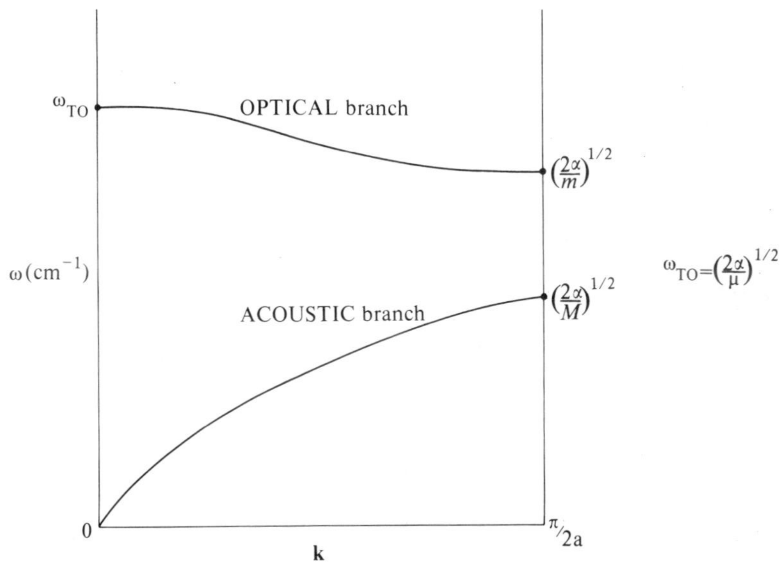

Fig. 3.4 The dispersion relation for the linear diatomic lattice, $M > m$.

Infrared and Raman studies determine the frequency at $\mathbf{k} = 0$ only. This limitation is due to the fact that when electromagnetic radiation interacts with the crystal lattice, absorption can occur only if momentum, $h\mathbf{k}$ (and hence the wave vector $\mathbf{k}$) and energy are both conserved. Consider infrared radiation of wavelength, λ, equal to 100 μ ($=10^{-2}$ cm). Then $\mathbf{k} = 2\pi/\lambda \sim 10^3\ \text{cm}^{-1}$. This is very small compared with $\mathbf{k}$ at the Brillouin zone boundary: $\mathbf{k}$ would have a value of about $10^8\ \text{cm}^{-1}$, the reciprocal of the nearest neighbour distance. Consequently when $\mathbf{k} \sim 0$ (i.e., at the Brillouin zone centre), energy and momentum are conserved and absorption can take place, subject to the gross selection rules for infrared and Raman activity. We shall now consider the application of the Born–von Kármán theory to the linear diatomic lattice. This one-dimensional model has been very useful in the study of polymer spectra and also gives a fairly realistic picture of the vibrations of a cubic lattice AB such as NaCl or CsCl.

Let the unit cell contain two particles of masses M, m with $M > m$, and let nearest neighbours be a distance a apart. Also let the particles M occupy odd numbered points u_{2n+1}, at distances $x = (2n + 1)a$ from the origin, while m occupy the even numbered points u_{2n} at distances $x = 2na$ from the origin. If the force constant between nearest neighbour atoms is α, then we have

$$M\ddot{u}_{2n} = \alpha(u_{2n+1} + u_{2n-1} - 2u_{2n}), \tag{3.9a}$$

$$M\ddot{u}_{2n+1} = \alpha(u_{2n+2} + u_{2n} - 2u_{2n+1}). \tag{3.9b}$$

These equations of motion may have solutions of the form

$$u_{2n} = A\, e^{i(\omega t - \mathbf{k}x)}, \qquad u_{2n+1} = B\, e^{i(\omega t - \mathbf{k}'x)}, \tag{3.10}$$

where $\omega = 2\pi\nu$, ν in sec^{-1}. From (3.9, 3.10) we have

$$\begin{aligned} A\, e^{i\mathbf{k}(2n+1)a} &= \frac{\alpha}{M\omega^2}\left(2A\, e^{i\mathbf{k}(2n+1)a} - B\, e^{i\mathbf{k}'(2n+1)a} - B\, e^{i\mathbf{k}'(2n-1)a}\right) \\ B\, e^{i\mathbf{k}(2n)a} &= \frac{\alpha}{M\omega^2}\left(2B\, e^{i\mathbf{k}'(2na)} - A\, e^{i\mathbf{k}(2n+1)a} - A\, e^{i\mathbf{k}(2n-1)a}\right). \end{aligned} \tag{3.11}$$

Solutions independent of n are obtainable only if $k = k'$; with this condition (3.11) reduces to

$$A = \frac{2\alpha}{M\omega^2}(A - B\cos ka). \qquad B = \frac{2\alpha}{m\omega^2}(B - A\cos ka), \tag{3.12}$$

which gives a nontrivial solution only if

$$\begin{vmatrix} \omega^2 - \dfrac{2\alpha}{M} & \dfrac{2\alpha}{M}\cos ka \\ \dfrac{2\alpha}{m}\cos ka & \omega^2 - \dfrac{2\alpha}{m} \end{vmatrix} = 0. \tag{3.13}$$

To get the desired dispersion relation, we require ω in terms of k.

The solutions are

$$\omega_1^2 = \alpha\left(\frac{1}{M} + \frac{1}{m}\right) - \alpha\left[\left(\frac{1}{M} + \frac{1}{m}\right)^2 - \frac{4\alpha^2}{mM}\sin^2 ka\right]^{1/2}, \quad (3.14a)$$

$$\omega_2^2 = \alpha\left(\frac{1}{M} + \frac{1}{m}\right) + \alpha\left[\left(\frac{1}{M} + \frac{1}{m}\right)^2 - \frac{4\alpha^2}{mM}\sin^2 ka\right]^{1/2}, \quad (3.14b)$$

where ω_1^2, ω_2^2 are the solutions corresponding to the acoustic and optical branches respectively.

As $\mathbf{k} \to 0$, we have ${\omega^2}_1 \to 0$, $\omega_2^2 = 2\alpha(1/m + 1/M)$. As $\mathbf{k} \to \pi/2a$, ω_1^2 increases to a limiting value of $2\alpha/M$, while ω_2^2 decreases to a limiting value of $2\alpha/m$. The dispersion relation is therefore represented by the curves shown in Fig. 3.4.

The reduced mass of the unit cell, μ, is given by

$$\frac{1}{\mu} = \frac{1}{m} + \frac{1}{M}.$$

It is ω_2 at $\mathbf{k} = 0$ which is observed by infrared or Raman studies.

In order to determine the form of the vibrations, we need to evaluate the ratio A/B, where A, B are the displacements of the particles M, m respectively.

Let the ratios corresponding to the solutions ω_1, ω_2 be A_1/B_1, A_2/B_2 respectively. Then

$$\frac{A_1}{B_1} = -\frac{(2\alpha/M)\cos ka}{\omega_1^2 - 2\alpha/M}. \quad (3.15)$$

As $\mathbf{k} \to 0$, $\cos ka \to 1$ and $\omega_1^2 \to 0$, giving $A_1/B_1 \to 1$. In such a mode, all particles are displaced equally, corresponding to a translation of the crystal as a whole.

As $\mathbf{k} \to \pi/2a$, $\omega_1^2 \to 2\alpha/M$ and thus $B_1 \to 0$; this corresponds to a mode in which the particles of mass m are stationary, while those of mass M move in antiphase.

For the optical branch, we have

$$\frac{A_2}{B_2} = -\frac{2\alpha/M\cos ka}{\omega_2^2 - 2\alpha/M}. \quad (3.16)$$

At $\mathbf{k} = 0$, we may write

$$\omega_2^2 = \frac{2\alpha}{\mu} = 2\alpha\left(\frac{1}{m} + \frac{1}{M}\right),$$

when (3.16) becomes $A_2/B_2 = -m/M$ or $MA_2 = -mB_2$; m and M move with equal amplitude and opposite phase. This is the only mode leading to absorption and will result in an infrared band if a change in dipole moment is associated with the motion. Finally, at $\mathbf{k} = \pi/2a$, $\omega_2^2 \rightarrow 2\alpha/m$, $A_2 \rightarrow 0$ giving a motion in which the particles of mass m move in antiphase while those of mass M are stationary. These four modes of motion are illustrated in Fig. 3.5.

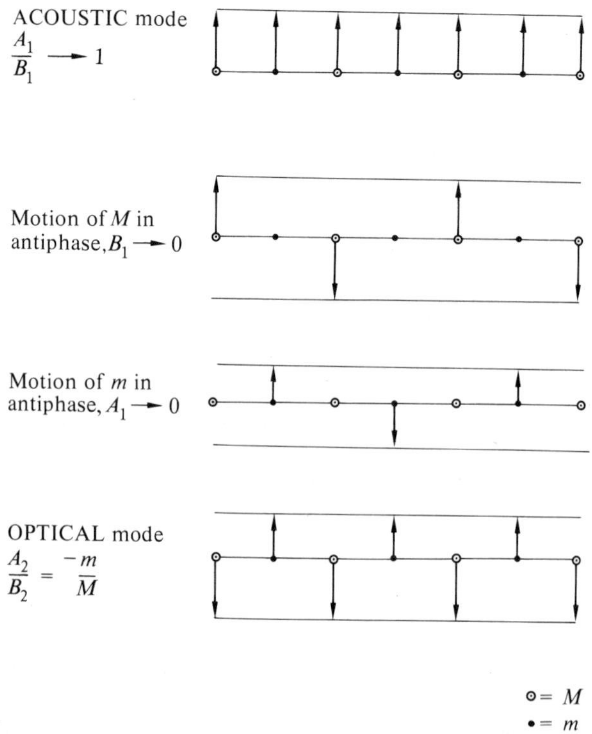

Fig. 3.5 Modes of motion of a linear diatomic lattice.

Blackman extended this theory to the two- and three-dimensional lattice; in the two-dimensional lattice he considered nearest neighbours at a distance a with force constant α, and second nearest neighbours at $2^{1/2}a$ with force constant γ. For the three-dimensional lattice, third nearest neighbours at $3^{1/2}a$ with force constant δ are also introduced. For the frequency leading to infrared absorption in the three-dimensional diatomic lattice, Blackman found

$$\underset{(\mathrm{cm}^{-1})}{\nu_{\mathrm{opt}}} = \frac{1}{2\pi c}\left[(2\alpha + 8\delta)\left(\frac{1}{m} + \frac{1}{M}\right)\right]^{1/2}, \tag{3.17}$$

and also showed the relations (3.18, 3.19) to exist between the force constants and the elastic stiffnesses:

$$C_{11}a = \alpha + 4\gamma + 4\delta, \tag{3.18a}$$

$$C_{12}a = 2\gamma + 4\delta, \tag{3.18b}$$

$$\frac{\gamma}{\alpha} = \frac{C_{12} + C_{44}}{4(C_{11} - C_{12} - C_{44})}. \tag{3.19}$$

Where reasonable values are obtained for γ/α, agreement between ν_{opt} calculated from (3.17) and ν_{obs} is quite good; but for LiF, MgO and ZnS, there is poor agreement and the values of γ/α are unrealistic. It is clear that in these cases the central-force field is inadequate. An alternative treatment due to Kellerman [739, 740] in which no relations between force constants and elastic constants are employed, gives better agreement with observation. The results obtained by the various methods are compared in Table 3.8. The frequencies obtained are those of the transverse optical mode; the longitudinal mode frequency is inactive in infrared absorption but can be obtained in a number of ways; the possibilities include neutron scattering, which gives the complete $\omega/\mathbf{k}$ curve rather than the value of ω at $\mathbf{k} = 0$, reflection measurements, where the reflection band is bounded approximately by the transverse (TO) and longitudinal (LO) frequencies, and dielectric constant measurements, which give ω_{LO} through the Lyddane–Sachs–Teller relationship

$$\frac{\omega_{LO}}{\omega_{TO}} = \left(\frac{\varepsilon_0}{\varepsilon_\infty}\right)^{1/2} \tag{3.20}$$

where ε_0 is the static, and ε_∞ the high-frequency dielectric constant.

The difficulties in the use of the Born–von Kármán theory referred to above limit its use to compounds, crystallizing in cubic ionic lattices, for which single-crystal elastic constant data are available. Few such compounds exist; nevertheless the Born–von Kármán theory has proved fairly effective in the calculation of the optically active translational frequency, as Table 3.8 shows.

The third method which has been applied to the calculation of lattice translational frequencies is the Wilson *FG* matrix method. In 1961, Shimanouchi, Tsuboi and Miyazawa [1291] published a paper giving the basis of the method and applying it to the one-dimensional diatomic lattice, and also to NaCl, diamond and CaF_2. Further studies were carried out on $CaCO_3$ [1442] and hexanitrocobalt(III) complexes [1052–3]. In these latter cases they used a Urey–Bradley force field and calculated UBFF constants from the observed frequencies. Since very few calculations have been made it has not been possible to determine the significance of the UBFF constants in relation to other physical properties of the lattice, such as elastic stiffness. The treatment given below is essentially that of Shimanouchi *et al.* [1291] with some alterations in the notation for consistency with the remainder of the text.

Table 3.8 Transverse optic mode frequencies calculated by Blackman's (B) and by Kellermann's (K) method; longitudinal frequencies calculated by Kellermann's method and by the Lyddane–Sachs–Teller relation.

	ν_{obs} (cm^{-1})	ν_{TO} (cm^{-1})		ν_{LO} (cm^{-1})	
		B	K	K	LST
LiF	307	—	405	887	674
NaCl	164	133	149	324	259
NaBr	134	107	141	272	202
KCl	142	127	140	253	210
KBr	113	108	112	201	161
KI	98	88	98	171	134
MgO	577	1090	516	718	1055
AgCl	103	103	96	306	184
AgBr	76	70	81	218	128
TlCl	86	76	—	235	214
TlBr	62	55	41	158	149

The G and F matrices for lattice translations are obtained by adding the G and F matrices for one Bravais cell (i.e., the smallest unit in which no two groups of atoms become equivalent as the result of a simple translation) to the sum of all the matrices representing interactions between that Bravais cell and its neighbours. The matrices may be set up in terms of either internal or Cartesian coordinates. Let the matrices relating to optically active vibrations be G_{opt}, F_{opt}. Then, if the Bravais cell is denoted by the suffixes i, j, k, and its internal coordinate vector is $\mathbf{I}_{ijk}$, we have

$$G_{opt} = \sum_{i',j',k'} G_{ijk,i'j'k'}, \tag{3.21a}$$

$$F_{opt} = \sum_{i',j',k'} F_{ijk,i'j'k'}, \tag{3.21b}$$

where the indices $(i'j'k')$ run over all cells interacting with the cell whose index is (ijk).

We can, further, define an optically active internal coordinate vector $\mathbf{I}_{opt}$ by

$$\mathbf{I}_{opt} = N \sum_{i,j,k} \mathbf{I}_{ijk}, \tag{3.22}$$

where N is the appropriate normalizing factor.

Consider the one-dimensional diatomic lattice of atoms with masses M, m, whose reciprocals are μ_1, μ_2, and let all nearest neighbour separations be a. For the ith Bravais cell and its interaction with the $(i-1)$th and $(i+1)$th Bravais cell we may consider the system shown in Fig. 3.6.

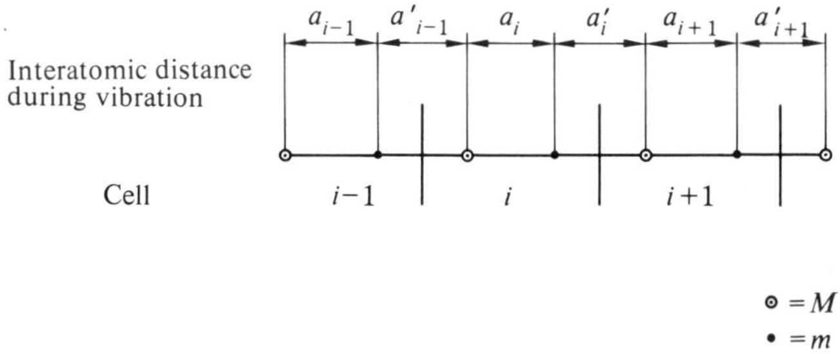

Fig. 3.6 The Bravais cell and its nearest neighbours in a linear diatomic lattice.

Then we can define internal coordinates

$$\begin{aligned} \Delta a_{i-1} &= a_{i-1} - a, & \Delta a_{i'-1} &= a_{i'-1} - a, \\ \Delta a_i &= a_i - a, & \Delta a_{i'} &= a_{i'} - a, \\ \Delta a_{i+1} &= a_{i+1} - a, & \Delta a_{i'+1} &= a_{i'+1} - a, \quad \text{etc.,} \end{aligned}$$

and the vectors $\mathbf{I}_i$ are

$$2^{-1/2}\begin{pmatrix}\Delta a_{i-1} \\ \Delta a_{i'-1}\end{pmatrix}, \qquad 2^{-1/2}\begin{pmatrix}\Delta a_i \\ \Delta a_{i'}\end{pmatrix}, \qquad 2^{-1/2}\begin{pmatrix}\Delta a_{i+1} \\ \Delta a_{i'+1}\end{pmatrix}.$$

We can determine the G_{opt} matrix elements in just the same way as for internal coordinates in a molecule. The only internal coordinates between which the G_{opt} elements are nonzero are those with one or more common atoms. There are only two types of G_{opt} element; principal and interaction stretching elements, of the Decius types g^2rr and g^1rr

$$G_{ii} = \mu_1 + \mu_2,$$

just as for a bond stretching element in a molecule.

$$G_{ii'} = \mu_c \cos\phi,$$

where μ_c is the mass of the common atom. Since the angle is 180°, we have $G_{ii'} = -\mu_c$. In this case we have m as the common atom, so $G_{ii'} = -\mu_2$. Where M is the common atom, as in $G_{i'-1,i}$ we have $G_{i'-1,i} = -\mu_1$. A portion of the complete G matrix is therefore as shown in Table 3.9, and

$$\begin{aligned} G_{\text{opt}} &= \sum_{i'=i-1}^{i+1} G_{ii'} \\ &= \begin{bmatrix} 0 & -\mu_1 \\ 0 & 0 \end{bmatrix} + \begin{bmatrix} \mu_1+\mu_2 & -\mu_2 \\ -\mu_2 & \mu_1+\mu_2 \end{bmatrix} + \begin{bmatrix} 0 & 0 \\ -\mu_1 & 0 \end{bmatrix} \\ &= \begin{bmatrix} \mu_1+\mu_2 & -(\mu_1+\mu_2) \\ -(\mu_1+\mu_2) & \mu_1+\mu_2 \end{bmatrix}. \end{aligned} \tag{3.23}$$

Table 3.9

	Δa_{i-1}	$\Delta a_{i'-1}$	Δa_i	$\Delta a_{i'}$	Δa_{i+1}	$\Delta a_{i'+1}$
Δa_{i-1}	$\mu_1+\mu_2$	$-\mu_2$	0	0	0	0
$\Delta a_{i'-1}$	$-\mu_2$	$\mu_1+\mu_2$	$-\mu_1$	0	0	0
Δa_i	0	$-\mu_1$	$\mu_1+\mu_2$	$-\mu_2$	0	0
$\Delta a_{i'}$	0	0	$-\mu_2$	$\mu_1+\mu_2$	$-\mu_1$	0
Δa_{i+1}	0	0	0	$-\mu_1$	$\mu_1+\mu_2$	$-\mu_2$
$\Delta a_{i'+1}$	0	0	0	0	$-\mu_2$	$\mu_1+\mu_2$

We can construct the F_{opt} matrix in the same way. If we have as a potential function

$$2V = f_r\,[\textstyle\sum (\Delta a_i)^2],$$

then all diagonal elements of F_{opt} will be f_r and all others will be zero, giving

$$\begin{aligned} F_{\text{opt}} &= \sum_{i'=i-1}^{i+1} F_{ii'} \\ &= \begin{vmatrix} 0 & 0 \\ 0 & 0 \end{vmatrix} + \begin{vmatrix} f_r & 0 \\ 0 & f_r \end{vmatrix} + \begin{vmatrix} 0 & 0 \\ 0 & 0 \end{vmatrix} \\ &= \begin{vmatrix} f_r & 0 \\ 0 & f_r \end{vmatrix} \end{aligned} \tag{3.24}$$

giving, since $|F_{\text{opt}}G_{\text{opt}} - E\lambda| = 0$, the two roots $\lambda = 0$ and $\lambda = 2f_r(\mu_1 + \mu_2)$. This is easily seen if we note that the sum of the roots is $\sum F_{\text{opt},ii}G_{\text{opt},ii} = 2f_r(\mu_1 + \mu_2)$, while the product of the roots, $|\det F_{\text{opt}}|\,|\det G_{\text{opt}}| = 0$.

Since $\lambda = 4\pi^2c^2\nu^2$ where ν is the frequency in cm^{-1}, we have

$$\nu = \frac{1}{2\pi c}\left(\frac{2f_r}{\mu}\right)^{1/2}, \qquad \text{where } \frac{1}{\mu} = \frac{1}{M} + \frac{1}{m} = \mu_1 + \mu_2. \tag{3.25}$$

This solution corresponds to ω_2 ($\mathbf{k} \to 0$) in the Born–von Kármán theory; similarly the zero solution corresponds to ω_1 ($\mathbf{k} \to 0$).

The evaluation of the G_{opt} matrix elements, though simple in the above case, in in general rather lengthy. It is possible to determine the G_{opt} elements more readily by using Cartesian rather than internal coordinates, since G_{opt} is diagonal in terms of Cartesian coordinates. Let the G_{opt} matrix so expressed be $G_{\text{opt(c)}}$, then

$$G_{\text{opt(c)}} = G_{(\text{c})ijk,\, ijk}, \tag{3.26}$$

its elements being the reciprocal masses of the atoms in the Bravais cell. $F_{\text{opt(c)}}$ is related to F_{opt} as follows:

There is a matrix B such that $I = BX$ where I, X are internal and Cartesian coordinates respectively. But

$$\begin{aligned} 2V &= I^{\dagger} F_{\text{opt}} I \\ &= X^{\dagger} B^{\dagger} F_{\text{opt}} B X \\ &= X^{\dagger} F_{\text{opt(c)}} X \\ \therefore\ F_{\text{opt(c)}} &= B^{\dagger} F_{\text{opt}} B. \end{aligned} \tag{3.27}$$

Analogous to $\mathbf{I}_{\text{opt}} = N \sum \mathbf{I}_{ijk}$ in internal coordinates, we can define

$$\mathbf{X}_{\text{opt}} = N \sum \mathbf{X}_{ijk} \tag{3.28}$$

as the Cartesian coordinates representing optically active vibrations; such coordinates must be totally symmetric with respect to the translational symmetry operations of the crystal. If n is the number of atoms in the Bravais cell, there will be $3n$ such coordinates.

Then we have

$F_{\text{opt}} = B^{\dagger}_{\text{opt}} F_{\text{opt}} B_{\text{opt}}$, where B_{opt} is defined by $I_{\text{opt}} = B_{\text{opt}} X_{\text{opt}}$.

But

$$\begin{aligned} \mathbf{I}_{\text{opt}} &= N \sum_{i,j,k} \mathbf{I}_{ijk} \\ &= N \sum_{i,j,k} \Big(\sum_{i'j'k'} B_{ijk,\, i'j'k'} X_{i'j'k'} \Big) \\ &= N \sum_{i',j',k'} \Big(\sum_{i,j,k} B_{ijk,\, i'j'k'} X_{i',j',k'} \Big) \end{aligned}$$

from which

$$B_{\text{opt}} = \sum_{i,j,k} B_{ijk,\, i'j'k'} \quad \text{or} \quad \sum_{i',j',k'} B_{ijk,\, i'j'k'} \tag{3.29}$$

The matrix elements B_{opt} are calculated only for those internal coordinates in terms of which the potential energy is expressed: for example Shimanouchi *et al.* [1291] used a UBFF involving only K and H for diamond, but included a repulsion term between the fluorine atoms in CaF_2. Just as with the secular determinant for internal vibrations, it is possible to reduce the order by a choice of symmetry coordinates. These must show the rotational symmetry of the lattice.

If such coordinates S_{opt} exist, they are related to the X_{opt} by

$$S_{opt} = UX_{opt} \tag{3.30}$$

and if the matrix elements derived from the S_{opt} are denoted by a superscript s, then

$$G^s_{opt(c)} = UG_{opt(c)}U^\dagger$$

$$F^s_{opt(c)} = UF_{opt(c)}U^\dagger = UB^\dagger_{opt}F_{opt}B_{opt}U^\dagger \tag{3.31}$$

$$= B^{\dagger s}_{opt}F_{opt}B^s_{opt}, \tag{3.32}$$

where

$$B^s_{opt} = B_{opt}U^\dagger, \qquad B^{\dagger s}_{opt} = UB^\dagger_{opt}. \tag{3.33}$$

We may reconsider the problem of the linear diatomic lattice, using the set of Cartesian coordinates Δx which represent the longitudinal displacements of the atoms; the superscripts (1), (2) refer to displacements of M, m respectively.

Then

$$\begin{aligned}
\Delta a_{i-1} &= -\Delta x^{(1)}_{i-1} + \Delta x^{(2)}_{i-1}, \\
\Delta a_{i'-1} &= -\Delta x^{(2)}_{i-1} + \Delta x^{(1)}_{i}, \\
\Delta a_{i} &= -\Delta x^{(1)}_{i} + \Delta x^{(2)}_{i}, \\
\Delta a_{i'} &= -\Delta x^{(2)}_{i} + \Delta x^{(1)}_{i+1}, \\
\Delta a_{i+1} &= -\Delta x^{(1)}_{i+1} + \Delta x^{(2)}_{i+1}, \\
\Delta a_{i'+1} &= -\Delta x^{(2)}_{i+1} + \Delta x^{(1)}_{i+2},
\end{aligned}$$

or in matrix form as shown in Table 3.10. Then

$$B_{opt} = \begin{vmatrix} 0 & 0 \\ 0 & 0 \end{vmatrix} + \begin{vmatrix} -1 & 1 \\ 0 & -1 \end{vmatrix} + \begin{vmatrix} 0 & 0 \\ 1 & 0 \end{vmatrix}$$

$$= \begin{vmatrix} -1 & 1 \\ 1 & -1 \end{vmatrix} \tag{3.34}$$

We may define symmetry coordinates

$$S_{opt} = 2^{-1/2}(\Delta x^{(1)} + \Delta x^{(2)}), 2^{-1/2}(\Delta x^{(1)} - \Delta x^{(2)}).$$

Coordinates of the form $(1/\sqrt{n}) \sum \Delta x^{(n)}$ are redundant; they depict a translation of the lattice as a whole, corresponding to an acoustic mode, and thus lead to a solution of the secular determinant with zero frequency.

As with internal vibrations, the U matrix may be constructed, where the redundancy is obvious, without including the redundant coordinate.

Table 3.10

	$\Delta x_{i-1}^{(1)}$	$\Delta x_{i-1}^{(2)}$	$\Delta x_i^{(1)}$	$\Delta x_i^{(2)}$	$\Delta x_{i+1}^{(1)}$	$\Delta x_{i+1}^{(2)}$	$\Delta x_{i+2}^{(1)}$
Δa_{i-1}	-1	$+1$	0	0	0	0	0
$\Delta a_{i'-1}$	0	-1	$+1$	0	0	0	0
Δa_i	0	0	-1	$+1$	0	0	0
$\Delta a_{i'}$	0	0	0	-1	$+1$	0	0
Δa_{i+1}	0	0	0	0	-1	$+1$	0
$\Delta a_{i'+1}$	0	0	0	0	0	-1	$+1$

Since $S_{\text{opt}} = UX_{\text{opt}}$, and S_{opt}, X_{opt} are respectively $2^{-1/2}(\Delta x^{(1)} - \Delta x^{(2)})$, $(\Delta x^{(1)} - \Delta x^{(2)})$, we have

$$U = [2^{-1/2} \quad -2^{-1/2}]$$

$$U^{\dagger} = \begin{bmatrix} 2^{-1/2} \\ -2^{-1/2} \end{bmatrix}$$

$$B_{\text{opt}}^{(s)} = BU^{\dagger} = \begin{bmatrix} -1 & 1 \\ 1 & -1 \end{bmatrix} \begin{bmatrix} 2^{-1/2} \\ -2^{-1/2} \end{bmatrix} = \begin{bmatrix} -2^{1/2} \\ 2^{1/2} \end{bmatrix}$$

$$\begin{aligned} F_{\text{opt(c)}}^{s} &= B_{\text{opt}}^{\dagger s} F_{\text{opt}} B_{\text{opt}}^{s} \\ &= [-2^{1/2} \quad 2^{1/2}] \begin{bmatrix} f_r & 0 \\ 0 & f_r \end{bmatrix} \begin{bmatrix} -2^{1/2} \\ 2^{1/2} \end{bmatrix} \\ &= [-2^{1/2} \quad 2^{1/2}] \begin{bmatrix} -2^{1/2} f_r \\ 2^{1/2} f_r \end{bmatrix} \\ &= [4f_r] \end{aligned} \tag{3.35}$$

$$G_{\text{opt(c)}}^{s} = UG_{\text{opt(c)}}U^{\dagger}$$

$$G_{\text{opt(c)}} = \begin{bmatrix} \mu_1 & 0 \\ 0 & \mu_2 \end{bmatrix},$$

since in Cartesian coordinates G_{opt} is diagonal and the elements are the reciprocal masses of the atoms in the Bravais cell.

$$\therefore\ G^{s}_{\text{opt(c)}} = [2^{-1/2} \quad -2^{-1/2}]\begin{bmatrix}\mu_1 & 0\\ 0 & \mu_2\end{bmatrix}\begin{bmatrix}2^{-1/2}\\ -2^{-1/2}\end{bmatrix}$$

$$= [2^{-1/2} \quad -2^{-1/2}]\begin{bmatrix}2^{-1/2}\mu_1\\ -2^{-1/2}\mu_2\end{bmatrix}$$

$$= \tfrac{1}{2}(\mu_1 + \mu_2). \tag{3.36}$$

Note that $F^{s}_{\text{opt(c)}}$ and $G^{s}_{\text{opt(c)}}$ are now 1×1 matrices, the redundancy leading to the zero root having been eliminated. We therefore have

$$\lambda = FG = 4f_r \times \tfrac{1}{2}(\mu_1 + \mu_2)$$
$$= 2f_r(\mu_1 + \mu_2)$$

as before. Therefore,

$$\nu = \frac{1}{2\pi c}\left(\frac{2f_r}{\mu}\right)^{1/2}. \tag{3.37}$$

The solutions for the two- and three-dimensional diatomic lattices are identical, doubly and triply degenerate modes of the frequency given by (3.37) being predicted.

Although the *FG* matrix method is of wide general applicability, its use in the determination of translational lattice frequencies is limited to those compounds whose crystal structures are known in detail. Without this knowledge it is not possible to construct the matrices. With this knowledge, the difficulties become similar to those of using the Born–von Kármán theory, namely

(i) the difficulty of finding a suitable force field, and
(ii) the lack of a theoretical basis whereby force constants can be related to some observable property of the crystal.

(c) The calculation of rotational mode frequencies

The rotational modes which have been observed are mostly restricted rotational (librational) modes, though in a few cases the existence of free rotation above some transition temperature has been suggested.

In theory, the fundamental librational frequency could be determined from the height of the potential-energy barrier restricting rotation, assuming a form for the variation in potential energy with angle of rotation. In practice, the problem has been frequently approached from the other direction; librational modes have been assigned either directly to low frequency bands, or by studying the fine structure of an internal fundamental and attributing this to combinations between the internal fundamental and successive levels of a librating species; the barrier height is then calculated from the librational frequencies. Two types of system have been studied; that in which the librating group is not attached to a framework (e.g., $CO_3^=$, NO_3^-, $SO_4^=$) and that in which one or more librating groups is attached to a framework (e.g., the XY_3 or XY_2 groups in systems such as XY_3—Y—XY_3, XY_2—Y—XY_2). In this second type of system the theory has been developed in a series of five articles by Miller *et al.* [454–7, 955], with special reference to organic rather than inorganic systems; in the first type only Lippincott and co-workers appear to have studied the problem to any great extent.

For an isolated librating group taking up n equivalent positions in the course of a rotation through 2π, Lippincott *et al.* [1248] have shown that a potential of the form

$$2V = V_0(1 - \cos n\alpha),$$

where V_0 is the barrier height and α the angle of rotation, gives V_0 in cm^{-1}:

$$V_0 = \frac{8\pi^2 Ic}{n^2 h} \nu_L^2 \tag{3.38}$$

where ν_L is the observed librational frequency in cm^{-1}.

They have calculated barrier heights from ν_L, itself deduced from studies of combinations with internal fundamentals, and compared V_0 calculated from (3.38) with the observed cut-off frequency. The measure of agreement obtained may well be coincidental since quite small values of the anharmonicity constant may strongly affect ν_L and hence the barrier height; nevertheless the values obtained are reasonable, being in the region of 200 cm^{-1} or 0·56 kcal $mole^{-1}$.

Table 3.11 Calculated barrier heights, and observed fundamental and cut-off frequencies (cm^{-1}) for some planar trigonal anions, assuming a threefold potential.

Anion	Compound	ν_L (cm^{-1})	Anharmonicity constant (calculated)	ν_0 (cm^{-1})	Observed cut-off frequency
NO_3^-	$NaNO_3$	32	3·0	155	238
	$Sr(NO_3)_2$	20	0·5	198	220
	$Ba(NO_3)_2$	19	0·5	179	191
	$Pb(NO_3)_2$	18	0·5	161	167
$CO_3^=$	$CaCO_3$ (calcite)	36	1·0	162	152

The first of Miller's articles dealt with the modes of a top with a threefold axis, attached to a rigid framework. Although he has dealt with CH_3 groups, there would be applications to systems such as M—NH_3 or

M—O—XO_3 in inorganic chemistry. One may write a potential

$$2V(\alpha) = V_3(1 - \cos 3\alpha) + V_6(1 - \cos 6\alpha) + \cdots \tag{3.39}$$

and, on substituting in the Schrödinger equation, we obtain Mathieu's equation. For each quantum number v there is one A and one E sublevel, and we may write

$$E_v = 2{\cdot}25Fb_{v\sigma} \tag{3.40}$$

where σ designates A or E, and b_v is an eigenvalue of Mathieu's equation:

$$F = \frac{h^2}{8\pi^2 rI_\alpha}; \qquad rI_\alpha = I_\alpha\left[1 - I_\alpha \sum_g \frac{\lambda_g^2}{I_g}\right],$$

where I_α is the moment of inertia of the top about its symmetry axis, λ_g is the cosine of the angle between the top axis and the gth principal axis of inertia of the molecule, and I_g is the moment of inertia about the gth principal axis of inertia. Now

$$\nu_{\text{obs}}(\text{cm}^{-1}) = \frac{\Delta E_{v\sigma}}{hc} \tag{3.41}$$

$$\Delta E = 2{\cdot}25F(\Delta b_{v\sigma}). \tag{3.42}$$

From $\Delta b_{v\sigma}$ it is possible to calculate a parameter s [645] such that

$$V_3 = 2{\cdot}25Fs. \tag{3.43}$$

A specimen calculation on CH_3CHO is given by Miller, as follows: $\nu_{0\to1}$ is observed at 150 cm^{-1} and assumed to represent both the A and E sublevels. Consider the A sublevel

$$\Delta E_{v\sigma} = E_{1(A)} - E_{0(A)} = 150\ \text{cm}^{-1} = 2{\cdot}25F(\Delta b_{v\sigma}).$$

But $F = 7{\cdot}696$ cm^{-1} from structural data.

$$\therefore\ \Delta b_{v\sigma} = b_{1(A)} - b_{0(A)} = \frac{150}{2{\cdot}25 \times 7{\cdot}696} = 8{\cdot}663.$$

From tables [644],

s	20	28
$b_{1(A)}$	12·10	14·52
$b_{0(A)}$	4·20	5·03
$\Delta b_{1\to0(A)}$	7·90	9·49

Linear interpolation gives $s = 23{\cdot}84$. From (3.43), $V_3 = 2{\cdot}25 \times 23{\cdot}84 \times 7{\cdot}696 = 1180$ cal mole^{-1}. Having found V_3, one may then reverse the calculation to find the frequencies of the other transitions.

Miller next considered the case of two or three rotating groups attached to the same atom. The two-rotor case, also considered by Pierce [1137] and by Swalen and Costain [1400], is applicable to systems such as $O_3P—O—PO_3$ or $O_2S—O—SO_2$.

Again considering a top with a threefold axis, we have

$$V(\alpha_1, \alpha_2) = \frac{V_3}{2}(2 - \cos 3\alpha_1 - \cos 3\alpha_2) + V_3' \cos 3\alpha_1 \cos 3\alpha_2 + V_3'' \sin 3\alpha_1 \sin 3\alpha_2 + \text{terms involving } 6\alpha. \quad (3.44)$$

The terms in V_3', V_3'' arise from interaction between the rotors and could be neglected to a first approximation. For a two-rotor system with C_{2v} symmetry there are two modes of motion, one of a_2 symmetry which is Raman active only, and one of b_1 symmetry which is both infrared and Raman active. These arise from the rotors having parallel and antiparallel directions of rotation respectively.

A simplification of the wave-mechanical problem leads to equations for V_3 analogous to (3.43); s values have to be found for the b_1 and a_2 species. If interaction between the rotors is neglected, each level will again be split into two sublevels. Reduction of the overall symmetry from C_{2v} to C_s causes both modes to become both infrared and Raman active, their species being $1a' + 1a''$.

The remaining articles by Miller deal largely with the differences in V_3 computed from various observed frequencies and with possible improvements in barrier height calculations, both by more accurate determination of s (using microwave techniques) and by considering terms in the potential function other than that in V_3. Miller's improvements to the potential mainly involve the inclusion of V_6, and in his last paper [955] he considers the situation where a barrier height V is given by

$$V = \sum_{n=1}^{3} n^2 V_n,$$

applying this to acrolein. Improvements to the simple cosine function for the threefold potential have also been well reviewed by Dellepiane and Zerbi [346]. These improvements, due to Scott and Scheraga [1262], Magnasco [899] and Mason and Kreevoy [914] involve allowing in various ways for interactions between non-bonded atoms. They all require a number of empirical or semi-empirical parameters which are not easy to evaluate, and Dellepiane and Zerbi found that none of the proposed potential functions reproduced the experimentally determined barrier heights over the series of molecules chosen for study; nor were they able to rationalize the disagreements with experiment.

From the point of view of the structural chemist, it is probably of secondary importance to determine barrier heights accurately. An approximate calculation based on the simple n-fold potential is sufficient to determine whether an observed frequency can, or cannot, reasonably be attributed to a librational fundamental.

The direct attack of determining V_0 independently of observed librational frequencies has been made by Leech, Powell and Sheppard [838]. They have deduced values of the barrier height by studying the variation of band shape with temperature. In general, the band consists of a central peak with wings: Bulanin and Orlova [210] have suggested that the intensity of the central peak is due to molecules not in rotation, while that of the wings is due to molecules in states characterized by a quantum number K such that the rotational levels lie above the potential-energy barrier. Then

$$\frac{I_{\text{wings}}}{I_{\text{total}}} = \frac{N_r}{N} = \frac{\int_{K^*}^{\infty} e^{-K^2h^2/8\pi^2IkT}\,\mathrm{d}K}{\int_0^{\infty} e^{-K^2h^2/8\pi^2IkT}\,\mathrm{d}K}. \tag{3.45}$$

From this, they find

$$V_0 = 2BhK^{*2}, \tag{3.46}$$

where B is the rotational constant and K^* the lowest quantum number corresponding to an energy greater than the barrier height.

An alternative procedure due to Rakov [1177] and also employed by Leech *et al.*, depends on the fact that the half-bandwidth Δ can be separated into a temperature-independent part Δ_0 and a temperature-dependent component proportional to $e^{-V/RT}$, where V is the barrier height for reorientation by Brownian motion. Put $\Delta = \Delta_0 + A_0\, e^{-V/RT}$. If $\Delta\nu_{1/2}$ is measured at a temperature so low that there is no rotation, this gives Δ_0 directly; also if the bandwidth at some higher temperature is Δ, we have

$$\log_e(\Delta - \Delta_0) = -\frac{V}{RT} + \log_e A; \tag{3.47}$$

thus V is found from a plot of $\log_e(\Delta - \Delta_0)$ against $1/T$.

In this procedure it is necessary to separate the observed band into the central peak and the rotational wings. The contour due to rotational motion can be determined from values of Coriolis constants ζ; the usual procedure is to plot the contour with different ζ values and take that which gives the best fit; the central peak is then fitted to a Lorentzian function.

$$\log_e d = \frac{a}{(\nu - \nu_0)^2 + b^2}, \tag{3.48}$$

where d = optical density

$a/b^2 = \log_e d_{max}$

$2b$ = half-bandwidth

ν_0 = frequency at which α is a maximum.

The integrated area of such a band is equal to $\pi b \log_e d_{max}$. The accuracy of estimating V by Rakov's procedure is very dependent on the separation of the band into its rotational and non-rotational components. In $Au_2I_2(CH_3)_4$ Leech *et al.* found values for the barrier height of 280 cm^{-1} at 180°K, and 305 cm^{-1} at 298°K, by Bulanin's procedure; the average over the range was 230 cm^{-1} using Rakov's method.

In conclusion we may say that it is relatively easy to determine whether a band can be reasonably attributed to a librational motion from a rough calculation of the potential-energy barrier, but difficult to obtain a very accurate value either of the librational fundamental frequency or the height of the potential-energy barrier.

Part Two

4. The spectra of compounds with no discrete molecular vibrating unit

4.1 Introduction

Although most of the spectroscopic information which exists deals with the internal modes of isolated vibrating groups, there is a growing interest in compounds which do not contain such groups. Work on such compounds has been largely concentrated on two classes. The first of these is the lattices AB, where the Born–von Kármán theory has provided a basis for the interpretation of observed absorption or reflection measurements in terms of the transverse optic modes at zero wave vector, and the Lyddane–Sachs–Teller relation is then applied to the calculation of longitudinal optic mode frequencies. The second class consists of compounds with stoichiometric formulae such as AB_2, AB_3, A_2B_3, ABX_3 and A_2BX_4. In such cases the spectra have been interpreted partly in terms of vibrations of the coordination polyhedra. Although these are not discrete vibrating units, such a treatment usually provides an adequate first approximation in terms of which the spectra may be usefully discussed. The adequacy of this approximation will naturally depend on the relative bond strengths in the compound. A good example of the value of such an approach is afforded by the spinels. In these compounds, which are formulated A_2BX_4, we may formally consider the structure as consisting of isolated ions of A and isolated BX_4 molecules. In (for example) Na_2WO_4, which has the spinel structure, this approximation is very good, since the interactions in the WO_4 group are much stronger than those in the NaO_6 coordination polyhedron. At the other extreme, in a compound such as Cr_2ZnO_4 the interactions within the coordination polyhedra of Zn and of Cr are similar in strength, and it is much less useful to discuss the spectra in terms of vibrations of isolated groups. Every gradation between these two situations is possible; the distinction between behaviour as a molecular crystal and as a 'giant molecule' crystal cannot be made on the basis of discrete physical properties of the crystal, but by consideration of the forms of the dispersion curves, and of the forms of the normal coordinates for crystal vibrations. In the following discussion, the selection rules for each compound are derived in terms of either factor-group or site-group analysis; the observed spectra are then tabulated and possible assignments considered in the light of the selection rules derived.

4.2 Compounds of the type AB

Most of the materials which have attracted interest have been discussed by Wilkinson [1510].

In Table 4.1, and in all other such tables, entries of the type $m(n)$ under

Table 4.1 Selection rules for diatomic lattices.

Lattice type	Space group or factor group	Z	Γ_{opt}				$\Gamma_{acoustic}$
			Activity R only	ir only	R and ir	ia	
NaCl	O_h^5	4(1)	—	$1F_{1u}$	—	—	$1F_{1u}$
CsCl	O_h^1	1	—	$1F_{1u}$	—	—	$1F_{1u}$
ZnS (Zincblende)	T_d^2	4(1)	—	—	$1F_2$	—	$1F_2$
ZnS (Wurtzite)	C_{6v}^4	2	$2E_2$	—	$1A_1 + 1E_1$	$2B_1$	$1A_1 + 1E_1$
HgS	D_{3h}	3	$2A'_1$	—	$3A''_2 + 5E'$	—	$1A''_2 + 1E'$
BN	D_{6h}^4	2	$2E_{2g}$	$1A_{2u} + 1E_u$	—	$2B_{1g}$	$1A_{2u} + 1E_{1u}$

Z imply that the content of the conventional unit cell is m formula units, while the factor-group analysis has been carried out for the non-redundant part of the unit cell (generally the primitive cell) containing n formula units.

Normal-coordinate analyses have been carried out for the one-dimensional diatomic lattice [1291] and for the wurtzite lattice [1444], using the Wilson *FG* matrix method; the first of these analyses is easily generalized to 3 dimensions and applied to the NaCl and CsCl structures. The selection rules for HgS are determined according to the factor group given by Wilkinson [1510]; but Wyckoff's compilation gives the space group as either D_3^4 or D_3^6, leading to the representations $\Gamma_{opt} = 2A_1(\text{R}) + 3A_2(\text{ir}) + 5E(\text{ir, R})$; $\Gamma_{acoustic} = A_2 + E$. According to Wilkinson, the unit cell contains no centre of symmetry; thus the LO modes will be Raman active. BN has been studied by Geick *et al.* [517]; originally, a graphite-like structure of factor group D_{3h} was proposed, but the spectroscopic data is in better accord with the tabulated structure.

Many compounds with the NaCl and CsCl structures have been studied;

the comprehensive study by Jones *et al.* [705] has been supplemented, and some values revised, by Hass [617], Berthold [136], Wilkinson [1510] and Jacobson and Nixon [694].

The values of f_r in Table 4.2 are calculated from the relation obtained

Table 4.2 Transverse and longitudinal optic mode frequencies (cm^{-1}) for compounds with the NaCl structure.

Compound	ν_{TO} (cm^{-1})	ν_{LO} (cm^{-1})	f_r (mdyn $Å^{-1}$)
LiF	305	685	0·14
LiCl	191	—	0·06
LiBr	159	—	0·05
NaF	245	422	0·18
NaCl	162	265	0·11
NaBr	135	202	0·10
NaI	117	180	0·08
KF	190	—	0·14
KCl	142	205	0·11
KBr	117	160	0·105
KI	102	128	0·09
RbF	156	—	0·11
RbCl	119	—	0·105
RbBr	90	—	0·10
RbI	75	—	0·085
CsF	115	—	0·065
AgCl	103	184	0·085
AgBr	80	128	0·085
MgO	403	717	0·45
CaO	286, 295	577, 587	0·27
SrO	227	487	0·20
MgS	251	—	0·245

by the Wilson *FG* matrix method (eq. (3.37)), namely

$$\nu = \frac{1}{2\pi c}\sqrt{\frac{2f_r}{\mu}},$$

where f_r is the force constant between nearest neighbours, and μ is defined by

$$\frac{1}{\mu} = \frac{1}{M_A} + \frac{1}{M_B}.$$

The frequencies of the LO modes at $\mathbf{k} = 0$ can be determined in one of three ways. Where the static and high-frequency dielectric constants ε_0

and ε_∞ are known, the Lyddane–Sachs–Teller relation (eq. (3.20)):

$$\frac{\omega_{LO}}{\omega_{TO}} = \left(\frac{\varepsilon_0}{\varepsilon_\infty}\right)^{1/2}$$

may be employed. A second method involves analysis of the reflection spectra to obtain the real (ε') and imaginary (ε'') parts of the dielectric constant; ω_{LO} is then the frequency at which ε' is zero. Calculations by Wilkinson for the compounds tabulated above show agreement to within a few cm^{-1}; slightly less good agreement is obtained by Jacobson and Nixon for CaO and SrO. The values given in Table 4.2 are those calculated from the reflection data; the Lyddane–Sachs–Teller relation gives ν_{LO} at 538 cm^{-1} for CaO and 445 cm^{-1} for SrO. Jacobson and Nixon also employed the technique due to Berreman [133–5] involving oblique incident radiation on a thin film of the material; this method gave the intermediate values of 564 cm^{-1} for CaO and 473 cm^{-1} for SrO. Berthold's value for ω_{TO} in MgS refers to the crystalline material; he found a band at 317 cm^{-1} in amorphous MgS.

The force constants f_r which are calculated from the Wilson *FG* matrix procedure differ somewhat from those tabulated by Wilkinson. Those constants are derived using the shell model due to Cochran [275]; the theory of Born and Mayer gives the expression

$$A = \frac{6R_0}{\beta}, \tag{4.1}$$

where A is the force constant, R_0 the lattice constant (i.e., the length of the cubic cell edge) and β the compressibility. The values of A are larger than those of f_r by a factor of about 5 (LiF, 0·78; NaCl, 0·40; KBr, 0·30 mdyn $Å^{-1}$). Materials with the CsCl structure whose spectra have been studied include the chlorides, bromides and iodides of Cs and Tl. As Table 4.1 shows, one infrared absorption of species F_{1u} is expected and there is no Raman one-phonon spectrum. The force constant A has been calculated for CsI as 0·28 mdyn $Å^{-1}$.

Table 4.3 Lattice frequencies in compounds with the CsCl structure.

Compound	ν_{LO}	ν_{TO}	f_r (mdyn $Å^{-1}$)
CsCl	—	99	0·08
CsBr	113	74	0·08
CsI	90	61·5	0·07
TlCl	100	63	0·035
TlBr	75	48	0·038
TlI	—	52	0·06

Unlike the alkali halides, ZnS and its isostructures crystallize in non-centrosymmetric space groups and some of the fundamental LO modes are therefore observable in the Raman effect. The wurtzite form of ZnS has attracted more attention from the theoretical viewpoint; normal coordinate analyses of this lattice have been carried out by Poulet and Mathieu [1161] and Tsuboi and Wada [1444]. The compounds studied are ZnO, ZnS, CdSe and CdS. In a number of cases the authors have

Table 4.4 Lattice frequencies in compounds with the wurtzite structure.

Mode / Compound	Class A_1		Class E_1		Class E_2		Ref.
	TO	LO	TO	LO	TO	LO	
ZnO	380	574	407	583	437	101	313
	420		438		383	180	974
	432		456		433	98	1444
	334		351		433	98	1444
	386		407		433	98	1444
ZnS	300						813, 977
	274	350	310				81
CdSe	185						977
	164	209	170	209			149
CdS	228		235		252	44	1444
	251		257		249	52	1444
	198		209		249	52	1444
	209		241		249	52	1444
	256	(325)			212	85	1161
	260						977
	241	304	265				81
	233·5	300	241	300			149

made no assignment, and their observations have been tabulated in accordance with the first assignment given in the table, for each compound. Values of LO mode frequencies in brackets have been calculated using the Lyddane–Sachs–Teller relation. The two sets of observed data for ZnO are in poor agreement; in addition to the tabulated frequencies, Mitra and Bryant found two others at 538 and 588 cm^{-1} which they did not assign. Tsuboi and Wada attempted to determine force constants corresponding to the values found by Damen *et al.* [313] in three different ways. For set (i) they used a simple valence-force field with Urey–Bradley constants K_1, K_2 denoting Zn—S 'stretching' and H denoting all types of 'bending'. The poor agreement with observation was attributed to the neglect of polarization implicit in this force field, and they attempted to correct it in two ways. To obtain set (ii) they considered the polarization as giving rise to an electric field which caused an additional force to act on each ion. If the field and polarization directions are parallel, the valence force constant K is effectively reduced by an amount k such that the observed frequency ν_{obs} will be lower than the calculated frequency ν_c on the basis of a field including chemical bonds and other short-range forces. They arrived at the relation

$$\frac{\nu_{obs}^2}{\nu_c^2} = \frac{K - k}{K}. \tag{4.2}$$

For both ZnO and CdS this approach causes an over-correction. To calculate the results of set (iii) an effective electric field E_{eff} is considered:

$$E_{eff} = eN_\gamma \tag{4.3}$$

and

$$k = eE_{eff} = e^2 N_\gamma, \tag{4.4}$$

from which

$$\frac{k}{K} = P_{ir}, \qquad \frac{\nu_{obs}^2}{\nu_c^2} = 1 - P_{ir}, \tag{4.5}$$

where

$$P_{ir} = \left[\left(\frac{\varepsilon - 1}{\varepsilon + 2}\right) - \left(\frac{n^2 - 1}{n^2 + 2}\right)\right] = \gamma \frac{Ne^2}{K}, \tag{4.6}$$

ε being the low-frequency dielectric constant.

Poulet and Mathieu's Raman data for CdS are in poor agreement with Tsuboi and Wada's observations; Poulet and Mathieu carried out a normal-coordinate analysis using a Urey–Bradley force field involving constants K (M—S stretching), H_1 and H_2 (SMS and MSM bending) and F_1, F_2 (M---M, S---S repulsions). They calculate ν_{TO} for the A_1 and E modes as 307 cm^{-1}, though the strong band in the infrared is at 258 cm^{-1}.

For the zincblende structure, one optically active mode is expected; this has been observed in absorption by Mitsuishi *et al.* [977] for ZnS, ZnSe, ZnTe, CdS and CdTe. Taylor [1424] has studied the Raman spectra of polycrystalline ZnSe and ZnTe, obtaining the TO and LO mode frequencies, and some assignments of the acoustic modes have been made from the multiphonon spectrum.

Geick *et al.* [517] have found in BN three TO modes at 1367 (ir), 1370 (R) and 783 (ir) cm^{-1}, and have determined LO modes corresponding to the first and last of these at 1610 and 828 cm^{-1}. The absence of LO modes from the Raman spectrum suggests a centrosymmetric space group and therefore eliminates structures (resembling graphite) based on the factor group D_{3h}.

Table 4.5 Lattice modes of compounds with the zincblende structure.

Compound \ Mode	TO	LO	TA	LA	Ref.
ZnS	310				977
ZnSe	205	250			1424
	215				977
ZnTe	174	203			1424
	190				977
CdS	270				977
CdTe	150				977
	139 or 142	148 or 152	35 or 37	97 or 143	173
	125	151	37	50	1357

Wilkinson [1510] has given the spectrum of cinnabar (HgS) as 36, 102, 336 cm^{-1} (A_2'') and 83, 92, 100, 280, 343 cm^{-1} (E') for the ir active modes, and 43, 256 (A_1''); 72, 88, 108, 282, 343 (E') for the Raman active modes, together with three unassigned frequencies observed in the Raman at 202, 290 and 351 cm^{-1}. The alternative factor group D_3 suggested by Wyckoff would lead to the same number of ir and Raman active modes.

4.3 Compounds of the type AB_2

The most frequently studied compounds are those with the fluorite structure and those with the rutile structure. The *FG* matrix method has been employed in the normal-coordinate analysis of fluorite by Shimanouchi *et al.* [1291] and of rutile by Tsuchida and Nakagawa [1445]. Other structures studied include those of cuprite, α-quartz and β-quartz.

Table 4.6 Selection rules for AB_2 lattices.

Lattice type	Space group or factor group	Z	Γ_{opt} Activity R only	ir only	ir, R	inactive	$\Gamma_{acoustic}$
CaF_2	O_h^5	4(1)	F_{2g}	F_{1u}	—	—	F_{1u}
TiO_2 (rutile)	D_{4h}^{14}	2	$A_{1g}+B_{1g}+B_{2g}+E_g$	$A_{2u}+3E_u$	—	$A_{2g}+2B_{1u}$	$A_{2u}+E_u$
Cu_2O	O_h^4	2	F_{1g}	$3F_{1u}$	—	$A_{2u}+E_u$	F_{1u}
α-quartz	D_3^4 or D_3^6	3	$4A_1$	$4A_2$	$8E$	—	A_2+E
β-quartz	D_6^4 or D_6^5	3	A_1+4E_2	$3A_2$	$4E_1$	$2B_1+3B_2$	A_2+E_1

A normal-coordinate analysis of the fluorite lattice has been carried out by Shimanouchi *et al.* [1291], the roots of the secular equation being given as

$$\lambda_{F_{1u}} = \tfrac{1}{3}(\mu_F + 2\mu_m)4K, \qquad \lambda_{F_{2g}} = \mu_F 4(F + \tfrac{1}{3}K)$$

where μ_F, μ_m are the reciprocal masses of fluorine and metal, and K, F are Urey–Bradley force constants for Ca—F stretching and F---F repulsion respectively. Of the six branches of the phonon dispersion curve corresponding to optical modes, TO_1 and LO_1 corresponding to the F_{1u} mode are strongly split, but TO_2 and LO_2 corresponding to the F_{2g} mode are degenerate at $\mathbf{k} = 0$. All frequencies have been observed by Wilkinson [1510] for the compounds studied; the TO_1 values are in fair agreement with Kaiser's earlier work [720], but Kaiser's calculations of ν_{LO_1} from the Lyddane–Sachs–Teller relation give systematically low values compared with observation.

Table 4.7 Observed frequencies for crystals with the CaF_2 structure.

Compound	Mode and method of observation			Force constants (mdyn Å^{-1})		Ref.
	TO_1(ir)	LO_1(ir)	$TO_2 = LO_2$(R)	K	F	
CaF_2	266	482	322	0·307	0·165	1510
	257	(463)				720
	259*					136
SrF_2	219	391	285·5	0·268	0·124	1510
	217	(374)				720
BaF_2	189	338	241	0·223	0·078	1510
	184	(326)				720
CdF_2	204	394	317	0·247	0·178	1510
PbF_2	103	337·5	256	0·072	0·144	1510
$SrCl_2$	141	228	184	0·167	0·120	1510

* This value is for the crystalline material; in amorphous CaF_2 Berthold [136] observed a band at 322 cm^{-1}.

Of the AB_2 structures, rutile has been most frequently studied. A normal-coordinate analysis has been carried out by Tsuchida and Nakagawa [1445], viewing the vibrations as essentially those of the AB_6 octahedron. A simple Urey–Bradley force field involving A—B stretching, BAB bending and B---B repulsion gave moderate agreement, for NiF_2, with Balkanski's observed transmission spectrum. Krishnamurti [798] in his study of TiO_2 noted that the site symmetry of Ti was D_{2h} but, like all other workers, analysed the vibrations in terms of the D_{4h} factor group. There is no suggestion of any splitting due to the site symmetry. Katiyar and Krishnan [727] have calculated, using a rigid-ion model, dispersion curves for MgF_2 in which the TO mode frequencies at $\mathbf{k} = 0$ are in very good agreement with Barker's observations [92].

As Table 4.8 shows, there are some pronounced disagreements in both observation and assignment. Eagels [422] calculated the A_{2u} mode frequency from the expression

$$1 - \varepsilon_{\infty c} = 4\pi\rho\left[1 - \left(\frac{\omega_{\mathrm{m}}}{\omega_{\mathrm{T}}}\right)^2\right]^{-1}, \tag{4.7}$$

where $\varepsilon_{\infty c}$ is the high-frequency dielectric constant in the c-direction, $4\pi\rho$ the absorption strength, ω_{m} the frequency of the reflection minimum on the high-frequency side of the band, and ω_{T} the A_{2u} (TO mode) frequency.

Table 4.8 The spectra of some crystals with the rutile structure.

Compound	A_{1g}	B_{1g}	B_{2g}	E_g	A_{2u}		E_u						Ref.
					TO	LO	TO	LO	TO	LO	TO	LO	
TiO_2	612	143	826	447									1155
					167	811	500	806	388	458	183	373	422
	610	143	825	445	333						256		798
SnO_2	634		777	475	465	704	605	757	284	368	243	273	1393
							610	670	292	323	252	263	929
MgF_2	295	73	405	154									802
	410	92	515	295	399	625	450	617	410	415	247	303	1155, 92
					265		435		405		280		685
					399·5		451		410		249·5		727
MnF_2	341	61	476	247									1155
FeF_2	340	73	496	257	440		480		320		200		78
CoF_2					360		420		270		205		78
NiF_2					375		445		285		225		78
	418	448	68	330	421		447		254		215		1445
ZnF_2	350	70	522	253	294	488	380	498	244	264	173	227	1155, 92

The figures for CoF_2 and NiF_2 are derived from the transmission spectra; calculations based on Kramers–Kronig analysis of the reflectance spectra for the E modes are in close agreement.

Raman data for SnO_2 from Wilkinson (private communication).

The equation was originally derived by Spitzer *et al.* [1355], and the values of the parameters (except for $\varepsilon_{\infty c}$) are taken from that work. Their calculated values for $\nu_{\mathrm{TO}}(A_{2u})$ and $\nu_{\mathrm{LO}}(A_{2u})$ are 189 and 870 cm^{-1} respectively. Krishnamurti's value of 333 cm^{-1}, however, seems to fit better into the general pattern of assignments. In two cases, inactive mode frequencies have been calculated; Krishnan [802] obtained the values 287 cm^{-1} (A_{2g}) and 428, 166 cm^{-1} (B_{1u}) for MgF_2, while Tsuchida and Nakagawa obtained 288 cm^{-1} (A_{2g}) and 414, 164 cm^{-1} (B_{1u}) for NiF_2. They also determined the form of the vibrations of the NiF_6 octahedron in NiF_2 by calculation both of the L matrix elements for the ir active modes, and a potential-energy distribution based on the Jacobian matrix elements which show the variation of frequency with force constant. For the NiF_6 octahedron of D_{4h} symmetry, we would expect six stretching modes. There will be two axial, of which one is symmetric and one antisymmetric with respect to inversion, and four equatorial. The remaining nine will be bending modes. Their Jacobian matrix elements suggest $\nu_1(A_{1g})$, $\nu_3(B_{1g})$

and $\nu_9(E_u)$ as partly axial and partly equatorial stretching, while $\nu_5(E_g)$, $\nu_6(A_{2u})$ and $\nu_7(B_{1u})$ appear to be dominated by the equatorial stretching force constant K_2. $\nu_2(A_{2g})$, $\nu_4(B_{2g})$, $\nu_8(B_{1u})$ and ν_{10}, $\nu_{11}(E_u)$ depend most strongly on the FNiF bending constant $H(\gamma)$. They themselves, on the basis of the L matrix elements, consider ν_6 and ν_9 as stretching and ν_{10} and ν_{11} as bending.

Both α- and β-quartz have been studied, and the changes in the spectrum as going through the transition at 573°K have been interpreted in terms of the change in factor group from $D_3(\alpha)$ to $D_6(\beta)$. The Raman spectrum of α-quartz has been obtained by Krishnan [800] and Krishnamurti [795–6], while the infrared frequencies have been obtained by Spitzer and Kleinman [1354] from reflection measurements. Elcombe [432] has considered both a Born–von Kármán model and a rigid-ion model and has obtained good agreement with the observed frequencies. Elastic constants calculated on the basis of the Born–von Kármán model are also in good agreement with experimental values. Similar calculations based on the Raman data for β-quartz due to Narayanaswamy [1059] have also proved successful. A recent Raman study due to Shapiro *et al.* [1281] shows very good agreement with previous observations for the totally symmetric frequencies in α-quartz, but an additional weak band is found at 147 cm^{-1}. This band becomes markedly stronger on heating from -77°K to 846°K, but is absent in the β phase. Only two totally symmetric modes are found in the β phase, at 459 cm^{-1} and at 207–162 cm^{-1} (the frequency decreasing with increasing temperature). No explanation of the additional

Table 4.9 The vibrational spectra of α- and β-quartz.

Class	A_1	A_2		B_1	B_2	E	
		TO	LO			TO	LO
α-quartz	1081	1080	1240			1163	1236
	466	777	790			1043	1161
	355	489	547			797	809
	207	364	384			697	699
	147*					450	510
						394	403
						265	270
						128	129
						E_2	E_1
β-quartz	453 (470)	(1075)		(610)	(1111)	1154 (1065)	1060 (1075)
	207–162*	(469)		(411)	(361)	686 (600)	792 (735)
					(74)	398 (366)	479 (417)
						250 (244)	94 (90)

* Observed by Shapiro only.
Values in brackets are calculated using Born–von Kármán model.

totally symmetric mode observed by Shapiro in each phase is satisfactory. Wilkinson, however, tabulates seven ir active modes whose positions at 100°K are 369, 504, 518 cm^{-1} (A_2) and 268, 384, 398, 458 cm^{-1} (E_1). For Cu_2O, he gives two F_{2u} frequencies (610, 148 cm^{-1}) at 100°K.

4.4 Compounds of the type A_2B_3

Spectroscopic studies of such compounds deal mainly with oxides $M_2^{III}O_3$, of which four structural types have been studied. Three of these types seem to form a series in which the structural type is connected with the radius ratio; these are the hexagonal La_2O_3 structure, the cubic bixbyite or Tl_2O_3 structure, and the well-known hexagonal corundum structure. Radius ratios $r_{M^{3+}}/r_{O^{2-}}$ of $>0{\cdot}87$, 0·60–0·87 and $<0{\cdot}60$ correlate with the La_2O_3, Tl_2O_3 and corundum structures respectively. All the rare-earth oxides exist in the La_2O_3 or Tl_2O_3 structure, but there is another, monoclinic form assumed by Sm_2O_3 and Gd_2O_3.

Table 4.10 Selection rules for M_2O_3 structures.

Structure	Space group	Z	Optical				Acoustic
			Activity R only	ir only	R and ir	ia	
La_2O_3	D_{3d}^3	1	$A_{1g}+2E_g$	$A_{2u}+2E_u$	—	$A_{2g}+A_{1u}$	$A_{2u}+E_u$
Corundum	D_{3d}^6	2	$2A_{1g}+5E_g$	$2A_{2u}+4E_u$	—	$2A_{1u}+3A_{2g}$	$A_{2u}+E_u$
Bixbyite	T_h^7	16	$9(A_g+E_g+3F_g)$	$32F_u$	—	$11(A_u+E_u)$	F_u
B—Sm_2O_3	C_{2h}^3	6	$14(B_g+2A_g)$	$15(A_u+2B_u)$	—	—	A_u+2B_u

Mitsuishi's transmission data [979] for Al_2O_3 and Fe_2O_3 are unassigned, but Brown [189] has attempted to assign the vibrations of Cr_2O_3 in terms of a CrO_6 octahedron, and his assignment is given. Force-constant calculations using a simple UBFF give K(Cr—O) = 1·81, H(OCrO) = 0·0, F(O---O) = 0·53 mdyn Å^{-1}. The activation of ν_6 is attributed to distortion of the CrO_6 octahedron, but could equally well be regarded as a site-symmetry effect since the metal atom occupies a C_3 site and all triply degenerate modes correlate $F \rightarrow A + E$, where all classes are both ir and R active. It is doubtful whether there is any significance in attempting to assign these modes in terms of vibrations of the CrO_6 coordination polyhedron, particularly since this would have 15 modes, whereas factor-group analysis shows 12 external (3 acoustic, $A_{2u} + E_u$; 3 translational, $A_{2g} + E_g$; 6 rotational, $A_{2g} + A_{2u} + E_g + E_u$) and 18 internal modes ($2A_{1g} + 2A_{1u} + A_{2g} + A_{2u} + 3E_g + 3E_u$). The variations between sets of observed data are again considerable and may well depend on the crystalline nature and previous history of the sample.

Table 4.11 The spectra of materials with the corundum structure, assigned according to factor-group analysis.

Compound	A_{1g}(R)	A_{2u}(ir)		E_g(R)	E_u(ir)		Ref.
		TO	LO		TO	LO	
Al_2O_3	645	583	871	751	635	900	1157, 90
	418	400	512	578	569	625	
				451	442	480	
				432	385	388	
				378			
		592			640		1510
		398			578		
					445		
					384		
Al_2O_3 containing 0·04 per cent Cr_2O_3		622			610		603
		560			570		
		479			510		
		432			455		
		385			380		
Cr_2O_3		613	759		613	766	1196
		538	602		532	602	
					444	446	
					417	420	

Table 4.12 Other spectral data for compounds with the corundum structure.

Compound	Assignment and description; descriptions apply to Cr_2O_3 only					
	ν_1 $\nu_{Cr-O(A_{1g})}$	ν_3 $\nu_{Cr-O(F_{1u})}$	ν_2 $\nu_{Cr-O(E_g)}$	ν_5 $\delta_{OCrO(F_{2g})}$	ν_4 $\delta_{OCrO(F_{1u})}$	ν_6 $\delta_{OCrO(F_{2u})}$
Al_2O_3	690	595	488	470	441	385
Fe_2O_3	571	476	444	385	323	235
Cr_2O_3	648(R)	635(ir)	482(R)	348(R)	306(ir)	244(R)

The effect of strain in inducing surface-active forbidden modes has been considered [1157].

Extensive studies of the rare-earth oxides, particularly those with the bixbyite structure, have been made by McDevitt and co-workers [99, 930]. They noticed that in these cubic oxides, a number of band frequencies gave straight lines when plotted against unit cell volumes, but did not interpret this. Owing to the complex nature of the unit cell no factor-group assignment has so far been made. The coordination polyhedron has not been definitely identified, but appears to be 4 + 2 oxygen atoms with the four nearest neighbours at about 2·0 Å from the metal. The other two rare-earth oxide types have been less extensively studied and no interpretation of any data has been given. The coordination polyhedron in the La_2O_3 structure involves four oxygens at about 2·30 Å and three more at about 2·7 Å, while in B—Sm_2O_3 there is an irregular octahedron with M–O distances between 2·28 and 2·57 Å.

Table 4.13 Spectra of rare-earth oxides with the bixbyite structure [930].

Metal \ Band	A	B	C	D	E	F	G	H	J	Others
Gd	535	440	391	355	318	302	280	130	124	
Dy	550	460	410	372	328	315	291	132	125	
Y	561	468	405	391	343	—	312	—	120	171, 182, 242
Ho	562	468	418	388	334	318	298	133	126	
Er	565	475	395	375	335	323	298	133	126	
Tm	568	481	405	382	346	324	309	131		
Yb	573	490	405	388	341	—	304	132	125	
Lu	580	495	405	390	350	337	309	133	125	439

This work was carried out on samples dispersed in a polythene matrix. Studies of these compounds as Nujol mulls [99] did not resolve bands B, C and D, but gave a broad absorption in the region 340–500 cm^{-1}. McDevitt and Davison thought that H and J might be combinations involving an external fundamental near 50 cm^{-1}. Petru and Muck [1131–3] investigated Y_2O_3 and Sc_2O_3, finding bands at 560, 435 cm^{-1} and 635, 435 cm^{-1} respectively. The effect of change in crystal structure is shown in the work of Goldsmith and Ross [548]; Sm_2O_3 and Eu_2O_3 have spectra intermediate between those of La_2O_3 and a bixbyite-type oxide, though according to Wyckoff these materials have the bixbyite structure. Possibly they are the B—Sm_2O_3 form of the oxide.

Table 4.14 Spectra of other rare-earth oxides [548].

Metal \ Band	A	B	C	D	E
La*†		430		350	290
Ce		450		330	
Pr		420			
Nd*		440		320	285
Sm*	520	460	380		290
Eu*	520		370	330	290

* Baun and McDevitt observed a band near 640 cm^{-1} in these compounds.
† Petru and Muck found two bands at 510 and 445 cm^{-1}.

4.5 Perovskites

A large number of crystal structures for materials of the perovskite type are given by Wyckoff, but only the ideal cubic structure has been thoroughly studied spectroscopically. The formula unit is ABX_3, where A is a large cation, B a small cation and X a large anion. In practice X is oxygen or fluorine. When X is oxygen, the sum of the charge of A and B must be 6, and I–V, II–IV and III–III perovskites are all known. Examples are $NaNbO_3$, $BaTiO_3$ and $GdFeO_3$. Only I–II fluoride perovskites, such as $KNiF_3$, are known. The relative sizes of the ions are determined by the fact that in the ideal structure the X ions are at face centres and the A cations at cube corners. If these are in contact, then for the B cations to fit into the octahedral hole at the body centre positions,

$$r_A + r_X = 2^{1/2}(r_B + r_X). \tag{4.8a}$$

In practice the ideal perovskite structure is still formed when the relation

$$r_A + r_X = 2^{1/2}t(r_B + r_X) \tag{4.8b}$$

is satisfied, where the Goldschmidt tolerance factor t is $>0{\cdot}89$. Distorted variants of the ideal structure occur for $0{\cdot}80 < t < 0{\cdot}89$, and with t less than 0·80 the hexagonal ilmenite ($FeTiO_3$) structure is favoured. This simplified view does not take into account the influences of polarizability and Jahn–Teller distortion on the structure.

The ideal structure is cubic, with the space group O_h^1, one molecule in the unit cell, and both A and B ions on sites of O_h symmetry. The selection rules are $\Gamma_{\text{optical}} = 3F_{1u}(\text{ir}) + F_{2u}(\text{inactive})$, while $\Gamma_{\text{acoustic}} = 1F_{1u}$. There is no first-order Raman spectrum. No factor-group analyses have been carried out on the distorted structures; selection rules have been derived

Table 4.15 Correlation table for vibrations of BX_6 octahedra of site symmetries found in distorted perovskites.

Factor group	Compound	Site group	Class and selection rules				
			ν_1	ν_2	$\nu_{3,4}$	ν_5	ν_6
O_h^1	$BaTiO_3$	O_h	A_{1g}(R)	E_g(R)	F_{1u}(ir)	F_{2u}(ia)	F_{2g}(R)
D_{6h}^4	$BaTiO_3$	(1) D_{3d}	A_{1g}(R)	E_g(R)	A_{2u}(ir) $+E_u$(ir)	A_{1u}(ia) $+E_u$(ir)	A_{1g}(R) $+E_g$(R)
		(2) C_{3v}	A_1(ir, R)	E(ir, R)	A_1+E	A_2(ia)$+E$	A_1+E
C_{2v}^{14}	$BaTiO_3$	C_{2v}	A_1(ir, R)	$2A_1$	$A_1+B_1+B_2$ (all ir, R)	$A_1+B_1+B_2$	A_2(R) $+B_1+B_2$
C_{2v}^8	$PbZrO_3$	C_2	A(ir, R)	$2A$	$A+2B$ (ir, R)	$A+2B$	$A+2B$
D_{2h}^{16}	$GdFeO_3$	C_i	A_g(R)	$2A_g$	$3A_u$(ir)	$3A_u$	$3A_g$

by correlation with lower space group or site symmetries. The distorted structures which have been studied spectroscopically include tetragonal $BaTiO_3$, whose space group is unknown; hexagonal $BaTiO_3$, and three orthorhombic structures, those of $BaTiO_3$, $PbZrO_3$ and $GdFeO_3$. The spectra have been analysed both in terms of lattice modes and in terms of vibrations of the BX_6 octahedron. Correlation tables can be constructed in terms of all the site symmetries of B atoms in the structures studied, but the factor group D_{6h}^4 of hexagonal $BaTiO_3$ does not correlate with O_h^1.

The hexagonal form of $BaTiO_3$ contains two non-equivalent sets of Ba, Ti and O atoms; there are two types of coordination polyhedron, TiO_6 and Ti_2O_9, formed by sharing faces between two TiO_6 units.

Fair agreement exists in both observation and assignment of the fluoride perovskites with the cubic structure. The descriptions of the three ir active modes in terms of vibrations of the BF_6 octahedron are fairly consistent between the various workers, being $\nu_1 = \nu_{B-F}$, ν_2 = FBF bending, ν_3 = lattice mode of A^+ against the BF_6 octahedron. Perry [1120] has termed our ν_1 and ν_2 equatorial and axial bending respectively; our ν_1, ν_2 and ν_3 correspond to ν_2, ν_3 and ν_1 in his notation. (The values of the L matrix elements in Nakagawa, Tsuchida and Shimanouchi's [1054] analysis support these descriptions.)

Table 4.16 Spectra of cubic perovskites ABF_3.

Compound	ν_1		ν_2		ν_3		Ref.
	TO	LO	TO	LO	TO	LO	
$KMgF_3$	450		295		140		685
	478		300		156		1445
	458	551	299	362	168	197	1124
$KNiF_3$	446		255		153		1445
	441	522	244	306	148	164	79
	441	501					80
$KZnF_3$	430		206		142		1445
	410	474	200	292	141	149	1124
$KCoF_3$	417	500	225	293	139	168	66
	430	512	224	293	140	158	1124
$KMnF_3$	395	479	193	268	115	143	66
	399	483	193	270	119	144	1124
$RbMnF_3$	369	452	197	273	113	124	66
	370	435	198	272	114	126	1124

Balkanski [79, 80] has observed other lines at 311 cm^{-1} and 501 cm^{-1} in $KNiF_3$. He has assigned the line at 501 cm^{-1} to the F_{2u} mode, but Nakagawa's *et al.* [1054] calculation gives 218 cm^{-1} for this frequency.

Alternatively, according to Balkanski, the line at 311 cm^{-1} could be the F_{2u} fundamental and that at 501 cm^{-1} a multiphonon process. Temperature dependence studies have failed to resolve this question but the oscillator strength of the 501 cm^{-1} line suggests a one-phonon process.

Of the cubic oxide perovskites, $BaTiO_3$ and $SrTiO_3$ have been frequently studied, and there are many disagreements, while $KTaO_3$, $BaZrO_3$ and $SrZrO_3$ have attracted less attention. A striking difference between the oxide and fluoride perovskites is the frequent appearance in the oxides of a band assigned to the inactive F_{2u} mode. This is designated ν_4 in the following tables, and Nakagawa's analysis suggests it to be an OBO equatorial bending, with one pair of trans oxygens moving upwards and the other downwards, with respect to the equatorial plane. The majority of the studies on $SrTiO_3$ give $\nu_1 \sim 550$ cm^{-1}, $\nu_2 \sim 180$ cm^{-1}, $\nu_3 \sim 100$ cm^{-1}, $\nu_6 \sim 90$ cm^{-1}. The values given are those of Matossi [923], Perry *et al.* [1123], who used ir, Raman and neutron scattering, and Nilsen and Skinner [1076] who deduced possible one-phonon energies from the observed second-order Raman spectrum. The $BaTiO_3$ data are due to Matossi [923], and Perry and co-workers [1120, 1122]. Other references are given in the table.

Table 4.17 Spectra of cubic perovskites ABO_3.

Compound	ν_1		ν_2		ν_3		ν_4		Ref.
	TO	LO	TO	LO	TO	LO	TO	LO	
$BaTiO_3$	515 (R)	780 (N)			16 calc.	230 (R)	340 (N)	340 (N)	1122
	495 (ir)								826
	450 (N)								1122
	505		210		115				1120
	510		183		33·8				923
$SrTiO_3$	555		185		100				1123
	544	815	178	265	265	473	90	173	92a
	419	809	185		342	516	114		
	499		195			518	126		1076
	518		211				130		
	546		178		87·5				923
$KTaO_3$	549	833	199	290	290	423	88	188	1121
$BaZrO_3$	505		210		115				1120
$SrZrO_3$*	522		240		143		325		1120

* This may be a distorted structure.
N = neutron scattering, ir = infrared, R = Raman.

The calculated value of 16 cm^{-1} is obtained from the variation of the Lyddane–Sachs–Teller relation due to Cochran and Cowley [276], and agrees well with the 13 cm^{-1} obtained by Ballantyne [82].

Four tetragonal perovskites have been studied, but they do not appear to form an isostructural set. $BaTiO_3$ (with which β-$PbTiO_3$ appears to be isostructural) has been assigned on the basis of a C_{4v} factor group, but tetragonal $SrTiO_3$ which exists below 110°K is not ferroelectric, suggesting one of the factor groups S_4, C_{4h}, D_{4h}, D_4. The Raman spectrum contains three new lines below 110°K (453, 148, 39 cm^{-1} at 79°K). These were interpreted by Schaufele and Weber [1244] as being the first-order Raman spectra of the polar TO modes. However, this would require the factor group to be C_{4v} as in $BaTiO_3$, and the non-ferroelectric nature of $SrTiO_3$ was considered by Nilsen and Skinner as evidence against this interpretation. Further, the TO mode energies were known to be 544, 170, 42 cm^{-1} in poor agreement with the Raman spectra. Nilsen and Skinner therefore consider that there is a non-polar tetragonal phase with no Raman-allowed single phonon processes, restricting the factor group to C_{4h} or D_{4h}. $PbZrO_3$ is classed by Perry as possibly tetragonal, and it is not clear whether the form of $PbTiO_3$ studied is β-$PbTiO_3$.

Table 4.18 Spectra of tetragonal perovskites.

Compound	ν_1		ν_2		ν_3	ν_4	Ref.
	TO	LO					
$BaTiO_3$	$502A_1$ $513E$		34	359	183	$416(B_1)$	690
	$515A_1$ 485 (R), 500 (ir)E	$720A_1$ $720E$	5 (R) 0–34 (ir)	$270A_1$, $305E$ } TO	$165A_1$, $183E$ } TO		1200
				$465A_1$, $460E$ } LO	$175A_1$, $180E$ } LO		1140
	$495A_1$ $517E$			$340A_1$ Calc.	220–245 (55–351°K)		826
	491		12		182		82
	491		12		184 175		1120
	518	722		LO 271, T-dependent		307	1122
$PbTiO_3$	530		83		172 220	400	1120
$PbHfO_3$	530		109		193		1120

None of the other perovskite structures have been extensively studied; the spectra seem rather insensitive to the crystal structure, particularly with the orthorhombic materials where the expected breakdown of $F_{1u} \rightarrow A_1 + B_1 + B_2$ is observed in no case.

The spectrum of $NaNiF_3$, which has the $GdFeO_3$ ($CaTiO_3$) orthorhombic structure has been obtained by Nakagawa; the complexity as compared with the other ABF_3 perovskites is attributed to nonlinearity of the FNiF system rather than the low crystal symmetry. Bands in the ir

Table 4.19 Spectra of other perovskites.

Compound	Space group and crystal system	ν_1 TO	ν_1 LO	ν_2	ν_3	ν_4	Ref.
$BaTiO_3$	D_{6h}^4 hexagonal (R) (180°K)	528	717	189 168	246	311	1122
	(ir)	530 490					
$BaTiO_3$	C_{2v}^{14} orthorhombic (R) (283°K)	519	717	195	245	309	1122
$KNbO_3$	C_{2v}^{14} (ir)	660 550		375			826
$NaNbO_3$	C_{2v}^{14} (ir)	675 560		375			826
$PbZrO_3$	C_{2v}^8 orthorhombic (R)	508		221	34	290	1120
$CaTiO_3$	D_{2h}^{16} orthorhombic (R)	549		179	148	443	1120
$SrHfO_3$	orthorhombic (R)	555		191	132	269	1120
$CaZrO_3$	space group unknown (R)	515		222	118	310	1120

spectrum of the material dispersed in polythene are 447, 314, 245, 196, 174, 136 and 100 cm^{-1} at room temperature; on cooling to liquid nitrogen temperature a band appears at 282 cm^{-1}. O'Shea *et al.* [1094] have studied the Raman spectrum of $SrTiO_3$ over a range of temperatures, including the transition points cubic-tetragonal (110°K) and tetragonal-orthorhombic (65°K). The Raman spectra should contain no first-order processes in any of these phases, but they found two lines in the room-temperature spectrum near 180 and 80 cm^{-1}. These are about 20 times stronger than the unusually strong second-order lines in KI, and are close in frequency to ν_2 and ν_3 as observed in absorption or reflection by other workers (see Table 4.17). These lines are therefore interpreted as first-order processes, activated by a reduction of the symmetry of $SrTiO_3$. X-ray data obtained by Lytle [X101] and the discovery that *cubic* $SrTiO_3$ is piezo-electric suggest that the compound has a pseudo-cubic, non-centrosymmetric structure.

4.6 Spinels

This class of compound takes its name from the mineral spinel, Al_2MgO_4, and the general formula A_2BO_4 embraces a wide variety of A and B atoms. The valencies of A and B may be 1 and 6, 2 and 4 or 3 and 2 respectively, the materials being customarily referred to as I–VI, II–IV or III–II spinels. The so-called *normal* spinel structure is cubic, space-group O_h^7, with eight molecules in the unit cell, and has the A atoms on octahedral sites of D_{3d} symmetry, and the B atoms on tetrahedral sites of T_d symmetry. The oxygen atoms occupy C_{3v} sites. The coordination polyhedron around A is a regular octahedron. In the inverse spinel structure, only half the A atoms occupy octahedral sites, the remainder lying on tetrahedral sites, as do the B atoms.

White and De Angelis [1508], in their very complete analysis of the spinel structure, show that Γ_{optical} for both structures is $A_{1g} + E_g + 3F_{2g}$ (all R) + $4F_{1u}$(ir) + $A_{2u} + E_u + F_{1g} + F_{2u}$ (all inactive), while $\Gamma_{\text{acoustic}} = 1F_{1u}$. This analysis refers to the primitive rhombohedral cell with $Z = 2$. In consequence, there are no gross differences to be expected between the spectra of normal and inverse spinels, and the published data are in agreement with this deduction. White and De Angelis deal with a number of other possible arrangements, but only one of these has been realized in simple A_2BO_4 spinels; this is the so-called 1 : 1 ordering on tetrahedral sites. This occurs when there are two different types of ion (B, B′) on the tetrahedral sites, such that B is surrounded entirely by B′ and vice versa. The space group of this structure is T_d^2, and with $Z = 2$ we have $\Gamma_{\text{optical}} = 3A_1 + 3E(\text{R}) + 7F_2(\text{ir, R}) + 3F_1$ (inactive); $\Gamma_{\text{acoustic}} = 1F_2$. A possible instance of this is the Ni_2SiO_4–Ni_2GeO_4 system studied by Tarte [1408–10].

A number of spinels have a tetragonal structure, with $Z = 4$ and the space group D_{4h}^{19}, both the A and B atoms occupying D_{2d} sites. Siratori and co-workers [1332–3] have analysed the vibrations according to the site group approximation, which gives $\Gamma_{\text{optical}} = 8B_2 + 11E(\text{ir, R}) + 6A_1 + 3B_1(\text{R}) + 3A_2$ (inactive); $\Gamma_{\text{acoustic}} = B_2 + E$. Owing to the centrosymmetric nature of the space group, one E and three B_2 modes do not appear in the ir. They wished to examine the possibility of systematic changes in the spectra with the extent of tetragonal distortion, but no such correlation was obtained.

The earliest attempt at an interpretation of the spectra of spinels was due to Waldron [1478]. In his study of six ferrites Fe_2BO_4 (B = Co, Fe, Mg, Mn, Ni, Zn) he considered the vibrations in terms of internal modes of the BO_4 tetrahedron, and modes involving the Fe(A) atoms, which he viewed as arising from the AO_6 octahedron. This should, if that view were correct, give a criterion for distinguishing between normal and inverse spinels, since the inverse spinels would contain both AO_4 and BO_4 tetrahedra, as well as AO_6 octahedra. Waldron further stated that there would be a reduction in local symmetry due to the vacancies on half the octahedral sites, which would result in total breakdown of the threefold degeneracy; but in view of the factor-group analysis by White this cannot be correct. Hafner and Laves [604] were unable to detect any systematic differences between the spectra of inverse and normal spinels; nor did they find that the force-constant values for tetrahedral and octahedral groups maintained a constant relation depending on whether the spinel was normal or inverse. The same conclusions were reached by Dasgupta and Sinha [317]. Waldron had originally interpreted his data to mean that the higher of the two frequencies he observed was characteristic of tetrahedra, the lower of octahedra.

It was then recognized by Tarte [1411] that such an interpretation was even approximately true only in special circumstances. When the co-ordination polyhedra round A and B have very different frequencies, the two sets of vibrations can be considered as independent of each other. This applies in (for instance) Ni_2SiO_4, Ni_2GeO_4, Na_2WO_4 and Na_2MoO_4. Where, however, the vibrational frequencies are similar in magnitude, the interactions between the groups becoming comparable in strength with those within the groups, the concept of separate sets of vibrations characteristic of AO_x and BO_y polyhedra becomes meaningless. The majority of spinels fall into this category. Tarte and Preudhomme [1417] suggested that the study of solid solutions of A_2BO_4, $A_2B'O_4$ would provide a criterion for distinguishing these situations; in the first case the bands characteristic of BO_4 would be replaced by those of $B'O_4$ on increasing the $A_2B'O_4$ concentration, while in the second case the spectrum would retain its original appearance but be shifted in frequency as a whole. White has referred to these two types of behaviour as ordered and continuous respectively. This second situation applies particularly to the III–II spinels investigated by Waldron and by Hafner and Laves. In a discussion of the spectra of aluminates [1415], Tarte further emphasized the need to distinguish between *isolated* and *condensed* coordination

Table 4.20 Vibrational assignments in tetrahedral AlO_4 and octahedral AlO_6.

Compound	Type of polyhedron	Characteristic range (cm^{-1})	Example	Observed frequency*
				5 bands
AlO_4	Condensed	900–700	$CaAl_2O_4$	880–780
	Isolated	800–650	Yttrium aluminium garnet	790, 730, 700
			Al_2NiO_4 (inverse spinel)	728
AlO_6	Condensed	680–500	α-Al_2O_3 (corundum)	630, 600
			Al_2CoO_4 (normal spinel)	683
	Isolated	530–400	Yttrium aluminium garnet	510, 480, 430
			Al_2MgO_4 (normal spinel)	510

* Estimated from printed spectra; no tabulation was made by Tarte.

polyhedra. It is well established that in such condensed species as $X_2O_7^{n-}$, vibrational interactions between the terminal XO_3 groups lead to an increase in the XO_3 stretching frequencies. This interaction, when occurring with octahedral groups, may raise the frequency to values equal to or greater than those normally associated with isolated tetrahedral groups. It is therefore important to discuss the spectra with respect to the isolated or condensed nature of the coordination polyhedron, and this may be illustrated by some data due to Tarte [1415] on the vibrations of coordination polyhedra of Al.

Dealing first with the spinels containing isolated BO_4 tetrahedra, we see that only II–IV and I–VI spinels are represented. In the II–IV cases, the evidence for isolated tetrahedra arises from the ordered behaviour of solid solutions. The pairs Ni_2SiO_4–Ni_2GeO_4 and Co_2GeO_4–Fe_2GeO_4 have been shown to behave in this way; this may also be true of the pairs $Co_2(Ni_2)GeO_4$–Mg_2GeO_4 though the interpretation here is less certain since it is not known whether Mg_2GeO_4 has the normal or inverse structure. In Na_2WO_4 and Na_2MoO_4 there is great similarity between the solid and solution frequencies [1508].

Table 4.21 Vibrations of BO_4 tetrahedra and AO_6 octahedra in spinels.

Compound	BO_4				AO_6	Ref.
	ν_1	ν_2	ν_3	ν_4		
Ni_2SiO_4	516		825		380	1167, 1410, 1418
Ni_2GeO_4	470		690		336	1167, 1410, 1418
Co_2GeO_4	426		680		321	1167, 1410, 1418
Fe_2GeO_4	402		688		319	1167, 1410, 1418
Mg_2GeO_4	450		694		485 274	1167, 1410, 1418
Na_2MoO_4 (R)	892	381	808	303		215
(ir)			838	325		271
Na_2WO_4 (R)	928	373	813	312		215
(ir)			835	320		271

The spinels in which the motions of the coordination polyhedra are not distinguishable are better discussed in terms of the factor-group analysis. These compounds should show four bands of species F_{1u} in the infrared. Since few of these have been studied below 300 cm^{-1}, in many cases three, and in some only two, have been observed.

Dupuis and Lorenzelli [402] have studied some IV–II inverse spinels where M^{IV} is Sn, and have classified the vibrations in terms of $M^{II}O_4$ tetrahedra, $M^{II}O_6$ octahedra and SnO_6 octahedra. They support their assignment by reference to some metastannates with the perovskite structure; these have bands near 650 cm^{-1} attributed to the SnO_6 octahedron.

Table 4.22 The spectra of III–II spinels.

Band	A	B	C	D	Ref.
Normal					
Mn_2MnO_4	550	474			604
Mn_2CuO_4	578				317
	553	465			
	531				
Fe_2MnO_4	550	392			1478
Fe_2ZnO_4	555	393		169	978
	550	415			1478
Al_2MgO_4	690	538			604
Al_2CoO_4	680	562			604
Al_2ZnO_4	667	562	510		604
Al_2FeO_4	671	562	510		604
Al_2MnO_4	649	529	485		604
			465		
Ga_2NiO_4	633	446			1508
V_2MgO_4	629	485			604
Cr_2MgO_4	645	535	370	190	604
		508			
Cr_2NiO_4	617	500	370	180	604
Cr_2CoO_4	627	526			604
Cr_2ZnO_4	625	532			604
Cr_2FeO_4	617	523			604
Cr_2CuO_4	602	521			604
Inverse					
Ni_2CuO_4	571	473			317
		458			
Fe_2CoO_4	575	374	320	181	978
	585	389			1478
Fe_2FeO_4	595	382			978
	570	375			1478
Fe_2NiO_4	587	396		196	978
	602	413			1478
Fe_2MgO_4	565	406			978
	581	433			1478
Fe_2CuO_4	592	420			1478
Ga_2MgO_4	617	469			604

A few tetragonal III–II spinels have been studied by Siratori *et al.* [1332–3]. The compounds Mn_2MnO_4, Mn_2ZnO_4 and C_2CuO_4 have two bands each possibly corresponding to bands C and D in the cubic spinels (247, 165; 265, 167; 194, 135 cm^{-1} respectively), and another band corresponding to a mode inactive under O_h symmetry (125, 131, 117 cm^{-1}

Table 4.23 The spectra of some inverse IV–II spinels.

	SnO_6			$M^{II}O_4$			$M^{II}O_6$	
M^{II} = Mg	630	582	520	450	418	400	360	340
Mn	630	565	515	435	—	400	335	310
Co	640	562	515	426	415	—	378	—
Zn	645	585	515	435	410	—	365	342

respectively). A number of spinels of more complex formula such as $LiCrGeO_4$, $LiCrTiO_4$ and $LiAl_5O_8$ have been studied by Tarte [1411–13]; the interpretation is broadly similar. Tarte has identified frequencies due to LiO_4 tetrahedra near 450 cm^{-1} by means of the Li^6–Li^7 isotopic shift.

5. The vibrational spectra of triatomic molecules

5.1 General

Triatomic molecules of the formula types X_3, XY_2 and XYZ have all been studied. Such molecules may be linear or nonlinear, and may be symmetric or asymmetric. The term asymmetric is used in either of the two following senses; (a) that the two bonds are of unequal length, as in CsI_3, and (b) that the atomic arrangement is XYY rather than YXY, as in N_2O.

Table 5.1 Point groups of triatomic molecules.

Formula type	Linear		Nonlinear	
	Sym	Asym	Sym	Asym
X_3	$D_{\infty h}$	$C_{\infty v}$	C_{2v}	C_s
XY_2	$D_{\infty h}$	$C_{\infty v}$	C_{2v}	C_s
XYZ	$C_{\infty v}$		C_s	

Table 5.2 Selection rules for point groups of triatomic molecules.

Assignment and description	ν_1	ν_2	ν_3
	ν_{sym}	δ	ν_{asym}
$D_{\infty h}$	Σ_g^+ (R)	Π_u (ir)	Σ_u^+ (ir)
$C_{\infty v}$	Σ^+	Π	Σ^+ all ir, R.
C_{2v}	A_1	A_1	B_1 all ir, R.
C_s	A'	A'	A' all ir, R.

5.2 Molecules of the type X_3

Only two classes of such molecules have been studied, the azides of monovalent and divalent cations, and some trihalide ions. The azides have been investigated by a number of workers, and the agreement between the various observations is good. The spectra are, in the main, those to be expected for linear symmetrical molecules, but in a number of compounds some additional features appear. For NaN_3, KN_3 and CsN_3, whose crystal structures are known, these features are interpreted in terms of correlation splitting, and a similar effect was suggested to explain the doubling of ν_2 and ν_3 in $Ba(N_3)_2$ [560]. Additional components in the ν_1 region for NaN_3 are attributed to $N^{14}N^{15}N^{14}$ (1350 cm^{-1}) and to sum and difference bands involving lattice modes (1208, 1260, 1449, 1501 cm^{-1}) from which the lattice frequencies are determined as 98 and 150 cm^{-1} [194]. A second phase of NH_4N_3 in which ν_3 was shifted from 2030 cm^{-1} to 2050 cm^{-1} was investigated by Dows *et al.* [384]; this was considered to be a defective version of phase I (stable at 68°K) and appeared to contain a species of composition NH, which may have been N_2H_2 with either $D_{\infty h}$ or C_{2h} symmetry. The doubling of ν_2 in NH_4N_3 was ascribed to hydrogen bonding.

Table 5.3 Internal modes of azides.

Compound	ν_1	ν_2	ν_3	Ref.
LiN_3	1369	635	2092	560
NaN_3	1358	639	2128	560
	1358	638	2043	194
		643·5, 628	2189	1108
		639·3, 628	2160	389
KN_3	1344	645	2041	560
	1343	647, 636	2036	158
NH_4N_3	1345	661, 650	2041	560
RbN_3	1339	642	2024	560
		643·5, 624·5	2008	1108
CsN_3	1329	635	2062	560
	1336, 1325	636, 634	2022	195
		640·0, 621·7	2015	1108
HgN_3	1322, 1273	647, 592	2080	329
$Ca(N_3)_2$	1381	638	2114	560
$Sr(N_3)_2$	1373	635	2096	560
$Ba(N_3)_2$	1354	650, 637	2123, 2083	560
$Pb(N_3)_2$	1350·6	628	2006	719

All the above data relate to solids; the only extensive study of solution spectra is that of Kahovec and Kohlrausch [719]. Their values of ν_1 do not differ from the tabulated values above by more than 8 cm^{-1}.

For NaN_3, KN_3 and CsN_3, external (lattice) modes have been identified. Apart from the two already mentioned, NaN_3 was found to have a rotatory Raman active mode of class E_g at 122 cm^{-1}. KN_3 and CsN_3 crystallize in the space group D_{4h}^{18} with $Z = 2$, so that their translatory modes are $1E_g$ (R) + $1A_{2u}$ + $2E_u$ (ir) + $1A_{2g}$ + $1B_{1u}$ (ia), and their rotatory modes $1B_{1g}$ + $1E_g$ (R) + $1A_{2g}$ (ia).

In CsN_3 these have been assigned as 42 (E_g), 118 (A_{2u}), 89, 38 (E_u) and 153 (B_{1g}), 108 (E_g) [195]. Bryant [193] found nine lattice modes in KN_3, assigning 100, 134, 228 and 253 cm^{-1} to translatory and 83, 101, 120, 147, 191 to rotatory modes.

A few compounds containing trihalide ions have been studied. Unlike the azides, these compounds are not all linear or symmetrical; Tasman and Boswijk [X149–50] have established that in CsI_3, the I—I distances are 2·83 and 3·03 Å, while the I—I—I angle is 175·3°. The Cl_3^- ion also appears to depart from the $D_{\infty h}$ structure in benzene solution. The force constants given in the table are calculated using a valence-force field with $f_{r\alpha}$ ($=F_{12}$) = 0.

Table 5.4 The spectra of trihalides.

Compound		ν_1	ν_2	ν_3	f_r	f_{rr}	Ref.
$(Pr_4N)^+Cl_3^-$	ir solid	268	(165)	242	0·96	0·55	442
	ir soln. C_6H_6	281					
$(Me_4N)^+Br_3^-$	R solid	162			0·91	0·32	1126
	R soln. $(CH_2Cl)_2$	196					
$(M^+Br_3^-)$ (M = H, K, NH_4, Rb, Sr)*	R soln. H_2O	166					
KI_3	R solid	114		148	0·68	0·12	902
CsI_3	ir solid	103	69	149			

* These compounds have a band at 258 cm^{-1} in the crystal. The assignment of ν_2 in Pr_4NCl_3 is based on a band at 330 cm^{-1} in acetonitrile solution which was interpreted as $2\nu_2$.

Nelson and Pimentel [1062] have studied the vapour phase ir spectrum of the Cl_3 radical, assigning a band at 374·6 cm^{-1} to ν_3. They calculate the Cl—Cl stretching and interaction force constants to be 1·28 and 0·31 mdyn $Å^{-1}$, and consider the two Cl—Cl bonds to be of unequal length.

5.3 Linear molecules of the type XY_2

A number of metal halides have been studied, using a wide variety of techniques. In particular, the matrix isolation method, using a rare gas (usually Ar) as matrix has proved valuable in determining isotopic shifts. The data are, on the whole, confused and inconsistent, and almost no complete assignments exist.

Table 5.5 Spectra of divalent metal halides.

Compound		ν_1	ν_2	ν_3	Ref.
BeF_2	ir vap.	—	825	1520	196
	ir (Ar matrix)	(680)	345	1555	1346
$BeCl_2$	ir vap.	—	482	1113	196
	ir (Ar matrix)	(390)	250	1135	1346
$BeBr_2$	ir (Ar matrix)	(230)	220	1010	1346
BeI_2	ir (Ar matrix)	(160)	(175)	873	1346
$MgCl_2$	ir vap	—	295	597	196
ZnF_2	ir (matrix)	—	150·5	758, 754, 750	869
$ZnCl_2$	ir (matrix)	—	103, 101	508·5, 505·5, 502, 498·5	869
$ZnBr_2$	ir (matrix)	—	76, 73	404, 399·5, 395·5	869
ZnI_2	ir (matrix)	—	62	346, 341·5, 337·5	869
$CdCl_2$	ν_1 from R aqueous solution; ν_3 ir solid	280	—	212	323, 1182
$CdBr_2$		186	—	122	
CdI_2		144	—	88	
NiF_2	ir (Ar matrix)	(606)	—	780	965
$NiCl_2$	ir (Ar matrix)	(351)	—	521	
$PbCl_2$	R solid	300	148	352	75
$HgCl_2$	R solid	316	126	375	1160
		310	100		
	ir solid	—	106	375	1183
$HgBr_2$	ir solid	—	66	232	1183
			43		
		187*	—	251	29
HgI_2	ir, R solid	111, 137	26	112	285

* Raman value from earlier work [1454a].
Values in parentheses are calculated using the simple valence force field.

Some early data largely obtained indirectly by Klemperer [768] has been omitted; in that work values of ν_1 and ν_2 of the mercuric halides were deduced from Raman and electronic spectra. Janz and McIntyre [697] studied $HgCl_2$ and $HgBr_2$ in the melt; their data are $\nu_1 = 313$, $\nu_2 = 382$ ($HgCl_2$); $\nu_1 = 206$, $\nu_2 = 135$ ($HgBr_2$). A later study by Janz and James [696] gives $\nu_1 = 313$, $\nu_3 = 376$ ($HgCl_2$); $\nu_1 = 195$, $\nu_3 = 271$ ($HgBr_2$). This is more consistent with the tabulated data and ν_1 for $HgBr_2$ agrees quite well with the early Raman value quoted by Adams [29]. Décamps and Hadni [330] have obtained the spectrum of $Hg(CN)_2$ and have carried out a factor-group analysis, but have not made any assignments to internal vibrations. Isotopic splittings have been observed in the zinc halides by Loewenschuss [869] and in the nickel halides by Milligan [965]; in both cases some absorptions attributed to cyclic dimers have been observed. In the zinc halides these are at 666 (ZnF_2), 425 ($ZnCl_2$), 326 ($ZnBr_2$) and 272 (ZnI_2), while in $NiCl_2$ there is a similar absorption at 439 cm^{-1}. Buechler [196] also observed bands at 857 and 608 cm^{-1} in the ir spectrum of $BeCl_2$, which he attributed to the dimer.

The rare gas halides KrF_2, XeF_2 and $XeCl_2$ have been studied: GVFF calculations on KrF_2 by Claassen *et al.* [258] indicate a negative value of the interaction constant f_{rr}. Coulson [290] attempted to explain this in terms of Pauling's theory of ionic–covalent resonance, suggesting a larger weight in KrF_2 than XeF_2 for the 'no bond' structure (III).

$$\underset{\text{I}}{F{-}X^{+}{-}F^{-}} \qquad \underset{\text{II}}{F^{-}{-}X^{+}{-}F} \qquad \underset{\text{III}}{\widehat{F\ X\ F}}$$

Table 5.6 Spectra of the rare-gas halides as vapours.

Compound	ν_1	ν_2	ν_3	f_r	f_{rr}	Ref.
KrF_2	449	232·6	588	2·46	−0·20	258, 1041
XeF_2	515	213·2	558	2·84	+0·13	258
$XeCl_2$	—	—	313	—	—	1061

Murchison *et al.* [1041] demonstrated the linear structure of KrF_2 from the intensity alternation in the rotational levels of ν_3; the lines corresponding to odd and even J respectively had intensity ratios 3:1. The Kr—F distance was calculated as $1{\cdot}875 \pm 0{\cdot}002$ Å, in fair agreement with the electron diffraction study due to Harshbarger [X64] which gave $1{\cdot}89 \pm 0{\cdot}01$ Å. Two types of complex halide ion have been studied; those of the type $(XHX)^-$ and those containing two different halogen atoms. The most frequently studied compounds are the hydrogen fluorides, particularly KHF_2 whose tetragonal structure is well known. Ketelaar has studied such compounds in a variety of alkali halide matrices [748], and Jones and Penneman investigated KDF_2 as well as KHF_2 [716]. Ikrami [691] has studied the bifluorides of Ca, Sr and Ba and found ν_1 to be active in the Sr and Ba compounds due to reduced symmetry of the HF_2^- ion.

The value of ν_1 for KHF_2 was found by Coté and Thompson [287] from combination bands. They also observed a splitting of ν_2 (1225, 1274 cm^{-1}) in crystals with random orientation. Newman and Badger [1069] using oriented crystals found ν_2 at 1230 and 1240 cm^{-1} with incident radiation respectively perpendicular and parallel to the C-axis. Both Jones

Table 5.7 Spectra of bifluorides.

Compound		ν_1	ν_2	ν_3	Ref.
$NaHF_2$	ir Nujol mull	—	1212	1600 vb	745
KHF_2		(610, 608, 594)	1230	1455 vb	745, 287
NH_4HF_2			1208	1600 vb	745
KHF_2	ir, KCl mixed crystal		1257	1570	745
KHF_2	ir solid		1233	1473	
	ir soln. H_2O		1206	1536	716
KDF_2	ir solid		888	1045	
	ir soln. D_2O		873	1102	
$Ca(HF_2)_2$					
$Sr(HF_2)_2$	ir solid	615	1150–1190	1530–1670	691
$Ba(HF_2)_2$		584			

and Penneman and Ikrami have found bands which they have attributed to $H_nF_{n+1}^-$, $n > 1$.

The first study of a hydrogen halide ion, containing a halogen other than fluorine, was made by Waddington [1471] who interpreted the spectrum of $(CH_3)_4$ $NHCl_2$ as showing the presence of linear, symmetrical HCl_2^-. McDaniel and Vallee [928] investigated compounds containing HCl_2^-, HBr_2^- and HI_2^-, obtaining results consistent with Waddington's conclusions. Evans and Lo [444], however, considered that not all such ions are symmetrical: they observed ν_1 in four compounds. Of the compounds tabulated below, they consider the tetraethyl ammonium salts and the tetra-*n*-butylammonium hydrogen and deuterium bromides to be symmetrical, and the others asymmetrical. The appearance of ν_1 in the tetraethylammonium salts is attributed to a site symmetry effect, and the lowering of ν_2 and ν_3 in these compounds to strong hydrogen bonding. All the salts classified as unsymmetrical show a broad, structured absorption in the region 1300–500 cm^{-1} which is assigned to $\nu_3 \pm n\nu_1$.

Nibler *et al.* [1073] found the strongest bands in the spectrum of $CsHCl_2$ at 1670, 1168·5, 660·5 and 602 cm^{-1}. They consider that the bands near 600 cm^{-1} should be assigned to ν_2 and that at 1168·5 to $2\nu_2$. The original assignment by Waddington, analogous to KHF_2, was not considered valid, since the H bonding is very much weaker in the other bihalides. The enhancement of the intensity of $2\nu_2$ depends on the strength of the H bond and is thus correctly predicted as decreasing in the order $HCl_2^- > ClHBr^- > ClHI^-$.

Recently Noble [1081] has observed $HCl_2\cdot$ and $DCl_2\cdot$ formed as a result of passing a mixture of HCl (DCl), chlorine and argon through a glow discharge and condensing at 14°K. In these radicals the position of ν_3 is very different from that in the anions ($HCl_2\cdot$, 696·4; $DCl_2\cdot$, 463·7).

Table 5.8 Infrared spectra of hydrogen halide ions.

Compound	ν_1	ν_2	ν_3	Ref.
$CsHCl_2$	—	602·0, 660·5	1670	1073
$(CH_3)_4NHCl_2$	—	1180	1565	1471
		1160	1565	928
	210	1185 } 1080 sh }	1575	444
$(CH_3)_4NDCl_2$	200	880 } 830 sh }	1200	444
$(C_2H_5)_4NHCl_2$		1050	730	444
$(C_2H_5)_4NDCl_2$		750	515	444
$(nC_4H_9)_4NHCl_2$		1150	1540	928
$(CH_3)_4NHBr_2$	126	1038	1420	443a
$(CH_3)_4NDBr_2$	123	752	1070	443a
$(C_2H_5)_4NHCl_2$		1170	1670	928
	160	940, 780, 630	1350	443a
$(C_2H_5)_4NDBr_2$	155		950	443a
$(nC_4H_9)_4HBr_2$		1170	1540	928
	(190)		700	443a
$(nC_4H_9)_4DBr_2$	(170)		490	443a
$(nC_4H_9)_4HI_2$		1165	1650	443a

Six trihalide species of the type XY_2^- have been studied. ClF_2^- is predicted to be linear, the structure being based on sp^3d hybridization with the apical positions occupied by fluorine atoms. The observed spectra of the Rb and Cs salts suggest a distortion or site symmetry effect, since ν_1 is observed. In $NOClF_2$ the formulation as $NO^+ClF_2^-$ is confirmed by the appearance of a band at 2279 cm^{-1} (attributed to NO^+); the alternatives NOF:ClF and $NOF_2^-Cl^+$ are expected to absorb in the region 1800–1900 cm^{-1} characteristic of N=O. Evans and Lo have prepared $(nC_3H_7)_4NClBr_2$, and the existence of two polarized Raman lines attributable to stretching modes suggests linear Cl—Br—Br as the atomic arrangement. The ICl_2^- ion has been shown by Gran and Romming [X58] to be asymmetric and nonlinear in $RbICl_2$, but has the Cl—I—Cl arrangement, while the IBr_2^- ion shows ν_1 in both the infrared and Raman spectra of the Rb and Cs salts, and is therefore probably non-centrosymmetric. According to Maki and Forneris [902] the IBr_2 ion in $CsIBr_2$ has the I—Br—Br arrangement.

Data for the nitronium ion, NO_2^+, in various salts, have been collected by Soulen and Schwartz [1350]; like the isoelectric CO_2, NO_2^+ is linear. CS_2 and CSe_2 have been studied by Wentink [1504]. Deb and Yoffe [328]

Table 5.9 Spectra of trihalide ions, XY_2^-.

Compound		ν_1	ν_2	ν_3	Ref.
$NOClF_2$	ir solid	—	—	635	241,242
$KClF_2$	R solid	475	—	—	249
	ir solid	470	—	661	249
$RbClF_2$	R solid	476	—	—	249
$CsClF_2$	ir solid	510, 478	—	636	249
$(nC_3H_7)_4NClBr_2$	R solid	184 (ν_{Br-Br})	—	235 (ν_{Br-Cl})	443
$CsBrCl_2$	ir solid	—	—	305	1126
$RbBrCl_2$	R soln. H_2O	280	—	—	1126
$CsBrI_2$	ir solid	117	84	168	902
$RbICl_2$	ir solid	282	140	218	902
	R solid	278	—	—	902
$CsICl_2$	ir, R solid	268	129	218	902
$KIBr_2$	R solid	158	—	—	902
$RbIBr_2$	ir, R solid	144	97	180	902
$CsIBr_2$	ir, R solid	143	98	178	902

have found ν_1 in the infrared spectra of five cyanamides although the $CN_2^=$ ion is linear and centrosymmetric; this is presumably a site symmetry effect. Considerable controversy exists over the geometry of the UO_2^{++} ion; Jones and Penneman [715] have determined the value of ν_3 for XO_2^{++} as perchlorates in $HClO_4$, where X is U (965), Np (969), Pu (962) and Am (939). A very early Raman study by Conn and Wu [282] gave ν_1 860–880 cm^{-1}, ν_2 199–210 cm^{-1}, ν_3 930–960 cm^{-1} for aqueous solutions of uranyl salts. Sutton [1399] ascribed the appearance of ν_2 and ν_3 in the Raman to polarization and to an asymmetric field due to the anion involved in the complex. As will be seen in the section on angular XY_2, the majority of workers consider UO_2^{++} and the analogous transuranyl ions to be nonlinear.

In an early study of the metaborates of Na, Ba and Pb^{II}, Duval and Lecomte [416, 418] found differences in the spectra depending on the age of the sample. They concluded that the free BO_2^- ion did not exist in the solid and that ageing led to the formation of $(BO_2)_n^{n-}$ ions which could be cyclic, n being most probably 2 or 3. The later work of Morgan and Staats [994] and Hisatsune and Suarez [666] on metaborate ion in alkali halide matrices shows the much simpler spectra attributable to the free ion. In a KBr matrix the values are $\nu_2 = 608$, $\nu_3 = 2029$ cm^{-1} for $(^{10}B\,^{16}O_2)^-$, and 587, 1959 cm^{-1} for $(^{11}B\,^{16}O_2)^-$. Only these two frequencies were observed in the matrix studies: freshly prepared $Ba(BO_2)_2$ was found to have bands at 860, 874, 956, 1000, 1138, 1158, 1310, 1351 cm^{-1}; on ageing, the spectrum became less complex, consisting of five bands at 836, 930, 1070, 1246 and 1400 cm^{-1}. $Pb(BO_2)_2$ behaved similarly.

Table 5.10 Vibrational spectra of $CN_2^{=}$, NO_2^+, CO_2, CS_2 and CSe_2.

Compound		ν_1	ν_2	ν_3	Ref.
$Na(CN_2)_2$	ir solid	1234	598	2120	328
$Tl(CN_2)_2$		1388	632	1857	
$Ag(CN_2)_2$		1281	631	1977	
$Hg(CN_2)_2$		1322	647	2075	
$CaCN_2$		—	620	2170	
$PbCN_2$		1307	626	1991	
$CdCN_2$		—	655	2140	
$HgCN_2$		—	682	2100	
$CuCN_2$		—	693	2050	
NO_2ClO_4*		1396	570	2360	1350
NO_2NO_3*		1400	538	2375	
CO_2	ir, R gas	1343	667	2349	1418a
	ir solid	—	660 653	2344	694a, 1093a
CS_2	ir, R gas	658	396·8	1532·5	1504
	ir, R liq	656·5	397	1510	
CSe_2	ir, R gas, liq	364	313	1303	1759

* ν_1 from Raman and ν_2, ν_3 from ir measurements on solids.

5.4 Angular molecules of the type XY_2

Changes due to bonding or structural effects will not cause alterations in the gross nature of the spectrum, since (as Tables 5.1 and 5.2 show) all vibrations are both infrared and Raman active. Investigations of angular XY_2 molecules have been largely directed towards effects manifested as frequency shifts, particularly complex formation.

By comparison of the spectra of a number of compounds $K_2M^{II}(NO_2)_4$ (M^{II} = Ca, Ba, Pb, Cu), and $K_4Ni(NO_2)_6$ with those of KNO_2 and $Ba(NO_2)_2, 2H_2O$, Le Postollec concluded that these did not contain complex ions in solution. In contrast, $Na_3M^{III}(NO_2)_6$ (M^{III} = Co, Rh, Ir) showed markedly different spectra, characteristic of the coordinated nitro group [842]. A further study established the existence of the coordinated nitro group in $M^I_2Pt(NO_2)_4$ (M^I = Na, K) and $Na_2Pd(NO_2)_4$. Some nitro complexes $[Co(NH_3)_5NO_2]X$ (X = Cl, Br, I) were studied by Siebert [1299]; Penland [1116] found that the nitrito isomer could be distinguished from the nitro form by the positions of the NO_2 stretching modes. Patil and Rao [1111] have found these modes in the region 1500–1000 cm^{-1} in rare earth nitrites, and have also attributed a band at 100–200 cm^{-1} to M—O stretching.

Very few halites have been studied: there is only one complete assignment for a chlorite. Djega–Mariadassoux [368] has calculated force constants for the bromite ion, using an SVFF.

Table 5.11 Vibrations of the nitro and nitrito groups.

Compound		ν_1	ν_2	ν_3	Ref.
Ionic NO_2^-					
$Na^{14}NO_2$	ir cryst.	1328	828·2	1261	1507
$Na^{15}NO_2$	ir cryst.	1302	823·6	—	1507
KNO_2	ir soln. H_2O	1320	—	1225	842
$Ba(NO_2)_2, 2H_2O$	ir soln. H_2O	1320	—	1230	842
$K_2Ba(NO_2)_4$	ir soln. H_2O	1320	—	1230	842
$K_4Ni(NO_2)_6$	ir soln. H_2O	1330	—	1240	842
NO_2 in complex					
$[Co(NH_3)_5NO_2]\ Cl_2$	ir KBr disc.	1310	821	1430	1299
	Br_2	1310	821	1429	1299
	I_2	1307	820	1429	1299
$K_2Pt(NO_2)_4$	ir solid	1393, 1386 1348, 1338	844 838	1440 1415	843
$Na_2Pd(NO_2)_4$	ir solid	1393, 1383 1344, 1332	838 832	1430 1405	843
$Na_2Pt(NO_2)_4$	ir, R solid	1364 A_{1g} 1325 B_{1g} 1325 E_u	847 A_{1g} 835 B_{1g} 832 E_u	1415 A_{2u} 1412 E_g	843
—ONO in complex					
$[Co(NH_3)_5ONO]Cl_2$	ir solid	1065	824	1468	1054a
$[Cr(NH_3)_5ONO]Cl_2$	ir solid	1048	839	1460	1054a
Organic NO_2		1375	900	1560	
Organic ONO		ν_{N-O} 820	—	$\nu_{N=O}$ 1620, 1660	

Table 5.12 Vibrations of the chlorite and bromite ions.

Compound		ν_1	ν_2	ν_3	Ref.
$NaClO_2$	R soln.	796	396	860 calc	916
$NaClO_2, 3H_2O$	R solid	786	402	844	916
$NaClO_2$	ir solid	—	—	860	
$LiBrO_2, H_2O$	ir solid	672	390	700	368
$NaBrO_2$	ir solid	680	340	698	368
	R solid	711	340	—	368
$Ca(BrO_2)_2, 2H_2O$	R cryst.	707	—	750	368
$Ba(BrO_2)_2, 2H_2O$	ir powder	630	360	680	368
	R powder	672	360	704	368

In section 5.3, it was noted that some workers considered the uranyl ion, UO_2^{++}, to be linear. The majority of the assignments, however, are based on an angular ion, though Lecomte and Freymann [836] have suggested that in UO_2Cl_2, $3H_2O$ the ion is linear but asymmetric. Sevchenko and Stepanov [1274] considered the ion to be angular in UO_2SO_4, $3H_2O$ and $KUO_2(SO_4)_2$, $2H_2O$, but possibly linear in the hexahydrated acetate and nitrate. According to Hoekstra [668], both uranates and diuranates contain the UO_2^{++} group, and the majority of the spectra are consistent with this being nonlinear. Jones and Penneman [715] have observed ν_3 in the perchlorates of NpO_2^+ and AmO_2^+ in $HClO_4$ solution at 824, 832 cm^{-1} respectively. The VO_2^+ group has been studied by Weidlein and Dehnicke [343, 1493] in various fluorocompounds. Griffith [568] has investigated the system MO_2 in peroxo compounds $M^I_3M^VO_8$ where M^I = K, NH_4, M^V = V, Nb, Ta, Cr. He has classified the vibrations as ν_1 (mainly $\nu(O{-}O)$ or $\delta(OMO)$), A_1; $\nu_2(\nu(M{-}O) + \delta(OMO))$, A_1; $\nu_3(\nu_{as}(MO))$, B_1. Strong bands in the region 810–880 cm^{-1} are assigned to ν_1. In a number of partially fluorine-substituted complexes such as $K_2TaO_2F_5$, ν_1 lies in the region 840–940 cm^{-1}, while other bands near 570, 600 or 650 cm^{-1} are ascribed to ν_2 or ν_3. Metal–fluorine stretching modes have also been identified near 450 cm^{-1} and in $K_2WO_3F_4$ and some other compounds containing M=O, the M=O stretching mode has been assigned near

Table 5.13 Vibrations of MO_2 systems.

Compound		ν_1	ν_2	ν_3	Ref.
The UO_2^{++} group					
$UO_2(SO_4)3H_2O$	ir solid	836	(200)	913	1274
$UO_2(NO_3)_2$	R soln. H_2O	860	200	909	1235
$UO_2(NO)_3)_26H_2O$	ir F, N mull	874	—	964	51
		869		937	
	ir solid	880	140	950	464
		850			
$3H_2O$	ir F, N mull	837	—	943	51
$2H_2O$	ir solid	885	196	950	464
		855	146		
anh.	ir solid	890	183	984	464
		845	152	975	
$K_2UO_2(NO_3)_2$	ir F, N mull	877	—	953	51
		873, 824			
UO_2Cl_2		—	—	905	1170
UO_2Br_2		852	—	930	1170
				905	
$UO_2Cl_23H_2O$		860	—	930	836
$UO_2(ClO_4)_2$	R soln. $HClO_4$	936	—	965	
$UO_2(ClO_4)_27H_2O$	R solid	888	—	958	
$5H_2O$		890	—	958	1453
$3H_2O$		892	—	967	
$1H_2O$		896	—	974	
anh.		900	—	996	

Table 5.13 (*cont.*).

Compound		ν_1	ν_2	ν_3	Ref.
Li_2UO_4		725	250	763	668
α-Na_2UO_4		737	270	775	
β-Na_2UO_4		705	280	745	
K_2UO_4		—	265	732	
Rb_2UO_4		—	270	726	
Cs_2UO_4		—	280	727	
$Cs_2U_2O_7$		—	265	754	
$K_2U_2O_7$		—	285	768	
$Rb_2U_2O_7$		—	280	761	
The VO_2^+ group					
VO_2F	ir solid	875	558	1024	343, 1493
$VO_2Sb_2F_{11}$	ir solid	888	572	1030	
VO_2SbF_6	ir solid	885	566	1028	
The MO_2 group					
K_3NbO_8		814	—	—	568
K_3TaO_8		813·5	—	—	
K_3CrO_8		875	—	—	
$(NH_4)_3VO_8$		846	—	—	
$(NH_4)_3CrO_8$		870	—	—	
			ν_2 or ν_3		
$K_2[TiO_2(SO_4)_2]$		862	667	—	
$K_2TiO_2F_5$		901	597	—	
$K_2TaO_2F_5$		867	634, 577	—	
$K_2NbO_2F_5$		929	588	—	
$K_2WO_3F_4$		905	617, 601	—	
$K_2MoO_3F_4$		939	599, 565	—	
$K_2MoO_5F_2$		869	649, 578	—	
		854			

950 cm^{-1}. An alternative interpretation of the spectra in Ti complexes was suggested by Vere and Patel [1463]; they considered that ν_3 was about 1000 cm^{-1}. However, this interpretation is unlikely to be correct. One would normally expect ν(O—O) (ν_1) to be greater than ν_{as}(M—O) (ν_3), but this would reverse the assignment. The bands near 1000 cm^{-1} are very weak and are not specific to the TiO_2 system. The assignment was supported by an analogous assignment of the bands in the region of 1000–1150 cm^{-1} in $Ti(OEt)_4$ to Ti—O stretching, but these bands are now known to be associated with the C—O linkages.

Although many divalent metal halides appear to be linear, Mann [907] has found MgF_2 to be angular ($\widehat{FMgF} = 150°$); all three bands were observed in the ir spectrum, and the lines due to $^{24,25,26}Mg$ were resolved. For $^{24}MgF_2$, he found $\nu_1 = 478{\cdot}04$, $\nu_2 = 241{\cdot}8$, $\nu_3 = 837{\cdot}41$ cm^{-1}. Donaldson *et al.* [376] have studied the orthorhombic (α) and monoclinic

(β) forms of SnF_2, finding bands at 575 and 370 cm^{-1} in the α form, and 446 and 350 cm^{-1} in the β form.

While the ClF_2^- ion was found to be linear, Christe and Sawodny [245–6] have found the ClF_2^+ ion to be angular. The spectrum of $ClF_2^+ AsF_6^-$ shows all three bands in the ir ($\nu_1 = 811$, $\nu_2 = 520, 558$, $\nu_3 = 818\ cm^{-1}$). The assignment given in Table 5.14 for FCl_2^+ differs from that of Christe *et al.* [1640], where the stretching modes are given as 529 and 586 cm^{-1}. The polyhalogen ions Cl_3^+, ClF_2^+, FCl_2^+ are all non-linear and the last of these has the atomic arrangement Cl—Cl—F. A number of compounds between oxygen or sulphur and a halogen have been studied and good agreement exists between different observations and assignments. The only doubt seems to be regarding Cl_2O. ν_2 is estimated as 320 cm^{-1} by Hedberg [630] on the basis of a band at 1245 cm^{-1} assigned as $\nu_2 + \nu_3$; but this may be an impurity due to HOCl, which is known to absorb at 1242 cm^{-1} [631]. A very full study by Rochkind [1211] resulted in considerable changes in the assignment; in particular the band at 969 cm^{-1} was assigned to $\nu_3 + \nu_4$ and the corresponding difference band was found at 379 cm^{-1}.

Table 5.14 Vibrations of polyhalogen ions and of halogen compounds with oxygen or sulphur.

Compound		ν_1	ν_2	ν_3	Ref.
F_2O	ir gas	928	461	831	130
ClO_2	ir gas	943·2	445	1110·5	1075
SCl_2	R liq.	514	208	535	1362
Cl_2O	ir gas	688	320	969	630
		640 centre	300 centre	677·2 P	1211
				685·9 Q	
				691·8 R	
	ir solid (77°K)	630·7	296·4	669·8 Q	
	ir N_2 matrix	639·9			
		637·0			
Cl_3^+	R, solid	493	225	508	1712
		485			
$ClO_2^+ AsF_6^-$		1043·7	521	1296·4	1644
		1038·3	517	1282·6	
$ClF_2^+ SbF_6^-$		807	387	830	1711
BrF_2^+	ir, R solid	705	362	702	1643
FCl_2^+		ν(Cl—F)	ν(Cl—Cl)		
		744	535, 528	299, 293	1712
$FN_2^+ AsF_6^-$		2370	803	1056	1876

The NO_2 monomer has been studied by Fateley *et al.* [453], using the matrix-isolation technique; two bands are reported, one weak band lying near 750 cm^{-1} and another band at 1611–1624 cm^{-1} whose position depends on the nature of the matrix. Moore [989] had previously found bands at 751 and 1616 cm^{-1} in NO_2 gas. From combinations he assigned $\nu_1 = 1244$ cm^{-1}, $\nu_2 = 749$ cm^{-1}; presumably the 1616 cm^{-1} band is ν_3.

The spectrum of H_2O and its isotopically substituted derivatives has been investigated many times. The assignments in the gas phase are well established.

Table 5.15 The gas-phase spectra of $H_2{}^{16}O$ and its isotopically substituted derivatives.

Compound	ν_1	ν_2	ν_3	Ref.
$H_2{}^{16}O$	3652	1595	3756	1272
$D_2{}^{16}O$	2671	1178	2788	123a
$HD{}^{16}O$	2727	1402	3707	123a
THO	—	1324	3720	1356
TDO	—	—	2735	1356
T_2O	—	996	2370	1356
$H_2{}^{18}O$	3647	1586	3744	551a
$D_2{}^{18}O$	2657	1169	2764	1138a

In the liquid, there is some disagreement about the stretching mode assignments. The work has been summarized by Senior and Thompson [1272]. They noted that both Schultz and Hornig [1249] and Walrafen [1482] assigned $\nu_1 = 3450$, $\nu_3 = 3615$ cm^{-1}; a much earlier study by Fox and Martin [482] allocated both bands to ν_1 and did not assign ν_3. Both sets of workers gave $\nu_2 = 1640$ cm^{-1}. The band at 3450 cm^{-1} is strong in both the ir and Raman spectra of liquid water, whereas that at 3615 cm^{-1} appears in the Raman only. This suggested to Senior and Thompson that ν_1 and ν_3 were coincident at 3450 cm^{-1}. The band at 3615 cm^{-1} was interpreted as a combination $(\nu_1 + \nu_H)$ where ν_H is the O---H---O frequency of the hydrogen bond. This mode does not combine with ν_3 (B_1), thus accounting for the absence of the 3615 cm^{-1} band from the ir spectrum. It would, however, combine with ν_1 (A_1), the combination appearing in the Raman. The same argument applies to the band at 3225 cm^{-1} which appears only in the Raman spectrum and is ascribed to $2\nu_2$.

The situation with regard to ice is even more confused, partly because of its polymorphism. The work by Ockman [1085–6] and Hornig and co-workers [680] on ice I led to broadly similar results; the one difference in interpretation was that Ockman preferred to assign 3140 cm^{-1} in H_2O to ν_1 and 3340 to $2\nu_2$ on the basis of the strong Raman polarization of the 3150 cm^{-1} line, though the assignment of 1580 cm^{-1} to ν_2 would imply a very large negative anharmonicity constant. Some librational and translational modes were identified in H_2O, D_2O and HOD, both on their own and in combination. Bertie and Whalley [137–9] studied the spectra of ices I_h, I_c, II, III, V, VI and IX, finding, for ices II, VI and IX, spectra consistent with crystallographic information [X79–81]. In their analysis of the spectrum of ice I they did not attempt to distinguish between ν_1 and ν_3; they noted that the hexagonal and cubic forms of ice I had the same spectra and attributed the differences between their results and those of earlier workers to those workers having used partly vitreous ice.

They attributed bands in the region 900–550 cm^{-1} (H_2O) and 675–425 cm^{-1} (D_2O) to librational fundamentals, on the basis of the isotopic shift ($\sim$1·35). Blue [161] had previously shown that, if the bending mode is symmetrical, such that the force constants for bending in two directions at right angles are the same, then the uncoupled librational modes (rocking and wagging) are coincident, the twisting mode about the C_2-axis being inactive. In principle, however, the existence of coupling, and the reduction of symmetry in the crystal, activated all types of libration, and it was considered essential to include the effects of coupling in making librational assignments. The study by Giguere and Arraudeau [530] was considered unsatisfactory in this respect by Bertie and Whalley.

In an attempt to calculate the librational heat capacity of ice, Blue gave expressions for the librational frequencies (in cm^{-1}) as follows:

$$\nu_x = \frac{1}{2\pi I_A c}[k_H r_H^2 \sin^2\phi + \tfrac{1}{2}k_O r_O^2]^{1/2}, \qquad (5.1)$$

$$\nu_y = \frac{1}{2\pi I_B c}[k_H r_H^2 \cos^2\phi]^{1/2}, \qquad (5.2)$$

$$\nu_z = \frac{1}{2\pi I_C c}[k_H r_H^2 + \tfrac{1}{2}k_O r_O^2]^{1/2}, \qquad (5.3)$$

where k_H, k_O are force constants such that, for a rotational displacement through an angle α about the C-axis, the restoring forces are $k_H r_H \alpha$, $k_O r_O \alpha$. Also, r_O is the oxygen displacement, r_H the sum of the O—H bond length and the hydrogen displacement, and ϕ is the angle made by the O—H bond with the x-axis, this being taken as the axis perpendicular to the C_2-axis (designated y), such that the xy-plane is the molecular plane. Thus, ν_x, ν_y and ν_z will correspond to wagging, twisting and rocking respectively. Van der Elsken and Robinson [1450] studied a number of alkaline-earth halide hydrates, and calculated values of k_H and k_O, but this approach has not been generally followed. Other studies of H_2O vibrations and librations in crystal hydrates have been carried out by

Table 5.16 Vibrational assignments in ice I.

Compound	ν_1	ν_3	ν_2	$2\nu_2$	$3\nu_L$	ν_L	ν_T	Ref.
H_2O	3140	3224	1580	3340	2235	846	160	1085
	3400	3220	1620	3140	2230	812	—	680
	3360	3210	1585	3125	2225	850	—	598
	3380–3330 3220 3150		1650 + 30 (also $2\nu_L$)	—	2266 + 20 (also $\nu_2 + \nu_L$)	900 840 770 660 555	—	137–9
D_2O	2520 2336	2432	1212	—*	1627	605	160	680 137–9
	2570 2495 2432	2440	1210	—	1635	—	—	
	2545 2485 2425 2395 2332 2240		1210 (also $2\nu_L$)	—	1650 (also $\nu_2 + \nu_L$)	675 640 425		
	ν_{O-H}	ν_{O-D}						
HDO	3310		1470		1925	600 810	—	1085
	3275	2416	1490	2875	1975	—	—	598

* In D_2O, $2\nu_2$ and ν_1 overlap.

Gamo [501–6], Prask and Boutin [1165], Ferraro and Walker [463] and Nakagawa and Shimanouchi [1050]. Prask and Boutin have studied a number of compounds by neutron scattering, and have attempted to relate librational frequencies to barriers restricting rotation. Calculations have been made using the cosine potential function

$$V = \tfrac{1}{2}V_0(1 - \cos 2\phi), \tag{5.4}$$

and also an empirically derived relation

$$V_0 = 0{\cdot}64 - 1{\cdot}6 \times 10^{-4}\nu + 0{\cdot}31 \times 10^{-4}\nu^2 + 0{\cdot}27, \tag{5.5}$$

where V_0 is the barrier height in kcal mole^{-1} and ν the librational frequency. They have also stated that, if the force constants f_ρ (rocking), f_τ (twisting) and f_ω (wagging) are approximately equal, the frequencies should fall in the order $\omega > \tau > \rho$. For cylindrical symmetry of the force constants about the hydrogen bonds, τ and ρ become degenerate. However, most studies show the frequencies to lie in the order $\rho > \omega > \tau$. In their study of $Ba(ClO_3)_2$, H_2O they found $\rho = 458$, $\omega = 390$, $\tau = 318$ cm^{-1}. For the twisting motion in $Ba(ClO_3)_2$, H_2O, independent determinations of V_0 as 5·0 kcal mole^{-1} gave 300 cm^{-1} by (5.4) and 360 cm^{-1}

by (5.5). The observed value is 318 cm^{-1}. A less satisfactory agreement is obtained with $Li_2SO_4H_2O$ where spin–lattice relaxation measurements gave $V_0 = 7{\cdot}3$ kcal $mole^{-1}$, corresponding to 365 cm^{-1} by (5.4). The observed librational frequencies are 832, 672 and 524 cm^{-1}.

In the study by van der Elsken and Robinson [1450], some doubling of the librational frequencies was observed, and this was attributed to the existence of two crystallographically distinct types of water, engaging in hydrogen bonds X---HOH---X and O---HOH---X, where X is a halogen. Since the rocking and wagging modes are ir active, four librational bands are expected in compounds containing two types of water. This expectation was realized in $NaBr_2.2H_2O$, and $BaCl_2.2H_2O$; there was no doubling in $SrCl_2.H_2O$, $BaCl_2.H_2O$ and $BaBr_2.H_2O$, all of which have only one type of water molecule (giving X---HOH---X bonds). Gamo [503–4] observed such doubling in $FeCl_2.4H_2O$ but not in $Na_2SO_4.10H_2O$, where two types of water also exist. He also found that, in a series of eleven hydrates, a linear plot of the H_2O rocking frequency against the metal–OH_2 distance had points corresponding to five of the compounds lying on it. These were all chlorides; there was no correlation between frequency and O—H---Cl distance.

Table 5.17 Librational frequencies in crystal hydrates.

Compound	ρ	ω	τ	Ref.
$NaBr.2H_2O$ (X—HOH—X)	590	405		
(O—HOH—X)	625	470		
$NaI.H_2O$ (X—HOH—X)	510	385		
(X—DOD—X)	385			
(O—HOH—X)	585	450		
(O—DOD—X)	430			
$NaI.2H_2O$ (X—HOH—X)	470	400		
(O—HOH—X)	615	470		
$SrCl_2.H_2O$	585	395		
$SrCl_2.D_2O$	435			
$SrCl_2.2H_2O$	710	480		1450
$SrCl_2.2D_2O$	520	365		
$BaCl_2.H_2O$	575	360		
$BaCl_2.2H_2O$ (X—HOH—X)	560	410		
$BaCl_2.2D_2O$ (X—DOD—X)	415			
$BaCl_2.2H_2O$ (O—HOH—X)	690	520		
$BaCl_2.2D_2O$ (X—DOD—X)	515	310		
$BaBr_2.H_2O$	550	335		
$BaBr_2.D_2O$	415			
$BaBr_2.2H_2O$	550	445		
$BaBr_2.2D_2O$	415			
$MgCl_2.6H_2O$	714			501
$CaCl_2.6H_2O$	719			501

Table 5.17 (*cont.*).

Compound	ρ	ω	τ	Ref.
$SrCl_2.6H_2O$	690			501
$FeCl_2.4H_2O$	750, 575, 470			503–4
$CoCl_2.2H_2O$	642	790		503–4
$CoCl_2.6H_2O$	776			501
$NiCl_2.6H_2O$	728	650		1165
	763			501
	755			1050
$AlCl_3.6H_2O$	790	570	300	1165
	844			501
$CrCl_3.6H_2O$	730	590	384	1165
	844			501
	800	541		1050
$Na_2SO_4.10H_2O$	—	720		506
$MgSO_4.7H_2O$	665	485	595	1165
	749	460		501; 1050
$FeSO_4.7H_2O$	712			501
$CoSO_4.7H_2O$	755			501
$NiSO_4.7H_2O$	760			501
$ZnSO_4.7H_2O$	758			501
$CuSO_4.5H_2O$	875			502
$1H_2O$	866, 813			502
$Al_2(SO_4)_3.18H_2O$	810	585	320	1165
	690	920		505
$KAl(SO_4)_2.12H_2O$	810	585	310	1165
$Al(NO_3)_3.9H_2O$	790	595	310	1165

The hydrides of S and Se have also been studied, and the spectrum of solid H_2S has been interpreted in relation to its crystal structure. H_2S, like H_2O, is polymorphic. The low temperature phase III (transition points 103·6°K in H_2S, 107·8°K in D_2S) is tetragonal, not orthorhombic as originally suggested [870]. The other two phases are disordered.

The assignment due to Reding and Hornig [1191] for H_2S and D_2S, phase III, is based on a D_{4h} factor group and a unit cell content of 8. A later study by Miller and Leroi [959] considers 8 possible structures; the only one in accordance with their observations involves a C_{4h} factor group with $Z = 8$, and all molecules on C_1 sites. Their observed frequencies are, for H_2S, 2521 (ν_1); 2546 (ν_3); 2540, 2544, 2552, 2556 (ν_1 or ν_3); 1163 and 1201 (ν_2, R); 1171 and 1186 (ν_2, ir). For D_2S they find 1831 and 1860 (ν_1); 1848 and 1853·5 (ν_3); 842·5 and 869 (ν_2, R); 847 and 857 (ν_2, ir). Reding and Hornig also observed some external modes. Absorptions at 93 cm^{-1} and 290 cm^{-1} in phase II were ascribed to translatory and rotatory modes respectively; corresponding frequencies in phase III were 91, 92 and 125, 170 cm^{-1}.

Table 5.18 Vibrational spectra of group VI hydrides.

Compound		ν_1	ν_2	ν_3	Ref.
H_2S	gas	2615	1183	(2627)	50a
D_2S		1892	934	2000	1074
HDS			1090		
H_2Se		2260	1074	2350	
D_2Se		1630	745	1696	218a
HDSe		1691	905	2352	
H_2S	solid, phase I	2557	(1170)	2557	
H_2S	solid, phase II	2550	(1170)	2562	
H_2S	solid, phase III	2521 A_{1g}			
		2554 E_g		2546 E_g	
		2523 A_{2u}	1171, 1186 (ir)	2544 E_u	959
		2532 E_u			
D_2S	solid phase III	1833 A_{1g}			
		1848 E_g		1853 E_g	
		1835 A_{2u}	847, 857 (ir)	1854 E_u	
		1843 E_u			

5.5 Molecules of the type XYZ

(a) Angular molecules

The angular molecules HOCl, DOCl have been studied in the gas phase [631], and recently the species HOF, DOF and $H^{18}OF$ have been prepared by photolysis of mixtures of fluorine with H_2O, D_2O and $H_2{}^{18}O$. These have been investigated using the matrix-isolation technique, N_2 being the matrix. The nitrosyl halides have been studied, with isotopic substitution in the cases of ONF and ONCl. Some controversy exists regarding the assignments in ONF. The original assignment by Woltz *et al.* [1524] gave the bending frequency as 521 cm^{-1} and ν_{N-F} as 766 cm^{-1}. A later study by Jones *et al.* [713] reversed this assignment on the basis of the potential-energy distribution, though the isotopic shifts would seem to support Woltz's original assignment. Similarly in ONCl, Burns and Bernstein [214] have assigned a band at 332 cm^{-1} to bending, but Eberhardt [424] has interpreted it as ν_{N-Cl}. Polo and Wilson [1154] have studied the unsymmetric $S^{16}O^{18}O$ and have carried out a normal-coordinate analysis. Their values for $S^{16}O_2$ and $S^{18}O_2$ are given for comparison.

(b) Linear molecules

HCN and a number of its derivatives have been studied, both molecules XCN and anions YCN^- (Y = O, S, Se). Complexes of the cyanate

Table 5.19 Vibrational frequencies of angular XYZ molecules.

Compound	ν_1	ν_2	ν_3	Ref.
HOF	3482	884	1393	1082
DOF	2573	835·5	1029·5	
$H^{18}OF$	3472	857	1395·3	
HOCl	3626	1242	739	631
DOCl	2674	911	739	
$^{16}O^{14}NF$	1843·5	765·8	519·9	713
$^{16}O^{15}NF$	1811·8	748·7	517·4	
$^{18}O^{14}NF$	1795·6	757·8	511·3	
$^{18}O^{15}NF$	1763·0	740·4	509·3	
$O^{14}N^{35}Cl$	1800	332	605	822
	1799	332	592	214
$O^{15}N^{35}Cl$	1769	331	590	822
$O^{14}N^{37}Cl$	—	325	—	424
ONBr	1801	265	542	214
$S^{16}O^{18}O$	1122	507	1341	1154
$S^{16}O_2$	1152	518	1360	
$S^{18}O_2$	—	—	1316	

(OCN^-) group invariably appear to be bound to the metal through the nitrogen; a number of M—N stretching frequencies have been identified. The group retains its linear configuration in these complexes. Both the C—O and C≡N frequencies increase on complex formation. Some fulminates and fulminato complexes have also been investigated, the complexes having the arrangement M—C≡N—O. Unlike the isocyanato complexes, these show a splitting of $\nu_{C\equiv N}$ due to Fermi resonance, and they do not show pronounced systematic shifts in the C≡N and N—O frequencies compared to the salts such as KCNO. A number of compounds XCZ where X, Z are O, S, Se, Te have been investigated.

The values of ν_1 given for the cyanates of Rb, Cs, Ag, Tl and NH_4 correspond to the bands observed by Waddington [1473] and ascribed

Table 5.20 Vibrational frequencies of linear molecules XYZ.

Compound		$\nu_1(\nu_{X-Y})$	$\nu_2(\delta)$	$\nu_3(\nu_{Y-Z})$	ν_{M-N}	Ref.
HCN	ir gas	3311	712	2097		50b
DCN	ir gas	2630	569	1925		50b
TCN	ir gas	2460	513	1724		1356a
FCN	ir gas	1078	451	2290		457a
ClCN	ir gas	784	380	2219		489a
		714				
BrCN	ir gas	586	342	2198		900
ICN	ir gas	470	321	2158		15

Table 5.20 (*cont.*).

Compound		$\nu_1(\nu_{X-Y})$	$\nu_2(\delta)$	$\nu_3(\nu_{Y-Z})$	ν_{M-N}	Ref.
NaOCN (C^{12})	ir alkali halide disc	1214·5	624·3	2226		901
(C^{13})	ir alkali halide disc	1197·2	608·7	2160		
KOCN (C^{12})	ir alkali halide disc	1207·3	628·0, 636·9	2165		
(C^{13})	ir alkali halide disc	1194·6	612·3, 620·0	2112		
RbOCN	ir mull	1205	627, 637	2210		1473
CsOCN	ir mull	1217	627, 635	2160		
AgOCN	ir mull	1203	605, 632	2145		
TlOCN	ir mull	1203	617, 627	2130		
NH_4OCN	ir mull	1293	640	2190		
KSCN	ir cryst.	749	470 (A'), 484 (A'')	2053		708, 709
	R soln. H_2O	743	470	2066		
KSeCN	ir solid	561	417, 426	2070		564
	R	564	—	2078		
NaSeCN	ir solid	535	426, 413	2079		
Isocyanato complexes (ir, mull and soln. CH_3NO_2) $M{-}N{\equiv}C{-}O$; O = X, N = Z						
$(Et_4N)_2M(NCO)_4$, M = Mn		1321	625	2196	325	478
Fe		1322	622	2199	325	
Co		1325	619	2206	345	
Ni		1321	619	2237	351	
Cu		1318	620, 615 sh	2198, 2222	338	
Zn		1326	624	2208	321	
$(Me_4N)_2Sn(NCO)_6$	ir mull	1307	622, 667	2183	383	479
Fulminates						
NaONC	ir, KBr disc	1106	471	2096		111, 113
KONC	ir, KBr disc	1064, 1057	473, 468	2043		
Fulminato complexes $M{-}C{=}N{-}O$; O = X, C = Z						
$Na_2Pd(CNO)_4$, $5H_2O$		1127	465	2188		113
$Na_2Ni(CNO)_4$, $5H_2O$		1126	—	2267, 2185		112
		1111	—	2242, 2167		
$(\phi_3P)_4Pd(CNO)_2$		1155	487	2172		113

Table 5.20 (*cont.*).

Compound	$\nu_1(\nu_{X-Y})$	$\nu_2(\delta)$	$\nu_3(\nu_{Y-Z})$	ν_{M-N}	Ref.
$[Fe(CNO)_6]^{4-}$	1049	—	2195		112
$[Co(CNO)_6]^{4-}$	1097	—	2161, 2241		112
$[Ni(CNO)_6]^{4-}$	1126	—	2185, 2267		112
OCS ir gas	2062	520	859		903
OCSe ir gas	2021	466	642		15
SCSe ir gas	1435	(355)	506		1504, 1504a
SCTe ir soln. CS_2	1347	(337)	423		1504, 1504a

by him to $2\nu_2$. He assigned ν_1 to the band observed in the region 1300–1350 cm^{-1}, by analogy with the isoelectronic CO_2 ($\nu_1 = 1337\ cm^{-1}$). He noted that in the isostructural K, Rb, Cs and Ag compounds, which are tetragonal with the space group D_{4h}^{18} (I4/mcm), $Z = 4$, the cyanate ion cannot lie along the principal axis, since the degeneracy of ν_2 is broken down. However, in the rhombohedral NaOCN and the tetragonal NH_4OCN, the ion does lie along the principal axis, the degeneracy being preserved. TlOCN is orthorhombic, with the space group D_{2h}^{26} (Ibam), $Z = 4$, and the degeneracy is, as expected, removed.

6. The spectra of tetratomic molecules

6.1 General

In this very large group of compounds we may consider two basic atomic arrangements; one in which all bonds radiate from a central atom, and one in which a chain is formed. There are also a few four-membered ring systems resulting from dimerization of alkali metal halides in the vapour phase.

Table 6.1 Selection rules for tetratomic molecules.

Formula	Arrangement	Point group	R only	Activity ir only	ir, R	In-active
XY_3	Trigonal	D_{3h}	A_1'	A_2''	$2E'$	—
	Pyramidal	C_{3v}	—	—	$2A_1 + 2E$	—
	T-shape	C_{2v}	—	—	$3A_1+2B_1+B_2$	—
	Nonlinear (as in HN_3)	C_S	—	—	$5A'+A''$	—
XY_2Z	Planar	C_{2v}	—	—	$3A_1+2B_1+B_2$	—
	Pyramidal	C_S	—	—	$4A'+2A''$	—
XYZW	Planar	C_S	—	—	$5A'+A''$	—
	Pyramidal	C_1	—	—	$6A$	—
X_2Y_2	Linear	$D_{\infty h}$	$2\Sigma_g^+ + \Pi_g$	$\Sigma_u^+ + \Pi_u$	—	—
	Trans planar	C_{2h}	$3A_g$	A_u+2B_u	—	—
	Cis planar	C_{2v}	A_2	—	$3A_1+2B_1$	—
	Non planar 'book'	C_2	—	—	$4A+2B$	—
	Y shape	C_{2v}	—	—	$3A_1+2B_1+B_2$	—
	Ring	D_{2h}	$2A_g+B_{1g}$	$B_{1u}+B_{2u}+B_{3u}$	—	—
	Ring	C_{2h}	$3A_g$	A_u+2B_u	—	—

6.2 Planar Molecules

(a) Molecules of the formula XY_3

A number of halides of the types $M^{II}Y_3^-$ (M = Zn, Cd, Hg) have been studied in solution, and the trihalides of B and Al have also been investigated. Morris *et al.* [1000] considered various structures for the species formed in solutions containing Zn^{++} and halide ion. For the zinc to halide ratios 2·5–4·5 (chloride) and 2·5–7·5 (bromide), in aqueous solution, they found a spectrum characteristic of ZnY_3^-. They considered four possibilities for the structure, these being the planar (D_{3h}), pyramidal (C_{3v}), T-shaped (C_{2v}) and a bridged dimer (D_{2h}) and chose the D_{3h} structure, as being that which led to the minimum repulsion between the halogen atoms. The $CdCl_3^-$ ion has been studied in the melt by Tanaka *et al.* [1402] and in tributylphosphate solution by Davies and Long [323]. The data obtained are quite different; Tanaka interpreted the data on the basis of a pyramidal ion, while Davies and Long considered the ion to be planar.

For $HgCl_3^-$, the Raman spectrum of the melt obtained by Janz and James [696] differs from the infrared and Raman spectra of the tetramethylammonium salt found by Deacon [327]. The stretching frequencies are in moderate agreement but the values for ν_4 are respectively 210 and 103 cm^{-1}. In view of the values for $CdCl_3^-$, the second assignment would seem more reasonable. Many studies of the boron halides have been made, particularly BF_3. The isotopic shifts have been investigated for all four halides. Some selection rule breakdowns have been observed; Steinhardt *et al.* [1381] have observed ν_1 in liquid BF_3, and have attributed this to association. The values of both the Trouton constant [X69] and the viscosity [X18] indicate that this could occur. Studies of solid BF_3 and BCl_3 by Dows *et al.* [381–2] suggest that the two compounds are not isostructural. Three studies of AlF_3 gave concordant results for the ir active modes; Snelson [1345] additionally observed a number of bands attributed to a dimer of D_{2h} symmetry. Klemperer [766] found a band at 610 cm^{-1} in the emission spectrum of $AlCl_3$ vapour at 900°C which he attributed to ν_3.

The largest group of planar molecules of the type XY_3 to be studied is that comprising the borates, carbonates and nitrates. A number of the initial studies, such as those by Vratny [1469] and Ferraro [461] on the nitrates, sought to relate deviations from the free-ion selection rules to the covalent character of the nitrate group; for instance Ferraro stated that 'an examination of the spectra of metallic nitrates has shown that in progressing from a monovalent to a tetravalent metallic nitrate, an increased lowering of the nitrate symmetry, and a transition from a D_{3h} to a C_{2v} point-group symmetry, occurs'. A large number of studies prior to about 1960 expressed the view that breakdown of the free-ion selection rules implied some form of covalent metal-to-anion bonding, and some studies suggested that the magnitude of the deviation from free-ion selection rules could be used to distinguish between monodentate and bidentate functions of the anion. It cannot be too strongly emphasized that, in fact, an observed breakdown of selection rules can lead only to a

Table 6.2 Vibrational frequencies of halides XY_3.

Compound		Assignment and description				Ref.
		ν_1	ν_2	ν_3	ν_4	
		$\nu_{sym}(A_1')$	$\pi(A_2'')$	$\nu_{as}(E')$	$\delta(E')$	
$ZnCl_3^-$	R soln. H_2O	286	—	—	—	1000
$ZnBr_3^-$	R soln. H_2O	182	—	—	—	
$CdCl_3^-$	R soln. TBP; also aqueous extract	265	—	287	90	323
$CdBr_3^-$		168	—	184	58	
CdI_3^-		124	—	161	51	
$CdCl_3^-$	R melt, $CdCl_2$/KCl	257	211	245	177	1402
$HgCl_3^-$	R melt	282	—	287	210	696
	ir and R, solid Me_4N salt	276	—	261	103	327
$^{10}BF_3$	ir and R, gas	888	720	1505	482	1381
	liquid	880	688	1488	477	
	solid	878	665	1476	477	
$^{11}BF_3$	gas	888	691	1454	480	
	liquid	880	656	1433	477	
	solid	878	632	1405	477	
$^{10}BCl_3$	ir gas	471	480	995	244	1263
$^{11}BCl_3$		471	460	956	243	
$^{10}BBr_3$	ir and R, gas	278	395	856	148	1505
	liquid	278	392	845	150	
$^{11}BBr_3$	ir, R gas	278	370	819	148	
	liquid	278	374	802	151	
$^{10}BI_3$	ir, R gas	190	352	737	—	
	soln. CS_2	190	—	724	100	
$^{11}BI_3$	ir, R gas	190	336	740	—	
	soln. CS_2	190	—	693	100	
AlF_3	ir gas	(800)	297	935	236	198
	ir Ar matrix	—	266	940	251	862
	ir	—	300	965	270	1345

determination of the effective symmetry of the absorbing species, and not to an identification of the causes of the breakdown. Many workers, such as Lippincott *et al.* [1248, 1499], Decius *et al.* [339] and Buijs and Schutte [206–9] considered the spectra in terms of selection rules imposed by the site symmetry of the anion, and a number of studies of the relation between the spectra of a substance and its polymorphism were carried out. These included work on NH_4NO_3 by Newman and Halford [1070–1], on alkaline earth carbonates by Lander [823], and on KNO_3 by Khanna *et al.* [752]. The structures of most compounds of the formula MXO_3 have been determined, and three structures are particularly common. These are calcite, aragonite and a third structure, exhibited by the rare-earth borates $SmBO_3$–$LuBO_3$. This is generally stated to be the vaterite

(μ-$CaCO_3$) structure, but the differences between the spectra of the rare-earth borates and of vaterite itself cast some doubt on this identification. The structure will be referred to in this work as the YBO_3 structure, this compound having been investigated in some detail by Bartram and Felten [X10].

Calcite is hexagonal, with the space group D_{3d}, $Z = 2$. The sites occupied by Ca, C and O have symmetries $C_{3i} = S_6$, D_3 and C_2 respectively, and, in terms of the site-group analysis, the free-ion selection rules should continue to be obeyed. Factor-group analysis also predicts the same activity for the internal modes. The correlation diagram is shown in Fig. 6.1. The most striking difference between prediction and observation is the appearance of ν_1 weakly in a number of carbonates with the

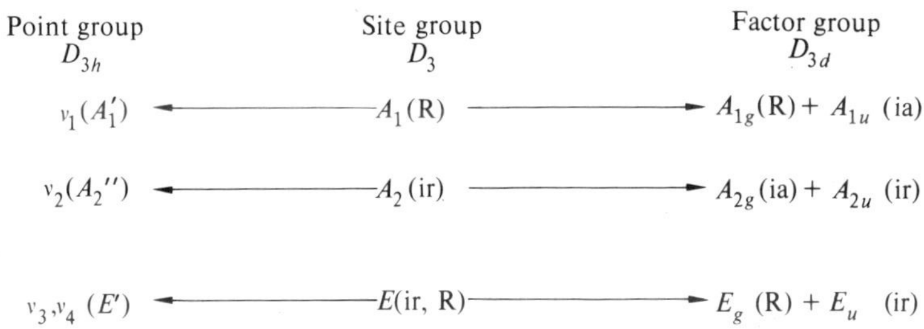

Fig. 6.1 Correlation diagram for calcite.

calcite structure. The lack of correlation of either the frequency or intensity of this band with the polarizing power of the cation and with the optical anisotropy was demonstrated by Goldsmith and Ross [548]; this would appear to rule out any explanation in terms of cation-to-anion bonding.

The appearance of ν_1 might suggest a distortion of the anion to C_{2v} or lower symmetry, but this would involve a breakdown of the degeneracy of ν_3 and ν_4, which is not observed. The observed spectra (see Fig. 6.2) suggest an effective C_3 symmetry of the carbonate ion, which could be brought about by rotational disordering. Such an explanation is consistent with the observations of Weir *et al.* [1499a] and Sterzel [1384] that deformation of the crystal activates ν_1. It is of interest that Porto *et al.* [1156] have found a nonzero intensity for α_{xy} in the $A_{1g}(\nu_1)$ mode, contrary to the selection rules. A second breakdown of the selection rules is seen in the almost invariable appearance of a doublet in the ν_2 region. The most generally accepted explanation is that the splitting arises through a combination with a lattice mode ν_L, the components of the doublet being ν_2 and $\nu_2 - \nu_L$. Buijs and Schutte [206–9, 1255] originally suggested that this might be a lattice translation, but Weir *et al.* considered a librational mode more probable. In a low-temperature study [1499] they

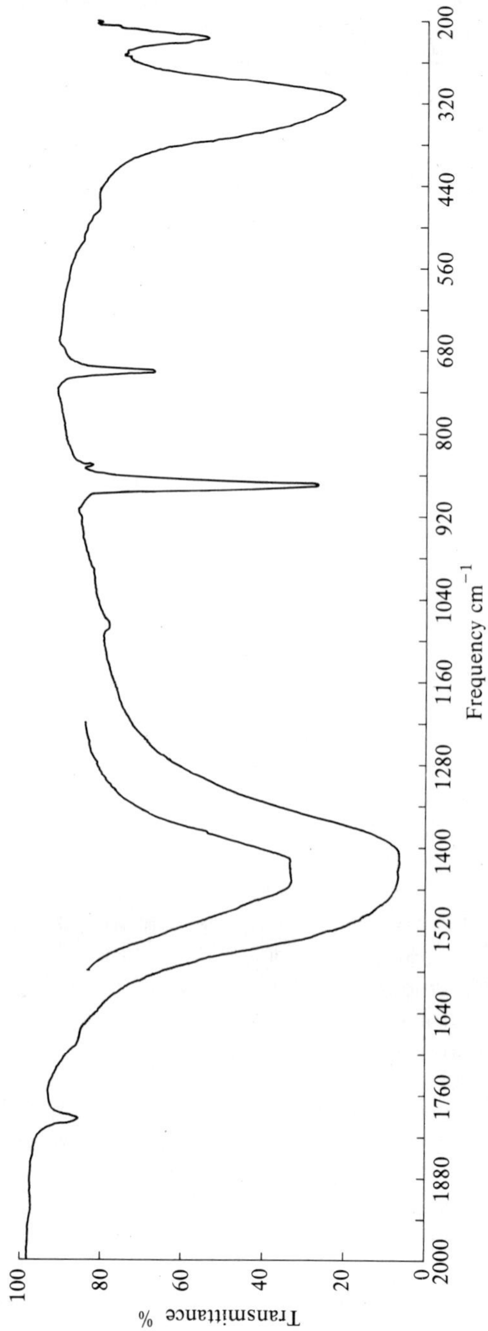

Fig. 6.2 Spectrum of calcite. The observed spectrum is not consistent with the D_3 site symmetry. The presence of the weak band near 1080 cm^{-1}, assigned to ν_1, may be due to rotational disordering, leading to a reduction in site symmetry to C_3. The degeneracy of ν_3 and ν_4 would be retained under C_3 symmetry.

observed fine structure associated with all the fundamental modes, interpreting this as combinations $\nu_{\text{int}} \pm n\nu_L$, calculating the value of ν_L and thus determining the barrier height restricting rotation (see chapter 3).

Table 6.3 Vibrational spectra of compounds with the calcite structure.

Compound		ν_1	ν_2	ν_3	ν_4	Ref.
$ScBO_3$	ir, NM, PF	—	770, 740	1240	640	970
$InBO_3$	ir, PF	—	768	1288	676	
			743			
$LuBO_3$	ir, PF	—	773	1275	630	1499
			748			
$MgCO_3$	ir, PF	—	892	1478	749	
$CaCO_3$	ir, PF	1087	874	1432	712	
			855			
$MnCO_3$	ir, NM	1081	864	1460	753	
			837			
$FeCO_3$	ir, NM	1081	862	1460	738	1218
$CoCO_3$	ir, NM	1081	854	1460	749	
$NiCO_3$	ir, NM	1073	831	1460	—	
$ZnCO_3$	ir, PF	—	873	1480	745	
$CdCO_3$	ir, PF	1075	862	1462	724	
			837			1499
$LiNO_3$	ir, PF	—	843	1420	738	
$NaNO_3$	ir, PF	1068	838	1395	727	

NM = nujol mull, PF = powder film.

In a number of these compounds, the isotopic shifts have been studied. Steele and Decius [1370] showed that in $ScBO_3$ (but not $InBO_3$) the isotopic shift was perturbed by a coupling effect. Sterzel [1385] found that ν_2 in $CaCO_3$ shifted from 877 cm^{-1} (^{12}C) to 850 cm^{-1} (^{13}C) and 826 cm^{-1} (^{14}C), in good agreement with calculation. He associated this with the absence of intramolecular coupling between vertical nearest neighbour $CO_3^=$ ions in the calcite structure, the distance apart being 4·26 Å.

The aragonite structure is displayed by a number of compounds MXO_3 with large cations; these crystallize in the space group D_{2h}^{16} with $Z = 4$, M and X atoms on C_s sites, and O atoms on C_s and C_1 sites. Under the C_s site symmetry, all internal modes become active and the degeneracy of ν_3 and ν_4 is completely broken down. The factor-group analysis predicts considerable correlation splitting, as shown in the correlation diagram (Fig. 6.3).

In compounds with the aragonite structure, ν_2 is split. Figure 6.4 illustrates this splitting. Decius [339] has suggested that this splitting is due to coupling between vertical nearest neighbour carbonate or nitrate

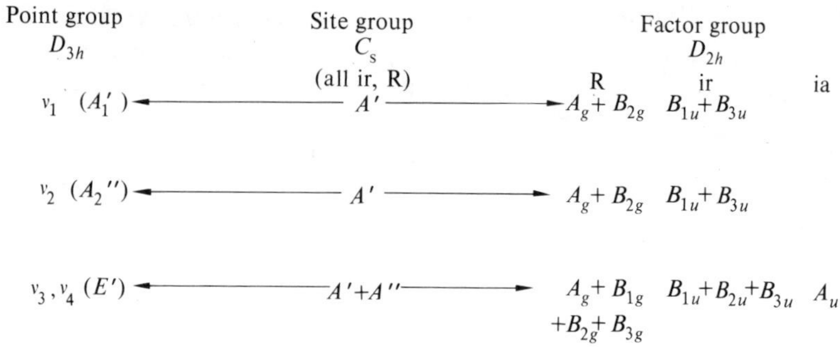

Fig. 6.3 Correlation diagram for aragonite.

ions. In aragonite, this anion–anion distance is 2·87 Å. Although the perturbation of the isotopic shifts observed in $LaBO_3$ and KNO_3 by Decius [333–5], and in divalent metal carbonates by Sterzel and Chorinsky [1385] is consistent with this explanation, the fact that the splittings are of very similar magnitude in aragonite and in calcite is not. For instance, in $CaCO_3$ (calcite) the splitting observed by Schroeder *et al.* [1248] was 19 cm^{-1}; in $SrCO_3$ it was 18 cm^{-1}. This similarity seems to point to combination with a librational fundamental as the most likely explanation for the splitting of ν_2. A number of compounds which have the aragonite structure at room temperature undergo phase transitions. Khanna *et al.* [752] have studied the behaviour of ν_1 and ν_2 of KNO_3 on varying the temperature. On heating to 130°C, there is a marked decrease in intensity of both ν_1 and ν_2, though not the total disappearance of ν_1 expected if the new phase (phase I) has the calcite structure. Again the explanation advanced is in terms of rotational disordering. On cooling phase I to 124°C, a further decrease in the intensity of ν_1 was found; the space group of the new phase (phase III) is said to be C_{3v}^5 with the nitrate ion on a C_{3v} site. ν_1 would remain ir active under these selection rules. Lander [823] found that $SrCO_3$ and $BaCO_3$ transformed from the aragonite to the calcite structure at 912°C and 803°C respectively, and that $BaCO_3$ underwent a further change to a cubic form with a structure resembling that of NaCl at 976°C. The existence of rotational disorder in the high temperature forms was suggested.

Weir and Lippincott [1499] observed splitting of ν_2 in powdered films of $SrCO_3$ and $BaCO_3$. The frequencies were 863, 845 cm^{-1} ($SrCO_3$) and 858, 845 cm^{-1} ($BaCO_3$). The values in square brackets are the calculated values associated with isotopic substitution.

$CaCO_3$ in the vaterite form has a spectrum characteristic of C_{2v} or lower symmetry, the observed frequencies [1499] being 1090 (ν_1); 878, 850 (ν_2); 1450 (ν_3); 747, 741 cm^{-1} (ν_4). As with some of the aragonite-type

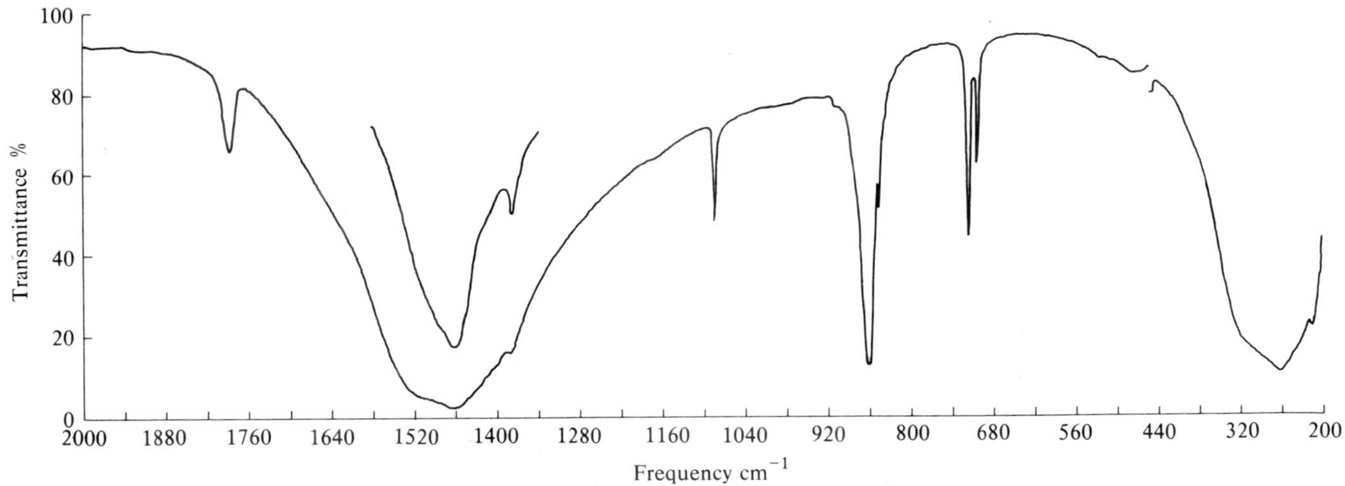

Fig. 6.4 Spectrum of aragonite. ν_1 is very much stronger than in calcite, and the degeneracy of ν_3 and ν_4 is now lifted. This is consistent with the C_s site symmetry. The additional band in the ν_2 region is much stronger than in calcite.

Table 6.4 Vibrational spectra of compounds with the aragonite structure.

Compound		ν_1	ν_2	ν_3	ν_4	Ref.
$LaBO_3$	ir P, mull	930	708	1310	606	1499
				1244	587	
$PrBO_3$	ir	944	788	1285	610	825
			713	1250	590	
$NdBO_3$	ir P, mull	951	791	1307	619	1499
			720		598	
$CaCO_3$	ir P	1083	857 (^{12}C)	1489	713	
				1411	700	
			[830] 846 (^{13}C)			
			[806] 828 (^{14}C)			
		1071	859 (^{12}C)	1462	706	1385
			[832] 845 (^{13}C)	1452	699	
			[808] 825 (^{14}C)			
		1060	860 (^{12}C)	1453	694	
			[833] 842 (^{13}C)			
			[809] 821 (^{14}C)			
$PbCO_3$	ir P	1053	840	1450	678	
			826		670	1248
KNO_3	ir P	1050	827	1420	714	

P = powder.

carbonates, ν_3 is not resolved and ν_2 is split. This general similarity between the spectra of the aragonite and vaterite forms of $CaCO_3$ is not reproduced when the rare-earth borate spectra are considered.

The spectra of the rare-earth borates with the YBO_3 structure have been studied by Weir and co-workers [1499, 1500] and by Laperches and Tarte [825]. In the first study, Weir and Lippincott found five principal regions of absorption; for $ErBO_3$, examined as a powder film, bands occurred at 1103, 931, 876, 716 and 561 cm^{-1}. In the second study, carried out on both mulls and films, absorptions of natural $ErBO_3$ were given as 1051, 1023, 1003; 921; 857; 707 and 559 cm^{-1} corresponding to the first set; also a strong band at 390 cm^{-1} and two weaker bands at 424 and 350 cm^{-1}. The observations of Laperches and Tarte on $ErBO_3$ were in moderate agreement with those of Weir; they found 1066, 1031, 980; 927; 871, 835; 711; 565; 401, 365, 267 cm^{-1}. There are two problems in the interpretation of these spectra; these are, first, the coordination of the B atom, and second, the nature of the vibrating unit. One version of the structure is that due to Bartram and Felten [X10]; in this, the boron atom is in 3-coordination, the BO_3^{3-} ion being T-shaped, with symmetry C_{2v}. The unit cell is hexagonal with $Z = 2$, and the space group D_{6h}^4, the anion site symmetry being D_{3d}. On the other hand, following Weir's suggestion of possible 4-coordination, Bradley *et al.* [X20] considered a structure containing B_3O_9 rings, whose symmetry would be D_{3h} if the ring were planar. This is again a hexagonal structure, with the space group D_{3h}^2 and the B atoms on C_s sites. Surprisingly for two such different structures, the spectroscopic evidence is insufficient to determine which (if either) is correct. The situation has been summarized by Ross [1217] who concluded that Bartram and Felten's structure was more likely to be correct but was not altogether satisfactory since it did not account for all the bands observed. The bands below 500 cm^{-1} do not display the correct isotopic shifts for lattice translations, and are too low to be accounted for as internal fundamentals. The spectra are tabulated in Table 6.5 without assignment.

Table 6.5 Vibrational spectra of rare-earth borates with the YBO_3 structure.

Compound and Reference									
$SmBO_3$	$GdBO_3$	$DyBO_3$	$HoBO_3$		$ErBO_3$		$TmBO_3$	$TbBO_3$	
LT [825]	LT	LT	LT	W [1500]	LT	W	W	LT	W
	1032	1050	1059	1045	1066	1051	1057	1073	1063
1013	1011	1020	1026	1020	1031	1023	1013	1041	1022
980	970	975	979	1002	980	1003	—	981	—
903	911	920	922	917	927	921	925	931	927
849	861	870	870	853	871	857	862	880	866
—	816	827	830	822	835	—	834	844	—
686	697	705	707	707	711	717	713	720	718
558	561	562	563	563	565	559	567	570	568
396	399	400	400	424	401	424	418	402	428
358	368	366	368	395	365	390	398	363	398
				348		350	355		357
							342		340
263	269	271	274	—	267	—	—	270	—

We shall now consider the remaining borates, carbonates and nitrates. Twelve other borates have been studied. Of these, $Mg_3(BO_3)_2$ and $Co_3(BO_3)_2$ are known to be isostructural, and the Ni and Cd borates are thought to be isostructural with these. The structure is orthorhombic, with the space group D_{2h}^{12}, $Z = 2$ and the anions on C_s sites [X14, X52]. In spite of this low symmetry, ν_1 is observed only in the Mg and Cd compounds. ν_3 is broad and unresolved, and this is also the case for the borates of Be, Ca, Sr and Ba. The last three of these may form an isostructural group; the similarity of the spectra of the Ca and Sr compounds is consistent with this. The borates of La, Nd, Sm and Eu display, in addition to their room-temperature structures, an unknown high-temperature structure which appears to be of low symmetry.

Table 6.6 Vibrational spectra of ortho-borates.

Compound		ν_1	ν_2	ν_3	ν_4	Ref.
$Mg_3(BO_3)_2$	ir, powder film and Nujol mull	1022	719	1290–1090	657 606	1499
Cd		1006	692	1290–1100	611 575	1499
Co		—	676	1250–1020	617 587	1499
Ni		—	691	1260–1080	619 598	1499
Be		928	672	1340–1240	643 600	1500
Ca		—	702	1310–1140	622 613	1500
Sr		—	715	1260–1090	612 600	1500
Ba		—	—	1260–1140	—	1500
HT forms:						
$LaBO_3$		952	757	1280, 1215	634 588, 582	1500
Nd		935	762 737	1390, 1310 1215, 1172	667 575, 563	1500
Sm		939	760 732	1388, 1338 1085, 1055	674 571, 561	1500
Eu		925	733 723	1360, 1210 1180, 1066	675 573, 564	1500

All the anhydrous divalent metal carbonates crystallize in one or other of the three $CaCO_3$ structures, and have been dealt with above; in addition, some alkali metal carbonates and rare-earth carbonates have been studied. All the carbonates of known structure crystallize in either the orthorhombic or monoclinic systems and should therefore exhibit no degenerate vibrations. The observed spectra are in good agreement with this prediction except that ν_3 is not split in the monoclinic K_2CO_3. Louisfert [886] has studied $CaCO_3 \cdot 6H_2O$ and 'amorphous' $CaCO_3$ of unknown structure.

Goldsmith and Ross [549] have studied the complete series of anhydrous and hydrated rare-earth carbonates; the data for the La compounds are typical, except for the anomalous Ce compounds which may contain some basic carbonate.

Many nitrates have been studied and in most cases deductions have been

Table 6.7 Infrared spectra of carbonates as nujol mulls.

Compound	ν_1	ν_2	ν_3	ν_4	Ref.
Li_2CO_3	1088	866	1495	741	206
			1437	712	
Na_2CO_3	1079	881	1451	701	
				694	
Na_2CO_3, H_2O	1068	868	1450	702	
				686	207
$Na_2CO_3, 10H_2O$	1068	890	1499	—	
			1379		
K_2CO_3	1060	880	1412	691	208
				685	
$CaCO_3$ 'amorphous'	1073	847	—	—	
$CaCO_3, 6H_2O$	1084	874	—	745	886
	1070	847			
$La_2(CO_3)_3$	1062	850	1490	718	
			1395	697	
$La_2(CO_3)_3, 8H_2O$	1060	847	1480	720	
			1385	609	549
$Ce_2(CO_3)_3, 7H_2O$	w 1086, 1073	870	1460	745, 717	
	s 1058	848	1400, 1350	677, 650	
$Ce_2(CO_3)_3$	1055	878	1550, sh 1460	720	
		852	1370	680	

drawn from the spectra as to the possible existence of metal-to-anion coordination, and whether (if it exists) the nitrate group is functioning as a monodentate or a bidentate ligand. In a few cases the relation between spectra and crystal structure has been discussed. From the structural data tabulated below, we may predict that in all cases except the anhydrous nitrates of Ca, Sr, Ba and Pb, all six vibrations will be observed. For these four compounds, the C_3 site symmetry allows ν_1 but does not lead to a breakdown of the degeneracy of ν_3 and ν_4. These compounds should therefore have four bands in the ir spectrum, with the possibility of some correlation splitting. This has been studied by Schutte [209, 1253–4]; the correlation diagram is shown in Fig. 6.5.

In general, the site-group approximation is adequate; in some cases ν_3 and ν_4 are not resolved or observed. The nitrates of divalent cations similarly appear to obey the site-group selection rules, but here, attempts have been made to establish the existence of cation–anion coordination, using the splitting of ν_3 as a criterion. In Table 6.11, the data are presented in terms of the free-ion vibrations ν_1–ν_4, for the sake of uniformity; however, it should be remembered that many of the assignments have been made on the basis that the species involved is M—O—NO_2 with C_{2v} symmetry, and the correlation between the free ion and the complex is

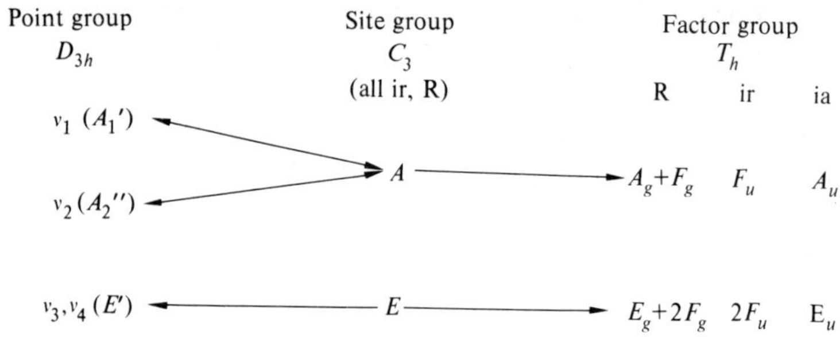

Fig. 6.5 Correlation diagram for barium nitrate.

$\nu_1(A_1') \rightarrow \nu_2(A'')$; $\nu_2(A'') \rightarrow \nu_6(B_2)$; $\nu_3(E') \rightarrow \nu_1(A_1) + \nu_4(B_1)$; $\nu_4(E') \rightarrow \nu_3(A_1) + \nu_5(B_1)$.

Of the anhydrous nitrates studied, those of Zn, Hg, Mn and Cu (the last in the β-form) have been classified as covalent by Addison and Gatehouse [37], and the vibrations assigned in terms of a nitrato group of C_{2v} symmetry; similarly anhydrous mercurous and nickel nitrates were classified as covalent by Vratny [1469], and anhydrous uranyl nitrate by Ferraro [464], who considered the nitrato group to be bidentate, and assigned the band at 1600 cm^{-1} to $\nu_1(\nu(N{=}O))$. All the bidentate nitrates considered so far are chelate, but the α-form of $Cu(NO_3)_2$ appears to have bridging nitrato groups. The spectrum shows bands at 1510, 1502, 1478, 1435, 1390, 1049, 1015, 796, 759, 746 and 712 cm^{-1} [463]. The bridging groups are arranged in a structure of parallel chains, with each non-bridging oxygen atom forming bonds to two copper atoms in other chains. Although it is clear in this case that the splitting of the vibrations arising from ν_3 in the free ion is greater in the β (chelate) than in the α (bridging) form, not enough comparable data exist to establish the value of this as

Table 6.8 Observed correlation splitting in strontium and barium nitrates.

Mode	Species	A_g	E_g	F_g	F_u
ν_1	Sr	1055	—	not obs.	1055
	Ba	1048	—	not obs.	1046
ν_2	Sr	(812)	—	not obs.	817
	Ba	(817)	—	not obs.	818
ν_3	Sr	—	736	715	741
	Ba	—	731	703	729
ν_4	Sr	—	1369	1403	1364
	Ba	—	1357	1387	1354

Table 6.9 Structural information on nitrates.

Compound	Crystal system	Z	Space group	Site symmetry	Ref.
α-$TlNO_3$, NH_4NO_3(I)	Cubic	4	—	—	W VIIa, 4
NH_4NO_3(II)	Tetragonal	2	C_{4v}^2	C_{2v}	W VIIa, 5
NH_4NO_3(III)	Orthorhombic	4	D_{2h}^{16}	C_s	X49, X92, X93
NH_4NO_3(IV)	Orthorhombic	2	D_{2h}^{13}	C_{2v}	
NH_4NO_3(V)	Tetragonal	8	C_4^3	C_1	W VIIa, 7
γ-$TlNO_3$	Orthorhombic	4	D_{2h}^{16}	C_s	X49, X92, X93
β-$TlNO_3$, $CsNO_3$ (room temperature)	Trigonal	9	C_{3v}^2	C_s	X46
$RbNO_3$, $CsNO_3$ (high temperature)	Cubic	8	T_h^6	C_3	X47
$M(NO_3)_2$ (M = Ca, Sr, Ba, Pb)	Cubic	4	T_h^6 or T^4	C_3	1254
$Ni(NO_3)_2$	Cubic	4	T_h^6	C_3	X167
α-$Cu(NO_3)_2$	Orthorhombic	4	C_{2v}^7	C_s or C_1	X164
$Sr(NO_3)_2$, $4H_2O$	Monoclinic	4	C_{2h}^6	C_2 or C_1	X123
$Cd(NO_3)_2$, $4H_2O$	Orthorhombic	8	C_{2v}^{19}	C_2	X102
$Ni(NO_3)_2$, $4H_2O$	Monoclinic	4	—	(point symmetry C_1)	X54
$Ni(NO_3)_2$, $6H_2O$	Triclinic	—	—	C_1	X53
$(UO_2)(NO_3)_2$, $2H_2O$	Monoclinic	2	—	—	X157–9
$(UO_2)(NO_3)_2$, $3H_2O$	Triclinic	—	—	C_1	X157–9
$(UO_2)(NO_3)_2$, $6H_2O$	Orthorhombic	4	C_{2v}^{12}	C_s	X157–9
$Al(NO_3)_3$, $9H_2O$	Monoclinic	—	C_{2h}^5	C_1	X82
$Sc(NO_3)_3$, $4H_2O$	Monoclinic	4	C_{2h}^1, C_2^1 or C_s^1	—	X121
$M(NO_3)_3$, xH_2O (M = La, Ce, Pr, Nd, Sm; $x \sim 6$)	Triclinic	2	C_i^1	C_1	X77–8, X129, X162

$LiNO_3$, $3H_2O$ and $AgNO_3$ are known to be orthorhombic [WN102], and $Co(NO_3)_2$, $6H_2O$ monoclinic [WN103].

a general test of the chelate or bridging nature of a nitrato group. Ferraro [464] and Gatehouse and Comyns [507] considered that the hydrated uranyl nitrates also contained bidentate nitrato groups, but Allpress and Hambly [51], comparing the observed spectra with that of $KUO_2(NO_3)_3$, which contains monodentate groups, found strong similarities, and therefore considered these compounds to contain monodentate groups. For $KUO_2(NO_3)_3$ they found (in terms of free-ion vibrations) ν_1, 1025; ν_2, 802; ν_3, 1555, 1521, (ν_{as}), 1271 (ν_{sym}); ν_4, 742, 715 cm^{-1}. Remarkably, the two sets of observed data for the dihydrate and hexahydrate are sufficiently different to support the differing assignments. The ambiguity

Table 6.10 Vibrational spectra of nitrates of monovalent cations.

Compound	Method and state	ν_1	ν_2	ν_3	ν_4	Ref.
$NO_2^+\ ^{14}NO_3^-$	ir, R solid (77°K)	1050	824	1413	722	1427–8
$NO_2^+\ ^{15}NO_3^-$		1050	802	1378	722	
$LiNO_3, 3H_2O$	R solid	1054	—	1416 1332	—	915
	ir, R soln, TBP	1054	828	1416 1332	717	730a
$AgNO_3$	Ref, melt	1029	800	1395 1310	695	1515
	ir, N mull	1046	806	1352	735 711	1248, 1469
$Rb^{14}NO_3$		1023	803	1536 1276	736 711	51
$Rb^{15}NO_3$		1019	782	1499 1244	736 711	
$Cs^{14}NO_3$		1025 1021	805	1519, 1504 1277	749, 739 713, 708	
$Cs^{15}NO_3$		1024 1020	784	1480, 1471 —	748, 738 712, 708	
$RbNO_3$	ir, N mull	1061	844	1370	726	1248, 1469
$CsNO_3$	ir, N mull	1053	844	1370	726	
NH_4NO_3(I)	ir film on AgCl/ IRTRAN 2	1047	832	1337	—	1430
(II)		1047	832	1365	—	
(IV)		1046	830	1348	715 713	
(VII)		1060	833	1363 sh 1340	—	
NH_4NO_3(III)	ir single crystal	1048	830	1530–1240	720	1070
(IV)		1046	830	1550–1250	715	
γ-$TlNO_3$	ir single crystal	1044	822	1410–1280	713 697	

Ref. = reflectance.

arises from the uncertainty of the position of the higher component of the free-ion ν_3 vibration. The transition metal nitrate dihydrates were considered by Hathaway [619] to contain a bidentate nitrato group, but the frequency of ν_{as} (NO_2) is more characteristic of a monodentate group. Walker and Ferraro [1479] have suggested 1480–1555 cm^{-1} as characteristic of a monodentate group, and $\sim$1600 cm^{-1} for ν(N=O) in a bidentate group.

All the compounds listed in Table 6.12 show a spectrum characteristic of a bound nitrato group, and the last three appear to contain a bidentate group. Walker and Ferraro [1479] have studied the anhydrous rare-earth nitrates and a representative selection of their results is given. They observed that two different types of spectrum were found for the nitrates

Table 6.11 Vibrational spectra of nitrates of divalent cations.

Compound		ν_1	ν_2	ν_3	ν_4	Ref.
$Mg(NO_3)_2$	ir, KBr disc	1045	825	1372	—	1469
$Ca(NO_3)_2$	ir, N mull	1064	815	1372	744	
$Sr(NO_3)_2$	ir, N mull	1055	817	1364	741	209,
$Ba(NO_3)_2$	ir, N mull	1046	818	1354	729	1248,
$Pb(NO_3)_2$	ir, N mull	—	836, 807	1373	726	500
		1058	817	—	731	
$Zn(NO_3)_2$	ir, N mull	1050	800	1546	—	
		1040	793	1300		
$Cd(NO_3)_2$	ir, N mull	1028	802	1366	751	
$Hg(NO_3)_2$	ir, N mull	1027	788	1495	750	
				1376		37
$Mn(NO_3)_2$	ir, N mull	1019	805	1553	759	
			799	1294		
$Co(NO_3)_2$	ir, N mull	—	799	1425	762	
			792	1350		
$Ni(NO_3)_2$	ir, N mull	1053	805	1470	758	1469
		1033		1300	744	
β-$Cu(NO_3)_2$	ir, N mull	1038	795	1565, 1546	770	37
		1016	787	1289, 1264		
$Mn(NO_3)_2, xH_2O$	ir, soln. H_2O	1117	812	1460, 1410	753	1469
		1035		1320		
$Mn(NO_3)_2, 2H_2O$	ir, N mull	1027	804	1555	760	
				1310	665	
$Co(NO_3)_2, 2H_2O$	ir, N mull	1055	812	1560–1540	760	
				1320		619
$Ni(NO_3)_2, 2H_2O$	ir, N mull	1061	810	1560–1540	760	
				1310		
$Cu(NO_3)_2, 2H_2O$	ir, N mull	1051	820	1515	770	
		1023	805	1285		
$Ni(NO_3)_2, 4H_2O$		1035	802	1460	745	499
				1310		
$Ca(NO_3)_2, 4H_2O$	ir, R soln. TBP	1052	825	1422	721	730a
				1353		
$UO_2(NO_3)_2$		1022	800	1600, 1560	758	464
				1355		
$UO_2(NO_3)_2, 2H_2O$		1028	801	1608–1510	745	464
				1350–1270		
$UO_2(NO_3)_2, 2H_2O$		1029	803	1540, 1515	752, 743	51
				1268	707	
$UO_2(NO_3)_2, 3H_2O$		1045	804	1540, 1506	751	51
		1032	800	1306, 1276	744	
$UO_2(NO_3)_2, 6H_2O$		1040 (1052)	802 (809)	1610–1418 (1536)	747 (788)	(51)
		1030 (1042)		1350–1280 (1497)	723 (750)	464
				(1333)		
				(1297)		
$Hg_2(NO_3)_2$	ir, N mull	1032	835	1355	797, 791	1469
		996, 962	808	1262	741, 700	

of Pr—Er and Tm—Lu respectively. $Tm(NO_3)_3$ shows a triplet at 1610, 1598, 1580 cm^{-1} presumably associated with $\nu(N{=}O)$, and two more groups of bands (1530, 1460, 1410; 1375, 1330, 1260) attributable to

Table 6.12 Vibrational spectra of nitrates of trivalent cations.

Compound		ν_1	ν_2	ν_3	ν_4	Ref.
$Al(NO_3)_3, 9H_2O$		1058 1040	—	1434 1388, 1321	719	921a
$In(NO_3)_3$	R, 4·2M soln. H_2O	1040	—	1510 1315	750	649
$Sc(NO_3)_3, 4H_2O$	ir solid	1035 970	805	1535 1335, 1315	760	773
$Sc(NO_3)_3, 4H_2O$	ir	1032	802	1511 1384, 1316	755	461
$Sc(NO_3)_3, 2H_2O$	ir	1025 970	805	1550 1335, 1280	760 720	773
$Sc(NO_3)_3$	ir	1015	805	1550 1280	765	773
$Y(NO_3)_3, 3H_2O$		1029	804	1495 1386, 1313	745	461
$Y(NO_3)_3$		1025	835 814	1470 1370, 1300	780 752	1469
$La(NO_3)_3, 5H_2O$		1034	809	1460 1369, 1319	—	461
$Ce(NO_3)_3, 6H_2O$		1044	826	1505 1417, 1330	735	1222a
$Sm(NO_3)_3, 6H_2O$		1031	806	1471 1333	—	461
$Nd(NO_3)_3$		1025	795	1495, 1455 1325, 1240	—	1479
$Sm(NO_3)_3$		1035 1030	801 794	1545, 1505, 1430 1320	—	1479
$Er(NO_3)_3$		1060 1030, 1015	804 800	1525, 1460, 1410 1375, 1330, 1300, 1260	— —	1479
$Tm(NO_3)_3$		1060, 1030 1000	800	see text	—	1479
$(NbO)(NO_3)_3$		1004	775, 770 783	1646 1213	—	470
$(VO)(NO_3)_3$		1014, 1007 995, 962	780	1635, 1612 1210, 1191	—	35

$\nu_{as}(NO_2)$ and $\nu_{sym}(NO_2)$. The spectra of the Yb and Lu compounds are very similar. A few nitrates of group IV metals have been studied; the tetrahydrated thorium, hafnium and zirconium nitrates have spectra rather similar to those of the hydrated M^{III} nitrates. The anhydrous nitrates of Sn, Ti and Zr, however, display the bands near 1600 cm^{-1} characteristic of $\nu(N{=}O)$ and contain bidentate nitrato groups.

The anhydrous nitrates appear to have symmetry nearly D_{2d} [X54a] and an assignment for the Zr compound has been made on this basis by Weidlein [1494], whose ir spectrum agrees well with that obtained by Field and Hardy [469]. Raman studies of the other two compounds by Addison *et al.* [35] show similar spectra.

Table 6.13a Infrared spectra of $M(NO_3)_4$, $4H_2O$.

Compound	ν_1	ν_2	ν_3	ν_4	Ref.
Zr	1020	798	1530, 1430 1360, 1270	—	461
Hf	1026	782	1520 1370, 1305	—	461
Th	1036 1030	811 805	1520, 1506 1323, 1292	756 744	37

Table 6.13b Vibrational spectra of $M(NO_3)_4$.

Compound	$\nu_{N=O}$	$\nu_{as}(NO_2)$	$\nu_{sym}(NO_2)$	Bending	ν_{as}(M—O)	ν_{sym}(M—O)
Sn ir	1657,1618	1228	—	802 783, 696	344	302
Ti ir	1663	1181	1002	780 680	450	310
Zr R	1650, 1610, 1576	1280, 1223	1028, 988	811, 780 774, 700	335 305	290 250
ir	1600, 1585	1275, 1225	1008, 986	790 763, 730	310	—

Although the existence of bands near 1600 cm^{-1} attributable to ν(N=O) serves to distinguish bidentate from monodentate nitrato groups in simple metal nitrates, it is much less easy to make the distinction in nitrato complexes. An alternative criterion is the separation between the two highest-frequency bands of the nitrato group. In monodentate complexes these are $\nu_4(B_1)$ and $\nu_1(A_1)$, the antisymmetric and symmetric stretches of the free NO_2 group. In a series of such complexes $[\phi_3MeAs]_2\ M^{II}(NO_3)_2$ the average value of this separation was 182 cm^{-1} (M^{II} = Co, Mn, Ni, Cu). In bidentate complexes the two highest frequencies are $\nu_1(A_1)$, ν(N=O) and $\nu_4(B_1)$, $\nu_{as}(NO_2)$. For two complexes [Ni en_2 NO_3]X the separations were 186 cm^{-1} (X = ClO_4) and 188 cm^{-1} (X = I). The data given in Table 6.14 show that no conclusive test can be based on the infrared spectra. A particularly good example is that of the complexes $M^I[UO_2(NO_3)_3]$; the K compound was assigned as a monodentate complex by Allpress and Hambly [51], while the Rb and Cs salts with virtually identical spectra were assigned as bidentate by Addison and Gatehouse [36, 37].

In the case of the complexed carbonate group, there is a far more clear-cut distinction. The highest stretching frequency for a monodentate complex lies in the range 1450–1500 cm^{-1}, compared with 1590–1640 cm^{-1} in the bidentate complexes. Further, the separation between the two highest stretching frequencies never exceeds 150 cm^{-1} in a monodentate

Table 6.14 Vibrational spectra of some nitrato complexes.

Monodentate	Assignment and description						Ref.
	ν_1	ν_2	ν_3	ν_4	ν_5	ν_6	
	$\nu_{sym}(NO_2)$	ν_{N-O}	$\delta(NO_2)$	$\nu_{as}(NO_2)$	$\delta(NO_2)$	π	
$[\phi_3MeAs]_2Cu(NO_3)_4$	1462	1022	—	1285	—	807	1389
$[Co^{III}(NH_3)_5NO_3]PtCl_4$	1481	1012	—	1269	—	796	
$Pd^{II}(H_2O)_2(NO_3)_2$	1502	988	—	1274	—	797 784	
cis $Pt^{II}(NH_3)_2(NO_3)_4$	1529	980 970	—	1264	—	781	509
trans $Pt^{II}(NH_3)_2(NO_3)_4$	1522	970	—	1253	—	821	
$K_2Ce^{IV}(NO_3)_6$	1531	1034	—	1290 1277	—	804 799	
$Co(\phi_3PO)_2(NO_3)_2$	1497	1024	—	1290	—	812	83a
Ni $en_2(NO_3)_2$	1420	(1035)	728 708	1305	—	818	306
$Ni(dien)(NO_3)_2$	1440	(1035)	736	1315	714	816	306
$Ni(Et_3P)_2(NO_3)_2$	1513	1018	—	1272	—	805	306
$KUO_2(NO_3)_3$	1555 1521	1025	742	1271	715	802	51
$RbUO_2(NO_3)_3$	1536	1023	736 711	1276	—	803	1433
$CsUO_2(NO_3)_3$	1519 1504	1025 1021	749 739	1277	713 708	805	37

Bidentate	Description						
	$\nu(N{=}O)$	$\nu_{sym}(NO_2)$	$\delta(NO_2)$	$\nu_{as}(NO_2)$	$\delta(NO_2)$	π	
$Co(Me_3PO)_2(NO_3)_2$	1517 1492, 1469	1024	—	1317 1304, 1282	—	812	288a
[Ni en_2 NO_3]ClO_4	1476	(1025)	746	1290	695	809	306
[Ni en_2 NO_3]I	1480	(1035)	740	1292	702	805	306

complex, and does not fall below 300 cm^{-1} for a bidentate complex. Both in-plane bending vibrations appear in a well-defined manner, in contrast to the nitrate spectra in which the lower vibration is often not seen.

The only other group of trigonal planar XY_3 molecules to have been studied consists of the trithio- and triselenocarbonates. After an early study of $BaCS_3$ by Duval, the correctness of which was later disputed [778], several further investigations were made. Little is known about the structures of any of these compounds; Krebs *et al.* [X94] found $BaCS_3$

to be hexagonal, with $Z = 6$, and possible space groups D_3^4, D_3^6, C_6^2, C_6^3. In the first two of these, the $CS_3^=$ ion would occupy sites of C_2 symmetry, leading to the activation of all six vibrations. Some difficulty arose regarding the assignments, as it was not originally considered likely that ν_1 and ν_2 were close together. Such an assignment is now generally accepted; in the selenocarbonate, $\nu_1 < \nu_2$.

Philippot and Maurin [1135] have recorded the spectra of the hexagonal and orthorhombic forms of $PbCS_3$; these two spectra are similar, but differ somewhat from that of Krebs *et al.* [787], ν_3 being unresolved. The bands near 830 cm^{-1} which occur in a number of spectra are now assigned as combinations $\nu_1 + \nu_4$, $\nu_2 + \nu_4$. Of the four complexes, Krebs and Mueller [779] considered that those containing Ni were possibly bidentate, and those containing Zn probably monodentate, with the Zn in tetrahedral coordination. The band at 840 cm^{-1} in these may again be $\nu_1 + \nu_4$ or $\nu_2 + \nu_4$.

SO_3 vapour has been assigned as planar by Stopperka [1387], with $\nu_1 = 1068$, $\nu_2 = 653$, $\nu_3 = 1332$, $\nu_4 = 530$ cm^{-1}.

Three interhalogen compounds, ClF_3, $BrCl_3$ and BrF_3, are unusual in having the *T*-shape with C_{2v} symmetry. This geometry arises by the occupation of two apical positions and one equatorial position of a trigonal bipyramid.

Only ClF_3 has been studied completely. The initial infrared and Raman work due to Jones *et al.* [704] established the existence of bands near 750, 710 and 510 cm^{-1} which were later assigned as the three stretching modes. Another early study by Schaefer and Wicke [1242] identified all six fundamentals but incorrectly assigned them according to a pyramidal

Table 6.15 Vibrational spectra of carbonato complexes.

Monodentate	Assignment and description						Ref.
	ν_1 $\nu_{sym}(CO_2)$	ν_2 $\nu(C—O)$	ν_3 $\delta(CO_2)$	ν_4 $\nu_{as}(CO_2)$	ν_5 $\delta(CO_2)$	ν_6 π	
$[Co(NH_3)_5CO_3]Cl$	1488	1062	741	1351	694	877 851	550
Br	1449	1072	738 727	1370	699	865 851	
I	1445	1065	778	1361	673	862 849	
$NO_3, \frac{1}{2}H_2O$	1471	1053 1047	738	1361	687	851	
ClO_4	1455	—	757	1370	681	852	495
$[Co(ND_3)_5CO_3]Br$	1471	1072	751	1369	687	854	
I	1467	1065	742	1360	687	853	
$NO_3, \frac{1}{2}H_2O$	1495	1053	745	1370	687	854	
ClO_4	1471	—	760	1370	690	855	

Table 6.15 (*cont.*).

Bidentate	Description						Ref.
	$\nu(C{=}O)$	$\nu_{sym}(CO_2)$	$\delta(CO_2)$	$\nu_{as}(CO_2)$	$\delta(CO_2)$	π	
$[Co(NH_3)_4CO_3]Cl$	1590	1043	768	1282	673	885	550
				1266	669	832	
ClO_4	1602	—	762	1284	672	836	495
Br	1595	1043	760	1282	671	848	550
				1263		833	
I	1595	1044	763	1284	673	840	
				1267	667	834	
NO_3	1630	1042	765	1282	677	851	
				1261	671	832	
SO_4	1610	1020	755	1280	678	850	
				1255		830	
$[Co(ND_3)_4CO_3]Cl$	1635	1031	753	1268	672	832	495
	1607						
ClO_4	1603	—	765	1292	676	835	
$[Co(en)_2CO_3]Br$	1628	—	759	1276	674	827	
	1615						
$K[Co(NH_3)_2(CO_3)_2]$	1623	1026	763	1265	673	839	
	1597		744				
$[Co(NH_3)_6Co(CO_3)_3]$	1590	1033	742	1287	677	831	

Table 6.16 Vibrational spectra of trithio- and triselenocarbonates.

Compound		ν_1	ν_2	ν_3	ν_4	Ref.
$SrCS_3$	ir	—	508	954	334	1160
				922	319	
$BaCS_3$	ir		518	943, 912	330, 321	1160
	R	510	516	920	325	1034
$PbCS_3$	ir	510	477	940	330	787
				875	318	
Tl_2CS_3	ir	520	495	900	312	787
				890		
$BaCSe_3$	ir	316	420	802	185	1010
$Ni(NH_3)_3CS_3$		505	490	1000	328	779
				860		
$Ni(ND_3)_3CS_3$		505	490	1001	328	
				860		
$Zn(NH_3)_2CS_3$		520	494	1010	328	
				867, 840(?)		
$Zn(ND_3)_2CS_3$		520	494	1010	328	
				867, 840(?)		

model. The microwave work by Smith [X141] favoured the planar *T*-shape and the data were reassigned by Weber and Ferigle [1490] on this basis. They gave the stretching frequencies as 750, 644, 508 cm^{-1}; in-plane bending as 428, 316 cm^{-1}; and out-of-plane bending as 395 cm^{-1}. The assignment tabulated below is due to Claassen *et al.* [266], who also identified two fundamentals of BrF_3 and calculated its spectrum using the same diagonal potential constants as in ClF_3. Haendler *et al.* [602] had previously observed some overtones of BrF_3 (1337, 1250, 1154 cm^{-1}) which are in approximate agreement with calculated binary combinations from Claassen's work. Nelson and Pimentel [1063] have studied $BrCl_3$, produced by passing a mixture of the halogens and an inert gas through a microwave discharge and depositing the product on a CsI window at 20°K. Using a Kr matrix they found a doublet at 364·3, 362·1 cm^{-1}.

Table 6.17 Vibrational spectra of ClF_3 and BrF_3. Assignment and description

Compound		$\nu_1(A_1)$	$\nu_2(A_1)$	$\nu_3(A_1)$	$\nu_4(B_1)$	$\nu_5(B_1)$	$\nu_6(B_2)$
		ν(XZ)	ν_s(XY)	δ	ν_{as}(XY)	δ	π
ClF_3	ir*	761	535	332	713	434	364
		741	518	319	703, 694		
	R*	753	529	—	—	—	—
		746					
	ir	752	529	328	702	442	328
BrF_3	ir*	674 (doublet)	(528)	(300)	613	(384)	(289)
	ir	675	552	242	612	350	242

* Ref. 266; other data, ref. 1872.

ν(XZ) refers to the stretching mode associated with the equatorial atom. Long and Jones [874] who studied the in-plane frequencies of ClF_3, calculated MUBFF constants and found a dependence on the inverse sixth power of the bond length. Their data is in good agreement with Claassen's.

One other group of planar, non-trigonal XY_3 molecules is XN_3 (X = H, D, F, Cl, Br). Thompson and Fletcher [1432] have carried out a normal-coordinate analysis of this system, which has C_s symmetry, and found the UBFF to give a rather poor fit for HN_3 and DN_3.

(b) Molecules of the type XY_2Z

As with the boron halides, their monosubstituted derivatives have been fully studied. All the compounds where Z is hydrogen or deuterium, and Y is F, Cl or Br, have been investigated by Lynds and co-workers [95, 892–6], and some of the mixed halides have been studied by Lindemann

Table 6.18 Vibrational spectra of molecules XY_3 with C_s symmetry.

Compound		Assignment and Description						Ref.
		ν_1	ν_2	ν_3	ν_4	ν_5	ν_6	
		ν_{X-N}	ν_{N-N}	ν_{N-N}	δ_{XNN}	δ_{NNN}	π	
HN_3	ir gas	3336	2140	1274	1150	522	637	1432
DN_3	ir gas	2480	2141	1183	955	498	638	1432
FN_3	ir matrix	869	2034	1086	654	503	—	964, 964a
ClN_3	ir matrix	723	2066	1144	—	520	—	964, 964a
BrN_3	ir matrix	687	2062	1160	—	530	—	964, 964a

and Wilson [860] and by Goubeau [556]. A more complete study has recently appeared [1915].

Some disagreement exists regarding the assignments in $HBCl_2$ and $HBBr_2$. Lynds [894] has argued that the assignment of Brieux de Mandirola and Westerkamp [185] is incorrect in that those workers have assigned ν_6 to the R-branch envelope of ν_5 and the P-branch to the ν_5 band centre. He points out that ν_6 is a perpendicular band with a strong central Q-branch, and the appearance of the band near 800 cm^{-1} is that of a class C ($\perp r$) band for an accidental symmetric top near the prolate limit.

A few compounds NO_2Z have been studied, where X is F, Cl, (OH) and (OD), treating these latter groups as point masses. The original assignment for NO_2F by Dodd *et al.* [370] has been revised by Bernitt [129], the band at 460 cm^{-1} being attributed to $FONO_2$ and that near 570 cm^{-1} resolved into two fundamentals. The original assignment of NO_2Cl by Ryason and Wilson [1222] has been twice revised, the latest version due to Bernitt [129] being tabulated. Force constants for the in-plane modes of $NO_2(OH)$ have been calculated by Rao and Sakku [1176].

In the same way that HNO_3 has been considered as $NO_2(OH)$, we may consider HCO_3^- as $CO_2(OH)^-$. The bicarbonates of Na, NH_4, K, Rb and Cs have been studied at 20°C and −180°C, and the effect of deuterium substitution has also been investigated. Crystallographic studies suggest that the bicarbonate ion is not monomeric in the solid; a chain polymer of C_{2v} symmetry was proposed for the Na compound, and a cyclic dimer of C_{2h} symmetry for the K, Rb and Cs salts. Comparison of the Raman and infrared spectra of $KHCO_3$ shows the correlation splitting expected for such a structure; the space group is C_{2h}^5, with $Z = 4$ [X112]. In Table 6.21, the notation of the assignment is not that adopted by Novak [1083] and Le Narvor [841]; theirs is given in brackets for comparison.

Table 6.19 Infrared spectra of substituted boron halides.

Compound	Assignment and description						Ref.
	$\nu_1(A_1)$	$\nu_2(A_1)$	$\nu_3(A_1)$	$\nu_4(B_1)$	$\nu_5(B_1)$	$\nu_6(B_2)$	
	$\nu_s(XY)$	$\nu(XZ)$	$\delta_s(YXY)$	$\nu_{as}(XY)$	$\delta_{as}(YXY)$ or $\delta(YXZ)$	π	
$H^{11}BF_2$	1162	2619	(550)	1461	1402	924	892
$H^{10}BF_2$	1193	2619	(555)	1506	—	945	
$D^{11}BF_2$	1125	1959	(544)	1394	1195	787	
$D^{10}BF_2$	1147	1959	(551)	—	1199	806·5	
$H^{11}BCl_2$	762	2625	289	1100	885	900	185
		2617		1100		795	893
$H^{10}BCl_2$	(556)	2628	(170)	1115	925	917	185
	740	2617	287	1089	892	784	893
$D^{11}BCl_2$	—	1969	285·5	1005	—	645	95
$D^{10}BCl_2$	—	1969	287·5	1050	—	661	95
$H^{11}BBr_2$	595	2600	—	1042	770	780	185
		2622		1036	772	731	895
$H^{10}BBr_2$	(532)	(2606)	(158)	1051	795	797	185, 895
	—	2622	—	1048	—	—	900
$D^{11}BBr_2$	585	1960	—	870	656	625	896
$D^{10}BBr_2$	—	—	—	907	—	—	896
$^{11}BF_2Cl$	1242	696	427	1421	366	604	860
$^{10}BF_2Cl$	1286	701	—	1470	366	629	
$^{11}BFCl_2$	—	1312	—	993	—	524	
$^{10}BFCl_2$	—	1355	—	1031	—	547	
$^{11}BF_2Br$	1208	632	—	1417	—	568	
$^{10}BF_2Br$	1247	637	—	1468	—	592	
$^{11}BFBr_2$	1302	—	—	862	—	—	
$^{10}BFBr_2$	1344	—	—	901	—	—	
$^{11}BCl_2Br$	878	405	218	951	217	430	
$^{10}BCl_2Br$	916	—	—	991	—	448	
$^{11}BClBr_2$	343	918	165	827	195	404	
$^{10}BClBr_2$	—	954	—	864	—	424	
$^{11}BF_2I$	1185	610	(270)	1410	318	529	1915
$^{10}BF_2I$	1220	620	(270)	1465	318	552	
$^{11}BFI_2$	366	1295	(125)	758	(250)	403	
$^{10}BFI_2$	370	1335	(125)	794	(250)	422	
$^{11}BCl_2I$	835	378	185	937	204	400	
$^{10}BCl_2I$	871	382	(185)	984	204	416	
$^{11}BClI_2$	285	882	119	718	174	350	
$^{10}BClI_2$	(287)	917	(119)	751	(174)	(367)	
$^{11}BBr_2I$	752	249	128	816	139	355	
$^{10}BBr_2I$	788	(249)	(128)	(852)	(139)	371	
$^{11}BBrI_2$	222	788	112	711	128	329	
$^{10}BBrI_2$	(222)	820	(112)	745	(128)	345	

Table 6.20 Vibrational spectra of compounds NO_2Z.

Compound		Assignment and description						Ref.
		$\nu_1(A_1)$	$\nu_2(A_1)$	$\nu_3(A_1)$	$\nu_4(B_1)$	$\nu_5(B_1)$	$\nu_6(B_2)$	
		$\nu_s(NO_2)$	$\nu(XZ)$	δ_s	$\nu_{as}(NO_2)$	δ_{as}	π	
NO_2F		1310	822	568	1792	560	742	129
NO_2Cl		1319	793	370	1685	408	652	129
$NO_2(OH)$	ir gas	1320	886	—	1710	583	—	278a
	R liq.	1300	925	680	1675	610	765	1192a
$NO_2(OD)$	ir gas	1313	888	—	1685	543	—	278a
	R liq.	1300	915	670	1645	580	764	1192a

Table 6.21 Vibrational spectra of bicarbonates.

Compound	Assignment and description						Ref.
	$\nu_1(\nu_1)$	$\nu_2(\nu_5)$	$\nu_3(\nu_3)$	$\nu_4(\nu_6)$	$\nu_5(\nu_4)$	$\nu_6(\nu_2)$	
	ν(C—OH) or (C—OD)	$\nu_s(CO_2)$	δ_s	$\nu_{as}(CO_2)$	δ_{as}	π	
HCO_3^- dispersed in KBr	960	1338	712	1697	579	835	128
	961	1338	715	1696	581	838	1788
$NaHCO_3$	1046	1448	650	1659	700	837	
	1035	1395, 1300		1616	690		
$NaDCO_3$	1040	1340	627	1625	691	835	
	990			1600	686		
NH_4HCO_3	1043	1398	648	1600	701	832	
	1026, 994	1367, 1290	640		688		1083
NH_4DCO_3	—	1390	640	1580	692	832	
		1355	616		670		
$KHCO_3$ (ir)	1005	1400	657	1650	698	831	
		1370		1625			
(R)	1029	1448	635	1680	676	832	
$KDCO_3$	994	1390	625	1620	692	830	
				1590			
$RbHCO_3$	1000	1394	670	1615	670	834	
		1362					
$RbDCO_3$	980	1380	635	1620	635	833	841
$CsHCO_3$	990	1398	657	1625	657	833	
		1350					
$CsDCO_3$	972	1370	635	1615	635	832	

Of the bands listed under ν_2 in the hydrogen carbonates, one is due to C—O—H in-plane bending. In the deuterium carbonates, this vibration is found in the region 1040–1080 cm^{-1}. In their studies of HCO_3^- in alkali halide matrices, both Bernitt *et al.* [128] and Maguire and Rubalcava [1788] give this band as 1211 cm^{-1}. They also assign the O—H stretching mode to a band at 3390 cm^{-1} and the torsional mode to a band at 660 [128] or 667 cm^{-1} [1788]. Their assignments for the species containing ^{13}C and D, however, differ in several respects. The out-of-plane bending mode lies close to ν_1 for C—O—H and for C—O—D is found at 710–720 cm^{-1}. The stretching modes are found in the region 2750–2250 cm^{-1} (O—H) and 2100–1850 cm^{-1} (O—D). The half-widths of these band systems are ~350 and 130 cm^{-1} respectively.

Formates have been studied with reference to metal–anion interaction and to crystal structure. Nakamoto [13] has considered four possible types of bonding in metal carboxylates, and Donaldson *et al.* [373–5] have studied a fifth type involving interaction of a lone pair with the carboxylate group. The effect on the stretching frequencies associated with these interactions is given in Table 6.22. The solution frequencies are taken as a datum.

Table 6.22 Metal–anion interactions in carboxylates and their effect on the COO stretching frequencies.

Type of interaction	ν_{as}	ν_{sym}	Remarks
Ionic	None	None	—
Monodentate	+	−	Effect is said to increase as the M—O bond strength increases
Bidentate (chelate) complex	Not known		
Bidentate (bridging) complex	+	+	—
Interaction with lone pair	—	—	Effect decreases in the order Sb(III) > Bi(III) > Sn(II) > Pb(II) > Tl(I)

The effect of crystal structure on the spectra of alkali metal formates has been studied by Newman [1068], by Harvey *et al.* [616] and by Charlton and Harvey [236]. Newman's work was concerned with the correct assignment of ν_2 $(\nu_{sym}(A_1))$ and $\nu_5(\delta_{CH}(B_1))$. His polarization studies on crystals of sodium formate, with parallel radiation normal to (010), suggested $\nu_2 > \nu_5$, but the generally accepted assignment, confirmed by the deuteration studies of Fonteyne [475a] and Harvey *et al.* [616], and the study by Hammaker and Walters using NaHCOO enriched with ^{13}C, [607] is that given in Table 6.23. A factor-group analysis of NaHCOO [616, 236] predicts six infrared-($3A_u + 3B_u$) and six Raman-active modes ($3A_g + 3B_g$). The infrared active modes were observed by

Table 6.23 Fundamental bands of formates [373].

Metal	Assignment and description					
	$\nu_1(\nu(C{-}H))$	$\nu_2(\nu_{sym}(CO_2))$	$\nu_3(\delta_{sym})$	$\nu_4(\nu_{as}(CO_2))$	$\nu_5(\delta(CH))$	$\nu_6(\pi)$
Li	2857	1387 1374	791	1587	1410 1399	1072
Na	2833	1364	775	1592	1381	1071
K	2833	1350	775	1590	1385	1064
Rb	2801	1350	774	1592	1383	1066
Cs	2801	1351	775	1595	1379	1099
Tl^{I}	2786	1355 1333	758	1590	1406 1387	—
NH_4	2825	1364	775	1592	1406	1072
Be	2907	1383	800 735	1613	1399	—
Mg	2907	1385 1374 1366	761	1603	1404 1391	1081
Ca	2874	1370 1351	803 800 781	1581	1408 1395	1075 1064
Sr	2874	1374 1364 1351	786, 781 763, 719	1572 1553	1404 1399	1087
Ba	2865	1362 1357	769 767 759	1585 1577	1403 1387	1075 1066
Pb^{II}	2833	1348 1333	763 758	1577	1389 1372	1075 1066
Sn^{II}	2874	1339 1311	781	1563	1385 1370	1064
Al^{III}	2907	1387 1366	792 774	1613	1420 1406	1121 1094
Ga^{III}	2849	1383	784 772 730	1597	—	1050
In^{III}	2874	1383 1368	800 794 728	1587 1582	1404 1391	1172
Sn^{IV}	2841	1368	719	1608	1404	—
Bi^{III}	2841	1333	788 772	1550	1399	1075
Sb^{III}	2841	1307 1325	769 740	1541	—	1026

Charlton and Harvey [236] at 2831, 1376, 782 cm^{-1} (A_u) and 1595, 1356, 1066 cm^{-1} (B_u). Three ir active translational modes ($A_u + 2B_u$) are also predicted and were observed at 57, 107 and 138 cm^{-1}. The alkaline earth formates have been investigated by Harvey *et al.* [616] and by Schutte and Buijs [1256]. The Sr, Ba and Pb compounds crystallize with the space group D_2^4, $Z = 4$, and the formate ions on two sets of fourfold general positions. These, according to Schutte and Buijs, give rise to the doubling of the fundamental frequencies. Only ν_3 shows any further correlation splitting. Two forms of calcium formate exist. The α-form is orthorhombic, D_{2h}^{15}, $Z = 4$, with all atoms on general positions, and the β-form tetragonal, D_4^4, $Z = 8$. Buijs and Schutte observed only the in-plane modes in this compound, finding (in the notation of Table 6.23) $\nu_1 = 2857$, $\nu_2 = 1359$, $\nu_3 = 790$, $\nu_4 = 1600$, $\nu_5 = 1383\ \text{cm}^{-1}$. The rare-earth formates have been studied by Saralidze *et al.* [1230].

The acetate group may also be treated as a tetratomic system, considering the CH_3 group as a point mass. Some frequencies of the $RCOO^-$ group are tabulated correspondingly to those of the formate group, and the same considerations apply to the determination of the type of metal–anion interaction. Edwards and Hayward [428] studied some anhydrous transition metal acetates and found $\nu_{as} - \nu_{sym} \sim 180\ \text{cm}^{-1}$ for three compounds which may be bridged. The hydrated Cu^{II} and Cr^{II} acetates are in fact $M_2^{II}(CH_3COO)_4,2H_2O$, and have been shown to be bridged [286].

Table 6.24 Some vibrational bands of acetates.

Metal	Assignment (4-atom system) and description				Ref.
	ν_1	ν_2	ν_3	ν_4	
	ν_{C-CH_3} $\nu_4(A_1)$	$\nu_{sym}(CO_2)$ $\nu_3(A_1)$	COO bending $\nu_5(A_1)$	$\nu_{as}(CO_2)$ $\nu_8(B_1)$	
Anhydrous					
Li	935	1441	661	1588, 1579	
Na	923	1435	645	1574	581
K	919	1402	644	1573	
Rb	925	1412	647	1567	374
Ag	—	1405	—	1520	
Mn^{II}	—	1410	—	1585	
Co^{II}	—	1405	—	1590	428
Ni^{II}	—	1415	—	1600	
Cu^{II}	—	1420	—	1595	
Zn	954	1446	694	1545	581
		1405			
Al	989	1472	645	1595	
		1417			
Sc	960	1462	673	1535	
		1412	639		
Y	955	1441	673	1536	
	943	1390	647		582
La	965	1465, 1449	668	1609	
	955, 945	1430, 1405	636	1525	
Ce	964	1466, 1449	668	1605	
	955, 943	1433, 1397	635	1523	
Nd	957	1455	670	1601, 1556	
	942	1436, 1408	640	1538, 1520	

Table 6.24 (*cont.*).

Metal	Assignment (4-atom system) and description ν_1	ν_2	ν_3	ν_4	Ref.
	ν_{C-CH_3} $\nu_4(A_1)$	$\nu_{sym}(CO_2)$ $\nu_3(A_1)$	COO bending $\nu_5(A_1)$	$\nu_{as}(CO_2)$ $\nu_8(B_1)$	
Tl^{I}	919	1412	648	1560	374
Pb^{II}	935	1410	665	1548	
Sn^{II}	930	1395	661	1530	
Bi^{III}	929	1410	640	1522	
Be	982	1480	650	1624	581
Mg	950	1427	666, 658	1580	
Ca	953	1452	668	1559	
	942	1411		1538	
Sr	946	1426	659	1558	
	934				
Ba	929	1412	664	1557	581
			647		
Cd	950	1446	685	1543	
		1410			
Hydrated					
Li, $2H_2O$	—	1471, 1453	693	1590	581
		1434, 1420	653		
Mg, $4H_2O$	949	1444	667	1535	
		1422	659		
Zn, $2H_2O$	955	1456	695	1550	
Cd, $2H_2O$	939	1409	670	1555	
La, $\frac{3}{2}H_2O$	948	1443	667	1557	582
		1398			
Nd, $4H_2O$	960	1459	667	1557	
	942	1448		1543	
Y, $4H_2O$	979, 968	1457	677	1543	
	956, 943	1406			
Cu^{II}, $1H_2O$	960	1422	693	1604	286
Cr^{II}, $1H_2O$	942	1423	698	1586, 1576	

According to van Niekerk and Schoening [X156], Co^{II} acetate is chelate. Some complexes of XO_2^{++} where X = U, Np, Am, Pu have been studied by Jones [714].

Far infrared studies have been made on the dihydrates of Li and Zn acetates by Ananthanarayanan [54], and of the acetates of Na, Pb, Zn, Cu, Co and UO_2 by Lorenzelli and Gesmundo [882]. In the first case some lattice modes were identified, and in the second, the C—C torsional mode was found in the region 160–190 cm^{-1}.

Seidel [1268] has found for $CS_2Se^{=}$ ν_{as}(C—S) 933, ν_{sym}(C—S) 844, ν(C—Se) 458, out of plane bending 486 cm^{-1}.

For $CSCl_2$ [1706], bands have been assigned to $\nu C = S$ (1135), ν_{sym} CCl_2 (500), δ_{sym} CCl_2 (295), ν_{as} CCl_2 (813), δ_{as} CCl_2 (305) and out-of-plane bending (471).

(c) Molecules of the formula X_2Y_2

Of the seven structures observed for molecules of this type, six are planar. The exception is the H_2O_2 'book' structure, of C_2 symmetry. Few inorganic molecules have the linear structure; the most frequently studied species is Hg_2Cl_2. This has been considered both as an isolated molecule of $D_{\infty h}$ symmetry, and also in terms of factor-group analysis. Hg_2Cl_2 crystallizes in the tetragonal system, space group D_{4h}^{17}, with $Z = 1$, giving rise to $2A_{1g} + E_g(\text{R}) + A_{2u} + E_u(\text{ir})$ modes. The Raman active modes are agreed to be 270 and 167 cm^{-1} (A_{1g}) and 42 cm^{-1} (E_g); these correspond respectively to ν_{sym}(HgCl), ν(Hg—Hg), and a bending mode. Three frequencies have been assigned as ir active fundamentals, at 260, 110 and 33 cm^{-1}. Adams [29], Goldstein [551] and Cooney [285] all agree that 260 cm^{-1} is a fundamental; this is an A_{2u} mode corresponding to ν_{as}(HgCl), and Goldstein assigns the 110 cm^{-1} band to the E_u bending mode. Hadni [600] gives the less probable assignment (having regard to the molecular motions involved) of 110 and 33 cm^{-1} as fundamentals. Goldstein gives the ν_{as}(Hg—X) modes as 180 cm^{-1} in the bromide and 135 cm^{-1} in the iodide, and Cooney gives for the A_{1g} modes, 221 and 132 cm^{-1} in the bromide and 192, 110 cm^{-1} in the iodide. Infrared and Raman studies on C_2N_2 give ν_{sym}(C—N), 851; ν(C=C), 2329; ν_{as}(C—N), 2157; δ_{sym}, 507; δ_{as}, 226 cm^{-1} [297a, 824].

The cis- and trans-planar forms, of C_{2v} and C_{2h} symmetry respectively, are both known; N_2F_2 and the hyponitrite ion $N_2O_2^{=}$ exist in both forms. Fateley *et al.* [453] have assigned the $(NO)_2$ dimer in both forms, but consider the cis-form to be the more stable. N_2F_2 was originally assigned as

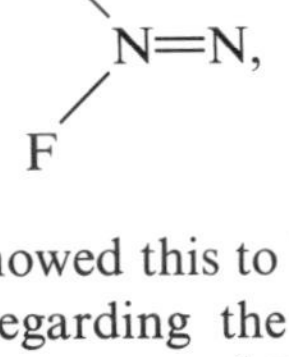

but microwave studies [15] showed this to be incorrect.

One disagreement exists regarding the assignments in trans-N_2F_2. Sanborn has given ν(N=N) as 1636 cm^{-1} [1229], but neither Shamir and Hyman [1275] nor King and Overend [760] observed a band at this frequency; they agree that the correct value is 1522 cm^{-1}. The three ir active mode frequencies are agreed by all workers. For trans-$N_2O_2^{=}$ the disagreement concerns $\nu_3(A_g)$, δ_{sym}. Millen *et al.* [945] gave this as

Table 6.25 Vivrational frequencies of cis-X_2Y_2 molecules.

Compound		Assignment and description						Ref.
		$\nu_1(A_1)$	$\nu_2(A_1)$	$\nu_3(A_1)$	$\nu_4(A_2)$	$\nu_5(B_1)$	$\nu_6(B_1)$	
		ν_s(N—Y)	ν(N—N)	δ_{sym}	τ	ν_{as}(N—Y)	δ_{as}	
N_2F_2	R, gas	884	1526	341	550?	945	729	760
$N_2O_2^=$	ir, Na salt	830	1314	584	—	1047	330	555
$(NO)_2$	ir, R liq.	1862	262	167	—	1770	196	453
	ir, CO_2 matrix	1862	—	—	—	—	—	453
	ir, Ar matrix	1866	—	—	—	1776	—	453

485 cm^{-1}, calculated from a combination frequency. Both Rauch and Decius [1188] and McGraw *et al.* [934] prefer 696 cm^{-1}. There is some variation in the observed frequency for $\nu_2(A_g)$, ν(N=N). Nachbaur [1043] attributes a band at 880 cm^{-1} in the ir spectrum of $Na_2N_2O_2$ to N—O stretching, but no-one else has found this.

Table 6.26 Vibrational frequencies of trans-X_2Y_2 molecules.

Compound	Assignment and description						Ref.
	$\nu_1(A_g)$	$\nu_2(A_g)$	$\nu_3(A_g)$	$\nu_4(A_u)$	$\nu_5(B_u)$	$\nu_6(B_u)$	
	ν_{sym}(N—Y)	ν(N—N)	δ_{sym}	τ	ν_{as}(N—Y)	δ_{as}	
N_2F_2	1010	1522	600	363·5	990	423	1275
$N_2O_2^=$	(1121)	(1419)	(696)	492	1031	371	1188, 934
$(NO)_2$	—	—	—	—	1740	—	453

Ring dimers of HCl and lithium halides have been studied; these have been assigned C_{2h} and D_{2h} symmetry respectively. Katz [730] has studied $(HCl)_2$ and $(DCl)_2$ in rare gas matrices and observed one of the B_u stretching modes, finding a steady decrease in frequency with increasing atomic mass of the matrix material. Observed values, for $(HCl)_2$ and $(DCl)_2$ respectively, are 240, 182 (Ne); 232, 173 (Ar); 208·5, 163 (Kr); 185, 147 (Xe). The vibrations of such a system are $3A_g(R) + 1A_u + 2B_u(ir)$. For the alkali halide dimers, the vibrations are $2A_g + 1B_{1g}(R) + B_{1u} + B_{2u} + B_{3u}(ir)$, the stretching modes being $A_g + B_{2u} + B_{3u}$. Berkowitz [125–6] has calculated all frequencies for X_2Y_2 (X = Li, Na, K; Y = F, Cl, Br, I) and for $(RbCl)_2$ and $(CsCl)_2$, but the agreement with experiment is not very good.

Table 6.27 Vibrations of alkali halide dimers.

Compound		Assignment and description							Ref.
		$\nu_4(B_{1u})$	$\nu_5(B_{2u})$			$\nu_6(B_{3u})$			
		π	stretching			stretching			
Li_2F_2	ir vapour	—		460			640		769
		300		570			655		1344
			Ar	Kr	Xe	Ar	Kr	Xe	
$^7Li_2F_2$	ir matrix	—	537	542	532	626	626	618	861
Li_2F_2	ir matrix	—	550			646			
			559	555	547	651	646	638	
$^6Li_2F_2$	ir matrix	—	569	573	565	661	661	652	
$^7Li_2F_2$		268	—	—	—	—	—	—	489
$^6Li_2F_2$		284	—	—	—	—	—	—	
Li_2Cl_2	ir vapour	—		335			460		769
$^7Li_2Cl_2$	ir matrix	174		—			—		489
$^6Li_2Cl_2$		186		—			—		
Li_2Br_2	ir vapour	—		295			413		769
$^7Li_2Br_2$	ir matrix	150		—			—		489
$^6Li_2Br_2$		162		—			—		
Li_2I_2	ir vapour	—		248			375		769

Both Linevsky [861] and Snelson [1344] have suggested the possibility of a planar trimer, and Snelson has ascribed bands at 230, 267, 509 and 736 cm^{-1} to this species. Abramowitz *et al.* [20] have suggested that Li_2F_2 might be linear, but Freiberg *et al.* [489] have pointed out that the bands near 250 cm^{-1} in this molecule have an intensity ratio of 1 : 3 : 3 : 1 rather than the 1 : 1 : 1 : 1 ratio demanded by the linear structure.

(d) Molecules of the formula WXYZ

There are three known shapes, all with C_s symmetry. In the compounds where the XYZ group is OBO, NCO, OCN or SCN, this group is linear, so the geometry is the same as that in HN_3. Where XYZ is ONO or NSO, the group is nonlinear and the molecule as a whole is either W-shaped (trans) or U-shaped (cis). HONO exists in both shapes, HNSO only in the U-shape.

In XNSO (X = halogen) the modes listed as ν_2 and ν_3 are approximately described as ν(N—S), ν(S—O) respectively. The P—R branch separations in the ir vapour spectra of these compounds indicate that they, like HNSO, have the *cis* arrangement [1695]. A somewhat different assignment for ClNSO had been proposed previously [1903].

Table 6.28 Vibrations of planar molecules, WXYZ.

Compound		Assignment and description						Ref.
		ν_1	ν_2	ν_3	ν_4	ν_5	ν_6	
		ν(W—X)	stretching		in-plane bending		τ	
HNCO	ir g	3531	2274	1327	797	572	670	645a
HNCS	ir g	3539	1980	851	470	542	615	83b
DNCS	ir g	2644	1944	854	374	483	566	83b
HOCN	ir matrix*	3506	2294	1098	1241	460	438	694b
DOCN	ir matrix*	2590	2292	1093	957	437	—	694b
ClSCN	ir soln†	520	2162	678	—	353	—	1075b
BrSCN		451	2157	676	—	369	—	
ISCN	ir soln†	372	2130	700	—	362	—	
trans HONO	ir g	3590	1700	793	1267	598	545	606a, 713a 379
cis HONO	ir g	3425	1639	855	1370	525	638	
trans DONO	ir g	2658	1695	739	1015	592	416	
cis DONO	ir g	2537	1620	817	1086	517	—	
HNSO	ir g	3345	1261	1090	911	453	759	1198a
DNSO	ir g	2480	1257	1055	757	410	594	
HOBO	ir g	3680	2030	—	1420	650	—	1507a
FNSO	liq	825	995	1230	600	228	395	1695
ClNSO	liq	526	989	1221	672	187	359	
		676	995	1248	532	187	359	1903
BrNSO	liq	451	1000	1214	624	161	342	1695
INSO	liq	372	1028	1247	602	154	330	

* The matrix materials used were Ar at 4°K, N_2 at 14° and 20°K, and CO at 20°K. There are small differences between the sets of data in the various materials.

† The solvents were C_2HCl_5 in the region 4000–1300 cm^{-1} and CS_2 from 1300 to 300 cm^{-1}.

6.3 Pyramidal molecules

(a) Molecules with the formula XY_3

The compounds XH_3, XD_3 and XT_3 have been studied, where X = N, P, As and Sb. The H_3O^+ ion has also been investigated in a variety of acid media. The data are fairly complete and the only problem in interpretation is the difficulty of distinguishing the two stretching modes, which lie very close together in the P, As and Sb compounds and in H_3O^+. Following the assignment in $(H_3O)_2PtCl_6$ [534a], the higher of the two stretching frequencies has been listed as ν_1.

A large number of halides has been studied, and, in general, the various observations and assignments are in good agreement. The only serious

Table 6.29 Vibrational frequencies of pyramidal hydrides.

Compound		$\nu_1(A_1)$	$\nu_2(A_1)$	$\nu_3(E)$	$\nu_4(E)$	Ref.
$^{14}NH_3$	ir gas	3337	968	3443	1626	123b–d
			932		1621	1000a
	ir, R liq.	3294	1054	3380	1628	762a
	ir solid (−82°C)	3223	1058	3375	1646	1189
$^{15}NH_3$		—	961	3335	1625	996
			926·3			
$^{14}ND_3$	ir gas	2421	749	2564	1191	123b–d
			746			
	ir solid (−82°C)	2318	817	2507	1196	1189
			815	2500		
$^{14}NT_3$	ir gas	2016	647	2163	1000	1396
PH_3	ir gas	2323	990	(2328)	1121	926
	calc.	2452	1041	2457	1154 }	500
PD_3	calc.	1761	756	1766	822 }	
PT_3		1398	623	1401	668	1396
AsH_3	ir vap.	2116	907	2123	1003	926
			904·4			
	ir solid	2092	894	—	979	633
AsD_3	ir vap.	1523	—	1529	714	926
AsT_3		1256	553	1258	590	1396
SbH_3	ir vap.	1891	782	1894	831	625, 1395
	ir solid	1843	769	—	812	633
SbD_3		1359	561	1362	593	625
SbT_3			466	1125	486	1395
$OH_3^+ClO_4^-$	R cryst.	3285	1182	—	1577	1423
NO_3^-		2780	1135	—	1680 }	1238
HSO_4^-		2840	1160	—	1620 }	
F^-		3150	1150	2468	1705 }	
Cl^-		3235	1150	2590	1700 }	
$OD_3^+Cl^-$		2350	1015 or 785	2245, 2000	1255 }	467
$OH_3^+Br^-$		3250	1150	2610	1705 }	
I^-		3380	1160	2635	1705 }	
$(OH_3)_2^+PtCl_6^{2-}$		3226	1070	2825	1695	534a

discrepancies relate to ν_4 of PF_3, which is variously stated as 344 cm^{-1} [1519] and 532 cm^{-1} [595], and to BiI_3. A few complexes $M^{IV}Y_3^-$ have been studied, and there is evidence to suggest that the tetrahalides of S, Se and Te are ionized in the solid state, giving $XY_3^+Y^-$.

NCl_3 has also been studied by Hendra and Mackenzie [641] whose results agree with those of Delhaye. A number of other workers have obtained concordant results on NF_3 [1241, 1519].

The structure of AsI_3 has been determined by Trotter [X154]; the space group is $C_{3i}^2(R\bar{3})$, and $Z = 2$. Hooper and Jakes [676a] have carried out

Table 6.30 Vibrational frequencies of pyramidal halides.

Compound		ν_1	$\nu_2(A_1)$	$\nu_3(E)$	$\nu_4(E)$	Ref.
$GeCl_3^-$	R soln.	320	162	253	139	360
$SnCl_3^-$	R soln.	297	128	256	103	1542
	R melt	268	—	220	105	273
$SnBr_3^-$	R soln.	211	83	181	65	1542
$PbCl_3^-$	R solid, $KPbCl_3$	218	170	244	203	75
NF_3	R vap.	1035	649	910	500	1276
	R liq.	1050	667	905	515	1099
	ir vap.	1031	642	907	497	
NCl_3	R soln. CCl_4	536	347	638	256	345
PF_3	ir vap.	892	487	860	344 or 532	595, 1519
PCl_3	ir vap.	504	252	482	198	880, 881, 884
PBr_3	R liq.	380	162	400	116	325
PI_3	R soln. C_6H_6	303	111	325	79	1363
AsF_3	ir liq.	715	343	644	274	184
$AsCl_3$	ir liq.	412	194	307	155	325
$AsBr_3$	ir, R liq.	(284)	128	275	98	946
AsI_3	R soln. CS_2	216	94	221	70	1363
$SbCl_3$	R vap.	360	165	320	134	1512
$SbBr_3$	R soln. CS_2	205	105	242	86	440
SbI_3	ir solid	177	89	146	71	906
$BiCl_3$	R	288	130	242	96	599
$BiBr_3$	ir solid	196	104	169	90	906
BiI_3	ir solid	145	90	115	71	
		—	111	174	77	599
$SCl_3^+AsF_6^-$	R solid	519	284	543	214	1239
$SeCl_3^+AsF_6^-$	R solid	437	200	390	168	
$TeCl_3^+AsF_6^-$	R solid	412	170	385	150	
$SeCl_3^+AlCl_4^-$	R solid	416	234	395	186	520–2
$TeCl_3^+AlCl_4^-$	R solid	391	185	367	139	
$SeCl_4$	R solid	388	206	346	164	
$TeCl_4$	R solid	374	191	342	143	
	ir solid	364	186	353	151	30
	ir soln. C_6H_6	362	181	347	—	
$TeBr_4$	ir solid	240	110	222	87	567
TeI_4	ir solid	140	88	172	62	

a factor-group analysis, finding the internal vibrations to be $2A_g + 2E_g(\mathrm{R}) + 2A_u + 2E_u(\mathrm{ir})$. The correlation with the internal vibrations is $\nu_1(A_1)$, $216 \rightarrow 185(A_g) + 201(A_u)$; $\nu_2(A_1)$, $94 \rightarrow 75(A_g) + 102(A_u)$; $\nu_3(E)$, $221 \rightarrow 205(E_g) + 216$ or $226(E_u)$; $\nu_4(E)$, $70 \rightarrow 50(E_g) + 74(E_u)$. Tramer [1438] has studied the Raman spectrum of $SbCl_3$, finding nine bands between 15 and 95 cm^{-1}, and has observed the splitting of the E internal modes (ν_3, 320, 315; ν_4, 150, 139). The group VI tetrahalides in the vapour phase have C_{2v} symmetry, arising from the occupation of two apical and two equatorial positions of a trigonal bipyramid; this leads to the selection rules $\Gamma_{int} = 4A_1 + 2B_1 + 2B_2$ (all ir, R) $+ 1A_2$(R). No interpretation of the spectra of the solids has been made in terms of this analysis, and even in benzene solution $TeCl_4$ appears to behave as $TeCl_3^+Cl^-$. As long ago as 1924, Voigt and Biltz [X161] showed that molten $TeCl_4$ is an electrical conductor, and although Schmitt and Zeil [X131] have interpreted NQR data to mean that solid $TeCl_4$ is an undissociated molecule, they are alone in this view. Adams and Lock [30] have observed a low-intensity band at 240 cm^{-1} which might arise from a weak bond Cl_3Te^+---Cl^-. The corresponding interaction in $TeCl_3^+$, $AlCl_4^-$ is said to be much weaker [567]. Greenwood has observed splitting in ν_3 of $TeCl_4$ (352, 344 cm^{-1}).

Many compounds in which Y is oxygen are known; X may be an element of group V (P, As); group VI (S, Se, Te); group VII (Cl, Br, I); or group VIII, where XeO_3 has been studied.

Most of the spectral data accumulated is due to Rocchiciolli; she has carried out studies on sulphites, selenites, chlorates and bromates. Dasent and Waddington [314–15] have examined a number of simple and complex iodates. Of the compounds whose spectra are listed in Tables 6.32–6.37, seventeen are of known crystal structure; optical crystallographic data are available for two other compounds. Only among the bromates (seven out of fifteen) and the iodates (six out of fifteen) has structural and spectroscopic work been carried out on fair-sized groups of isostructural compounds. The effect of crystal structure on the spectra of these compounds is confined to the splitting of the *E* vibrations when the anion occupies a site of less than twofold symmetry. The crystal structure of ten of the compounds of known structure leads to the prediction that ν_3 and ν_4 should be split in these cases; in fact the prediction is certainly borne out in only five cases, those of $AgClO_3$, $AgBrO_3$, $Sr(ClO_3)_2$, $Ba(BrO_3)_2$, H_2O and $Pb(IO_3)_2$. Only ν_3 of lead iodate was observed. Conversely, there are seven compounds for which neither the site nor the point symmetry allows splitting of the degenerate vibrations, but only three of these, $NaBrO_3$, Na_2SO_3 and $Co(BrO_3)_2$, $6H_2O$, display the predicted spectra. The situation is particularly complicated here, since there are compounds isostructural with these (see Table 6.31) which do not give the predicted spectra. $LiIO_3$ is a special case; structural work has shown [X184] that this compound does not contain discrete IO_3^- ions but IO_6 octahedra. Since the I atom occupies a D_3 site, there should be seven ir active vibrations ($2A_2 + 5E$) and eight Raman active ($3A_1 + 5E$). The study did not go below 400 cm^{-1} and only two vibrations were observed.

Newman and Powell in their study of the double salts do not actually assign the bands; the table follows Rocchiciolli's assignments of the simple sulphites, in which $\nu_1 \geqq \nu_3$.

The two asterisked bands in the Zn and Cd double salts are broad and

Table 6.31 Structural data for compounds whose spectra have been determined.

Compound	Optical or crystal class	Z	Space group	Anion: Site symmetry	Anion: Point symmetry	Ref.
Na_2SO_3	Hexagonal	2	C_{3i}^1	C_3	C_{3v}	X185
$MgSeO_3, 6H_2O$	Rhombohedral	1	C_3^4	C_3	C_{3v}	X45
†$NaClO_3$	Cubic	4	T^4	C_3	C_{3v}	X182
*$KClO_3$ (LT form)	Monoclinic	2	C_{2h}^2	C_S	—	X183
†$AgClO_3$	Tetragonal	8	C_{4h}^5	C_S	C_S	X110
$KBrO_3$	Rhombohedral	1	C_{3v}^5	C_{3v}	—	X183
$Ba(BrO_3)_2H_2O$	Monoclinic	4	C_{2h}^6	C_1	—	X83–6
‡$Zn(BrO_3)_2, 6H_2O$	Cubic	4	T_h^6	C_3	C_{3v}	X166
$LiIO_3$	Hexagonal	2	D_6^6	D_3	see text	X184
§$NaIO_3$	Orthorhombic	4	D_{2h}^{16}	C_S	C_S	X109
$Pb(IO_3)_2$	Orthorhombic	4	D_{2h}^{10}	—	—	X144
$Sr(ClO_3)_2$	Orthorhombic	—	—	—	—	WN109
KIO_3	Monoclinic	—	—	—	—	WN108

* $TlIO_3$ is isostructural with this compound.
† Isostructural with corresponding bromate.
‡ The Mg and Co salts are isostructural with this compound.
§ The NH_4 salt is isostructural with this compound.

unresolved. The two frequencies assigned to ν_1 in Ag_2SO_3 by Rocchiciolli raise a problem; it seems possible that the band at 953 cm^{-1} is a component of ν_3, as it would be reasonable to expect a splitting in both ν_3 and ν_4. Another possibility is that one of the two bands is $2\nu_4$ (or $\nu_4' + \nu_4''$). There is no immediately obvious explanation for the splitting of ν_2 in some of the double salts.

Two other studies have been made of Na_2SO_3 in various states; in all cases ν_1 and ν_3 were resolved, in contrast to Newman and Powell's results tabulated below.

Loehr and Plane [868] have investigated alkaline solutions containing As(III), and attributed bands at 752(A_1) and 680(E) to stretching, and at 340 cm^{-1} to bending.

Duval and Lecomte [419] have observed bands in the spectrum of the PO_3^- ion at 1150, 1080, 920, 820, 590 and 460 cm^{-1}.

A Raman study of aqueous solutions of Na_2SeO_3 has been made by Siebert [1295–6]; the data given in Table 6.34 is taken from the infrared study of the solids as Vaseline mulls, by Rocchiciolli [1206].

The Mg, Ca, Ba, Cu, Ag and Pb compounds also have one or two weak bands between ν_2 and ν_3; Rocchiciolli has suggested that these are due

Table 6.32 Vibrational frequencies of sulphites.

Compound	ν_1	ν_2	ν_3	ν_4	Ref.
Na_2SO_3	970	633	970	498	1066
K_2SO_3	965	619	965	478	
$CaSO_3$	980	649	941	517	
				487, 445	
$CaSO_3, 2H_2O$	970	655	945	524	
				495, 452	
$SrSO_3$	958	640	930	519, 497	
$BaSO_3$	951	630	930	508, 493	
Ag_2SO_3	998	631	915	476, 494	1205
	953				
$CdSO_3$	965	649	937	527	
			892	475	
$PbSO_3$	969	625	897	483	
				472	
UO_2SO_3	972	633	921	483	
			894		
$(NH_4)_2Mg(SO_3)_2, H_2O$	973	679	918	444	
$(NH_4)_2Zn(SO_3)_2, 2H_2O$	952–857*	650	*	486	
$(NH_4)_2Cd(SO_3)_2$	958–870*	646	*	479	
		615			
$(NH_4)_2Fe(SO_3)_2, 2H_2O$	959	663	895	487	1066
$(NH_4)_2Co(SO_3)_2, 2H_2O$	957	672	890	490	
		616			
$(NH_4)_2Ni(SO_3)_2, 2H_2O$	958	674	880	495	
		632			
$(NH_4)_2Mn(SO_3)_2$	964	655	907	482	

* This absorption is due to both ν_1 and ν_3 but the individual bands cannot be identified.

Table 6.33 The vibrational spectrum of Na_2SO_3.

State		ν_1	ν_2	ν_3	ν_4	Ref.
Saturated aqueous solution	R	967	620	933	469	441
		966	612	952	471	1326
Solid	ir	1002	632	954	—	441
	R	983	632	947	494	1326
Solid (nujol mulls)	ir	1010	633	961	496	441
Calculated values for $^{34}SO_3$		972	623	937	457	

either to an impurity or to a combination between a lattice mode (175–220 cm^{-1}) and ν_4. Weak bands also occur above 830 cm^{-1} in the spectra of the Mg, Ca, Cu and Zn compounds.

While ν_1 is greater than ν_3 in sulphites and selenites (where the bands are distinguishable) the reverse situation obtains where the central atom

Table 6.34 Vibrational frequencies of selenites.

Compound	ν_1	ν_2	ν_3	ν_4
Na_2SeO_3	788	449	740	387
$MgSeO_3, 6H_2O$	806	457	740	412
				396
$CaSeO_3, H_2O$	784	430	713	418
				375
$BaSeO_3$	802	453	765	419
			737	376
$CuSeO_3$	774	461	714	427
				353
Ag_2SeO_3	790	441	727	386
$ZnSeO_3, H_2O$	799	455	767	—
$PbSeO_3$	790	457	759	442
			734	394

is a halogen. On the whole, the chlorates and bromates seem to give well-resolved spectra, splittings of ν_3 and ν_4 being observed in the majority of cases. All the data quoted are taken from the work of Rocchiciolli on chlorates and bromates; the infrared spectra are those of the solids as Vaseline mulls. Rocchiciolli's interpretation of the splitting of ν_3 in $NaClO_3$ as due to anion distortion is disputed by Hollenberg and Dows [671]. The external modes of $NaClO_3$ and $KClO_3$ have been studied by Shantakumari [1277, 1279]. For $NaClO_3$ six Raman lines have been observed, at 70, 83, 103, 123, 131 and 179 cm^{-1}. $KClO_3$ has the space group C_{2h}^2, with $Z = 2$, and nine Raman active lattice modes are predicted. These are $4A_g + 2B_g$ (translational) and $A_g + 2B_g$ (rotational). Lines were observed at 54, 82, 98, 127 and 145 cm^{-1}. Ten Raman active lattice modes of $Ba(ClO_3)_2, H_2O$ have been observed by Shantakumari [1278].

Raman studies on aqueous solutions of the bromates show that their spectra are very similar to those of the chlorates. Again the most comprehensive spectroscopic study is that of Rocchiciolli [1207], of the infrared spectra as Vaseline mulls. A feature not found in the chlorates is a band in the region 460–490 cm^{-1} of the dihydrates of Mg, Co, Ni and Zn bromates; this was attributed to either a lattice mode or metal–oxygen stretching. No analogous band appears in the spectra of the anhydrous, monohydrated or hexahydrated salts. The assignment to ν_1 and ν_3 is not given explicitly by Rocchiciolli; in the table, the lowest frequency band in the stretching region is assigned to ν_1, in line with the Raman studies. It is possible that one band in this region is $\nu_2 + \nu_4$.

Dupuis [405] gives different values for the stretching frequencies in $Ba(BrO_3)_2H_2O$, and the interpretation is also different; she has $\nu_1 =$

Table 6.35 Vibrational frequencies of chlorates [1204].

Compound	ν_1	ν_2	ν_3	ν_4
$NaClO_3$	933	619	989	478
			968	
$KClO_3$	933	619	989	479
			968	
$Mg(ClO_3)_2$	915	609	975	499
			958	478
$Ca(ClO_3)_2$	938	626	984	493
			959	478
$Sr(ClO_3)_2$	921	610	1028	505
			951	471
$Ba(ClO_3)_2$	921	605	972	503
			955	483
$Cr(ClO_3)_3$	915	610	971	495
			954	481
$Nd(ClO_3)_3$	912	607	970	498
			952	478
$Co(ClO_3)_2$	892	610	989	498
			950	479
$Ni(ClO_3)_2$	935	609	987	478
			965	
$Cu(ClO_3)_2$	935	609	988	480
			967	458
$AgClO_3$	935	612	987	493
			967	470
$Pb(ClO_3)_2$	935	602	987	499
			966	467
$Zn(ClO_3)_2$	934	607	988	498
			965	478
$Cd(ClO_3)_2$	934	607	987	501
			968	478

815 cm^{-1}; $\nu_3 = 790, 780\ cm^{-1}$. On heating, an anhydrous compound was obtained ($\nu_1 = 815$; $\nu_2 = 450$; $\nu_3 = 790$; $\nu_4 = 400, 360\ cm^{-1}$) and on further heating a more complex spectrum was obtained, which Dupuis considered as due to a second anhydrous form; though it has been suggested that the spectrum at this stage is due to the formation of perbromate ($4BrO_3^- \rightarrow 3BrO_4^- + Br^-$). The values are $\nu_1 = 845$; $\nu_2 = 440, 430$; $\nu_3 = 790$; $\nu_4 = 390, 380, 360\ cm^{-1}$.

The iodates have mainly been studied in the stretching region; a band in the region 400–450 which appears in the majority of the spectra is listed as ν_2 but could be M—O stretching. In some cases a definite assignment to M—O stretching has been given and is tabulated as such. There are three bands in the stretching region in most cases; this may arise from the breakdown of ν_3 or the extra band may be $2\nu_2$.

Table 6.36 Vibrational frequencies of bromates [1207].

Compound	ν_1	ν_2	ν_3	ν_4	Other bands
$NaBrO_3$	802	444	826	366	
$KBrO_3$	780	429	825 792	358	
$AgBrO_3$	762	435	799 788	386 354	
$Ba(BrO_3)_2, H_2O$	784	435	816 810	387 361	
$Sr(BrO_3)_2, H_2O$	776	442	831 820	398 367	
$Pb(BrO_3)_2, H_2O$	798	442	829	364	
$Co(BrO_3)_2, 6H_2O$	778	418	797	357	
$Ni(BrO_3)_2, 6H_2O$	779	430	820 796	357	
$Mg(BrO_3)_2, 6H_2O$	807	432	823	399 356	
$Zn(BrO_3)_2, 6H_2O$	805	422	829 823	358	
$Cu(BrO_3)_2, 6H_2O$	799	—	823	—	
$Ni(BrO_3)_2, 2H_2O$	779	429	829	358	486
$Mg(BrO_3)_2, 2H_2O$	775	—	849 816	370	465
$Co(BrO_3)_2, 2H_2O$	777–88	425	823	360	465
$Zn(BrO_3)_2, 2H_2O$	768	—	823 806	—	470

The structural evidence suggests that the Na, NH_4 and Tl compounds could show splitting of ν_3, but only the Na salt shows three bands in the stretching region. Dasent and Waddington have assigned the doublet to ν_1.

Both the infrared [1337] and Raman [259] spectra of XeO_3 have been recorded. XeO_3 is orthorhombic, space group D_2^4, $Z = 4$, with all atoms on general positions [X152]. The expected splitting of ν_3 and ν_4 was not observed, the infrared spectrum of the solid being ν_1, 770; ν_2, 311; ν_3, 820; ν_4, 298 cm^{-1}. The corresponding Raman values for a 2M aqueous solution are 780, 344, 833, 317 cm^{-1}. The pyramidal XY_3 group forms far fewer complexes than the planar XY_3 group. Among the pyramidal XY_3 groups, the most extensive investigations are those on sulphito complexes; a few compounds which may be iodato complexes have also been studied. Monodentate sulphito complexes bound through oxygen and through sulphur have been found. Binding through oxygen lowers the symmetry of the group from C_{3v} to C_s, thus splitting the two degenerate vibrations; a lowering of one of the S—O stretching vibrations is also expected [1066]. Conversely, on binding through sulphur, the C_{3v} symmetry is

Table 6.37 Vibrational spectra of iodates.

Compound		ν_1	ν_2	ν_3	ν_4	ν_{M-O}	Ref.
IO_3^-	aqueous soln.	779	390	826	350	—	
$NaIO_3$	solid	774 760	—	796	—	—	
KIO_3		750	—	795	—	—	
$RbIO_3$		757 741	—	795	—	—	
$CsIO_3$		763 748	—	773	—	—	
NH_4IO_3		730	—	790	—	—	
$N(CH_3)_4IO_3$		780 757	—	796	—	—	
$AgIO_3$		705 695	—	762, 752 742, 728	—	—	314–15
$TlIO_3$		705	—	743	—	—	
$LiIO_3$		778	404	778	—	—	
$Hg(IO_3)_2$		683	—	778 757	—	436	
$Hg_2(IO_3)_2$		650	—	794, 776 719	—	448	
$Pb(IO_3)_2$		690	—	770 720	—	423	
$Ni(IO_3)_2$		765	440 407	765	—	—	
$Fe(IO_3)_3$		697	—	808 757	—	451	
$Lu(IO_3)_3$	anhydrous $3H_2O$	827–775 831–812	405	827–775 775	325	—	612
$Zr(IO_3)_4$		670	406	790 split 742	320 290	450	326
$Hf(IO_3)_4$		680	414	790 split 738	325 290	450	
$Th(IO_3)_4$		770	410	770	—	—	314

retained and an increase in the S—O stretching frequency predicted [1066]. Spectra of iodates, $M^I_2M^{IV}(IO_3)_6$ have been interpreted on the basis that these compounds are monodentate complexes bound through oxygen. It has not so far been possible to differentiate between the suggested structures for bidentate sulphito complexes, though Newman and Powell [1066] have found two spectral types which are presumably characteristic of different structures. Table 6.38 shows that, in spite of the C_{3v} symmetry attributed to monodentate complexes bound through sulphur, ν_3 is split in the majority of cases, and ν_4 in several. Presumably a structural effect is superimposed on that due to complex formation.

The two complexes marked with an asterisk in Table 6.39 may, in fact, be bound through sulphur [69, 70].

Table 6.38 Vibrations of monodentate complexes bound through sulphur.

Molecule	$\nu_1(A_1)$	$\nu_2(A_1)$	$\nu_3(E)$	$\nu_4(E)$	Ref.
$K_2[Hg(SO_3)_2]$	996	640	1102	510	187a
$(NH_4)_2[Hg(SO_3)_2]$	967	643	1117	514	1066
$[Co(NH_3)_5SO_3]Cl$	985	633	1110	519	1066
$[Co(en)_2SO_3Cl]$	984	625	1117	—	77
			1075		
$K_6[Pt(SO_3)_4]$	977	654	1100	559, 534	70
	967	637	1084	522, 508	
$K_6[Pt(SO_3)_4]2H_2O$	964	660	1082	540	1066
			1057		
trans $Na_2[Pt(SO_3)_2(NH_3)_2]$	980	636	1112	539	70
			1093	527	
			1076	511	
trans $K_2[Pd(SO_3)_2(NH_3)_2]$	986, 958	646	1074	525	1066
				511	
cis $K_2[Pd(SO_3)_2(NH_3)_2]$	995, 977, 958	655	1093–1056	568	1066
		622		522, 506	
$K_4[Ir(NO_2)_5(SO_3)]$	996	643	1140	529	69
			1100		
			1085		
$Na_3(NH_4)_2[IrCl_4(SO_3)_2]$	980	648	1122	527	69
	922		1058		
$K_3[Rh(SO_3)_3(NH_3)_3]3.5H_2O$	954	638	1107	524	1066
			1056		

Deabriges [326] has studied complexes $M_2Zr(IO_3)_6$, where M = K, NH_4, Rb, Cs. A detailed assignment was not given, but bands were attributed to IO_3 stretching (800, 770, 700, 665, 610 cm^{-1}); Zr—O stretching (415) and IO_3 bending (395, 380, 355, 310, 270 cm^{-1}). These values are for the Rb compound; the other spectra are similar.

It seems likely that all these compounds except the last have the arrangement

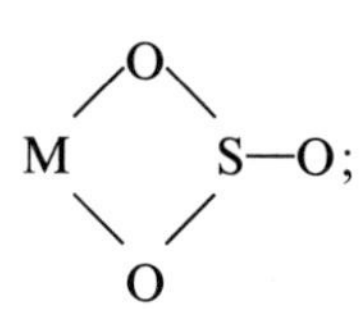

Baldwin [77] has found the Co complex to be monomeric.

Table 6.39 Vibrations of monodentate complexes bound through oxygen.

Molecule	ν_1	ν_2	ν_3	ν_4	ν_{M-O}	Ref.
$Na_7[IrCl_2(SO_3)_4]$*	922	656	1175 1125 1069	507	—	69
$(NH_4)_5[IrCl_4(SO_3)_2]$*	972	650 613	1167, 1073 1036	501	—	69
$Tl_2[Cu(SO_3)_2]$	862	673 602	989 902	506 460	—	1066
$K_2Mn(IO_3)_6$	630	—	786 755	—	480	314
$(NH_4)_2Mn(IO_3)_6$	640	—	789 758	—	479	314
$K_2Pb(IO_3)_6$	695	—	783 757	—	420	314
$K_2Ti(IO_3)_6$	656	—	787 757	—	443	314

Table 6.40 Vibrations of bidentate complexes.

Molecule	$\nu_1(A')$	$\nu_2(A')$	$\nu_3(A'+A'')$	$\nu_4(A'+A'')$	Ref.
$K_4[IrCl_3(SO_3)_2]$	927	652 612	1164 1078(1054)	506	69
$[Coen_2SO_3]Cl$	989	649 625	1119 1093, 1036	—	69
$K_2[Pt(SO_3)_2]2H_2O$	977	—	1166 1092, 1036	—	1066
$K_3[Rh(SO_3)_3]2H_2O$	939	690 647	1157 1113, 1058	527	314
$Na_3[Co(SO_3)_3]4H_2O$	968	661	1129 1070, 1000	520	314
$K_2[Pd(SO_3)_2]$	933 904	664 637	1157 1099, 1056	563 516, 497	314

(b) Molecules with the formula XY_2Z

A large number of such compounds has been studied, and chemically these fall into two classes; mixed halides and halogen-substituted hydrides constitute the first class, and compounds of the type MOZ_2 or MO_2Z, where Z is a halogen or (OH) treated as a point mass, form the second. Four halogen-substituted hydrides have been studied, and only one, NF_2H, has been completely assigned.

Only two mixed halides of nitrogen, NF_2Cl and $NFCl_2$, have been studied. Mueller [1030] has given Raman data for eleven mixed halides

Table 6.41 Vibrations of mixed halide–hydrides.

Compound	ν_1	ν_2	ν_3	ν_4	ν_5	ν_6	Ref.
	(ν(X—Z))(A')	($\nu_{sym}(XY_2)$)(A')	(δ)(A')	(δ)(A')	($\nu_{as}(XY_2)$)(A'')	(δ)(A'')	
NF_2H	3193	1307	972	500	1424	888	280
NF_2D	(2333)	1008	972	(500)	1042	888	
$NClH_2$	686	—	1553	1032	3380	—	990
NCl_2H	3279	687	1002	—	695	1295	

of phosphorus, some of which also have been studied by Delwaulle and Francois [358–9]. The only disagreement in assignment concerns $PFBr_2$. Mueller gives $\nu_s(PBr_2)$ as 420 and $\nu_{as}(PBr_2)$ as 395 cm^{-1}; Delwaulle gives the reverse assignment.

Table 6.42 Vibrations of mixed halides of group V elements.

Compound		Assignment and description						Ref.
		ν_1	ν_2	ν_3	ν_4	ν_5	ν_6	
		ν(X—Z)	ν_s(X—Y)	δ	δ	ν_{as}(X—Y)	δ	
NF_2Cl	ir vap.	697	930	556	377	855	—	280
	ir, Ar matrix	692	918	552	366	842	382	
						837	377	
$NFCl_2$*	ir vap.	825	615	409	273	692	344	661
PF_2Cl	R liq.	529	857	411	304	829	257	1015, 1029, 1030
PF_2Br		448	853	391	244	825	215	
PF_2I		412	850	378	190	845	170	
PCl_2F		823	522	199	326	498	266	
PCl_2Br		400	495	230	167	495	149	
PCl_2I		340	485	222	144	485	135	
PBr_2F		816	420	123	256	395	220	
PBr_2Cl		480	390	123	197	390	153	
PBr_2I		331	380	147	108	380	102	
PI_2Cl		470	324	87	165	324	120	
PI_2Br		380	320	88	132	320	105	
$AsCl_2Br$		282	380	178	129	380	142	
$AsBr_2Cl$		380	280	103	154	280	130	

*See footnote to Table 6.44.

The data shows that ν_3 and ν_6 are primarily XYZ bending whereas ν_4 is XY_2 bending.

Paetzold and Aurich [1103] noted that the Se—F stretching frequency was abnormally low, the usual range for it being 540–630 cm^{-1}; they thought that the compound might be a fluorine-bonded polymer. Walrafen [1480–1] did not assign the bending modes in $SeO_2(OH)$ except to point out that the line at 345 cm^{-1} must be ν_3 or ν_4 (A') since it is polarized.

Table 6.43 Vibrations of compounds XY_2Z, where Y is oxygen.

Compound		ν_1	ν_2	ν_3	ν_4	ν_5	ν_6	Ref.
VO_2Cl	ir solid	503	857	273	537	997	112	1493
KSO_2F*	ir solid	595	1105	375	495	1175	270	1264
$SeO_2(OH)^-$	R, soln. H_2O	615	855	345	410	790	320	1480–1
$KSeO_2F$	ir, R solid	440	882	348	415	909	282	1103
ClO_2F	ir vap, R liq.	630	1106	402	547	1271	367	1342

* The Rb and Cs salts have very similar spectra to this.

The tabulated assignment is made by analogy with that for the SeO_2F^- ion.

The compounds XY_2Z where Z is oxygen are more numerous, and disagreements exist, concerning both observation and assignment. There is some divergence between the results of the infrared study of SOF_2 due to O'Loane and Wilson [1090] and those of the Raman work due to Bender and Wood [123]. However, neither of them have observed the band at 326 cm^{-1} which Yost [1554a] quoted from the work of Best and Trampe. Nevertheless Gillespie and Robinson [536] consider that the band at 326 cm^{-1} does exist and is ν_4. O'Loane and Wilson have suggested that a band observed at 940 cm^{-1} is $\nu_3 + \nu_4$, and have assigned an unobserved band at 410 cm^{-1} to ν_4. Bender and Wood were unable to detect such a band in the Raman spectrum, though they point out that the band at 393 cm^{-1} is broad. The corresponding band in the thio analogue (SSF_2) may be the one observed at 334 cm^{-1}; Seel [1265] did not attempt to assign this band definitely and it may be ν_3. A complete assignment was given by Brown [192] who did not find the band at 533 cm^{-1} observed by Seel; he attributed it to an impurity, probably SF_4 or SOF_2. The data for $SeOCl_2$ is that of Paetzold [1100]; an earlier study by Rolfe and Woodward [1213] gave ν_1 (νSe—O) as 995 cm^{-1}. The values found by Walrafen [1480–1] for the bending modes differ from the tabulated figures; he gives 342 as either ν_3 or ν_4 and the other two lines as 396 and 314 cm^{-1}. He also finds a line at 900 in $SeO(OH)_2$ attributed either to a torsional mode of OH or the antisymmetric stretching of the two Se=O groups in a dimer.

6.4 Non-planar molecules, X_2Y_2

Since the non-planar structure of H_2O_2 was determined by Penney and Sutherland [X115], several other compounds have been shown to possess similar structures. These include H_2S_2, $S_2(Hal)_2$ and $Se(Hal)_2$. The dimer of NO may also have a non-planar structure, though Fateley *et al.* [453] have assigned its spectrum on the basis of planar cis and trans isomers.

Table 6.44 Vibrations of compounds XY_2Z, where Z is oxygen or sulphur.

Compound		Assignment and description						Ref.
		ν_1	ν_2	ν_3	ν_4	ν_5	ν_6	
		ν(X—Z)	$\nu_s(XY_2)$	δ(ZXY)	$\delta(XY_2)$	$\nu_{as}(XY_2)$	δ	
SOF_2	ir vap.	1333	808	530	(410)	748	390	1090
	R liq.	1308	801	529	—	720	395	123
$SOCl_2$	ir vap.	1251	492	344	194	455	284	913
	R liq.	1229	490	394	194	443	284	536, 931
$SOBr_2$	R liq.	1121	405	267	120	405	223	1364
$SeOF_2$	R liq.	1005	659	368	271	601	305	1100
$SeOCl_2$	R liq.	950	390	278	162	348	250	
$SeO(OH)_2$	R solid	859 831	702 or 690	430	336	702 or 690	364	
$SeO(OH)_2$	R soln. H_2O	882	691	407	308	691	354	1311, 1313
$SeO(OD)_2$	R solid	814	689	426	—	689	—	
SSF_2	R vap.	719	761	364	411	692	338	192
	ir vap.*	714	757	334	—	689	415	1265

* Values given are Q branch positions for the A' bands and the mean of the P and R branch positions for the A'' bands.

Considerable discrepancies in observation, and differences in assignment, exist regarding most of these compounds, though in many cases the early studies included bands due to impurities. An additional difficulty with H_2O_2 is the sensitivity of the torsional mode $\nu_4(A)$ to the physical state of the compound. Miller and Hornig [962] have carried out a normal-coordinate analysis of H_2O_2 and have also calculated the expected splittings of the various modes in the crystal. H_2O_2 has the space group D_4^4 ($P4_12_1$) with $Z = 4$, and the correlation splitting of the molecular modes is A $\rightarrow A_1 + B_1 + E$, B $\rightarrow A_2 + B_2 + E$. The A_2 and E modes are ir active, while all except the A_2 modes are Raman active. Translational and librational modes of species A_2 and E have been assigned [532].

The major disagreement concerns the position of the torsional mode, ν_4. In the vapour phase, Chin and Giguere [240] observed a system with a centre at 314 cm^{-1}, maxima at 242·76 and 370·70 cm^{-1}, the doublet representing a double minimum in the potential function. Miller and Hornig [962] had predicted a frequency of about 230 cm^{-1}. A band at 575 cm^{-1} previously attributed to ν_6 by Bain and Giguere [74] was re-interpreted [531] as $2\nu_4$. The corresponding value for D_2O_2 in the vapour phase has not been determined. In the liquid, Taylor [1420] assigned ν_4 as 635 in H_2O_2 and 538 in D_2O_2, at 243°K. Simon and Kreigsmann [1305] found a band at 228 cm^{-1} in liquid H_2O_2 at 278°K which they assigned to ν_4, and made the same assignment for the solid. However, most workers have assigned bands in the 650–700 cm^{-1} region in the solid to ν_4. Miller and Hornig, working at 80°K and 203°K, found such bands at 650 (A_2), 690 (E) and 635 (A_2), 668 (E) at the respective temperatures. Giguere and Harvey [532a], working at −173°C, found the band at 690 cm^{-1}, and in a later study Giguere and Chapados [532] confirmed this assignment and found the corresponding frequency in D_2O_2 as 511 cm^{-1}. They attributed the band at 650 cm^{-1} to a librational mode of species A_2. Their other main disagreement with Miller and Hornig concerns a band at 809 cm^{-1}. Miller and Hornig had assigned this as a combination of the A_2 torsional mode with a libration, whose frequency would thus be 159 cm^{-1}. Giguere and Chapados point out that the observed libration at 151 cm^{-1} is of species E and, even if it were of the correct symmetry class, would require a negative anharmonicity constant. They assigned the band to a librational fundamental, though the earlier study by Giguere and Harvey gave it as a combination. Miller and Hornig argued that an assignment such as that of Giguere and Chapados would be unlikely, since it implies that intermolecular forces play a greater part than intramolecular forces in hindering the rotation of the OH groups about the C_2-axis. Giguere and Chapados, however, consider that this is in fact the case, and is shown to be so by the increase in ν_4 from 314 cm^{-1} in the vapour to 690 cm^{-1} in the solid. Catalano and Sanborn [225] found ν_4 at 385 cm^{-1} for H_2O_2 in a solid nitrogen matrix.

Some attempts have been made to relate the torsional frequency to the height of the barrier restricting rotation. Microwave measurements by Massey *et al.* [X104a, b] gave a value of 113 cm^{-1}, but a re-interpretation of these by Blomgren [159] gave a revised value of 980 cm^{-1}. Chin and Giguere, taking ν_4 and $2\nu_4$ as 314 and 575 cm^{-1} respectively, arrived at a figure of approximately 875 cm^{-1} for the barrier height; an earlier estimate by Zumwalt and Giguere [1561] was 495 cm^{-1}, while calorimetric entropy measurements [X55a] suggested roughly 1225 cm^{-1}. Redington *et al.* [1192] and Hunt *et al.* [688] both gave the trans- and cis-barrier heights and the equilibrium value of the dihedral angle; these values were 300 cm^{-1}, 1300 cm^{-1}, 109·5° and 386±4 cm^{-1}, 2460±25 cm^{-1} and 111·5°, respectively. Busing and Levy have shown that in the solid, the dihedral angle is reduced to about 90°, which favours the view of Giguere and Chapados about the relative importance of inter- and intramolecular forces.

In the complete assignments, ν_1, ν_2 and ν_3 are Raman values for the liquid, ν_5 and ν_6 infrared values for the vapour, and ν_6 the infrared value for the solid.

The analogous sulphur compounds have been less frequently studied and the only major disagreement again concerns the torsional mode.

Table 6.45 Vibrational assignments in H_2O_2 and D_2O_2.

Assignment and description	H_2O_2					D_2O_2	
	Complete [531]	ir Solid [962]	ir Solid [532, 532a]	ir, R Liquid [962, 1420]	Vapour [1192]	Complete [531, 532]	Vapour [1192]
ν(O—H) ν_1	3610	3285	3292	3400	3774	2670	2750
δ(HOO) ν_2	1315	1407	1410	1400	1380	1000	1040
ν(O—O) ν_3	877	882	880	880	880	877	880
Torsion ν_4	660	690 658	690	635	—	480	—
ν(O—H) ν_5	3614	3192	3182	3400	3788	2680	2765
δ(HOO) ν_6	1266	1385	—	1350	1266	947	947
Translations			109 (A_2) 151 (E) 192 (E)			105 (A_2) 148 (E) 188 (E)	
Librations		159	236 (A_2) 656 (A_2) 219 (E) 259 (E) 809 (E)			230 (A_2) 482 (A_2) 210 (E) 244 (E) 585 (E)	

Zengin and Giguere [1555] originally suggested a value of 202 cm^{-1} in solid H_2S_2, derived from a combination at 1070 cm^{-1} attributed to $\nu_4 + \nu_6$ (868 cm^{-1}). Winnewisser and Winnewisser [1520] recently examined H_2S_2, D_2S_2 and HSSD in the liquid and the vapour, and found for the liquids 420, 300 and 391 cm^{-1}, and for the vapours 416·5, 306 and 365 cm^{-1} respectively. For the other frequencies all four studies [458, 1518, 1520, 1555] are in fair agreement, and the values given are those of Zengin and Giguere [1555] for the ir spectrum of the solid. These are ν_1, 2495; ν_2, 868; ν_3, 501; ν_5, 2480; ν_6, 868 cm^{-1}.

The sulphur and selenium halides have been the subject of numerous conflicting interpretations, mostly concerning the relative positions of ν(S—S) and the S—X stretching frequencies. Bradley [179] has pointed out that ν(S—S) is near 500 cm^{-1} in H_2S_2 and $(CH_3)_2S_2$, and that this frequency should be relatively insensitive to substitution. This suggests that, as far as ν_1 and ν_2 are concerned, the interpretation due to Matutano [925] is correct. Similarly with S_2Cl_2, all workers except Stammreich [1360] (whose interpretation, however, is given in the compilations of both Nakamoto [13] and Siebert [15]) agree that ν(S—S) lies near 540 cm^{-1} rather than 450 cm^{-1}. Additional support for the majority view comes from the calculations of Cyvin [309], whose force constants, on this assignment, agree well with mean square amplitudes of vibration derived from electron diffraction measurements by Hirota [660]. His other suggested change, however, which involves reversing ν_2 and ν_5 so that $\nu_5 > \nu_2$, is not generally agreed to and is contradicted by the Raman polarizations observed by Hendra and Park [643]. For S_2Br_2, Stammreich is again in disagreement with the other five studies [178, 483, 643, 660, 676], regarding the positions of ν_1, ν_2 and ν_5. In addition, both Stammreich and Bradley give ν_{sym}(S—Br) as 302 cm^{-1}, whereas the other four studies give values near 355 cm^{-1}. Hendra and Park consider the band at 302 cm^{-1} to be spurious, as they do that at 418 cm^{-1} in Se_2Cl_2. This is assigned by Stammreich as ν_5 and he reverses the positions of ν_1 and ν_2 as given by Frankiss [483] and Hendra and Park [643]. These two sets of workers also agree regarding the assignment of Se_2Br_2; Stammreich has ν_1, 265; ν_2, 205 (not observed by anyone else); ν_5, 292 cm^{-1}.

Table 6.46 Vibrational assignments in sulphur and selenium halides.

Compound		Assignment and description						Ref.
		ν_1	ν_2	ν_3	ν_4	ν_5	ν_6	
		ν(S—S)	ν_s(S—X)	δ	τ	ν_{as}(S—X)	δ	
S_2F_2	ir vap.	526	745	—	—	807	—	925
	ir vap.	616 R	718 R	—	—	684 R	333 R	
		607 P	708 P			677 Q	326 Q	1265
						668 P	319 P	
S_2Cl_2	R liq.	540	446	205	102	434	238	179
S_2Br_2	ir liq.	531	302	175	66	355	198	179
	R liq.	534	356	171	68	353	200	643
Se_2Cl_2	liq.	289	357	133	91	349	150	483
Se_2Br_2	R liq.	286	265	105*	63*	265	119	643
	soln. C_6H_6	290	265	105*	63*	265	119	483

* These two frequencies are those of the pure liquid.

Smith, Keller and Johnson [1335] have obtained the Raman spectrum of the $(NO)_2$ dimer, which they consider may have C_{2v} or C_2 symmetry. The observed lines are at 1861, 1760, 262, 196, 167 and 487 cm^{-1}; the first two were also observed in the ir spectrum of both the liquid and the solid.

7. The spectra of pentatomic molecules

7.1 General

The great majority of these are based on the tetrahedral arrangement; other possible shapes include square planar, tetragonal pyramidal and trigonal bipyramidal with either an apical or (more commonly) an equatorial position unoccupied. All these shapes correspond to formulae XY_4 or its substituted derivatives, but there are some other formulae. The most common is X_2Y_3 (linear, as in C_3O_2, L-shaped, as in H_2F_3, and a structure like that of ethylene but with one terminal atom missing, as in N_2O_3). The monomer of formic acid is the only molecule of formula X_2Y_2Z which is not tetrahedral to have been studied, and there is one unusual XY_4 structure. This is ascribed to the $CS_4^{=}$ ion and is planar Y-shaped:

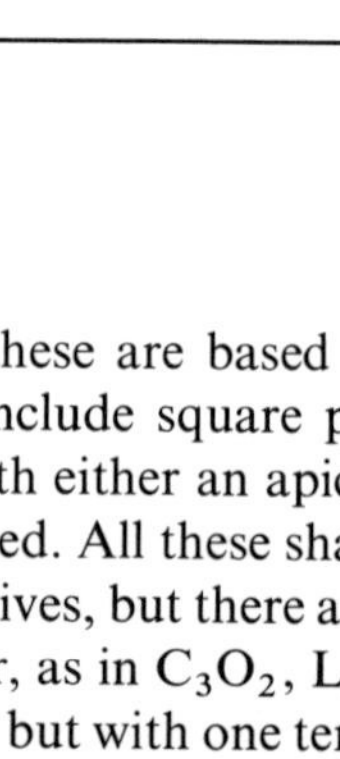

[1022].

There are no five-membered rings corresponding to any of the above formulae.

7.2 Molecules of the formula XY_4

(a) Tetrahedral molecules

(i) Hydrides. As well as the hydrides of group IV, those of groups III (MH_4^-) and V (MH_4^+) have been studied. The borohydrides fall into two groups, those which contain isolated BH_4^- ions and those with bridged structures. The alkali metal hydrides fall into the first class and the boro-

Table 7.1 Selection rules for pentatomic molecules.

Formula	Point group	Activity			Inactive
		R only	ir only	R and ir	
Tetrahedral					
XY_4	T_d	A_1+E	—	$2F_2$	—
XY_3Z	C_{3v}	—	—	$3A_1+3E$	—
XY_2Z_2	C_{2v}	A_2	—	$4A_1+2B_1+2B_2$	—
XY_2ZW	C_S	—	—	$6A'+3A''$	—
XYZWY	C_1	—	—	$9A$	—
Square planar					
XY_4	D_{4h}	$A_{1g}+B_{1g}+B_{2g}$	$A_{2u}+2E_u$	—	B_{1u}
XY_3Z	C_{2v}	$2A_2$	—	$3A_1+2B_1+2B_2$	—
XY_2Z_2 (trans)	D_{2h}	$2A_g+B_{1g}$	$B_{1u}+2B_{2u}+2B_{3u}$	—	A_u
XY_2Z_2 (cis)	C_{2v}	A_2	—	$4A_1+2B_1+2B_2$	—
Trigonal bipyramidal (apical position vacant)					
XY_4	C_{3v}	—	—	$3A_1+3E$	—
Trigonal bipyramidal (equatorial position vacant)					
XY_4, *XY_2Z_2	C_{2v}	A_2	—	$4A_1+2B_1+2B_2$	—
Tetragonal pyramidal					
XY_4	C_{4v}	$2B_1+B_2$	—	$2A_1+2E$	—
Planar Y-shaped					
XY_4	C_{2v}	—	—	$4A_1+2B_1+3B_2$	—
Linear X_2Y_3	$D_{\infty h}$	$2\Sigma_g^+ + \Pi_g$	$2\Sigma_u^+ + 2\Pi_u$	—	—
L-shaped X_2Y_3	C_{2v}	A_2	—	$4A_1+3B_1+B_2$	—
N_2O_3 structure	C_S	—	—	$7A'+2A''$	—
Formic acid structure	C_S	—	—	$7A'+2A''$	—

* This refers to the arrangement in which equivalent atoms occupy equivalent positions, such as XeO_2F_2, in which the two oxygen atoms occupy the equatorial, and the two fluorine atoms the axial, positions.

hydrides of Be and Al into the second. Vibrations in the vicinity of 2000 cm^{-1} and 1500 cm^{-1} are characteristic of the bridge system

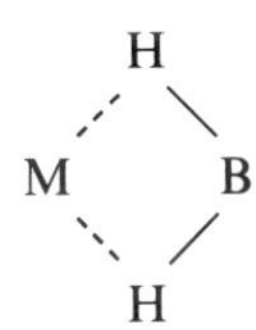

where M is Be or Al.

$LiBH_4$ is known to have the space group D_{2h}^{16}, with $Z = 4$ and the B atoms on C_S sites [X63]. This is consistent with the activation of ν_2 observed by Schutte [1251–2]. Schutte also studied thin films of $NaBH_4$ on NaCl and KBr plates, and concluded that, in the high temperature phase, the BH_4 ions occupied O_h sites, but in the low temperature phase a tetragonal distortion took place with the BH_4^- ions on D_{2d} sites. Splitting of both ν_3 and ν_4 into two components ($B_2 + E$) under these conditions was observed. Partial assignments of the ir spectra of $Be(BH_4)_2$ and $Al(BH_4)_3$ were given by Price [1168–9], and an almost complete assignment of the latter compound by Emery and Taylor [435]. The bridged structure assigned to $Al(BH_4)_3$ is confirmed by an NMR study [X112a].

Table 7.2 Vibrational spectra of ionic borohydrides.

Compound		$\nu_1(A_1)$	$\nu_2(E)$	$\nu_3(F_2)$	$\nu_4(F_2)$
$Li^{11}BH_4$	R soln. liq. NH_3	2265	1202	—	1082
$Li^{10}BH_4$		2268	1206	—	1085
$Na^{11}BH_4$		2264	1210	2244	1080
$Na^{10}BH_4$		2270	1208	2250	1093
$K^{11}BH_4$		2266	1213	—	1084
		2270	(1065)	2270	1080
$Li^{11}BD_4$		1569	854	1698	822
$Li^{10}BD_4$		1570	—	1704	833
$Na^{11}BD_4$		1570	855	1696	823
$Na^{10}BD_4$		1571	858	1706	834
$K^{11}BD_4$		1572	864	1694	827

All the data in Table 7.2 is due to Emery and Taylor [434] except for the second set for $K^{11}BH_4$, given by Woodward and Roberts [1536]. Ketelaar and Schutte have studied the $^{11}BH_4^-$ ion in a variety of alkali halide matrices, finding a steady decrease in both ν_3 and ν_4 with increasing cation and anion radii. Neither frequency is split, so the local symmetry is considered to remain as T_d [747]. Toyuki *et al.* [1435a] have calculated SUBFF constants from Emery and Taylor's data. Brearley and Sundaram [182] have calculated the frequencies of the $^{11}BT_4^-$ ion, using symmetry force constants in agreement with those of Emery and Taylor; their values are ν_1, 1334; ν_2, 712; ν_3, 1474; ν_4, 703 cm^{-1}.

For the Be compound, Price observed bands at 940–1090, 1530, 1985, 2060, 2165, 2515 and 2630 cm^{-1} in the ir spectrum.

The only complete assignment of an aluminohydride is that of Lippincott [865], who gave ν_1, 1790; ν_2, 799; ν_3, 1741; ν_4, 769 cm^{-1} for a 1·28M solution of $LiAlH_4$ in diethyl ether. This Raman data is very similar to that obtained by D'Or and Fuger [378], who also studied the ir spectrum of the solid, finding bands at 1782, 1647, 875 and 704 cm^{-1}. Roszinski [1219] found a similar ir spectrum, with an additional band at 800 cm^{-1}. In $LiAlD_4$ he found bands at 1285, 1182, 644, 569 and 510 cm^{-1}. Dautel and Zeil [320] found bands at 426 and 416 cm^{-1} for $LiAlH_4$ dissolved in $(C_2H_4)_2O$. More recent work on $LiAlH_4$ gives ν(Al—H) as 1780, 1650

Table 7.3 Vibrational assignment for $Al(BH_4)_3$ with D_{3h} symmetry.

Class		Assignment and approximate description	^{11}B	^{10}B
A_1' (R)	ν_1	ν_{sym}(terminal)	2473	2476
	ν_2	Bridge expansion	2069	2071
	ν_3	Bridge stretch (sym, ‖)	1495	1499
	ν_4	δ_{sym}(terminal)	1116	1123
	ν_5	ν_{sym}(Al—B)	510	523
A_2'' (ir)	ν_{11}	Bridge stretch (asym, ‖)	2154	—
	ν_{12}	Bridge shear	—	—
	ν_{13}	BH_2 out of plane rock	978	—
	ν_{14}	$\pi(AlB_3)$	—	—
E' (ir, R)	ν_{15}	ν_{asym}(terminal)	2549	2561
	ν_{16}	ν_{sym}(terminal)	2493	—
	ν_{17}	Bridge expansion	2010	2014
	ν_{18}	Bridge stretch (sym, ‖)	1521	1505
	ν_{19}	δ(terminal)	1116	1123
	ν_{20}	BH_2 in-plane rock	1392	1398
	ν_{21}	ν(Al—B)	602	612
	ν_{22}	Bridge deformation	—	—
	ν_{23}	AlB_3 in-plane bending	318	324
E'' (R)	ν_{24}	Bridge stretch	—	—
	ν_{25}	Bridge torsion	1149	1151
	ν_{26}	BH_2 out-of-plane rock	976	982
	ν_{27}	Bridge shear	—	—
	ν_{28}	BH_2 torsion	—	—

The vibrations ν_1–ν_4, ν_{11}, ν_{12}, ν_{13} are in phase; ν_{15}–ν_{20}, ν_{22}, ν_{24}–ν_{28} are out of phase. The (inactive) $2A''$ and $3A_2'$ modes are not listed.

cm^{-1}; for $NaAlH_4$, ν(Al—H) and Al—H bending are given as 1650, 895 cm^{-1} respectively [1672]. The hydrides of group IVA have been studied and the assignments are well established except for SnH_4 where some disagreement exists.

Table 7.4 Vibrational spectra of gaseous hydrides of group IVA elements.

Compound	ν_1	ν_2	ν_3	ν_4	Ref.
CH_4	2917	1534	3019	1306	15
CD_4	2109	1092	2259	996	15
SiH_4	2187	975	2190	914	15
SiD_4	(1545)	(689)	1597	681	776a, 936
GeH_4	2106	931	2114	819	15
GeD_4	1504	665	1522	596	15
SnH_4	—	—	1860	703	15
	—	758	1902	678	856
SnD_4	—	539	1368	487	856

The hydrides of the group VA elements N, P and As have been studied in the form of various salts. Mathieu and Poulet [922] have made an extensive study of the stretching vibrations of ammonium salts in relation to the site symmetry of the NH_4^+ ion. We may consider three possible effects.

(i) ν_1 remains inactive and ν_3 is not split. This, corresponding to the free-ion selection rules, is to be expected for site symmetries O_h and T_d.
(ii) ν_1 remains inactive and ν_3 is split. This should occur with the tetragonal site groups D_{2d} and S_4.
(iii) ν_1 is activated and ν_3 is split. This should occur if the NH_4^+ group lies on a site of threefold or lower symmetry.

There are also some cases where the site occupied by the NH_4^+ ion is a subgroup of O_h rather than T_d; examples are ammonium alum (S_6), $(NH_4)_2PtCl_4$ (D_{2h}). Presumably the NH_4^+ ion is considered to be in free rotation with effective O_h symmetry.

Since Mathieu and Poulet apparently did not attempt to find ν_1 in the infrared, but only in the Raman spectrum, only the predictions about ν_3 can be tested.

Table 7.5 The effect of site symmetry on the splitting of ν_3 in ammonium salts.

Case I Splitting predicted and found	Case II Splitting predicted and not found	Case III No splitting predicted or found	Case IV Splitting found not predicted
NH_4SCN (C_1) NH_4NO_3 (C_{2v}) $(NH_4)_2BeF_4$ (C_S)	NH_4F (C_{3v}) NH_4HF_2 (C_{2v}) $(NH_4)_2SO_4$ (C_S) $NH_4H_2PO_4$ (S_4) NH_4ClO_4 (C_S) NH_4BF_4 (C_S)	NH_4Cl (T_d) NH_4Br (T_d) $(NH_4)_2SnBr_6$ (T_d) $(NH_4)_2PtCl_6$ (T_d) $(NH_4)_3IrCl_6$ (T_d) $(NH_4)_2SnCl_6$ (T_d) $(NH_4)_2SiF_6$ (T_d)	$(NH_4)_3FeF_6$ (O_h, T_d) $(NH_4)_3AlF_6$ (O_h, T_d) $(NH_4)_3Co(NO_2)_6$ (T_h, T)

The anomalies in case IV may be attributed to the existence of two distinct sets of NH_4^+ ions, in each compound. There is no obvious explanation for the case II anomalies. Some more detailed analyses have been carried out. Couture and Mathieu [292] have predicted that for NH_4Cl, coupling between 4 ions of symmetry C_{3v} should give the correlation splittings $\nu_1(A_1) \rightarrow A + F$, $\nu_2(E) \rightarrow E + 2F$, ν_3, $\nu_4(F_2) \rightarrow A + E + 3F$. Bands observed in the Raman spectra of the solid were ν_1, 3042; ν_2, 1711, 1771, 1985; ν_3, 2818, 3133; ν_4, 1403, 1428, 1444 cm^{-1}. Martinez and Wagner [912] and Durig *et al.* [408] have carried out site-group and factor-group analyses of PH_4I respectively. This compound is tetragonal, space group D_{4h}^7, with $Z = 2$. The correlation diagram is as shown in Fig. 7.1, and the assignment is $\nu_1(A_{1g})$, 2295; $\nu_2(A_{1g})$, 1026, (B_{1g}), 1086; $\nu_3(E_u)$, 2372, (E_g), 2366; $\nu_4(A_{2u})$, 954, (E_u), 932, (B_{2g}), 974, (E_g), 919 cm^{-1}. Durig *et al.* [408] have also observed some of the external modes. The ir

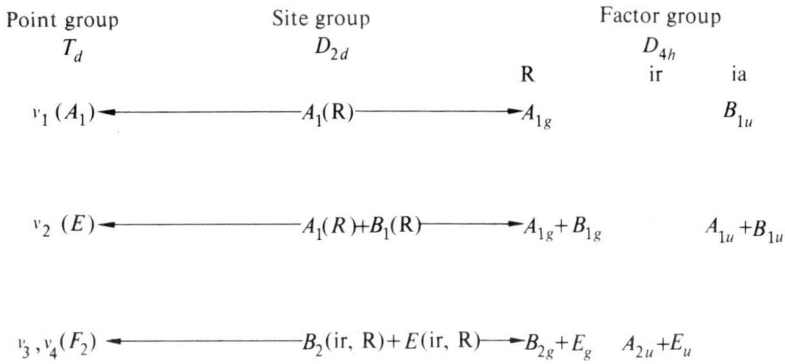

Fig. 7.1 Correlation diagram for phosphonium iodide.

active modes consist of $A_{2u} + E_u$(translational) and E_u(librational); these were observed at 107, 122 and 342 cm^{-1} respectively. The Raman active modes are $A_{1g} + B_{2g} + 2E_g$(translational) and E_g(librational). Lines were observed at 80 cm^{-1} (B_{2g}), 112 cm^{-1} $(E_g$, translational) and 318 cm^{-1} $(E_g$, librational). Gutowsky *et al.* [X59] developed two equations for the calculation of the barrier restricting rotation of a tetrahedral NH_4^+ ion about a C_4 cubic axis. Assuming electrostatic forces only, in a CsCl type lattice, they found

$$V_0 = 46{\cdot}8 \times (\tfrac{1}{4}\,e)^2\,\frac{r^4}{d^5}, \tag{7.1}$$

and more generally

$$16V_0 = \left(h\nu + \frac{5\hbar^2}{2I}\right)^2 \left(\frac{\hbar^2}{2I}\right)^{-1}. \tag{7.2}$$

In (7.1), r and d are the bond length and length of the cell edge respectively. Calculations by Rush [X130] on PH_4I using these equations give respectively 7·5 and 7·1 ± 0·5 kcal mole^{-1}; the observed Raman frequency gives 6·4 kcal mole^{-1} in PH_4I.

Other studies of solids containing MH_4^+ include those on NH_4F and ND_4F [1153], NT_4Cl and $^{15}NH_4Cl$ [995], PH_4Cl and PH_4Br [634–5],

PH_4I [1538], and AsH_4Cl, AsH_4Br [633–5]; these two compounds have been partially studied in the vapour.

Table 7.6 Vibrational frequencies for some group V hydride ions.

Compound		ν_1	ν_2	ν_3	ν_4	Ref.
NH_4^+ in NH_4Cl	R solid	3033	1685	3134	1397	292
NH_4^+ in $^{15}NH_4Cl$	ir solid (−190°C)	3054	1656	3121	1348	995
ND_4^+ in ND_4F	ir solid (−195°C)	2035	1257	2148	1120	1153
NT_4^+ in NT_4Cl	ir solid (20°C)	1889	976	2022	913	995
PH_4^+ in PH_4I	ir, R solid			See text		
PD_4^+ in PD_4I	ir, R solid	1654	772 725	1732	672	408
AsH_4^+ in AsH_4I	ir vapour	—	—	2151(Q)	842(Q)	633
	ir solid	2146	982	2174	842	

(ii) Halides. Some halide complexes of the form $(MHal_4)^{2=}$ have been studied, but in only one case, that of K_2BeF_4, has an analysis been made in terms of the structure. This compound is orthorhombic, space group D_{2h}^{16}, with $Z = 2$, and the $BeF_4^=$ ions on C_S sites. Under this site symmetry we have $A_1 \rightarrow A'$, $E \rightarrow A' + A''$, $F_2 \rightarrow 2A' + A''$; considering the unit cell as a whole, $A' \rightarrow A_g + B_{1g} + B_{2u} + B_{3u}$, $A'' \rightarrow A_u + B_{1u} + B_{2g} + B_{3g}$. Only the A_u mode is inactive, so the full spectrum should consist of 15 ir active bands and 18 R active bands. Lorenzelli and Gesmundo [883] observed 5 bands in the ir and 8 in the Raman. Lecomte *et al.* [835] studied some other fluoride complexes, identifying ν_3 and ν_4 and observing another band 30–40 cm^{-1} below ν_4, which they did not assign. Most of the possible halide complexes of Zn, Cd and Hg have been studied (excepting the fluorides); shifts on going from the solid Cs salts to the $(ZnHal_4)^=$ ion in solution were interpreted in terms of tetrahedral co-ordination of Zn in the solid. The assignment given for $HgCl_4^=$ is that due to Deacon [327]; an earlier study of the Raman spectrum of the melt [696] appears to lead to an incorrect assignment. The anomaly in the force constants of $MnBr_4^{2-}$ previously reported [429] is now shown not to exist [1689].

Halide complexes of trivalent cations have also been studied; the majority of the difficulties in interpretation involve the boron and aluminium complexes. A study of MBF_4 ($M = NH_4$, Na, K) by Coté and Thompson [288] showed ν_3 and ν_4 as doublets. The ratios of the component frequencies were slightly lower than 0·960, the product-rule ratio for the ^{10}B:^{11}B shift. Hoard and Blair [X71] had shown the NH_4 and K salts to have the $BaSO_4$ structure, with the space group D_{2h}^{16}, and C_S site symmetry; Klinkenberg [X89] had suggested D_{2h}^{17} for $NaBF_4$ with C_{2v}

Table 7.7 Vibrational spectra of halide complexes of divalent cations.

Compound		ν_1	ν_2	ν_3	ν_4	Ref.
K_2BeF_4	ir cryst.	620	186	804	391 383	883
	R cryst.	—	164 208	766, 780 832, 850	369 375	
	ir solid	—	—	833 820, 770	383–394(350)	835
K_2ZnF_4	ir solid	—	—	750	481(428)	
K_2CoF_4	ir solid	—	—	747	481(436)	
$(NH_4)_2CuF_4$	ir solid	—	—	821	471(447)	
Cs_2ZnCl_4	R solid	288	116	298	130	1172
Cs_2ZnBr_4	R solid	178	80	212	89	
Cs_2ZnI_4	R solid	130	60	172	70	
$ZnCl_4^{=}$	R soln. H_2O	275	79	306	104	1000, 351–2
$ZnBr_4^{=}$	R soln. H_2O	172	66	208	88	
$ZnI_4^{=}$	R soln. H_2O	122	44	170	62	
$CdCl_4^{=}$	R soln. TBP	260	—	281	92	323
$CdBr_4^{=}$	R soln. H_2O	167	62	185	53	1212
$CdI_4^{=}$	R soln. H_2O	117	45	147	37	
$(Me_4N)_2HgCl_4$	ir, R solid	264	—	225	111	327
$HgI_4^{=}$	R soln. H_2O	126	35	140 calc.	41	873
$(Et_4As)_2HgI_4$	ir solid	—	—	118	55	327
K_2PbCl_4	R solid	$257A_1$	$149A_1$ $199A_2$	$352A_1$ $340B_1$ $325B_2$	$236A_1$ $171B_1$ $210B_2$	75
$(Et_3NH)_2MnCl_4$	ir, R solid	256	—	301 278	120	429
$(n\text{-}Bu_4N)_2MnBr_4$	ir, R solid	195	65	221 209	89	
$(Et_4N)_2MnI_4$	ir, R solid	108	—	193 188	56	

The frequencies in brackets were unassigned by Lecomte *et al.* The value 140 cm^{-1} for ν_3 of $HgI_4^{=}$ was calculated using an MUBFF with H = 0. K_2PbCl_4 was also studied in the melt and the frequencies obtained were very similar. The correlation between T_d and C_{2v} symmetry tabulated was not given expressly by Balasubramanyam and Nanis [75]. The $MgCl_4^{=}$ complex has been studied by Adams [25], who found ν_1 in the Raman spectrum of a solution in $SOCl_2$ at 252 cm^{-1}, and ν_3 in the ir spectrum of the $(Et_4N)_2$ salt at 344 cm^{-1}.

site symmetry. In all cases the degeneracy of ν_3 and ν_4 should be totally broken down and six components observed when the isotopic shift is taken into account. Greenwood [563] has observed six components in the ν_3 region of the ir spectrum of KBF_4, but the situation is complicated by the possibility of Fermi resonance with $2\nu_4$. The fluoroaluminates present a number of unresolved problems. $LiAlF_4$ has two bands in the regions 850–800, 785–775 cm^{-1}. These have been ascribed to a dimer with C_{3v} symmetry containing Li bridges. The spectrum of cryolite, Na_3AlF_6, as well as having a band at 554 cm^{-1} attributed to ν_1 of AlF_6^{3-}, has another band at 633 cm^{-1} which has been interpreted as ν_1 of AlF_4^-. $KAlF_4$ has been studied by Lecomte *et al.* [835], who assigned the bands at 747 and 533 to ν_3 and ν_4 of AlF_4^-, did not assign the band at 394 cm^{-1}, and attributed the bands at 646 and 480 cm^{-1} to AlF_6^{3-}. Balasubramanyam and Nanis [76] have assigned the Raman spectrum of liquid $KAlCl_4$ on the basis of C_{2v} point symmetry of $AlCl_4^-$. The chloride, bromide and iodide complexes of Ga, In and Tl have all been studied. Woodward and co-workers have given two different versions of the $GaCl_4^-$ spectrum [1535, 1541]. Davies and Long [322] have found a band at 101 cm^{-1} in the Raman spectrum of $TlCl_4^-$ which no-one else has observed; they interpret it as ν_2 or ν_4, neither of which agrees with Spiro's values [1352]. An unusual complex is postulated by Hathaway [621] to explain the observed spectrum of Me_3SnBF_4. This compound has bands at 758 (ν_1), 353 (ν_2), 1100 (triplet, ν_3) and 530 (split, ν_4). A band at 562 due to ν_{as}(Sn—C) obscures the splitting of ν_4. There is an unassigned band at 447 cm^{-1}. The structure suggested (see Fig. 7.2) is not (I) but (II). The band at 482 cm^{-1} in $KSbF_4$ is unassigned.

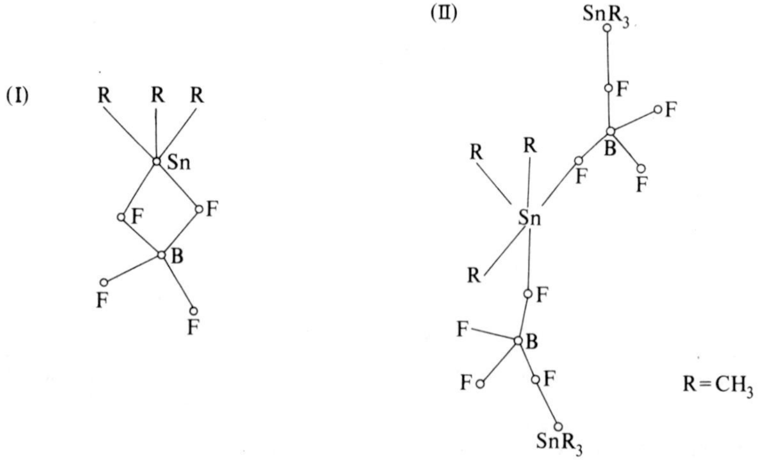

Fig. 7.2 Possible structures for the complex $(CH_3)_3SnBF_4$, involving 5-coordinate tin.

The spectra of the halides of C, Si, Ge and Sn have all been studied, except for SnF_4. Fournier *et al.* [481] have investigated the phase change in solid CF_4 and concluded that the form stable above 76·2°K is disordered, but the low temperature form (investigated at 70°K) has a well-defined structure, factor group S_4, with the molecules on S_4 sites. This

Table 7.8 Vibrational spectra of halides of trivalent cations.

Compound		ν_1	ν_2	ν_3	ν_4	Ref.
KBF_4	ir solid	775	—	1158, 1135 1115, 1086 1073, 1063	542	563
$NaBF_4$	R soln.	769	353	1016 (^{10}B) 984 (^{11}B)	529 (^{10}B) 524 (^{11}B, calc.)	553
R_4NBCl_4	R soln.	405	190	670	274	257
R_4NBBr_4	liq. SO_2	243	117	605	166	257
$NaAlCl_4$	R liq.	349	145	580	183	76
$KAlCl_4$	R liq.	353 A_1	168 A_1 155 A_2	562 A_1 505 B_1 592 B_2	136 A_1 196 B_1 182 B_2	76
$GaCl_4^-$	R soln. H_2O	346 396	114 134	386 453	149 172	1535, 1541
	R melt Ga_2Cl_4	346	115	380	153	1532
$GaBr_4^-$	R soln.	210	71	278	102	1080
	R solid Ga_2Br_4	209	79	288	107	1533
GaI_4^-	R soln.	145	52	222	73	1540
$InCl_4^-$	R soln.	321	89	337	112	1541
$InBr_4^-$	R soln.	197	55	239	79	1529
InI_4^-	R soln.	139	42	185	58	1540
$\phi_4AsTlCl_4$	R solid	312	78	296	60	1352
$TlBr_4^-$	R solid Tl_2Br_4	182	56	198	74	1353
$KTlBr_4$	R solid	182	58	196	58	1353
$RbTlBr_4 1H_2O$	R solid	180	59	193	59	1353
$CsTlBr_4$	R solid	184	58	201	69	1353
$FeCl_4^-$	R soln.	330	106	385	133	1541
$KSbF_4$	ir solid	—	—	747	550 (482)	835

accounts for the appearance of ν_2 in the infrared spectrum and the non-coincident doublets representing ν_3 and ν_4 in the ir. Liquid CF_4 is one of the few liquids in which vibrational selection rules are broken; ν_1 and ν_2 appear weakly in the ir spectrum due to the anisotropic field set up by neighbouring molecules. None of the other spectra given in Table 7.9 show any remarkable features.

In CI_4, Stammreich [1369] also found two other bands at 270, 360 cm^{-1}, which he assigned to impurities. Hadni [599] has identified bands at 26, 33 and 92 cm^{-1} in SnI_4 as lattice modes. A few transition metal halides have been studied and in most cases there are some disagreements. One of these concerns $TiCl_4$. In the vapour phase, Hawkins and Carpenter [622] found ν_3 at 498·5 cm^{-1}; in the ir spectrum of the liquid two bands are found, at 490 and 506 cm^{-1}. Delwaulle and Francois [356a] attributed this splitting to Fermi resonance between $\nu_3(F_2)$ and $\nu_1 + \nu_2$. Griffiths [577] contradicts this for two reasons: firstly, $\nu_1 + \nu_2$ is of species E and cannot interact with $\nu_3(F_2)$, and secondly, $\nu_1 + \nu_2$ is inactive in the ir. Dove *et al.* [380] have suggested the ternary combination

Table 7.9 Vibrational spectra of the tetrahalides of C, Si, Ge, Sn and Pb.

Compound		ν_1	ν_2	ν_3	ν_4	Ref.
CF_4	R vapour	908·5	435	1263 L*	631·2	230, 481, 982
				1283 T*		
	R liquid	907·5	435	1250 T	630	
				1266		
	ir liquid	908	434	1222	630·3	
	R solid HT phase	907·7	436·6	1246	630·6	
	R solid LT phase	907·2	438·5	1230·1, 1243·7 T	631·2	
				1321 L	629·6	
	ir solid LT phase	—	436·0	1237	631·9	
				1229	629·8	
CCl_4	R liquid	459	217	791	315	876
				762		
CBr_4	R liquid	268	128	673	182	
CI_4	ir, R solid	178	90	555	123	1369
SiF_4	ir, R vapour	800	268	1031	391	701a
$SiCl_4$	R liquid	424	150	610	221	361
$SiBr_4$	R liquid	249	90	487	137	361
SiI_4	R liquid	168	63	168	94	361
GeF_4	ir, R vapour	738	205	800	260	227
$GeCl_4$	R liquid	396	132	453	172	361; 1064
$GeBr_4$	R liquid	235	80	327	112	361
GeI_4	R soln. CS_2	159	60	264	80	1364, 1369
$SnCl_4$	R liquid	367	104	403	129	361
$SnBr_4$	R liquid	220	54	279	63	361
SnI_4	R soln. CCl_4	149	47	216	63	1364
						1369
	ir solid	—	—	208	64·5	599,
					60·5	518
$PbCl_4$	R liquid	327	90	348	93	1064

* L and T refer to longitudinal and transverse components.

$\nu_3 + \nu_4 - \nu_2 = 505$ cm^{-1} ($A_1 + A_2 + 2E + 2F_1 + 2F_2$) which would satisfy the objections raised by Griffiths; however he considers that the spectrum should be interpreted in terms of a chlorine-bridged dimer. Some support from this comes from the presence of the band in the Raman spectrum of liquid $TiCl_4$ and its absence from the ir spectrum of a solution of $TiCl_4$ in CCl_4. Against this, the two most likely structures are centrosymmetric, so the coincidence in the ir and Raman spectra of the liquid at 506 cm^{-1} is forbidden. Griffiths says, however, that it may not be a true coincidence since the line widths are very large and the frequency measurement is thus not exact. A doublet at 383, 397 cm^{-1} was found in the R spectrum of $TiBr_4$ [354]. Weidlein [1494] has similarly interpreted the spectrum of solid $ZrCl_4$ as that of a bridged structure of symmetry D_{2h}.

There are four interpretations of the spectrum of VCl_4, all of which are tabulated. MUBFF calculations by Creighton [300] suggest that ν_2 should lie between 83 and 129 cm^{-1} (assuming $H > 0$), with the likeliest value 105 cm^{-1}. Dove [380] suggested 128 cm^{-1}, Grubb and Belford [585] 83 cm^{-1}, both on the basis of combinations assigned to $\nu_2 + \nu_3$. Freymann [491] found bands at 93, 104, 141, 180, 341 and 384 cm^{-1} in the spectrum of UCl_4, and at 110 and 145 cm^{-1} in that of UBr_4, but did not assign them. Aleksandrovskaya and Godnev [48] have calculated the fundamental frequencies of gaseous Hf halides by two methods, their own (the results of which are tabulated) and one due to Hildebrand [657]; the results are in good agreement but no experimental data exists.

Table 7.10 Vibrational spectra of transition metal tetrahalides.

Compound		ν_1	ν_2	ν_3	ν_4	Ref.
$TiCl_4$	vapour	388	119	498·5	141	622
$TiBr_4$		229·5	74	383	90·5	948
$ZrCl_4$	ir vapour	(388)	(102)	421	(112)	1494
VCl_4	vapour	383	(105)	487	129	300
VCl_4	R soln. CCl_4	383	128	475	128	380
		383	128	475	159	380
VCl_4	ir	383	(83)	487	128	585
$HfCl_4$	calc.	347	88	378	93	48
$HfBr_4$	calc.	211	58	255	46	48
HfI_4	calc.	142	39	185	42	48
UF_4	ir, R solid	614	340	420	180	1770

A few species involving complexes containing $(MHal_4)^+$ have been studied, and most assignments have been made by analogy with the isoelectronic $MHal_4$ compounds. Mueller and Fadini [1009] have carried out some force-constant calculations.

Table 7.11 Vibrational spectra of tetrahalide cations of group V elements.

Compound			ν_1	ν_2	ν_3	ν_4	Ref.
NF_4^+	in $NF_4^+AsF_6^-$	R solid	813	488	1159	625	243
PCl_4^+	in PCl_5	R solid	451	173	627	201	523
PBr_4^+	in PBr_5	R solid	227	72	474	140	526
$AsCl_4^+$	in $AsCl_4^+AsF_6^-$	R solid	422	156	500	187	1492
$SbCl_4^+$	in $SbCl_4F$	R solid	353	143	399	153	342

(iii) Hydroxo- and oxo-species. A few tetrahedral hydroxo-species have been studied. Fordyce and Baum [477] have assigned the Raman solution spectrum of the $Zn(OH)_4^{=}$ ion as ν_1, 484; ν_2, 285; ν_3, 430; ν_4, 322 cm^{-1}. The $B(OH)_4^-$ spectrum has been studied in solution and in the solid.

Goubeau [554] assigned the Raman spectrum of an aqueous solution of the Na salt as ν_1, 754; ν_2, 379; ν_3, 947; ν_4, 533 cm^{-1}. The $B(OH)_4^-$ ion in the solid has been studied in the mineral teepleite, $Na_2[B(OH)_4]Cl$, and in the material sometimes formulated $Sr(BO_2)_24H_2O$, though more correctly $Sr[B(OH)_4]_2$. In this compound, the assignment given by Kessler and Lehmann is ν_1, 730–740; ν_2, 622; ν_3, 918, 976; ν_4, 507, 529, 552 cm^{-1} [743]. The structure of teepleite has been determined by Fornaseri [X51] and a factor-group analysis carried out by Klee [765]. The factor group is D_{4h} and the site group D_{2d}. An early Raman study by Lippincott [865] of alkaline aluminate solutions revealed lines at 615 cm^{-1} (ν_1) and 310 cm^{-1} (ν_2, ν_4). He calculated ν_3 as 720 cm^{-1} but did not find it. A re-study by Carreira [222] suggests that the isolated tetrahedral $Al(OH)_4^-$ ion does not exist in such solutions. Electrode potential measurements by Plumb and Swaine [X119] are interpreted to mean that at pH > 12·5 the predominant species is linear AlO_2^-, and between 8 and 12 either planar $Al(OH)_4^-$ or an octahedrally coordinated polymer. At pH 13, Carreira found a line at 628 cm^{-1} attributed to ν_1 of AlO_2^-, and at pH 10·2 lines at 830 ($\nu_1(A_{1g})$); 625 ($\nu_2(E_g)$); 1240 ($\nu_3(F_{1u})$) and 540 ($\nu_5(F_{2g})$) attributed to the octahedral polymer. Deuteration, NMR studies and vapour-pressure measurements confirm the high-pH species as AlO_2^-.

The oxo-anions form a very large group, and the central atom may be a member of any of the groups of the Periodic Table. The only species containing a group I element is the LiO_4 tetrahedron, which, according to Tarte [1412–3] exists in Li_2CO_3, the spinel $LiFeCr_4O_8$, and Li_2MoO_4, Li_2WO_4 which have the phenacite (Be_2SiO_4) structure. Bands assigned to the LiO_4 tetrahedron on the basis of ^7Li–^6Li shifts occur at 497, 435, 416, 397 cm^{-1} (Li_2CO_3); 461 (spinel); 471, 452 (Li_2WO_4); and 464, 432, 391 (Li_2MoO_4).

Bands due to the BeO_4 tetrahedron have been assigned in the region 900–700 cm^{-1} by Funck [497] and Launer [828]. Funck [498] calculated the frequencies of the BO_4 group as ν_1, 750; ν_2, 380; ν_3, 1030; ν_4, 520 cm^{-1}.

In BPO_4, Gesmundo and Lorenzelli [528] assigned BO_4 stretching frequencies as 1170, 1110, 1065 cm^{-1} and bending frequencies in two groups, 645, 635, 610 cm^{-1} and 236, 206 cm^{-1}. Tarte [1415] has assigned frequencies to the AlO_4 tetrahedron as 900–700 cm^{-1} where the tetrahedra are condensed, as in $CaAl_2O_4$, and 800–650 cm^{-1} in isolated tetrahedra such as exist in the 3:5 rare-earth aluminates with the garnet structure, e.g., $Al_5Y_3O_{12}$.

A few silicates and germanates contain isolated tetrahedra and their frequencies have been assigned by Tarte [1408–10, 1414]. Tarte has also suggested that isolated TiO_4 tetrahedra exist in mixed crystals $Zn(Si, Ti)O_4$

which retains the willemite ($ZnSiO_4$) structure up to about 10 per cent Ti. Tarte also considers Ba_2TiO_4, investigated by Bobovich [167] to contain isolated TiO_4 tetrahedra. ν_3 of the TiO_4 group is assigned at 745 cm^{-1}. Fortnum and Edwards [480] found bands at 935, 777, 607 and 448 cm^{-1} in the Raman spectrum of aqueous silicates; all but the 777 cm^{-1} band correspond approximately to vibrations in the solid.

Table 7.12 Vibrational spectra of silicates and germanates.

Structure	Compound	ν_1	ν_2	ν_3	ν_4	Ref.
Willemite	Zn_2SiO_4	872	—	978, 934, 905	462 396, 380	1408
	Zn_2GeO_4	798	—	808, 772, 746	325 311, 290	1408
Olivine	Mg_2SiO_4	842	422	1001, 982	615, 510	1409
	Ca_2SiO_4	819	340	953, 859	565, 520, 496	1409
	Ni_2SiO_4	830	392	988, 882	589, 524, 498	1409
	Fe_2SiO_4	832	363	963, 879	566, 509, 481	1409
	Co_2SiO_4	829	373	973, 875	575, 513, 484	1409
	Mn_2SiO_4	819	350	958, 863	567, 509, 486	1409
	Mn_2GeO_4	700	320	780	475, 390	1409
	Cd_2GeO_4	700	320	780	480, 390, 360	1409
Monticellite	$CaMgSiO_4$	833	425 ?	982, 890	597, 517	1409
	$CaCoSiO_4$	826	—	964, 886	571, 518, 498	1409

Very many oxo-anions where the central atom is a member of groups V, VI or VII have been studied, and structural similarities exist between them. For instance, a number of rare-earth phosphates have the monazite ($LaPO_4$) structure, which is also assumed by $SrSeO_4$. The scheelite

Table 7.13 Correlation table for T_d and site groups which appear as subgroups of T_d.

Site group	ν_1	ν_2	ν_3, ν_4
T_d	A_1(R)	E(R)	F_2(ir, R)
T	A(R)	E(R)	F(ir, R)
S_4	A(R)	A(R) + B(ir, R)	B(ir, R) + E(ir, R)
D_{2d}	A_1(R)	A_1(R) + B_1(R)	B_2(ir, R) + E(ir, R)
C_{3v}	A_1(ir, R)	E(ir, R)	A_1(ir, R) + E(ir, R)
C_3	A(ir, R)	E(ir, R)	A(ir, R) + E(ir, R)
C_{2v}	A_1(ir, R)	A_1(ir, R) + A_2(R)	A_1 + B_1 + B_2(all ir, R)
D_2	A(R)	$2A$(R)	B_1 + B_2 + B_3(all ir, R)
C_2	A(ir, R)	$2A$(ir, R)	A + $2B$(all ir, R)
C_s	A'(ir, R)	A'(ir, R) + A''(ir, R)	$2A'$ + A''(all ir, R)
C_1	A	$2A$	$3A$

Table 7.14 Structural information concerning compounds with tetrahedral oxo-anions.

Site group	Space group	Compounds	Ref.
T_d	O_h^7	SPINELS Na_2MoO_4, Na_2WO_4, Ag_2MoO_4	WVIIIb, 5
T	T_d^4	Ag_3PO_4	X66
C_{3v}	D_{3d}^5	$Sr_3(PO_4)_2$, $Ba_3(PO_4)_2$, $Ba_3(VO_4)_2$; and possibly $Sr_3(CrO_4)_2$, $Ba_3(CrO_4)_2$, $Ba_3(MnO_4)_2$	X132
	C_{6v}^4	$LiMnO_43H_2O$, $LiClO_43H_2O$	X87, X171
D_{2d}	D_{4h}^{19}	ZIRCON structure HT and LT forms of rare-earth phosphates from Dy to Lu; $CaCrO_4$	X133–7
S_4	S_4^1 or S_4^2	$BeSO_4$	WVIIIa, 17
	C_{4h}^6	SCHEELITE structure $M^{II}MoO_4$, $M^{II}WO_4$ (M^{II} = Ca, Sr, Ba, Pb). $CdWO_4$ M^IReO_4, M^ITcO_4 (M^I = K, NH_4, Rb, Ag) $NaReO_4$, $NaIO_4$, $KRuO_4$.	WVIIIa, 13, 19
C_3	T^4	$Sr(MnO_4)_2$, $3H_2O$	X44
	C_6^6 or C_{6h}^2	$Ba(ClO_4)_2$, $3H_2O$	X104
C_{2v}	D_{2h}^{17}	$M^{II}SO_4$ (M^{II} = Mg, Ca, Mn, Co, Ni, Fe)	X34–5
	D_{2h}^{17}	Na_2CrO_4	X111
D_2	D_{2h}^{24}	Na_2SO_4, Ag_2SO_4	X22, 23, 36, 68, 107
C_2	D_3^4	LT forms of rare earth phosphates La—Tb	X106 X168–9
	C_{2h}^6	KIESERITE structure $M^{II}SO_4$, H_2O (M^{II} = Mg, Co, Ni, Mn, Zn)	 X113
	C_{2h}^6	$Nd_2(SO_4)_3$, $5H_2O$; $CaSO_4$, $2H_2O$; $CdSO_4$, $\frac{8}{3}H_2O$; NO_2ClO_4; $Ca_3(CrO_4)_2$	X143 X3 X179, X98
	D_4^4	$NiSO_4$, $6H_2O$	X12
	D_{2h}^{24}	$Zr(SO_4)_2$, $4H_2O$; $Ba(MnO_4)_2$.	X139, X151 X61–2

Table 7.14 (*cont.*).

Site group	Space group	Compounds	Ref.
C_s	D_{2h}^{16}	Li_3PO_4, and possibly Li_3VO_4, Li_3CrO_4, Li_3MnO_4	X186–7
C_s	D_{2h}^{16}	α-$M^{II}SO_4$ (M^{II} = Cu, Co, Fe, Zn)	X90, X34–5
	D_{2h}^{16}	β-K_2SO_4 structure $(NH_4)_2SO_4$; $\beta M^I_2SO_4$ (M^I = K, Rb, Cs, Tl), K_2CrO_4, K_2MnO_4, possibly Rb_2MnO_4, Cs_2MnO_4	X55 WVIIIb, 11
	D_{2h}^{16}	BARYTES structure $BaSO_4$, $SrSO_4$, $PbSO_4$, β-$BaSeO_4$, $BaCrO_4$, $BaFeO_4$, $KMnO_4$, $CsReO_4$, $TlReO_4$, M^IClO_4 (M^I = K, NH_4, Rb, Cs, Tl), possibly M^IMnO_4 (M^I = Rb, Cs, NH_4)	WVIIIa, 29
	C_{2v}^{7}	$M^{II}(ClO_4)_2$, $6H_2O$ (M^{II} = Mg, Cd, Zn)	X173
C_1	D_3^1	PHENACITE structure Li_2MoO_4, Li_2WO_4	X181
	D_2^4	$M^{II}SO_4$, $7H_2O$ (M^{II} = Mg, Ni, Zn)	X8, X13
	D_{6h}^4	$ScPO_4$, $2H_2O$	781
	C_{2v}^9	$SnSO_4$; $SrCrO_4$; Nd and Eu $(ReO_4)_3$, $4H_2O$	X122, X177, X120
	C_{2h}^5	$FeSO_4$, $7H_2O$; $CoSO_4$, $7H_2O$; Li_2SO_4; Na_2SO_4, $10H_2O$; Tutton's salts; Na_2ZnSO_4, $4H_2O$; $AgMnO_4$; La and Ce $(ReO_4)_3$, $4H_2O$ MONAZITE structure HT forms of rare-earth phosphates La—Tb.	X11, X1, X126, X72–4, X120 X155
	C_{2h}^6	$NiSO_4$, $6H_2O$	X146
	C_i^1	$CuSO_4$, $5H_2O$; Er and Lu$(ReO_4)_3$, $4H_2O$	X5, X57, X120
	D_{2h}^{15}	$Na_2MoO_42H_2O$ and possibly the corresponding tungstate.	X176

($CaWO_4$) structure is exhibited not only by tungstates and molybdates, but by perrhenates and periodates. Consequently these compounds will be dealt with primarily according to their structure and secondarily according to their chemical composition. In a number of cases, factor-group analyses have been carried out (e.g., zircon [963], monazite (chapter 3 of this book), scheelite [91], spinel [1508]), but the site group approximation will be used here.

The three compounds with the spinel structure, and also Ag_3PO_4, should display the free-ion selection rules. This appears to be so for Ag_3PO_4 (ν_3, 965; ν_4, 555 cm^{-1}) [1379] and Na_2WO_4 (ν_3, 835; ν_4, 300 cm^{-1}) [271]. However, Na_2MoO_4 seems to show a splitting of ν_4 (313, 290 cm^{-1}) though ν_3 is unsplit (830 cm^{-1}), and Ag_2MoO_4 shows four bands altogether (830, 420, 350, 285 cm^{-1}) [271].

The materials with the zircon structure should show four bands only, two components each of ν_3 and ν_4. Hezel and Ross have studied the rare-earth phosphates and found these selection rules obeyed in all cases. A representative set of results is that for YPO_4 (ν_3, 1050 and 990 cm^{-1}; ν_4, 641 and 521 cm^{-1}). Doyle and Eddy [386] found ν_3, unsplit, in $CaCrO_4$ at 880 cm^{-1}. The hexagonal rare-earth phosphates were studied crystallographically by Mooney [X106] who concluded that the space group was either D_4^6 or D_3^4. These imply site symmetries D_2 and C_2 respectively. Steger and Schmidt [1379] observed the three components of ν_3 and ν_4, and did not find ν_1; they thus preferred the first assignment. In a later study, Hezel and Ross [653] observed ν_1 and thus considered the second assignment correct. Weigel's X-ray work [168, 169] supports this conclusion. For $NdPO_4$, $\frac{3}{2}$ H_2O, Hezel and Ross found ν_1, 969; ν_3, 1064, 1042, 1026; ν_4, 619, 570, 539 cm^{-1}. The spectra of the other members of the group are similar except that ν_3 is not always resolved. Figure 7.3 shows the spectra of some phosphates and arsenates with the zircon,

---

Fig. 7.3 (a) Spectrum of anhydrous holmium phosphate; (b) spectrum of hydrated lutetium phosphate. Both these spectra show two doublets, one in the stretching and one in the bending region, consistent with the D_{2d} site symmetry of the phosphate ion in a phosphate with the zircon structure. The water in the lutetium phosphate appears to be uncoordinated water, as no librational modes are observed.

(c) Spectrum of the high temperature form of praseodymium arsenate; (d) Spectrum of the low temperature form of praseodymium arsenate. These have the monazite and $LaPO_4$, xH_2O structures respectively.

(e) Spectrum of barium molybdate. This compound has the scheelite structure, and shows five bands which can be assigned to fundamentals. The doublet near 850 cm^{-1} is assigned to ν_3, but there is some controversy over the assignment of the bending modes. The author assigns the band at 374 cm^{-1} to ν_2 and the two bands at 324,292 cm^{-1} to ν_4, corresponding to the assignments of the solution spectra (ν_2 = 407 cm^{-1}, ν_4 = 320 cm^{-1}).

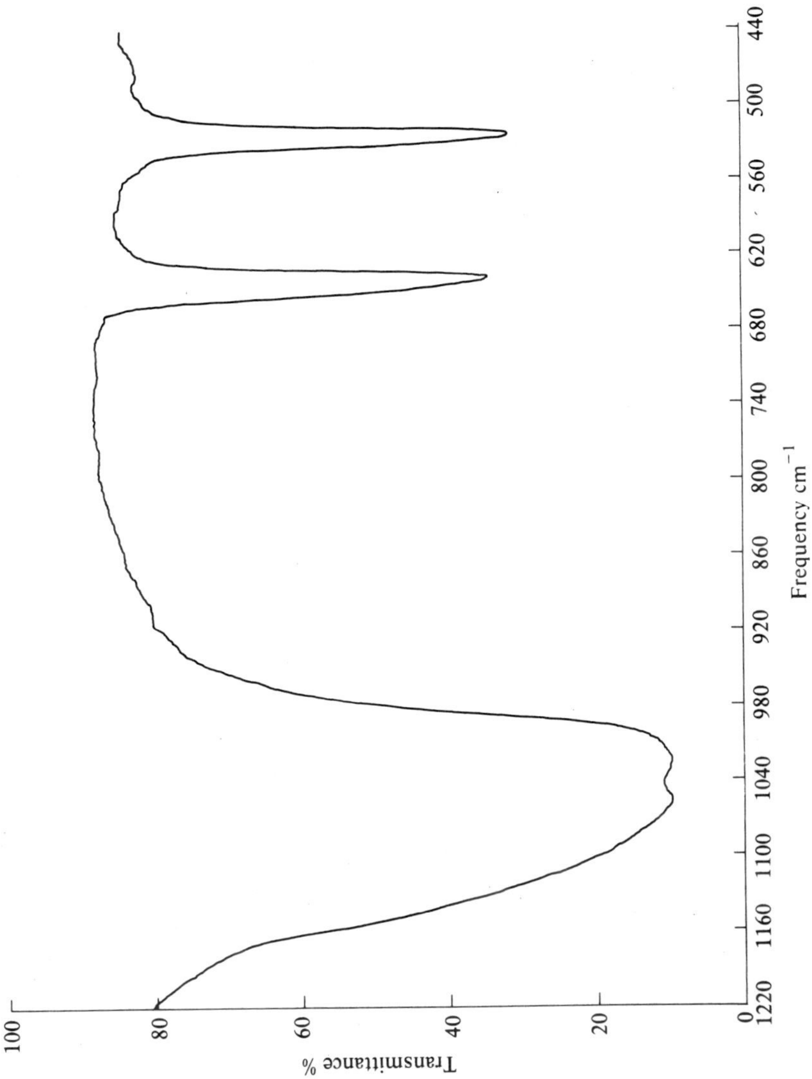
Transmittance %
100
80
60
40
20
0
1220
1160
1100
1040
980
920
860
800
740
680
620
560
500
440
Frequency cm^{-1}

Fig. 7.3(a)

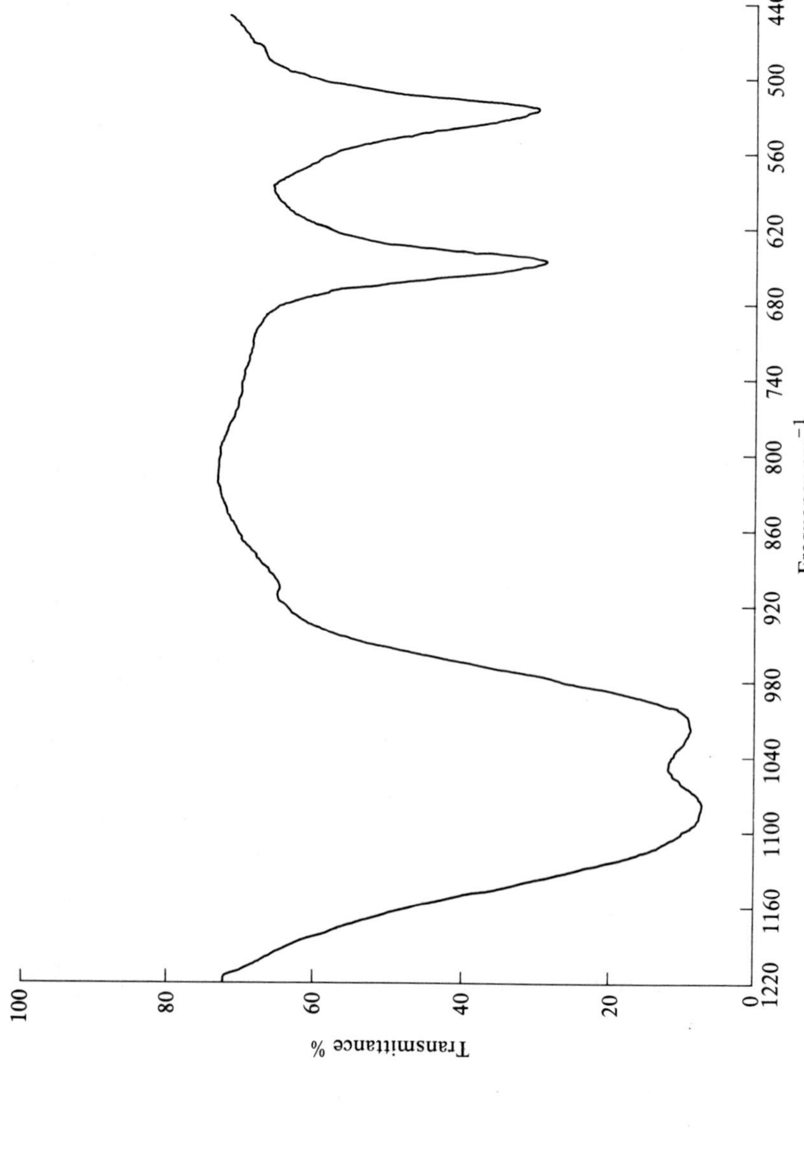
Transmittance %
100
80
60
40
20
0
1220
1160
1100
1040
980
920
860
800
740
680
620
560
500
440
Frequency cm^{-1}

Fig. 7.3(b)

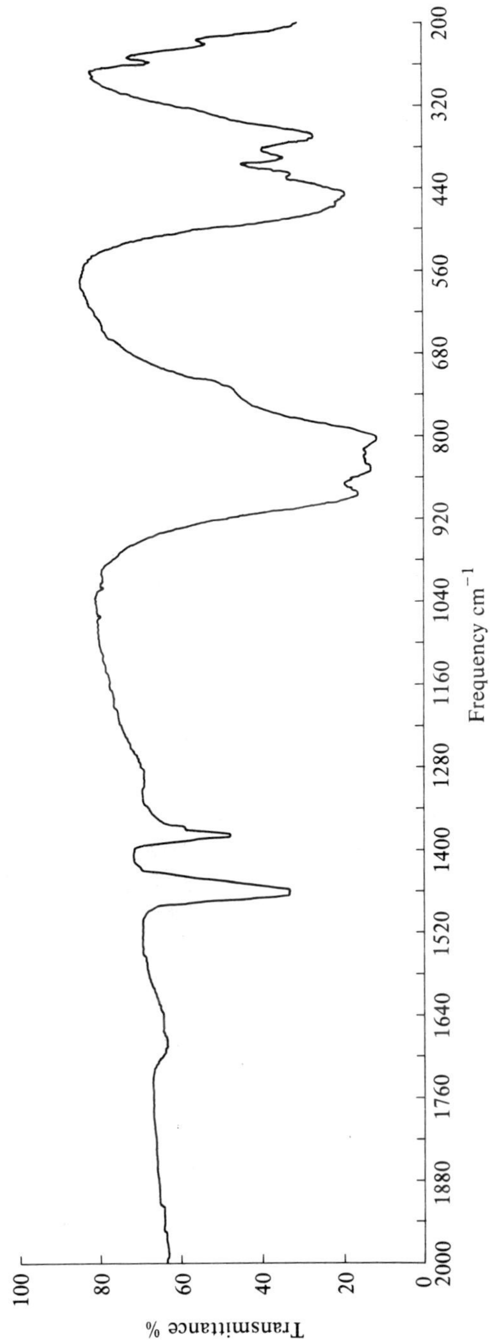
Transmittance %
100
80
60
40
20
0
2000
1880
1760
1640
1520
1400
1280
1160
1040
920
800
680
560
440
320
200
Frequency cm^{-1}

Fig. 7.3(c)

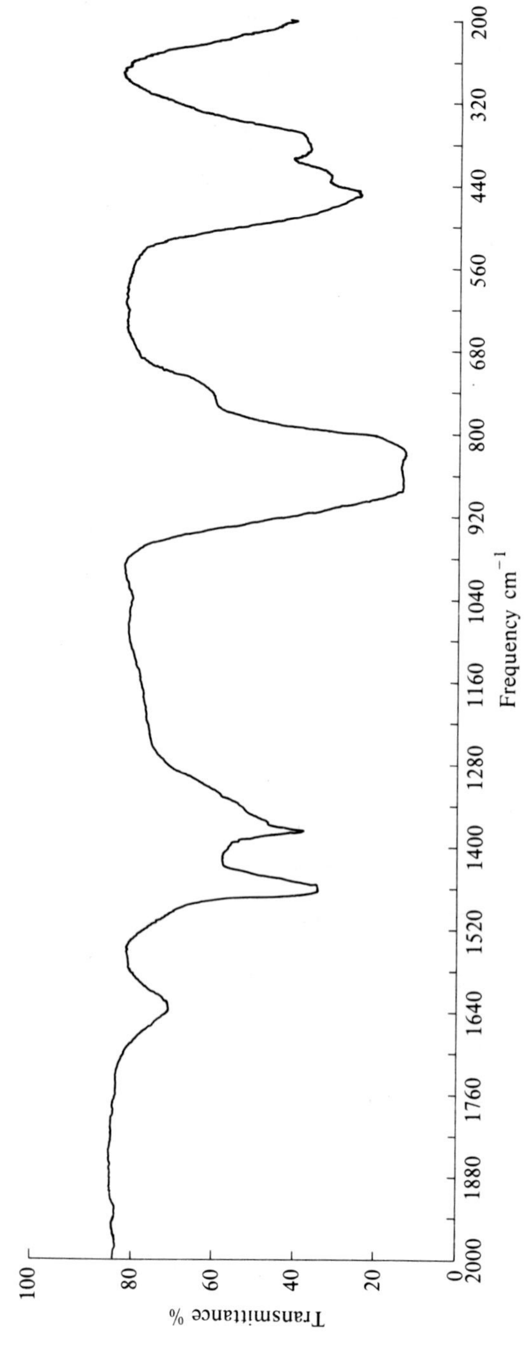
Transmittance %
100
80
60
40
20
0
2000
1880
1760
1640
1520
1400
1280
1160
1040
920
800
680
560
440
320
200
Frequency cm^{-1}

Fig. 7.3(d)

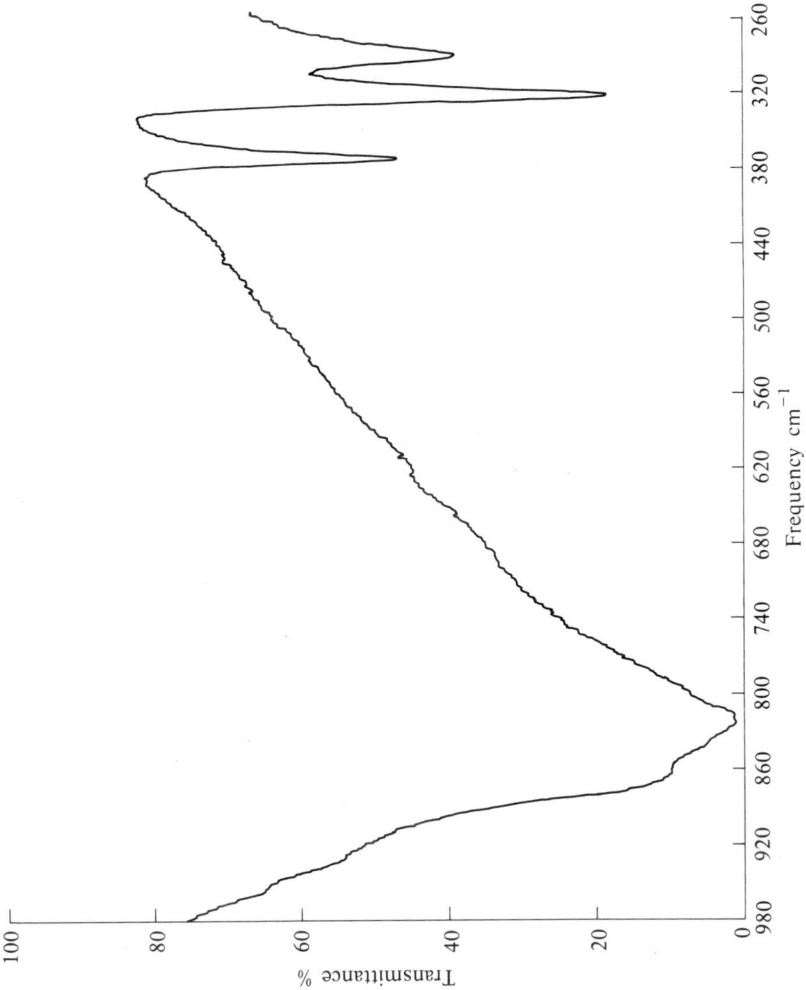
Transmittance %
100
80
60
40
20
0
980
920
860
800
740
680
620
560
500
440
380
320
260
Frequency cm^{-1}

Fig. 7.3(e)

monazite and hexagonal $LaPO_4$, xH_2O structures. The sulphate ion in Na_2SO_4 does occupy a site of D_2 symmetry; ν_1 and ν_2 were not observed by Steger and Schmidt, but ν_3 and ν_4 were found as doublets (ν_3, 1135, 1100; ν_4, 640, 620 cm^{-1}) rather than triplets as predicted. In the isostructural Ag_2SO_4, Baran [86] found vibrations at 1102, 1054, 973 cm^{-1} which could be interpreted as the three components of ν_3, though 973 cm^{-1} is more characteristic of ν_1. The selection rules for S_4 site symmetry are not adhered to by $BeSO_4$; Steger and Schmidt's assignment is ν_1, 980; ν_2, 575, 510; ν_3, 1150, 1090; ν_4, 920, 770, 740 cm^{-1}. The assignment of the 920 cm^{-1} band seems improbable. The numerous scheelites, in which the anion site symmetry is also S_4, do obey the selection rules, though there have been some disagreements about assignments, particularly among the tungstates and molybdates. The situation has been reviewed by Brown *et al.* [192a] whose assignments for tungstate, molybdate and periodate are given; that for the perrhenate is due to Ulbricht and Kreigsmann [1446–7] and for the pertechnate to Busey and Keller [215] and Mueller and Rittner [1032].

The scheelite structure is one of the few for which the correlation splitting of the internal modes has been observed. The selection rules in the factor-group approximation are ν_1, A_g(R) + B_u(inactive); ν_2, A_g + B_g(R) + A_u(ir) + B_u; ν_3, ν_4, B_g + E_g(R) + A_u(ir) + E_u(ir).

Table 7.15 Vibrational assignments in scheelites.

Assignment and group	$PbMoO_4$	$CaWO_4$*	KIO_4	$KTcO_4$	$KReO_4$
ν_1 A_g	872	912	792	913	965
B_u‡	859	893 (reflection)	796 vw.	—	—
ν_2 A_g	321	336	278	350, 348	348
B_g	354	401	293		
A_u	376	408	264	351	—
B_u‡	412	428 vw	—	—	—
ν_3 B_g	824	838	858	920, 887	924
E_g	772, 748	797	844		896
A_u	820	769	870	910	932
E_u	770	823	851		912
ν_4 B_g	296	336	—	327	330
E_g	384	409	—		336
A_u	268	290 †	314, 338	328	—
E_u	304	330	398	313	—

* Raman data from Scott [1260].

† One of these is an E_u rotational mode.

‡ Although these modes are inactive, it has been suggested [1158] that they are activated by strain in the sample. They appear very weakly. An analogous band at 870 cm^{-1} in the ir spectrum of $KReO_4$ was attributed to an unspecified combination involving lattice modes.

In the site groups C_{3v} and C_3, both ν_1 and ν_2 are activated and ν_3, ν_4 split into two components each. Mooney *et al.* [988] have made some calculations for strontium and barium phosphates, assuming that the P—O bond along the trigonal axis in the crystal is stronger than the others. Their calculations are in poor agreement with the study by Petreanu *et al.* [1127], whose observations (using normal $Ba_3(PO_4)_2$ and $Ba_3(PO_4)_2$ containing 75 per cent ^{18}O) gave ν_1, 980; ν_2, 392, 367; ν_3, 1082; ν_4, 590, 540, 489, 435 cm^{-1}.

Guerchais [589] assigned two bands to ν_3, at 1051 (split) and 1000 cm^{-1}; in the corresponding vanadate, chromate(V) and manganate(V), which may be isostructural with the phosphate, he gave for ν_3 respectively 862, 809; 847, 773; 828, 758 cm^{-1}. Baran and Aymonimo [87] agreed with these values for barium chromate(V), and gave 839, 783 cm^{-1} for ν_3 in the strontium compound. Assignments by Campbell [219] and Doyle and Eddy [386] of chromates(V) have been criticized by Mueller [1016], who gives ν_1, 830; ν_2, 330; ν_3, 765; ν_4, 330 cm^{-1}. He points out that the increase in the negative charge on $(XO_4)^{n-}$ correlates with an increase in ν_1 and a decrease in ν_3. The assignment for K_2CrO_4 solution [1358] is ν_1, 849; ν_2, 348; ν_3, 884; ν_4, 368 cm^{-1}.

The isostructural trihydrated lithium perchlorate and permanganate have been studied. Mathieu and Couture-Mathieu [918] found all nine lines in the Raman spectrum of the perchlorate (ν_1, 936; ν_2, 432, 464; ν_3, 1075, 1101, 1105; ν_4, 623, 635, 639 cm^{-1}). Doyle and Kirkpatrick [387] found, as predicted, three bands in the stretching region for the permanganate, at 930, 895, 845 cm^{-1}. For the strontium salt, however, they found six bands in this region at 945, 930, 920, 900, 890 and 845 cm^{-1}. The expected spectrum of $Ba(ClO_4)_2$, $3H_2O$ was not found by Ross [1215], as ν_3 was split into three rather than two components (1143, 1111, 1089 cm^{-1}); ν_1 was found at 941, ν_4 at 627, 637 cm^{-1}.

The anhydrite ($CaSO_4$) structure has been analysed by Krishnan and Shantakumari [805]; their labelling of the irreducible representations is incorrect. They have identified all nine of the Raman active internal modes, and Decius [338], who studied alkali halides doped with $CaSO_4$, has found all eight of the ir active modes when KBr was used as the host crystal. The assignment (Raman active modes given first) is ν_1, 1018; 976; ν_2, 497, 416; 496; ν_3, 1160, 1128, 1108; 1180·5, 1150·5, 1080·5; ν_4, 674, 628, 608; 648, 627, 612 cm^{-1}. The other sulphates with this space group and site group show similar ir spectra.

As Tables 7.13 and 7.14 show, there are very many compounds in which the site symmetry is sufficiently low to activate all nine vibrations in the infrared. The major deviations from the selection rules observed in these compounds are, firstly, the absence of ν_2, and secondly, an incorrect number of bands in the ν_3 region. Work on the rare-earth phosphates with the monazite structure [653] has demonstrated the great weakness of ν_2 relative to ν_1; the expected extinction coefficient of ν_2 in most compounds containing tetrahedral XO_4 anions is less than 5. In the ν_3 region, the vibrations are sometimes unresolved; but in the monazites there are extra bands which probably arise from combinations involving ν_4 and ν_2. Correlation splitting has been observed in gypsum by Hass and Sutherland [618] and in $BaSO_4$ by Ramdas [1181]. The spectra of anhydrous transition-metal sulphates have been studied by Pannetier [1107] in both the α- and β-forms; although the structures differ, the selection rules in the site-group approximation are the same.

Table 7.16 Vibrational spectra of compounds containing tetrahedral anions, on sites of low symmetry.

Compound		ν_1	ν_2	ν_3	ν_4	Ref.
Li_3PO_4	ir	—	535, 515	1090	610	1379
			485, 445, 405	1050	590	
$SmPO_4$	ir	962	488	1099, 1062	635, 580	653
				1040, 1018	568, 539	
				1000		
$SmPO_4, \frac{3}{2}H_2O$	ir	975	—	1070	621	653
				1040, 1026	570, 540	
$ScPO_4, 2H_2O$	ir	—	420	1180	620	772
				1050	560	
K_2SO_4	ir	983	458	1152, 1132	619	1246
			448, 438	1109, 1098	612	
				1092	605	
$CaSO_4, 2H_2O$	ir A_u	1000	—	1131	602	618
	B_u	—	—	1142, 1118	674, 604	
	R A_g	1006	492, 413	1144	621	
	B_g	—	—	1138, 1117	669, 624	
$SrSO_4$	ir	974 A_1	491 A_1	1147 A_1	627 A_1	338
				1081 B_1	613 B_1	
				1179 B_2	642 B_2	
$BaSO_4$	ir	980	470, 439	1205, 1136	641	1181
				1100, 1042	633, 614	
	R A_g	988	453	1145, 1105	630, 622	
	B_{2g}	967	462, 453	1167, 1138	647	
	B_{1g}, B_{3g}	—	—	1084	622	
α-$CoSO_4$	ir	988	474	1200, 1129, 1042	690, 677, 593	1107
β-$CoSO_4$	ir	991	480	1200, 1145, 1057	679, 604, 595	
$NiSO_4, H_2O$	ir	1017	—	1149, 1121, 1087	671, 628, 602	653
$NiSO_4, 6H_2O$	tetragonal R	987	466, 436	1131, 1088, 1050	639, 620, 596	792
		971	422, 374			
	monoclinic R	984	472, 439, 397	1148, 1080	620	
$ZnSO_4, 7H_2O$	R	990	461, 445	1141, 1121, 1094	620, 609	1280
				1062, 1057		
	ir	982	—	1142, 1111, 1093	652, 629, 609	653
$CdSO_4, \frac{8}{3}H_2O$	R	1006	462, 456	1157, 1115	626	803
				1101, 1076		
	ir	991	—	1124, 1087	635, 620	653
$Ce_2(SO_4)_3, 5H_2O$	ir	960	468, 418	1145, 1100, 1025	670, 618, 595	1159

Table 7.16 (*cont.*).

Compound		ν_1	ν_2	ν_3	ν_4	Ref.
$Na_2Zn(SO_4)_2, 4H_2O$	R	990	477, 452	1191, 1163 1104, 1071	621, 609	804
$K_2Zn(SO_4)_2, 6H_2O$	R	990	462, 450	1157, 1128 1112, 1084	634 612	57
	ir	985	461, 441	1142, 1100	642 632, 615	
β-$BaSeO_4$	ir	840	357, 344	926, 884, 862	452, 416	1416
K_2CrO_4	ir	853	397 or 384	915 split, 893	397 or 384	
$BaCrO_4$*	ir	860	—	949, 894, 873	419 389, 375	
$SrCrO_4$	ir	840	—	905, 890, 875	—	386
Li_2MoO_4†	ir	—	440	830	370	271
Li_2WO_4†	ir	—	415	835	330, 290	
$Na_2MoO_4, 2H_2O$	ir	898, 903	—	855, 820, 805	—	215
	R	897, 905	—	843, 836, 805	—	
$Na_2WO_4, 2H_2O$	ir	931	—	855, 825, 808	—	
	R	931	—	893, 838	355, 276	
$KClO_4$	R	938	461	1130, 1080	627	278
	ir	941	464‡	1143, 1111 1087, 1085	637 629, 627	653
NO_2ClO_4	R	938	460	1143, 1082	630	944
$Zn(ClO_4)_2, 6H_2O$	R	935	465	1131, 1115 1086, 1073	634, 625	1494a
	ir	941	—	1143, 1111, 1093	637 628, 625	653
$BaFeO_4$*	ir	790	—	870, 818, 783	362, 340, 306	1416
K_2MnO_4	ir	—	—	837	—	1209
$KMnO_4$	ir	844	385	910	407	1139
	ir	850	—	910	406, 395, 387	639
	R	841	345	—	420–415	
$Ba(MnO_4)_2$	ir	840	—	940, 930 920, 885	—	387
$Mg(MnO_4)_2, 6H_2O$	ir	859	—	904	—	
$CsReO_4$*	R	969	—	932, 923, 899	323, 331 335, 340	1446–7
	ir	967	—	928, 914, 902	—	
$La(ReO_4)_3, 4H_2O$	ir	973	—	934, 920 908, 872	(332)	
	ir	—	—	940–910 880	(325)	
	ir	965	—	930, 874	(315)	
$KRuO_4$	ir	826	—	848	282	572

* In these compounds all the bending modes are listed under ν_4. For $CsReO_4$, Mueller [1020] has suggested the assignment ν_2, 320, 315; ν_4, 335 cm^{-1}.

† Clark and Doyle did not assign the bending modes; they are tabulated to correspond with the assignments in Table 7.16.

‡ Taken from Schroeder *et al.* [1246].

The above table is by no means exhaustive, but contains a member of each of the structural groups listed in Table 7.14. A large number of other compounds containing tetrahedral anions, but whose crystal structure is not known, has also been studied. The only species of which all the compounds studied are of unknown structure is ruthenate(VI); Griffith [572] has found ν_1, 808; ν_2, 323; ν_3, 837, 790; ν_4, 323 in K_2RuO_4; for $BaRuO_4$ his values were 811, 335, 840, 330 cm^{-1}.

Table 7.17 Raman spectra of tetrahedral oxo-anions in solution.

Compound	ν_1	ν_2	ν_3	ν_4
$PO_4^{\equiv}$	938	420	1017	567
$AsO_4^{\equiv}$	810	342	810	398
$VO_4^{\equiv}$	874	345	855	345
$SO_4^{=}$	981	451	1104	613
$SeO_4^{=}$	837	345	873	314
$CrO_4^{=}$	847	348	884	368
$MoO_4^{=}$	894	407	833	320
$WO_4^{=}$	931	405	833	318
ClO_4^-	928	460	1128	645
BrO_4^-	801	331	878	410
IO_4^-	791	256	853	325
MnO_4^-	838	355	921	429
TcO_4^-	912	—	912	325
ReO_4^-	971	331	918	331

Some compounds have been studied in which the XO_4 group acts as a monodentate or bidentate ligand; in the latter case, both bridging and chelate ligands are known. In principle, monodentate and bidentate ligands may be differentiated by the splitting of ν_3 and ν_4 into 2 components under C_{3v} symmetry, 3 under C_{2v} symmetry. However, further reductions in symmetry due to crystal structure have been observed for monodentate sulphato groups [656]. A suggestion that the forbidden ν_1 should be stronger, and the splittings of ν_3 and ν_4 greater, for bidentate compared with monodentate ligands [1055], has been disproved as far as the cobaltammine sulphato complexes are concerned [656]. In some cases the two sets of criteria are contradictory, as for $Cu(en)_2(ClO_4)_2$ which is assigned as containing monodentate ClO_4^- since there are no bands close to 1200 cm^{-1} (i.e., the splitting of ν_3 is relatively small), but shows ν_3 as a triplet, suggesting a bidentate group [935].

Eskenazi [436] has investigated seven sulphato-complexes of palladium; in each of the four bidentate chelate complexes there is a component of $\nu_3 > 1200\ cm^{-1}$. A number of sulphato complexes involving trivalent and tetravalent metals have been studied [446, 754] but no clear distinction between those assigned as monodentate (e.g., $Rh_2(SO_4)_3$, $14H_2O$) and bidentate ($Rh_2(SO_4)_3$, $4H_2O$) is apparent from the results. Warrier [1487] has suggested that $Th(ClO_4)_4$, 4 and 8 H_2O, $ZrO(ClO_4)_2$, $2H_2O$ and

Table 7.18 Spectra of some tetrahedral ligands XO_4.

Structure		ν_1	ν_2	ν_3	ν_4	Ref.
MONODENTATE						
$[Co(NH_3)_5SO_4]I$		956	470	1149, 1124, 1029	644, 608, 593	656
$Cu(ClO_4)_2, 3H_2O$		920	480	1158, 1030	648, 620, 605	619
$Cu(bipy)_2(ClO_4)_2$		934	412	1111, 1087, 1069	663, 644	935
$[Co(NH_3)_5CrO_4]Cl$	ir	875	340	915, 895	373, 358	284
	R	872	338	910	388	
$[Co(NH_3)_5MoO_4]Cl$	ir	910	—	877, 833	350	
	R	908	335	—	350	
BIDENTATE CHELATE						
$TiOSO_4, 3H_2O$		917	466	1199, 1142, 1005	621, 609, 592	1463
$PdSO_4, 2H_2O$		960	—	1235, 1089, 995	—	436
$NaIn(SO_4)_2, 2H_2O$		1017	486, 446, 406	1202, 1165 1120, 1088	643, 619 603, 553	754
$[In(SO_4)_2]^-$	aqueous solution†	1000	—	1125	625	649
$[NH_4)_3In(SO_4)_3$		1011 999	462	1174, 1115, 1050	668, 659, 653 614, 601	754
$[Co(NH_3)_4CrO_4]NO_3$		850	350	944, 925, 893	398	284
$[Co(NH_3)_4MoO_4]NO_3$		920	—	868, 845, 795	371	
BIDENTATE BRIDGING						
$Co(NH_3)_4PO_4$		894	—	1106, 1055, 1040, 917	643 576, 531	656
$Co(NH_3)_4PO_4, 2H_2O$		895	—	1109, 1043, 918	644	1299
$Co(NH_3)_4PO_4, 3H_2O$		901	—	1083, 1060, 912	641 575, 539	656
$Th(SO_4)_2, 4H_2O$		1040 970	420 400	1265, 1227 1135, 1100	642, 615 608, 590	447
$PdSO_4$		996	—	1160, 1105, 1035	—	436
RCl_3*		996	452 422	1159, 1106, 1064	650, 636, 604	656
$Mn(ClO_4)_2, 2H_2O$		945	467 453	1210, 1138, 1030	666, 635, 613	619
$Cu(bipy)_2SO_4, 2H_2O$		971	—	1166, 1096, 1053–1035	643, 604, 595	935

* R = 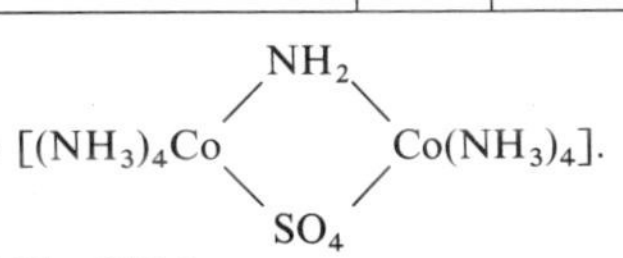

† 2.1M $In_2(SO_4)_3$.

$TiO(ClO_4)_2, 6H_2O$ contain a monodentate ClO_4^- group, as has Vdovenko [1453] for $UO_2(ClO_4)_2, 3H_2O$. He has also studied the anhydrous compound and the mono-, penta- and heptahydrates. Hathaway [619] considers that the dihydrates of Co and Ni perchlorates contain a bridging group, as in the Mn compound.

Some incomplete assignments for RuO_4 and OsO_4 have been given [369, 624, 1093]; the first complete assignment of OsO_4, by Levin and

Abramowitz [854] involved a band in the vapour at 353 cm^{-1}, ascribed to ν_2, which has not been found by subsequent workers. McDowell [932] gave for OsO_4 vapour ν_1, 970; ν_2, 333; ν_3, 960·5; ν_4, 326 cm^{-1}. Raman studies of the liquid have been made by Griffith [572] and Davidson [321]; Griffith's values are ν_1, 964; ν_2, 338; ν_3, 952; ν_4, 334 cm^{-1}. Davidson found ν_2 and ν_4 to coincide at 338 cm^{-1}. For liquid RuO_4, Griffith gives ν_1, 882; ν_2, 323; ν_3, 914; ν_4, 334 cm^{-1}. Davidson's values agree closely. Selig [1270] has found ν_3, 877 and ν_4, 306 cm^{-1} in the ir spectrum of XeO_4 vapour.

(iv) Thio-anions. Very few of these have been studied, and there is little structural information. Gattow [511] has shown $(NH_4)_2MoS_4$ to be isomorphous with β-K_2SO_4, but ν_3 does not show the splitting characteristic of an orthorhombic structure.

Table 7.19 Vibrational spectra of tetrahedral thio-anions.

Compound		ν_1	ν_2	ν_3	ν_4	Ref.
Na_3PS_4	ir solid	420	210	540	280	851
	R	413	218	560	277	
Na_3VS_4	ir solid	404	—	470	200?	1021, 1027
$MoS_4^{=}$	ir	460	—	480	155	511
$WS_4^{=}$	ir	450	—	466	145	
$WS_4^{=}$		485	178	460	185	851

Mueller and Krebs [1021] state that transition metal thio-anions have more M=S character than the corresponding main group compounds. Gattow [511] found bands, which he left unassigned, at 220, 195 cm^{-1} in the thiomolybdate and 215, 187 cm^{-1} in the thiotungstate. The reversal of his assignment by Leroy and Kaufmann [851] is based on calculations employing both a UBFF and a GVFF.

(b) Square planar molecules

Most of the compounds studied are halide complexes of Au(III), Pd(II) and Pt(II); the interhalogen compounds $KICl_4$ and $RbClF_4$ have been investigated, and a complete assignment exists for XeF_4. Some inconsistencies exist in the labelling of the modes; the convention used here is that of Claassen [255]. The inactive out-of-plane bending mode (ν_4,(B_{1u})) in XeF_4 may be represented by the line at 231 cm^{-1} in the Raman spectrum of the solid. Hyman [689] has suggested that the band at 123 cm^{-1} in the ir spectrum of XeF_4 is a rotational line of HF; the calculation due to Yeranos [1549] gives 182 cm^{-1}. The assignment for $RbClF_4$ is unusual in that the stretching mode ν_5 is below the bending frequencies ν_2, ν_7; Christe and Sawodny [248] have stated that this situation can arise because of the semi-ionic nature of the bonds (three-centre, four-electron), combined with the small size of the central atom and the small (i.e., near 90°) bond angles.

Some recent studies by Christe *et al.* [1641, 1642] have led to a reassignment of the frequencies of the ClF_4^- ion in which ν_5 is above ν_7 but below ν_2; for BrF_4^-, ν_5 lies above both ν_2 and ν_7.

Table 7.20 Vibrational spectra of square planar XY_4 groups.

Compound	Assignment and description						Ref.
	$\nu_1(A_{1g})$	$\nu_2(A_{2u})$	$\nu_3(B_{1g})$	$\nu_5(B_{2g})$	$\nu_6(E_u)$	$\nu_7(E_u)$	
	ν_{sym} in phase	π	δ_{sym}	ν_{sym} out of phase	ν_{as}	δ_{as}	
K_2PtCl_4	333	168	196	306	321	191	636–7
K_2PtBr_4	205	135	125	190	232	135	636–7
K_2PtI_4	142	105	—	126	180	127	636–7
K_2PdCl_4	310	—	198	275	336	193	636–7
K_2PdBr_4	192	130	125	165	260	140	636–7
$AuCl_4^-$	347	—	171	324	356	173	1224
$AuBr_4^-$	212	—	102	196	252	100	1224
$KAuI_4$	148	—	75	110	192	113	636–7
ICl_4^-	288	—	128	261	258	107	1126
XeF_4	543	291	235	502	586	123	254–5
$RbClF_4$	505	486	288	—	748	430	248
ClF_4^-	505	425	288	417	680–500	237	1641
BrF_4^-	523	317	246	449	542	183	1642

(c) Trigonal bipyramidal and other molecules

Microwave [X19, X153] and the electron diffraction [X88] measurements have shown that SF_4 has the C_{2v} symmetry arising from the non-occupation of one equatorial position. The two assignments for SF_4, due to Dodd [370a] and Levin and Berney [855] disagree regarding ν_2, ν_3, ν_5 and ν_9.

Levin and Berney considered the band at 715 cm^{-1} to be either an overtone of ν_3 or a hot band transition involving ν_8. They assigned the weak band at 460 cm^{-1} to $2\nu_9$. The assignment of ν_5 was based on observed combination frequencies. Rolfe *et al.* [1214] studied the Raman spectrum of liquid SeF_4, and found a strong polarized band at 744 cm^{-1}, strong (probably depolarized) bands at 545 and 360 cm^{-1}, and weak bands at 1009, 713, 400 and 194 cm^{-1}. Aynsley [67] studied SeF_4 in a CCl_4

Table 7.21 Vibrational assignments in SF_4.

Reference			[370a]	[855]				[370a]	[855]
A_1	ν_1	ν_{sym}(eq)	889	892	B_1	ν_6	ν_{as}(axial)	867	867
	ν_2	ν_{sym}(axial)	715	558		ν_7	δ	532	532
	ν_3	δ	557	353	B_2	ν_8	ν_{as}(eq)	728	728
	ν_4	δ	235	171		ν_9	δ	463	226
A_2	ν_5	Torsion	401 or 645	645 or 401					

matrix; both this spectrum and the spectrum of the solid suggested the existence of association between a fluorine atom of one molecule and an Se atom of another. NMR and Trouton constant measurement support the idea of association.

The compound $(NH_4)_2CS_4$ has been studied by Mueller and Krebs [1022]. They observed bands at 955, 825, 519 and 485 cm^{-1} in the ir spectrum. The higher frequencies (attributed to stretching) are greater than those in $CS_3^{=}$, and it was thus concluded that the $CS_4^{=}$ ion could not be tetrahedral. They suggested the planar arrangement

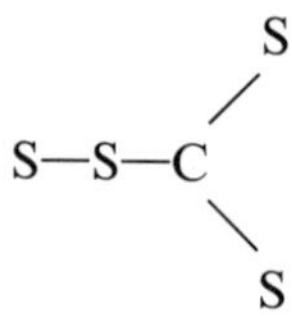

7.3 Molecules of the formula XY_3Z

(a) Hydrides or deuterides where Z is a halogen

The compounds studied are those in which X is C, Si or Ge. All the data given is for the gas phase. The Ge compounds have also been studied in the solid phase; at 77°K the A_1 frequencies are all lowered, while the E frequencies showed random shifts from the gas-phase values. No splittings were observed.

(b) Halides and thio-anions where Z is oxygen

For NOF_3, Skiens and Cady [1334] give the unassigned ir spectrum of the vapour as 1720 (doublet), 1430, 1300vs, 925 triplet, 820 triplet, 730 cm^{-1}. Mitra [971] has studied $AsOF_3$ and gives for the ir spectrum of the vapour 811, 784 (ν(As—O)) and 703 cm^{-1} (ν(As—F)). The assignment of two bands to ν(As—O) seems incorrect unless they represent the P and R branches of the same fundamental. Slight differences in the assignments of the phosphorus oxyhalides exist, particularly regarding ν_3 and ν_5 of

Table 7.22 Vibrational frequencies of monosubstituted hydrides.

Compound	Assignment and description						Ref.
	$\nu_1(A_1)$	$\nu_2(A_1)$	$\nu_3(A_1)$	$\nu_4(E)$	$\nu_5(E)$	$\nu_6{}^*(E)$	Ref.
	ν(X—Z)	(X—H)	δ	ν(X—H)	δ(HXH)	δ(HXZ)	Ref.
CH_3F	1049	2965	1464	3006	1467	1182	15
CH_3Cl	732	2967	1355	3042	1455	1015	15
CH_3Br	610	2972	1306	3060	1449	954	15
CH_3I	533	2970	1251	3062	1439	881	15
SiH_3F	872	2206	990	2196	943	728	15
SiH_3Cl	551	2201	949	2195	954	664	15
SiH_3Br	430	2200	930	2196	950	633	15
SiH_3I	362	2192	903	2206	941	592	15
GeH_3F	689	2121	859	2132	874	643	488
GeH_3Cl	423	2121	848	2129	874	602	488
GeH_3Br	305	2116	833	2127	871	578	488
GeH_3I	248	2112	812	2121	854	558	488
GeD_3F	618	1525	696	1535	635	466	488
GeD_3Cl	421	1522	614	1530	630	434	488
GeD_3Br	304	1515	597	1530	621	420	488
GeD_3I	248	1510	582	1525	603	404	488

* This mode is sometimes described as XY_3 rocking.

Table 7.23 Vibrational frequencies of oxygen-substituted halides and thio-anions.

Compound		Assignment and description						Ref.
		$\nu_1(A_1)$	$\nu_2(A_1)$	$\nu_3(A_1)$	$\nu_4(E)$	$\nu_5(E)$	$\nu_6(E)$	Ref.
		ν(X—O)	ν(X—Y)	δ	ν(X—Y)	δ	δ	
$Zn(BF_3O)$, H_2O	ir solid, R soln. H_2O	1048	768	464	1376	530	384	872
NOF_3		1691	743	528	883	558	400	304
POF_3	R	1415	873	473	990	485	345	595
$POCl_3$	R liq.	1290	486	267	581	249	193	355, 358
$POBr_3$	R	1261	340	173	488	267	118	355, 358
$VOCl_3$	ir, R liq.	1035	408	165	504	249	129	953
$VOBr_3$	ir	1025	271	120	400	212	83	946
Cs_2MoOS_3	ir solid	857	467	190	477	195	270	849
Cs_2WOS_3	ir solid	873	473	185	459	194	260	849
$[AsOF_3]$	R soln. H_2O of K salt	833	705	376	705	376	305	911
SOF_3^+	R solid, AsF_6^- salt	1538 1529	911	513	1063 1057	497	391	1623

POF_3. Delwaulle [358] gave both frequencies as 476 cm^{-1}; Ziomek and Piotrowski [1559] reversed the tabulated assignment. They also gave a lower value ($\sim$1390 cm^{-1}) for ν_1 than Gutowsky and Liehr [595]. Leroy and Kaufmann [849] have studied the $M^{VI}OS_3^=$ ion in the Raman as well as in the ir; the values obtained are similar. They found the spectra of the tetramethylammonium salts to resemble closely those of the Cs salts.

(c) Halides and oxo-anions where Z is sulphur

There is much disagreement about the assignment for PSF_3. The original Raman study by Delwaulle and Francois [356] gave ν_1, 695; ν_2, 874; ν_3, 440; ν_4, 940; ν_5, 402, ν_6, 276 cm^{-1}. This assignment (with minor variations in the value of ν_2) is given in both Nakamoto's and Siebert's books, and was also accepted by Horn *et al.* [677]. Ziomek [1559] had an assignment, in which ν_2 ($\nu_S PF_3$)) and ν_1 (ν(P=S)) were reversed in magnitude (ν_1, 940; ν_2, 695 cm^{-1}). Doubt was cast on all these assignments by three later studies, due to Cavell [228], Durig and Clark [410] and Mueller [1025]. Durig and Clark found a weak band at 880 cm^{-1} in the ir, which they assigned to $2\nu_3$; it was absent from the Raman. Cavell felt that the band (which he found at 882 cm^{-1}) might be an impurity, but could not identify it. Mueller did not mention a band in that region at all. Of these three studies, Cavell's gave ν_1, 983; ν_2, 698 cm^{-1}, while the other two reversed that assignment. Arguments based on intensities, correlations with other compounds containing P=S, electronegativities, potential energy distributions and treatment of the group as a diatomic 'molecule' were advanced in these studies. The potential energy distribution, which would have provided the strongest argument, was not given by Mueller, who referred to it as unpublished work by Krebs. A recent calculation by Shurvell [1292a] leads to the conclusion that ν_1 and ν_2 are strongly mixed, and this is shown by his PED values in Table 7.24. His fit to the observed

Table 7.24

Assignment	Force constant and mode			
	F_1 (PF stretch)	F_2 (PS stretch)	F_3 (FPF bend)	F_4 (SPF bend)
ν_1 PF stretch	0·453	0·396	0·114	0·037
ν_2 PS stretch	0·529	0·433	0·029	0·009
ν_3 sym. bend	0·171	0·171	0·612	0·199

A_1 frequencies is not particularly good; he obtained 1015, 644, 417 cm^{-1} as against the observed values of 983, 698, 442 cm^{-1}, and the PED above does not include a stretch–stretch interaction constant. He states, however, that inclusion of other interactions does not greatly affect the PED, and in the author's opinion, Shurvell's conclusion about the assignment of $\nu(P{=}S)$ and $\nu_S(PF_3)$ should be adopted. Marsult-Herail and Tartar [910] have studied $PSCl_3$ (and $POCl_3$) in the solid down to 200 cm^{-1} and found splittings of all the bands they observed. They attribute these partly to isotope effects and partly to coupling. Although $TlReO_3S$ has the space group D_{2h}^{16}, with the Re on a C_S site, no splitting of the E vibrations was found.

Table 7.25 Vibrational frequencies of sulphur-substituted halides and oxo-anions.

Compound		Assignment and description						
		$\nu_1(A_1)$	$\nu_2(A_1)$	$\nu_3(A_1)$	$\nu_4(E)$	$\nu_5(E)$	$\nu_6(E)$	Ref.
		$\nu(X{-}S)$	$\nu_s(X{-}Y)$	δ	$\nu_{as}(X{-}Y)$	δ	δ	
NSF_3	ir vap.	1515	775	521	811	429	342	1198b
PSF_3	ir vap.	983* 698		442	951	404	275	1292a
$PSCl_3$	ir vap., R liq.	770	431	252	847	248	174	410
$PSBr_3$	R liq.	726	299	165	431	175	133	1049
PO_3S^{3-}	R soln. H_2O	436	961	610	1037	519	367	1376
AsO_3S^{3-}	ir mull, Na salt.	441	791	378	441	390	312	911
ReO_3S^-	ir, Tl salt.	504	938	330	891	310	240	846
$S_2O_3^=$	R soln. H_2O Na salt.	446	995	669	1132	541	335	1295–6

* See text.

Newman [1065] has studied some complexes of $S_2O_3^=$ in an attempt to determine whether the ion, when functioning as a bidentate ligand, is chelate or bridging. $Ni(thiourea)_4S_2O_3, 4H_2O$ shows three S—O stretching modes and is assigned as probably chelate.

$(NH_4)_2M_2I_2(S_2O_3)$, where M=Ag or Cu, show splitting of ν_{as} and ν_s(S—O) which could be interpreted as belonging to a chelate ligand though the spectrum is complicated by possible Fermi resonance.

(d) Other substituted XO_4 groups

These include substituted phosphate, sulphate, selenate and perchlorate ions, where Z is H, OH or a halogen; the halochromates; the osmiamate $KOsO_3N$ and the corresponding rhenium compound K_2ReO_3N; the rhenium oxyhalides; and perchloryl fluoride and oxyfluoride. Few of the assignments call for comment. Simon and Schmidt [1321] have established that $RbHSO_3$ and its Cs analogue contain the ion

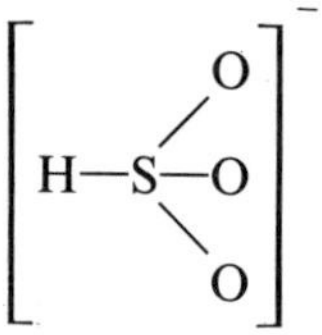

rather than

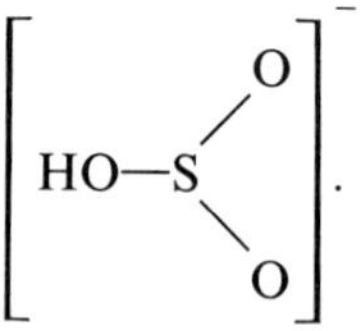

For $NaSO_3Cl$, Ciruna and Robinson [252] assign the SO_3 bending frequencies such that $\nu_5 > \nu_3$; Steger [1372] reverses this. A more recent Raman study of several solid chlorosulphates and of HSO_3Cl and $NaSO_3Cl$ in solution has led to a further revision in which $\nu_3 > \nu_5$ and the positions of ν_4 and ν_6 are substantially altered [1892]. An early study [398] of the fluorochromate ion involves a band at 460 cm^{-1}: Stammreich considers this to be an impurity [1367]. Mueller [1026] has revised the assignment given by Miller and Carlson [949] for ν(Re—Cl), $\delta(ReO_3)$ in ReO_3Cl. They gave these as 293, 435 cm^{-1} respectively. Mueller has shown that this leads to force constants of 1·60, 2·06 mdyn $Å^{-1}$ for Re—Cl stretching and ReO_3 bending; the reversed assignment gives the more realistic values 3·25, 1·01 mdyn $Å^{-1}$. He has not suggested a corresponding alteration in the assignment for ReO_3Br. The O—F frequency in $ClO_3(OF)$ has been given by Agahigian [45] as 885 cm^{-1}.

7.4 Molecules of the formulae XY_2Z_2, XY_2ZW and XYZWV

Compounds of the formula XY_2Z_2 which have been studied include substituted hydrides of group IV elements and a variety of molecules where Y is oxygen and Z is H, OH or a halogen. Since these molecules are of low symmetry, there is little or no structural information to be obtained from breakdown of selection rules. Some information about bonding has been deduced in a few cases. Mueller and Dehnicke [1035] have studied $Be(PO_2Cl_2)_2$ and $M(PO_2Cl_2)_3$ where M = Al, Ga, In, Fe. They concluded that the Be compound was ionic, since no vibrations characteristic of the beryllium–oxygen coordination polyhedron were observed. They did assign the F_{1u} stretching and bending modes of MO_6 in the other compounds (Al, 542, 337; Ga, 498, 246; In, 481, 190; Fe, 492, 273 cm^{-1}). They could not determine whether the PO_2Cl_2 group was chelate, leading

Table 7.26 Vibrational frequencies in other compounds containing monosubstituted XO_4 groups.

Compound		Assignment and description						
		$\nu_1(A_1)$	$\nu_2(A_1)$	$\nu_3(A_1)$	$\nu_4(E)$	$\nu_5(E)$	$\nu_6(E)$	Ref.
		$\nu(X-Z)$	$\nu_s(XO_3)$	δ	$\nu_{as}(XO_3)$	$\delta(OXO)$	$\delta(OXZ)$	
$RbHSO_3$	ir, R solid	2616	1039	509	1195	625	1135	1321
K_2HPO_4	R soln. H_2O	891	970	528	1084	528	389	1373
	ir solid	858	989	535	1080	535	—	
$NaHSO_4$	R solid	887	1051	594	1200	594	429	1297
$NaHSeO_4$	ir, AgCl disc	702	870	444	948	444	—	1101
		732	850	or 410	918	or 410	—	
	R solid*	680	843	443	938 893	409	323	
$HClO_4$	ir, R anhydrous liq.	742	1036	577	1230	577	430	15
$NaSO_3F$	R solid	786	1082	566	1287	592	409	1297
KSO_3F	ir Nujol mull	732	1073	565	1299	583	—	1282
$NaSO_3Cl$	R solid	416	1050	585	1195	535	220	1297
$NaSO_3Cl$	ir solid	(395)	1060	529, 524	1245	598	(240)	252
NMe_4SO_3Cl	R solid	381	1042	601	1300	553	312	1892
$KCrO_3F$	R soln. HF; ir mull and soln. acetone	637	911	338	955	370	261	1367
CrO_3Cl^-	R soln.	438	935	295	954	365	209	1368
$NaOsO_3N$	R soln. H_2O	1021	897	309	871	309	312	1531
K_2ReO_3N	ir solid	1022	878	315	830	273	380	785
ReO_3Cl	ir, R liq.	435	1001	293	960	344	196	1026
ReO_3Br	ir liq.	195	997	350	963	332	168	949
ClO_3F	ir vapour	715	1069	549	1315	589	405	858
$ClO_3(OF)$	ir vapour	666	1049	—	1298	—	—	45
BrO_3F	ir vapour	605	875	354	974	376	296	1646

* Arbitrary assignment of ν_3, ν_5, ν_6.

to a monomeric, or bridging, leading to a polymeric, structure. H_2SO_4 has been assigned as $SO_2(OH)_2$ and some disagreements exist; these arise partially through attempts to correlate the frequencies with the corresponding values in SO_2F_2. Both SO_2F_2 and SO_2Cl_2 have been frequently studied. For SO_2F_2, two early studies [123, 1118] gave the A_2 mode ν_5 as near 385 cm^{-1}; this was later confirmed and found to be almost coincident with ν_4. The band originally assigned to ν_4 at 274 cm^{-1} by Hunt and Wilson [686] was later agreed to be due to an impurity [687, 854]. For SO_2Cl_2, Siebert [15, p. 73] has given an assignment due to Gillespie and Robinson [536] and Martz and Lagemann [913]; a more recent assignment by Hunt and Wilson [687] differs from this in several respects. Studies of MoO_2Cl_2, MoO_2Br_2 and WO_2Cl_2 suggest that, unlike the sulphuryl and chromyl halides, these are bridged dimers [24]. Assignments of $MoO_2S_2^=$ and $WO_2S_2^=$ agree fairly well [1024, 848]. Pitts *et al.* [1152]

have studied the spectrum of $IO_2F_2^-$ as the K and Ag salts, but did not interpret it. They were attempting to resolve the disagreement existing over the structure of IO_2AsF_6; this was formulated as $IO_2^+AsF_6^-$ by Schmeisser and Lang [X132a], but as $AsF_4^+IO_2F_2^-$ by Aynsley and Sampath [X4]. The results favour the first formulation, since the compound does not show the bands at 855, 844, 806, 704 cm^{-1} found in $IO_2F_2^-$. Whereas the XY_2Z_2 molecule gives rise to $4A_1 + A_2 + 2B_1 + 2B_2$ vibrations, the WXY_2Z molecule gives $6A' + 3A''$. In some published assignments the bending frequencies are not assigned to individual modes, or even, in a few instances, to symmetry classes. Descriptions of the bending modes vary greatly; in Table 7.28 those given by Durig and Clark [410] for SPF_2Cl are listed; the most common variant is to refer to ν_9 as XY_2 rocking. The formulae are written so that the atoms are in the order WXY_2Z throughout. The Raman spectra of many of the compounds were studied by Delwaulle and Francois over the period 1946–9; later ir work has been generally in good agreement with the Raman data. Some assignments by Gillespie and Robinson [539] for FSO_2Cl and FSO_2Br have been modified by Toyuki and Shimizu [1436] and Reed and Lovejoy [1193] respectively. Delwaulle and Francois [359] have given the Raman spectra of POFClBr, PSFClBr as ν(P═O), (P═S) 1319, 729; ν(P—F), 890, 892; ν(P—Cl) 590, 535; ν(P—Br) 495, 414; bending 372, 311, 298, 173, 233 and 267, 245, 160, 314, 175 cm^{-1}.

7.5 Molecules of the formula X_3Y_2

The most frequently studied of these is C_3O_2. For many years it has been debated whether this molecule is linear or bent. Lafferty, Maki and Plyler [815] have interpreted the pure rotation spectrum as that of a linear molecule, and calculated that the lowest bending mode would lie between 25 and 70 cm^{-1}. From entropy measurements McDougall [X105] deduced a frequency of 61·6 ± 2·6 cm^{-1}; this had not been observed in the study (a) by Long *et al.* [875] or (b) by Miller and Fateley [954], but Miller *et al.* [956], (c) found a band at 63 ± 2 cm^{-1}. Both the low frequency and the low intensity of this mode are connected with the possibility of contributions from the structures

$$O^+{\equiv}C{-}C^-{=}C{=}O, \qquad O{=}C{=}C^-{-}C{\equiv}O^+.$$

in which ═C$^-$— is isoelectronic with N and would have a bent configuration around it at lowest energy [X118]. Since only ν_7 involves mainly bending at the central atom, it will be the only frequency so affected. The bending modes are shown in Fig. 7.4.

The spectrum of N_2O_3 has been obtained by Hisatsune and Devlin [662] in the ir and Raman, with normal N_2O_3 and N_2O_3 enriched with

Table 7.27 Vibrational frequencies of tetrahedral molecules XY_2Z_2.

Compound		Assignment and description									Ref.
		$\nu_1(A_1)$	$\nu_2(A_1)$	$\nu_3(A_1)$	$\nu_4(A_1)$	$\nu_5(A_2)$	$\nu_6(B_1)$	$\nu_7(B_1)$	$\nu_8(B_2)$	$\nu_9(B_2)$	
		$\nu_s(XY)$	$\nu_s(XZ)$	$\delta(XY_2)$	$\delta(XZ_2)$	Torsion	$\nu_{as}(XY)$	δ	$\nu_{as}(XZ)$	δ	
SiH_2F_2	ir vap.	2246	897	982	322	—	2251	730	981	903	15
SiH_2Cl_2	R liq. ir vap.	2224	954	527	188	710	2237	602	590, 587	876	250
SiD_2Cl_2	ir vap.	1608	695	519	187	(506)	1637	466	566, 563	663	250
SiH_2Br_2	R liq. ir vap.	2200	400	940	122	688	2220	556	465	832	15
NaH_2PO_2	(−190°K) ir film	2361	1167	1073	472	926*	2318	812	1187	1090	891
NaD_2PO_2	(−190°K) ir film	1685	838	1052	467	680*	1610	673	1182	802	
KPO_2F_2	R melt, ir cryst.	1145	517	824	277	365	1311	512 or 481	857	512 or 481	199
$Be(PO_2Cl_2)_2$	ir, Nujol mull	1147	590	418	350	—	1259	518	708	334	1035
$Al(PO_2Cl_2)_3$	ir, Nujol mull	1147	587	424	311	—	1258	462	630	311	
$KPO_2(OH)_2$†	R, soln. H_2O	1071	877	—	—	1027	1152	936	—	—	1373
$H_2SO_4 = SO_2(OH)_2$	ir, R liq.‡	1137	910	563	392	422	1368	563	973	563	1297,
		1195				392				422	539
$D_2SO_4 = SO_2(OD)_2$	ir, R liq.	1170	907	522	356	356	1340	560	980	395	539
SO_2F_2	ir, Ar matrix	1269	848	544	384	388	1502	553	885	539	859
SO_2Cl_2		1205	406	577	209	363	1434	391	582	284	687
SeO_2F_2	R liq.	971	702	360	277	277	1059	360	702	340	1104
CrO_2F_2		1006	727	304	182	(422)	1016	274	789	259	667
CrO_2Cl_2	ir, R liq.	981	Mixed 465, 356		140	224	995	211	496	257	952
$(NH_4)_2MoO_2S_2$		820	473	310	200	270	800	248	504	270	1024
$(NH_4)_2WO_2S_2$		850	475	310	196	280	810	235	475	280	

* Observed in spectrum of Ca salt.

† Three unassigned bending modes were found at 512, 393, 355 cm^{-1}.

‡ Where there are two values, the upper line gives Siebert's assignment, the lower line Gillespie and Robinson's.

Table 7.28 Vibrational frequencies of tetrahedral molecules WXY_2Z.

Compound		Assignment and description										Ref.
		$\nu_1(A')$	$\nu_2(A')$	$\nu_3(A')$	$\nu_7(A'')$	$\nu_4(A')$	$\nu_5(A')$	$\nu_6(A')$	$\nu_8(A'')$	$\nu_9(A'')$	Completely unassigned bending modes	
		$\nu_s(XY_2)$	$\nu(X—W)$	$\nu(X—Z)$	$\nu_{as}(XY_2)$	$\omega(XY_2)$	$\delta(XY_2)$	$\delta(WXZ)$	$\tau(XY_2)$	$\pi(WXZ)$ or $\rho(XY_2)$		
OPF_2Cl	ir vap.	900	1384	623	960	419, 412 unassigned A'			—	—		1015
	R liq.	895	1358	618	948	410 unassigned A'			424 unassigned A''			359
OPF_2Br	ir vap.	884	1380	561	948						412, 317, 243	1029
	R liq.	883	1344	552	947						411, 320, 241	359
$OPCl_2F$	ir vap.	546	1358	907	626	382	330	(205)	374	(247)		1015
	R liq.	547	1331	804	620	386	330	207	374	254		359
$OPBr_2F$	ir vap.	316	1337	888	536	316	279	136	291	232		1029
	R liq.	306	1303	880	538	306	273	134	291	220		359
SPF_2Cl	ir vap.	946	738	541	920	395	361	198	317	—	}	410
	R liq.	939	727	536	913	394	359	207	314	251	}	
SPF_2Br	ir, R	938	719	477	911	389, 288, 175 unassigned A'			297, 175 unassigned A''			1029, 357
$SPCl_2F$	ir vap.	478	753	912	575	331	268	192	319	192	}	410
	R liq.	474	737	902	567	327	268	193	317	193	}	
$SPBr_2F$	ir	377	713	887	470	274	218	129	254	162		1027
FSO_2Cl		1224	823	627	1455	505	430	308	480	308		1436
FSO_2Br	ir vap.	1226	812	305	1455	601	458	244	489	262		1193
	R liq.											
$(OH)SO_2F$		1230	960	850	1445	560	560	405	391	555	}	539
$(OH)SO_2Cl$		1209	920	416	1408	482	623	200	312	513	}	

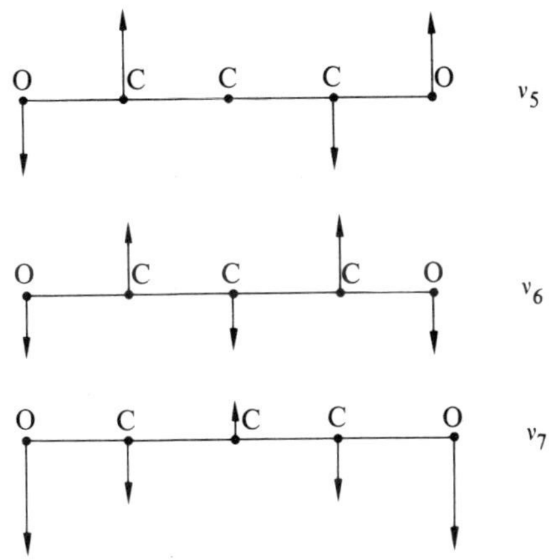

Fig. 7.4 Bending modes of C_3O_2.

Table 7.29 Assignments in C_3O_2 and C_3S_2 with $\mathbf{D}_{\infty h}$ symmetry.

Assignment and description				C_3O_2 (a)	(b), (c)	C_3S_2 [367]
Σ_g^+	ν_1	ν_{sym}	$\nu(C{=}Y)$	2200	2200	485
	ν_2	ν_{sym}	$\nu(C{=}C)$	832	830	1683
Σ_u^+	ν_3	ν_{as}	$\nu(C{=}Y)$	2258	2258	1030
	ν_4	ν_{as}	$\nu(C{=}C)$	1575	1573	2089·5
Π_g	ν_5	Bending		578	577	470
Π_u	ν_6	Bending		549	·550	502
	ν_7	Bending		(198)	63	94

N^{15}. They have assigned the frequencies using a UBFF. Yeranos and Joncich [1554] have also carried out UBFF calculations on this molecule. Hisatsune's assignments for $^{14}N_2O_3$ and $^{15}N_2O_3$ are ν_1 (ν(N=O)) 1863, 1830; ν_2 ($\nu_{as}(NO_2)$) 1589, 1546; ν_3 ($\nu_s(NO_2)$) 1297, 1277; ν_4 $\delta(NO_2)$ 783, 771; ν_5 (NO rock) 407, 397; ν_6 (NO_2 rock) 313, 308; ν_7 (ν(N—N)) 253, not given; ν_8 (NO_2 wag) 627, 614; ν_9 (NO_2 twist), 35, not given. The vibrations are $7A' + 2A''$. The L-shaped $H_2F_3^-$ ion has been studied by Azman [68] as KH_2F_3; KD_2F_3 was also investigated. The ir spectra of mulls gave ν(H—F) 1770; δ(FHF) 1082, 1050, 1020; and a band designated ν(F—F), though there is no F—F bond as such, at 455 cm^{-1}. This is presumably analogous to the O—O stretching in hydrogen-bonded O---H—O systems. Corresponding values for the deutero-ion were 1300–1400; 785, 745; 450 cm^{-1}.

7.6 Other pentatomic molecules

NO_2OF, NO_2OCl and HCOOH are all molecules of C_S symmetry. The first two molecules have been studied by Miller *et al.* [961] who have corrected a previous assignment, which involved impurity bands due to NO_2F. Millikan and Pitzer [966] have studied all the possible deuterated forms of HCOOH, as well as the normal molecule. Their assignment is to a slight extent arbitrary since there is considerable mixing of ν_4, ν_5 and ν_6; the shift of ν_7 (δ(OCO)) on deuteration is also interpreted as indicating coupling with ν_6 (δ(OH)).

Table 7.30 Infrared spectra of NO_2OF and NO_2OCl.

Compound	Assignment and description				
	$\nu_1(A')$	$\nu_2(A')$	$\nu_3(A')$	$\nu_4(A')$	$\nu_5(A')$
	$\nu_{as}(NO_2)$	$\nu_s(NO_2)$	ν(O—Hal)	$\delta(NO_2)$	ν(N—O)
NO_2OF	1759	1301	928	804	633
NO_2OCl	1735	1292	809	781	560

Compound	Assignment and description			
	$\nu_6(A')$	$\nu_7(A')$	$\nu_8(A'')$	$\nu_9(A'')$
	ρ(NOHal)	$\rho(NO_2)$	π	τ
NO_2OF	455	303	709	152
NO_2OCl	434	270	711	(121)

Table 7.31 Vibrational spectra of HCOOH and its deuterates*

Assignment	Description	HCOOH	DCOOD	DCOOH	HCOOD
$\nu_1(A')$	ν(O—H)	3570	2632	3570	2632
ν_2	ν(C—H)	2943	2232	2220	2948
ν_3	ν(C=O)	1770	1742	1756	1772
ν_4	ν(C—O)	1105	1171	1143	1178
ν_5	δ(CH)	1387	945	970	(1360)
ν_6	δ(OH)	1229	1040	(1220)	990
ν_7	δ(OCO)	636	538	629	541
$\nu_8(A'')$	δ(CH)	1033	873	(870)	~1000
ν_9	δ(OH)	636	489	629	512

* Values in brackets calculated from the product rule.

8. The vibrational spectra of hexatomic molecules

8.1 General

Many of these molecules have structures in which all bonds radiate from a central atom. The two such arrangements most commonly encountered are the trigonal bipyramid and the tetragonal pyramid. Molecules of the formulae XY_5 and XY_4Z are known with both arrangements; some trigonal bipyramidal molecules XY_3Z_2 have been studied. Many molecules X_2Y_4 exist, and in these the bonds do not radiate from a central atom; the structure may be simply $Y_2X—XY_2$ or may be bridged:

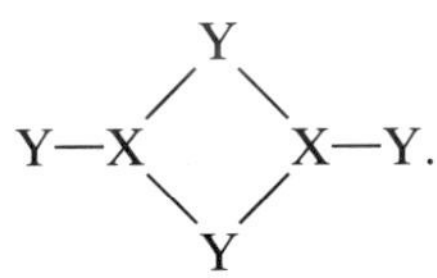

A few inorganic six-membered rings are known; for instance the cyclic trimers $(HF)_3$, $(LiF)_3$ have been postulated, and the S_6 molecule has a well-established structure.

8.2 Molecules of the formula XY_5

(a) Trigonal bipyramidal molecules

The majority of the compounds studied are halides MX_5, though some anionic species MX_5^- (M = Ge, Sn, Ti) have been investigated. Only one of these has been studied more than once; the two assignments for $SnCl_5^-$ [102, 299] agree closely. The halides of the pentavalent elements, however, have been the source of many disagreements. The first complete assignment of PF_5 [578] involved a band at 126 cm^{-1}, assigned to ν_7 (PF_2 axial bending); this was soon afterwards shown to be a rotational line of HF [681]. Three subsequent assignments agreed on a value near 305 cm^{-1} [281, 574, 1452] but a recent far infrared study [344] showed

Table 8.1 Selection rules for hexatomic molecules.

Type		Point group	Activity			
			R only	ir only	ir, R	Inactive
XY_5	Trigonal bipyramidal					
	XY_5	D_{3h}	$2A_1' + E''$	$2A_2''$	$3E'$	—
	XY_4Z (Z axial)	C_{3v}	—	—	$4A_1+4E$	—
	XY_4Z (Z equatorial)	C_{2v}	A_2	—	$5A_1+3B_1+3B_2$	—
	XY_3Z_2 (both Z axial)	D_{3h}	$2A_1'+E''$	$2A_2''$	$3E'$	—
	XY_3Z_2 (both Z equatorial)	C_{2v}	A_2	—	$5A_1+3B_1+3B_2$	—
	XY_3Z_2 (one Z of each type)	C_S	—	—	$8A'+4A''$	—
	Tetragonal pyramidal XY_5 XY_4Z (Z axial)	C_{4v}	$2B_1+B_2$	—	$3A_1+3E$	—
X_2Y_4 (not bridged)	System X—X(Y)(Y) non-planar Relative position of the Y / Y other group.					
	Trans	C_{2h}	$4A_g+2B_g$	$3A_u+3B_u$	—	—
	Gauche	C_2	—	—	$7A+5B$	—
	Cis	C_{2v}	$3A_2$	—	$4A_1+3B_1+2B_2$	—
	System X—X(Y)(Y) planar Relative position of the Y / Y other group.					
	Co-planar	D_{2h}	$3A_g+2B_{1g}+B_{2g}$	$B_{1u}+2B_{2u}+2B_{3u}$	—	A_u
	In plane at 90°	D_{2d}	$3A_1+B_1$	—	$2B_2+3E$	—
	Any other position	D_2	$4A$	—	$2B_1+3B_2+3B_3$	—
X_2Y_4 (bridged)	Ring X_2Y_2 planar Positions of terminal Y atoms relative to the ring.					
	Co-planar	D_{2h}	$3A_g+2B_{2g}+B_{3g}$	$2B_{1u}+2B_{2u}+2B_{3u}$	—	—
	Trans	C_{2h}	$4A_g+2B_g$	$2A_u+4B_u$	—	—
	Cis	C_{2v}	$2A_2$	—	$5A_1+3B_1+2B_2$	—
	Ring X_2Y_2 non-planar Terminal Y atoms					
	Trans	C_S	—	—	$8A'+4A''$	—
	Cis	C_{2v}	$2A_2$	—	$5A_1+3B_1+2B_2$	—

that this band decreased in intensity on reducing the temperature, behaviour characteristic of difference bands. It is plausibly assigned to ν_3 (ν_{as}(PF_2, axial))–ν_2(ν_s(PF_2 axial)). This problem is thus unresolved. PCl_5 has been investigated eight times, and only two of the eight vibrations have been assigned to the same frequency by all workers. These are ν_1(ν_{sym}(PCl_3)) at 394 and ν_7 at 100 cm^{-1}. Only two of the eight assignments agree throughout. The earliest assignment was that of Siebert [1293], who studied the Raman spectrum. He gave ν_2, 271; ν_5, 495; ν_6, 190; ν_7, 100 and ν_8, 190 cm^{-1}; he also calculated the A_2'' frequencies ν_3, ν_4 as 473, 124 cm^{-1} respectively, and observed a line in the Raman at 449 cm^{-1} which he assigned to ν_3. Six years later, Wilmshurst and Bernstein [1514] produced a markedly different assignment. ν_1 and ν_2 were considered to be accidentally degenerate; ν_3 was found at 465, ν_5, 592, ν_6, 335 and ν_8, 280 cm^{-1}. ν_4 was calculated as 176 cm^{-1} on the basis of two combinations (264, $\nu_4 + \nu_7$; 441, $\nu_4 + \nu_8$). In 1963, both Carlson [221] and Taylor and Woodward [1419] tackled the problem. Carlson disagreed with Wilmshurst and Bernstein about ν_4 and ν_6, giving the values 299, 273 cm^{-1} respectively. He noted that the values of the thermodynamic properties calculated using the Wilmshurst–Bernstein assignment were correct; thus the sum of the frequencies in that assignment should be correct. Taylor and Woodward disagreed with Carlson about ν_2 and ν_8; they gave ν_2 as 370 and ν_8 as 261, compared with 393, 280 cm^{-1}. Nagarajan [1047–8] followed Taylor and Woodward's assignment except for ν_8 which was listed as 161 cm^{-1}. Van der Voorn *et al.* [1452] restored Taylor and Woodward's assignment, but the two most recent studies, by Condrate and Nakamoto [281] and Beattie [102] give values for ν_2 close to that of Siebert. They also disagree over ν_8, giving 282 and 261 cm^{-1} respectively. Condrate and Nakamoto ascribed the band found by Siebert at 495 cm^{-1} to $POCl_3$, and Beattie assigned Nagarajan's 161 cm^{-1} band as ν_8–ν_7. Nakamoto's book gives an assignment resembling Wilmshurst and Bernstein's, while Siebert's book quotes three references and gives Taylor and Woodward's assignment. Beattie's assignment, as the most recent, is given in Table 8.2. The assignments for SbF_5 and $SbCl_5$ show much better agreement; that of Condrate and Nakamoto is given. The assignments of ν_3 (294 cm^{-1}) by Gaunt and Ainscough [514] and ν_4 (517 cm^{-1}) by Krishnapillai [807] are clearly incorrect. Olie [1088] has studied solid $SbCl_5$ and finds two different types of spectrum; that of the high temperature form closely resembles that of the liquid, so this form is presumably monomeric; the low temperature form has bands at 382, 337, 181, 171 in the Raman and 407, 374 and 250 cm^{-1} in the ir. None of these frequencies are characteristic of $SbCl_4^+$ or $SbCl_6^-$ (with the possible exceptions of 407 and 171 cm^{-1}), and Olie has concluded that the LT form of $SbCl_5$ may be a bridged dimer similar to that postulated by Walton and

Brisdon [1486] for $NbCl_5$ and $TaCl_5$. The interpretation of these spectra is uncertain since it is probable that the studies by Carlson [221] (deposited film) Gaunt and Ainscough [515] (solid) and Bader and Huang [71] (solution in CCl_4) refer to the dimer. On the other hand the ir study by Werder [1506] using the matrix-isolation technique, with N_2 and cyclohexane as matrices, may refer to the monomer and is assigned as such in Table 8.2. VF_5 has been completely assigned as the result of study of the ir spectrum of the vapour and the Raman spectra of the liquid [262]. Selig [1271] has obtained the Raman spectra of liquid NbF_5 as 767, 683, 315 (?), 253, 226, 136 (?) cm^{-1}, and of liquid TaF_5 as 756, 713, 690, 312 (?), 245, 213, 121 cm^{-1}, and has suggested that these relate to the monomer with D_{3h} symmetry, but has not assigned the frequencies.

(b) Tetragonal pyramidal molecules

Most work has been carried out on the interhalogen compounds, ClF_5, BrF_5 and IF_5; a few ionic species are also known. Spiro [1352] has found bands at 300, 270, 230, 194, 152 and 126 in K_2TlCl_5, $2H_2O$. Szymanski [1401] has studied the $SbCl_5^{=}$ ion and has given the assignments ν_1, 445; ν_2, 285; ν_3, 180 (A_1); ν_4, 420, ν_5, —; (B_1); ν_6, 117? (B_2); ν_7, 300; ν_8, 255; ν_9 90 (E). Selig [1269] has observed that the addition product between XeF_6 and BF_3 or AsF_5 shows peaks at 612, 620 cm^{-1} in the Raman spectrum of the solid, which are not characteristic of any of the reactants, and ascribes them to XeF_5^+. Edwards *et al.* [427] have found peaks in the ir spectra of $CsTeF_5$ and $RbTeF_5$ at 642, 627, 475 cm^{-1}; the isoelectronic IF_5 absorbs at 645 cm^{-1} but not near 475 cm^{-1}. All three interhalogen compounds have been assigned by Begun *et al.* [119]; in addition, assignments for BrF_5 have been proposed by McDowell and Asprey [933], Stephenson and Jones [1383] and Khanna [751], and for IF_5 by Lord *et al.* [877], Khanna [751] and Gillespie and Clase [535]. The spectrum of BrF_5 was originally reported by Burke and Jones [212], without assignment. Gatti [510] has determined some combination frequencies in the ir spectrum of ClF_5. The descriptions of the modes are taken from [933, 1383]. The assignments disagree regarding ν_5, ν_6 and ν_9 of BrF_5 and ν_2, ν_5, ν_6 and ν_9 of IF_5.

For BrF_5, Stephenson and Jones had originally given ν_5, 315 and ν_6, 481. Khanna assigned 315 to ν_9 and 244 to ν_5. The assignment quoted restores Stephenson and Jones' value of ν_9 and ascribes the band at 312 cm^{-1} to ν_6; Begun *et al.* did not find a band at 481 cm^{-1}. In Lord's original assignment of IF_5, two bands at 710, 693 cm^{-1} were assigned to ν_1 and ν_2 respectively, and ν_5, ν_6 were given as 275, 572 cm^{-1} respectively. Both Begun *et al.* and Gillespie and Clase observed a band near 595 cm^{-1} which Lord did not find, and assigned it to ν_2. Begun did not observe a doublet

Table 8.2 Vibrational frequencies of trigonal bipyramidal molecules XY_5.

	Assignment and description								
	$\nu_1(A_1')$	$\nu_2(A_1')$	$\nu_3(A_2'')$	$\nu_4(A_2'')$	$\nu_5(E')$	$\nu_6(E')$	$\nu_7(E')$	$\nu_8(E'')$	Ref.
Molecule	$\nu_s(XY_3eq)$	$\nu_s(XY_2ax)$	$\nu_{as}(XY_2ax)$	$\pi(XY_3)$	$\nu_{as}(XY_3eq)$	$\delta(XY_3)$	$\delta(XY_2ax)$	Rocking	
$SnCl_5^-$	338	256	330	148	351	160	59	—	
$SnBr_5^-$	202	154	208	106	257	111	—	103	299
$TiCl_5^-$	—	—	346	170	417	212	83	—	
					385				
$GeCl_5^-$	348	236	310	200	395	200	—	—	102
PF_5	817	640	945	576	1026	534	See text 305?	514	1452
PCl_5	393	282	448	300	580	273	100	261	102
SbF_5	718	672	724	212	758	265	93	224	281
$SbCl_5$	357	307	371	154	395	172	72	165	299
VF_5	719	608	784	331	810	282	200	350	262
$NbCl_5$	(349)	(293)	396	126	444	159	99	(139)	1506
$NbCl_5$*	—	—	394	151	245	172	105	(183)	71
$TaCl_5$*	—	—	382	147	238	166	107	(198)	

* These are the values for the ir spectrum in CCl_4 solution; they are probably due to the dimer but are formally assigned as for the monomer, for comparison with the matrix isolation study of $NbCl_5$. Normal coordinate analyses with some errors have been given [1047, 597]; corrected versions have been given by Krynauw *et al.* [809] and Van der Voorn *et al.* [1452].

Table 8.3 Vibrational spectra of interhalogen compounds XY_5 [119].

Compound	Assignment and description								
	$\nu_1(A_1)$	$\nu_2(A_1)$	$\nu_3(A_1)$	$\nu_4(B_1)$	$\nu_5(B_1)$	$\nu_6(B_2)$	$\nu_7(E)$	$\nu_8(E)$	$\nu_9(E)$
	ν_{ax}	$\nu_{s,eq}$	π	$\nu_{as,eq}$	π	δ_{eq}	$\nu_{as,eq}$	δ_{ax}	δ_{eq}
ClF_5	709	541	486	480	—	375	732	—	302
BrF_5	683	587	369	535	(281)	321	644	415	237
IF_5	710	593	318	575	(257)	273	608	372	189

near 700 cm^{-1}; Gillespie and Clase did (705, 694 cm^{-1}) and assigned one of the two components to the ternary combination $2\nu_9 + \nu_3$ $(A_1 + B_1 + B_2)$ in Fermi resonance with $\nu_1(A_1)$. They also agreed in assigning the band at 275 cm^{-1} to ν_6. Khanna gives Lord's assignment with ν_5 and ν_9 reversed.

8.3 Molecules of the formula XY_4Z

(a) Trigonal bipyramidal

$PFCl_4$ and $SbFCl_4$ have been assigned according to C_{3v} symmetry. Dipole moment and nuclear quadrupole resonance measurements on PF_4Cl support the alternative structure with C_{2v} symmetry; SOF_4 has the same atomic arrangement. Dehnicke and Weidlein [342], who studied $SbFCl_4$ in $CHCl_3$ and CCl_4 solution, gave the assignment ν_1 (ν(Sb—F)), 545; ν_2 (ν_{axial}(Sb—Cl)). 306; ν_3 (ν_{eq}($SbCl_3$)), 363; ν_4 (rocking), 137 (all A_1); ν_5 (ν_{eq}), 400; ν_6 (δ_{eq} ($SbCl_3$)) 177; ν_7 (δ_{axial}(ClSbF)), 445; ν_8 (torsion) 166 (all E). Griffiths' [578] assignment for $PFCl_4$, with the corresponding descriptions, are ν_1, 778, ν_2, 388; ν_3, 427; ν_4, 265; ν_5, 601; ν_6, 339; ν_7, 297; ν_8, 110 cm^{-1}. The original study of the ir spectrum of PF_4Cl by Carter and Holmes [224] contained no assignment, but Holmes [673–5] later carried out a normal-coordinate analysis and made a full assignment. An early incomplete study by Dudley *et al.* [391] of SOF_4 was followed by Goggin [546] who made a complete assignment. Sathianadan [1232] used this assignment to calculate GVFF force constants for the molecule.

The descriptions are those of Holmes; the A_1 modes of SOF_4 are tabulated to some extent arbitrarily since Goggin's descriptions are less definite. This assignment is arbitrary as between ν_1 and ν_2 and as between ν_3 and ν_5.

(b) Tetragonal pyramidal

$XeOF_4$ has been studied by Claassen *et al.* [255] and Hyman [689]; normal-coordinate calculations, using a UBFF, have been made by

Table 8.4 Vibrational spectra of $PClF_4$ and SOF_4 (XZY_4).

Class	Assignment and description	$PClF_4$	SOF_4
A_1	ν_1 $\nu_{sym}(XY_2)$	895	741
	ν_2 $\nu_{sym}(XYY')$	691	820
	ν_3 $\delta(XY_2)$	(501)	309
	ν_4 $\nu(X—Z)$	434	1379
	ν_5 $\delta(XYY')$	144	570
A_2	ν_6 torsion	356	189
B_1	ν_7 $\nu_{as}(XY_2)$	921	927
	ν_8 $\delta(XY_2Z)$	(490)	640
	ν_9 $\delta(XYY')$	144	552
B_2	ν_{10} $\nu_{as}(XYY')$	903	927
	ν_{11} $\pi(XY_2Z)$	560	481
	ν_{12} rocking	356	272

Yeranos [1547]. The assignments agree well. Yeranos gives ν_1 (ν(Xe—O)), 920; ν_2 (ν_s(XeF)), 567; ν_3 (δ(FXeO)), 285; ν_4 (ν_{as}(XeF)), 527; ν_5 (δ(FXeO)). 230; ν_6 ($\delta_{as}(XeF_2)$), 233; ν_7 ($\nu(XeF_2)$), 608; ν_8 (δ(FXeO)), 365; ν_9 ($\delta(XeF_2)$), 161 cm^{-1}. His calculated values do not agree very well with these observations. Blanchard [157] has studied WOF_4 and $MoOF_4$, and assigns for WOF_4, ν_1 1064, 1055, 1047; ν_3 745, 735, 720; ν_7 705, 700, 690; ν_8 560. Corresponding values for $MoOF_4$ are 1045 (ν_1), 680 (ν_3), 720 (ν_7), 530 (ν_8).

8.4 Molecules of the formula XY_3Z_2

PCl_3F_2 and SbF_3Cl_2 have D_{3h} symmetry, the three Y atoms occupying the equatorial positions, whereas PF_3Cl_2, PF_3Br_2 and PF_3H_2 have C_{2v} symmetry, with the Y atoms on one axial and two equatorial positions. For PCl_3F_2, the assignments of Griffiths [578] and Van der Voorn [1452] agree. In assigning the E' bending modes, Van der Voorn noted that the value of ν_6, the XY_3 bending, should be similar to that in PCl_5, while that of ν_7, the axial XY_2 bending, should be similar to that in PF_5.

Table 8.5 Vibrational spectra of PCl_3F_2 and SbF_3Cl_2.

Class	Assignment and description	PCl_3F_2 [1452]	SbF_3Cl_2 [341]
A_1'	ν_1 $\nu_s(XY_3)$	387	609
	ν_2 $\nu_s(XZ_2)$	633	393
A_2''	ν_3 $\nu_{as}(XZ_2)$	867	395
	ν_4 $\pi(XY_3)$	328	—
E'	ν_5 $\nu(XY_3)$	625	655
	ν_6 $\delta(XY_3)$	122	455
	ν_7 $\delta(XZ_2)$	404	145
E''	ν_8 rocking	357	294

Griffiths *et al.* [578] made an assignment for PF_3Cl_2 which was later revised by Holmes [674–5] on the basis that 488, 338 were improbably high for the PF_2 axial bending modes (the axial mode in PF_5 is at 301 cm^{-1}); the reassignment was 338 (A_1), 124 (B_1). The band at 488 was reassigned to the B_1 equatorial bending mode and the band at 368 formerly assigned to this was reassigned to ν_{12} (B_2 rock) and ν_6 (A_2 twist).

Salthouse and Waddington's assignment [1228] differs in a number of ways; in particular they considered that replacement of Cl by F should increase the P—Cl stretching frequencies; thus ν_7 ($\nu_{as}(PCl_2)$) was assigned as 667 cm^{-1} rather than 427, as the P—Cl frequencies in PCl_5 are 581, 395 cm^{-1}. They also reversed Holmes' assignment of ν_1 and ν_{10} and gave ν_8 as 562 cm^{-1}.

Table 8.6 Vibrational spectra of PF_3Cl_2, PF_3Br_2 and PF_3H_2.

Class	Assignment and description		PF_3Cl_2 [675]	PF_3Cl_2 [1228]	PF_3Br_2 [1228]	PF_3H_2 [552]
A_1	ν_1	$\nu_{eq}(X—Y)$	893	925	913	995
	ν_2	$\nu_{ax}(XY_2)$	663	650	607	820
	ν_3	$\delta_{ax}(XY_2)$	338	429	420	471
	ν_4	$\nu(XZ_2)$	407	407	318	2538
	ν_5	$\delta(XZ_2)$	124	—	—	1292
A_2	ν_6	twisting	368	—	—	655
B_1	ν_7	$\nu(XZ_2)$	427	667	587	2465
	ν_8	$\delta_{eq}(XYZ_2)$	488	562	510	765
	ν_9	$\delta_{ax}(XY_2)$	124	337	335	389
B_2	ν_{10}	$\nu_{ax}(XY_2)$	925	895	878	1026
	ν_{11}	$\pi(XYZ_2)$	500	490	458	867
	ν_{12}	rocking	368	—	—	329

Goubeau [552] has observed a much more complex spectrum in solid PF_3H_2, with 11 ir active bands (the strongest at 713, 960 and 1118 cm^{-1}) and 14 Raman active bands. He considers that in the solid there may be a fluorine-bridged dimer.

8.5 Molecules of the formula X_2Y_4

All the molecules studied have the simple $Y_2X—XY_2$ structure, and five of the six arrangements listed in Table 8.1 have been observed. The planar D_{2h} arrangement has been suggested for B_2F_4, B_2Cl_4, N_2O_4 and $S_2O_4^=$. For the two boron compounds this is the arrangement in the crystal; in the vapour, they have the staggered configuration with D_{2d} symmetry. The oxalate ion in solution has been assigned by Begun and Fletcher [118] on the basis of D_{2d} symmetry, though Siebert [1299] and Murata and

Kawai [1037–8] give an incomplete assignment on the basis of D_{2h} symmetry. A further assignment of the solution data, according to D_2 symmetry, has been given by Hester and Plane [648].

Mann and Fano [908] have carried out a normal-coordinate analysis of the X_2Y_4 system for the general case of planar

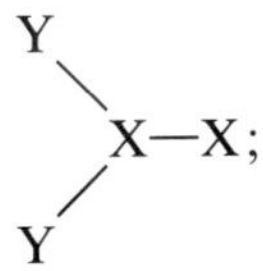

this is easily specialized to D_{2d} or D_{2h} by setting the dihedral angle equal to 90° or 0°. Gayles and Self [516] have studied B_2F_4 in the solid and the vapour phase and assigned the frequencies on the basis of D_{2h} symmetry. Begun and Fletcher [117] have studied N_2O_4 with both ^{14}N and ^{15}N and their assignments agree well with the predictions based on the product rule except for the B_{2u} modes, where the calculated and observed ratios are 1·024 and 1·035 respectively.

Table 8.7 Vibrational spectra of molecules X_2Y_4 with D_{2d} symmetry.

Class	Assignment and description	B_2F_4 [516] ir, vap		B_2Cl_4 [908] ir vap, R liq	$C_2O_4^{=}$ [118] ir, R soln
A_1	ν_1 ν_s(X—Y)	(1382)	1377	1131	1486
	ν_2 ν(X—X)	(624)	—	401	900
	ν_3 $\delta_s(XY_2)$	(301)	315	225	449
B_2	ν_4 ν(X—Y)	1151		730	1579
	ν_5 $\delta(XY_2)$	541		291	524
E	ν_6 ν(X—Y)	1375		917	1310
	ν_7 π	657		617	761
	ν_8 $\delta(XY_2)$	325		180	301

For B_2F_4, the A_1 values in brackets in Table 8.7 were observed for the $^{10}B\,^{11}B$ compound; the other A_1 values are calculated for the $^{11}B\,^{11}B$ compound. Finch's observations [473] agree well with those tabulated, but his calculation fits a model of D_{2h} symmetry better than one of D_{2d} symmetry. He also points out that the signs of $G_{9,\,10}$ and $G_{10,\,11}$ in Mann and Fano's analysis should be changed. In Table 8.8, which deals with the molecules of D_{2h} symmetry, the classifications and descriptions are those given by Gayles and Self. Begun and Fletcher's assignment for N_2O_4 has been tabulated to correspond to this.

Simon and Kuechler [1308] have suggested the D_{2h} structure for $Na_2S_2O_4$; in their Raman study of a 20 per cent aqueous solution they found bands at 1070, 998, 581, 465, 229 and 176 cm^{-1}. Duval and Lecomte

[414], however, have assigned the spectrum of the solid in terms of C_{2v} symmetry.

Table 8.8 Vibrational spectra of molecules with D_{2h} symmetry.

Class	Assignment and description	B_2F_4 ir, R solid	$^{14}N_2O_4$	$^{15}N_2O_4$	$C_2O_4^=$ [1037–8]
			ir vap, R liq and solid		
A_g	ν_1 $\nu_s(XY_2)$	1408	1380	1360	1481
	ν_2 $\nu(X—X)$	513	266	264	902
	ν_3 $\delta(XY_2)$	323	808	797	421
B_{2g}	ν_4 $\nu_{as}(XY_2)$	1340	1712	1672	1647
	ν_5 $\delta_{as}(XY_2)$	395	482	478	550
B_{3g}	ν_6 π	666	672	654	—
A_u	ν_7 torsion	—	—	—	—
B_{1u}	ν_8 $\nu_s(XY_2)$	1138	1262	1251	1330
	ν_9 $\delta_s(XY_2)$	537	750	739	774
B_{2u}	ν_{10} π	646	429	420	—
B_{3u}	ν_{11} $\nu_{as}(XY_2)$	1332	1748	1707	1640
	ν_{12} $\delta_{as}(XY_2)$	384	381	377	295

Snyder and Hisatsune [1348] have studied solid N_2O_4; its space group is T_h and the N_2O_4 molecules occupy D_{2h} sites. They have calculated a barrier to rotation of 2·9 kcal mole^{-1} and estimated a torsional frequency near 50 cm^{-1}. A number of other studies [117, 453, 561, 1397–8, 1509] also agree with that quoted in Table 8.8.

The remaining molecules studied have the two Y_2 groups in different planes. The C_{2h} (trans) configuration has been established for liquid P_2Cl_4 and for solid P_2I_4, Baudler and Fricke [96] have suggested the C_2 (gauche) configuration for P_2I_4 in CS_2 solution, but both Frankiss [485–6] and Carroll and Cowley consider that the trans configuration is retained. Additional bands found by Baudler and Fricke in the ir spectrum are attributed by Carroll and Cowley [223] to PI_3 formed by the reaction $3P_2I_4 \rightarrow 4PI_3 + \frac{1}{2}P_4$. They claim to have identified both these products using ^{31}P nmr, and PI_3 additionally from the uv spectrum. Studies of N_2F_4 suggest the existence of a trans as well as a gauche isomer; the Raman spectrum of the liquid and the ir spectrum of the vapour have been obtained by Durig and Clark [411] and by Koster and Miller [774], whose assignments differ in several respects. Both for the trans and gauche forms of N_2F_4, the most recent work [1824] is in conflict with previous assignments. Oskam *et al.* base their new assignment largely on the observed band contours in the ir, although, as they point out, there are some bands for which the Raman depolarization ratios and the ir band contours are not consistent. The assignments and descriptions are those of Frankiss for P_2Cl_4.

Table 8.9 Vibrational spectra of N_2F_4, P_2Cl_4 and P_2I_4.

Class	Assignment and description		N_2F_4 [411]	[774]	[1824]	P_2Cl_4 [484]	P_2I_4 [486]
A_g	ν_1	$\nu(XY_2)$	1010	1036	1035	494	319, 307
	ν_2	$\nu(X—X)$	600	600	600	410	
	ν_3	$\nu(XY_2$ wag)	719	719	719	238	117
	ν_4	$\nu(XY_2$ bend)	354	494	354	149	78
A_u	ν_5	$\nu(XY_2)$	997	993	946	501	330
	ν_6	$\nu(XY_2$ rock)	464	250	252	202	52
	ν_7	$\nu(XY_2$ twist)*	131	96	131	91	89
B_g	ν_8	$\nu(XY_2)$	867	867	867	480	328
	ν_9	$\nu(XY_2$ rock)*	494	300	423	189	98
B_u	ν_{10}	$\nu(XY_2)$	945	944	999	508	301
	ν_{11}	$\nu(XY_2$ wag)	542	539	541	275	112
	ν_{12}	$\nu(XY_2$ bend)	254	466	466	139	62?

* Referred to as torsion and XY_2 twist respectively in [1824].

As stated earlier, a number of assignments, involving different symmetries, have been given for the oxalate ion. In the solid state, both the D_{2h} and D_2 arrangements have been found, the first for $Na_2C_2O_4$ [X78b] and the second for $Ag_2C_2O_4$ [X58a] and $(NH_4)_2C_2O_4$, H_2O [X78a]. The assignment by Hester and Plane [648] of the Raman frequencies of aqueous $K_2C_2O_4$ solution is ν_1 $(\nu_s(C{=}O))$ 1490, ν_2 $(\nu(C—C))$ 905, ν_3 $(\delta(OCO))$ 460 (A); ν_5 $(\nu_{as}(C{=}O))$ 1647, ν_6 $(CO_2$ wag) 1450, ν_7 $(CO_2$ rock) 1310, ν_8 $(\nu_{as}(C{=}O))$ 1600, ν_9 $(CO_2$ wag) 1390, ν_{11} $(\nu_s(C{=}O))$ 1315 (B). ν_4 (A) and ν_{10}, ν_{12} (B) were not found. The oxalate ion in complexes of the type $K_2M^{II}(COO_2)_2$ and $[Co(NH_3)_4(COO)_2]^{+}X^{-}$ has been assigned on the basis of C_{2v} symmetry by Fujita *et al.* [494] and Siebert [1299] respectively. In the trisoxalato complexes $K_3M^{III}(COO_2)_3$ where M^{III} = Ga or Al the overall symmetry of the molecule becomes D_3 and Hester and Plane [648] have made an assignment on that basis. Bis- and trisoxalato complexes of In have been studied by Kharitonov [754].

N_2H_4, N_2D_4, N_2F_4 and P_2H_4 have all been assigned the C_2 (gauche) structure in the vapour. N_2H_4 and N_2D_4 appear to retain this configuration in the liquid and solid. By analysis of the external modes Baglin *et al.* [72] have shown that the space group C_{2h}^2 suggested for N_2H_4 [X37] cannot be correct, since only three ir active modes are predicted, whereas seven are observed. The alternative C_2^2 suggested by Busing *et al.* [X28] requires nine ir active external modes and is favoured by Baglin. On the other hand, since the unit cell contains two molecules, correlation splitting of every vibration is expected if the C_2^2 space group is correct, whereas no splitting is expected for the C_{2h}^2 space group due to the operation of the rule of mutual exclusion. Durig *et al.* did not observe any splitting of the internal modes in the solid [409]. They have also given the previous assignments due to Giguere and Liu [533] (ir vapour), Catalano [226] (ir matrix) and Ziomek and Zeidler [1560] (Raman, liquid). The assignments listed agree well on the whole, except for the torsional mode ν_7 where Catalano gives 377 and Durig 627 cm^{-1}.

Table 8.10 Vibrational assignments for molecules X_2Y_4 with C_2 symmetry.

Class	Assignment and description		N_2H_4 ir solid [409]	N_2D_4 ir solid [409]	R liq [411]	N_2F_4 R liq [774]	ir vap [411]	ir vap [1824]	P_2H_4 ir vap [1079]
A	ν_1	$\nu_s(XY_2)$	3200	2347	1036	1025	1031	1009	—
	ν_2	$\nu_{as}(XY_2)$	3310	2407	1025	1010	1023	934	—
	ν_3	XY_2 wag	1304	1041	733	733	738	740	633
	ν_4	$\nu(X-X)$	1126	938	588	588	586	590	—
	ν_5	XY_2 twist	883	744	423	—	—	300	—
	ν_6	XY_2 bend	1603	1184	300	354	295	284	—
	ν_7	torsion	627	495	122	122	116	123	—
B	ν_8	$\nu_{as}(XY_2)$	3310	2479	960	960	958 }	1016	2312
	ν_9	$\nu_s(XY_2)$	3310	2469	923	923	933 }	962	
	ν_{10}	XY_2 wag	1350	1004	515	515	519	518	801, 792 788, 780
	ν_{11}	XY_2 twist	1066	814	284	423	284	494	835, 827, 823
	ν_{12}	XY_2 bend	1655	1213	242	284	242	242	1091, 1081, 1071

Baudler and Schmidt observed bands in the Raman spectrum of P_2H_4 at 2286, 1070, 856, 653, 456, 437, 412, 161 cm^{-1}, and of P_2D_4 at 1661, 774, 574, 432 cm^{-1} [97–8].

I_2O_4 was first considered by Wise and Hannan [1523] to be $IO^+IO_3^-$ but Dasent and Waddington [316] showed that the band at 1105 cm^{-1} was due to moisture; they expected, in any case, a band at a frequency similar to that found in the isoelectronic TeO (796 cm^{-1}). They found bands at 658, 622 (ν_1), 408 (ν_2), 825, 783, 745 (ν_3) where these refer to the iodate group, and another band at 578 cm^{-1} due to ν (I—O—I) in a polymer chain.

8.6 Six-atomic rings

The only such inorganic molecule to have been fully studied is S_6. The ring has D_{3d} symmetry and the vibrations are $2A_{1g} + 1A_{1u} + 1A_{2u} + 2E_g + 2E_u$. The stretching modes were identified by Nimon *et al.* [1078] as 471 (A_{1g}), 448 (E_g) and 463 (E_u), the bending modes as 262 (A_{1g}), 313 (A_{2u}), 202 (E_g) and 180 (E_u). The forbidden A_{1u} stretching mode was identified as a weak band at 390 cm^{-1} in the ir spectrum of the Nujol mull. It was not observed in the spectrum of a solution in CS_2, or in the Raman spectrum of the solid or solution. Berkowitz [127] gives a spectrum in good agreement with Nimon's. Nimon and Neff [1077] also carried out a normal coordinate analysis, using both a GVFF and a UBFF. Snelson in his study of Li_2F_2 also found four bands at 736, 509, 267, 230 cm^{-1} which he attributed to Li_3F_3 [1344]. A band in the spectrum of gaseous HF at 390 cm^{-1} originally observed by Rothschild [1219a] was considered to be possibly due to a ring trimer [670]; Smith [1340] however, showed this band to be due to SiF_4.

9. The vibrational spectra of heptatomic molecules

9.1 General

Many molecules based on the octahedral XY_6 arrangement have been studied. The molecules treated in this chapter include halides and halide complexes, and the skeletal vibrations of other complexes where the ligand is treated as a point mass (as for NH_3, H_2O, NO_2^-, OH^-) or where only the atom of the ligand which is bonded to the metal is considered as part of the skeleton (as for the oxygen atoms in acido complexes involving $CO_3^=$, $SO_4^=$, $PO_4^{\equiv}$, etc.). The other 7-atomic molecules to have been studied are mostly of the type X_2Y_5. Such molecules may be $Y_3X—XY_2$ or $Y_2X—Y—XY_2$. We shall be concerned only with the latter type, as there is no spectroscopic information about any inorganic molecule known to be of the former type. Simon *et al.* [1331] have given diagrams of most of the possible structures which a $Y_2X—Y—XY_2$ molecule can assume. Apart from those with no symmetry elements at all, one can visualize ten possible arrangements, though many of these may be discounted. If the XYX bridge is linear, there are four arrangements with symmetry higher than C_1. Where at least one XY_2 group is coplanar with the bridge, the other may be staggered (I), coplanar (II) or gauche (III) with respect to it. These structures have point groups D_{2d}, D_{2h}, D_2 respectively. If neither group is coplanar with the bridge, but the bridge bisects the two terminal angles (the groups being trans to each other), the symmetry is C_{2h} (IV). When the bridge is angular, the highest possible symmetry is C_{2v}. This can be achieved if all seven atoms lie in one plane (V) or if the plane of the bridge bisects the two terminal angles, the XY_2 groups being eclipsed (VI). If the groups are staggered, however, the symmetry is reduced to C_s (VII). This symmetry is also found if the plane of the bridge contains one terminal angle and bisects the other (VIII). There are two lower-symmetry modifications of (V). If one Y atom from each terminal group lies in the plane of the bridge, and if the other Y atoms form X—Y bonds making equal angles with the plane of the bridge, the symmetry will be C_s if they are on the same side (IX) and C_2 if they are on opposite sides (X).

Table 9.1 Selection rules for heptatomic molecules.

	Point group	Activity R only	ir only	ir and R	In-active
Octahedral					
XY_6	O_h	$A_{1g} + E_g + F_{2g}$	$2F_{1u}$	—	F_{2u}
XY_5Z	C_{4v}	$2B_1 + B_2$	—	$4A_1 + 4E$	—
XY_4Z_2 trans	D_{4h}	$2A_{1g} + B_{1g} + B_{2g} + E_g$	$2A_{2u} + 3E_u$	—	B_{2u}
XY_4Z_2 cis	C_{2v}	$2A_2$	—	$6A_1 + 3B_1 + 4B_2$	—
X_2Y_5					
Structure					
I*	D_{2d}	$3A_1 + B_1$	—	$3B_2 + 4E$	—
II	D_{2h}	$3A_g + B_{2g} + 2B_{3g}$	$3B_{1u} + 3B_{2u} + 2B_{3u}$	—	A_u
III	D_2	$4A$	—	$3B_1 + 4B_2 + 4B_3$	—
IV	C_{2h}	$4A_g + 2B_g$	$4A_u + 5B_u$	—	—
V	C_{2v}	$2A_2$	—	$6A_1 + 5B_1 + 2B_2$	—
VI	C_{2v}	$3A_2$	—	$5A_1 + 3B_1 + 4B_2$	—
VII	C_s	—	—	$9A' + 6A''$	—
VIII	C_s	—	—	$10A' + 5A''$	—
IX	C_s	—	—	$8A' + 7A''$	—
X	C_2	—	—	$8A + 7B$	—

* Roman numerals refer to the descriptions in the text.

9.2 Octahedral molecules with the formula XY_6

(a) Halides

The earlier assignments of fluorides of hexavalent metals have in many cases been based on combination frequencies observed in the infrared; later studies, generally using both infrared and Raman have, in the main, confirmed these assignments. In only two cases, MoF_6 and ReF_6, have substantial changes been made. In the work on MoF_6 by Gaunt [512] and by Burke, Smith and Nielsen [213], ν_6 was calculated as 240 or 234 cm^{-1} from the combination at 882 (879) assigned to $\nu_2 + \nu_6$. Claassen *et al.* [263] revised this to $\nu_6 = 190\ cm^{-1}$, fitting $\nu_2 + \nu_6$ to 832 and $\nu_2 - \nu_6$ to 453 cm^{-1}. For ReF_6, Gaunt [513] had originally given $\nu_4 = 393$, $\nu_6 = 170\ cm^{-1}$. Claassen, however, gave ν_4 as 257 and ν_6 as 193 cm^{-1} [260]. This value of ν_4 was based on direct observation, whereas Gaunt's value was based on a combination frequency which Claassen considered to be due to ReF_7 or $ReOF_5$. Claassen's assignment of ν_6 is derived from $\nu_2 - \nu_6 = 403\ cm^{-1}$, whereas Gaunt's was based on $\nu_2 - \nu_6 = 413\ cm^{-1}$. An alternative is $\nu_6 = 157$ (from $\nu_5 + \nu_6 = 403$).

In OsF_6 the great breadth of ν_2 and its binary combinations was ascribed to a possible Jahn–Teller effect, the $5d^2$ non-bonding electrons giving a fivefold ground state; a similar effect was observed in ReF_6, where the $5d^1$ electron gives a fourfold ground state. Although the same effect could occur in IrF_6, it was not observed. Weinstock *et al.* [1495–8] considered that there was no static Jahn–Teller effect in RhF_6, which also displays splitting of ν_2, but a dynamic coupling involving this mode only.

Table 9.2 Vibrational frequencies of octahedral halides XY_6.

Compound	Assignment and description						Ref.
	$\nu_1(A_{1g})$	$\nu_2(E_g)$	$\nu_3(F_{1u})$	$\nu_4(F_{1u})$	$\nu_5(F_{2g})$	$\nu_6(F_{2u})$	
	$\nu_{sym}(XY)$	$\nu(XY_2)$	$\nu_{as}(XY_2)$	$\delta(XY_2)$	$\delta(XY_2)$	$\delta(XY_2)$	
SF_6	775	644	932	613	524	(344)	
SeF_6	708	662	780	437	405	(260)	512
TeF_6	701	674	752	327	313	(197)	
MoF_6	741	643	741	264	306	(190)	263
WF_6	769	670	712	256	322	(216)	512
WCl_6	(408)	(312)	367	165	(206)	(97)	445
UF_6	667	535	626	186	202	(144)	512, 265, 493
NpF_6	(648)	(528)	624	199	(206)	(164)	493, 905
PuF_6	(628)	(523)	615	206	(211)	(171)	
TcF_6	(705)	(551)	748	265	(255)	—	263
ReF_6	755	596	715	257	246	(193)	260
RuF_6	(675)	—	735	275	(262)	—	1495
RhF_6	(634)	(595)	724	283	(269)	(192)	
OsF_6	733	632	720	268	252	230	1496
IrF_6	(705)	(644)	720	276	(260)	(208)	264, 924
PtF_6	(658)	(601)	705	273	(242)	(211)	1496

Kim *et al.* [758] have computed orbital valence force field constants for the hexahalides. They claim that this force field is superior to the UBFF in three respects. First, the redundancies among the angle bending co-ordinates are eliminated; secondly, the bending frequencies calculated in this way give better agreement with observation; and, lastly, that least squares calculations lead to more rapid convergence.

Smith [1338] has examined the ir, and Hyman [689] the Raman spectra of XeF_6. In the ir spectrum of the vapour there are two bands in the stretching region, at 612 and 520 cm^{-1}, and in the Raman spectrum three bands in this region (655, 635, 582 in HF solution, 660, 600, 550 in the solid). This would suggest that XeF_6 is not octahedral, since such symmetry requires the ir and Raman spectra to show one and two stretching modes respectively. Malm *et al.* [904] and Weaver *et al.* [1488a] have suggested that the overtone ir spectrum is characteristic of O_h symmetry, and the band at 520 cm^{-1} is due to an impurity. However, Bartell *et al.* [X9] have carried out an electron diffraction study of XeF_6 vapour and find the results incompatible with a regular octahedral structure. Siebert [15] has suggested a structure of C_{5v} symmetry; this would be a pentagonal bipyramid with one axial position occupied by the lone pair. The selection rules for such a structure would be $\Gamma_{vib} = 3A_1$ (ir, R) + $3E_1$ (ir, R) + $3E_2$ (R), and the stretching modes would contribute $2A_1 + E_1 + E_2$ to this: we should thus expect three bands in the ir spectrum and four in the Raman spectrum, in the stretching region.

Adams [1] has tabulated a large number of frequencies for octahedral anions, $(MHal_6)^{n-}$, but there are few complete assignments. Of the 13 such ions tabulated by Siebert [15], five have complete assignments given (excluding ν_6) by direct observation. Some attempts have been made to study the effects of crystal structure, particularly by Griffiths and Irish [579] on fluorogermanates (also quoted by Adams [1],) Begun and Rutenberg [120] on various fluoro-anions, and Brown *et al.* [190] on fluoro-anions of group IV metals. Evans and Long [439] have interpreted the spectra of SF_4SbF_5, SeF_4SbF_5 and SeF_4AsF_5 in terms of $M^{VI}F_3^+ + M^VF_6^-$, but did not definitely adopt this interpretation of TeF_4SbF_5, which they thought might contain a fluorine bridge. However, the frequencies derived for SbF_6^- on the basis of the first structure agree well with the corresponding values from the S and Se complexes. The compound $IAsF_{12}$ has been formulated as $IF_6^+AsF_6^-$ and, in the original assignment, Christe and Sawodny [246] gave ν_1 of IF_6^+ as 708, ν_2 as 732 cm^{-1}. Hardwick and Leroi [611] carried out an MUBFF calculation resulting in a reversal of this assignment, but, in a further paper, Christe and Sawodny [247] stated that a trend exists with increasing oxidation number and mass of the central atom to decrease the separation of ν_1 and ν_2 and, in the limit, to reverse their relative positions. They also abandoned their original assignment of 404 cm^{-1} for ν_4 of IF_6^+, replacing this by 343 cm^{-1}. Christe [244] has also investigated IF_6^-, isoelectronic with XeF_6, and concluded that this ion is non-centrosymmetric. Possible fundamentals in the ir spectrum are 653, 592, 534, 485, 322, 285 cm^{-1} and in the Raman 673, 591, 546, 322, 274 cm^{-1}. Adams [28] has noted a dependence of ν_3 in halide complexes with the K_2PtCl_6 structure on cation size; the same dependence (a decrease) was also found in the series M_2WCl_6, M_2WBr_6, M_2MoCl_6 where M = K, Rb, Cs. In this case Adams suggested the variation as due to $d\pi$—$d\pi$ bonding, which is favoured by the smaller cation. Brown *et al.* [190], who have studied $M^I_2M^{IV}Hal_6$, where M^I = K, Rb, Cs, M^{IV} = Ti, Sn, Pb, Pd, Os, Pt, Ir, Hal = Cl, Br, do not agree with this interpretation, considering that π-bonding is not important in non-transition metal halides. The information from these two and other partial studies, such as that of Peacock and Sharp [1114] on the ir active frequencies of fluoro complexes, is not tabulated here; only reasonably complete assignments are tabulated, unless the complex is the only one of its type to have been studied.

The crystal structures of a number of fluoro-complexes are known, and, as stated earlier, some spectra–structure relations have been discussed. K_2SiF_6 is cubic, with the space group O_h^5 ($Fm3m$), and the anions on O_h sites. This is also true of Cs_2SiF_6, Cs_2GeF_6, $(NH_4)_2GeF_6$ (cubic), and KPF_6. All except the last have the K_2PtCl_6 structure [W IX, c 10]. Brown has reported a splitting in the ν_3 region for Cs_2SiF_6 (shoulder 740; 726), but apart from this, all the above compounds obey the free-ion selection rules, as expected. $BaGeF_6$ has the space group D_{3d}^3 ($P\bar{3}m1$) with the anions on D_{3d} sites. Under this symmetry, the correlation with O_h is $A_{1g} \rightarrow A_{1g}$ (R), $E_g \rightarrow E_g$ (R), $F_{1u} \rightarrow A_{2u} + E_u$ (both ir), $F_{2g} \rightarrow A_{1g} + E_g$, $F_{2u} \rightarrow A_{1u}$ (inactive) $+ E_u$. Consequently ν_3, ν_4 and ν_5 should all be split, and ν_6 should show one ir active component. Other crystals with this arrangement include $CsNbF_6$ and $CsTaF_6$, while the β-ammonium fluorosilicate structure, which has the same space group and anion site symmetry, is represented by γ-K_2TiF_6, γ-Rb_2TiF_6, K_2GeF_6 and the low-temperature form of Rb_2GeF_6.

Table 9.3 Vibrational spectra of octahedral fluoro-complexes.

Compound		Assignment						Ref.
		$\nu_1(A_{1g})$	$\nu_2(E_g)$	$\nu_3(F_{1u})$	$\nu_4(F_{1u})$	$\nu_5(F_{2g})$	$\nu_6(F_{2u})$	
$(NH_4)_3GaF_6$	solid R	536	359	—	—	271	—	1911
Na_2SiF_6	solid	663	477	741	483	408	—	120
H_2GeF_6	soln. H_2O	627	454	—	—	318	—	579
K_2GeF_6	solid	624	471	603	359	335	—	
					339			120
Na_2SnF_6	solid	592	477	559	300	252		
K_2TiF_6	solid ir	—	—	585	312	—	278	190
KPF_6	solid	751	580	830	558	477	—	
$CsAsF_6$	solid	685	576	699	392	372	—	120
$LiSbF_6$	solid	668	558	669	350	294	—	
$CsNbF_6$	solid R	683	562	—	—	280	—	736–7
$CsTaF_6$	solid R	692	581	—	—	272	—	
CsU^VF_6	solid	506	—	503	150	145	100	1195

Griffiths and Irish found all five ir active components in the spectrum of $BaGeF_6$ (ν_3, 602 (A_{2u}), 635 (E_u); ν_4, 337 (A_{2u}), 367 (E_u); ν_6, 452 (E_u)). Keller [736, 737] did not observe the predicted splitting of ν_5 in the Raman spectra of $CsNbF_6$, $CsTaF_6$. For K_2TiF_6, Rb_2TiF_6 and K_2GeF_6, a splitting has been found for ν_4 of about 20 cm^{-1}. No definite splitting of ν_3 has been observed, though Griffiths and Irish observed that the band was asymmetric and considered that it might be composed of two components at 602 and 635 cm^{-1}. They also observed a weak band at 464 cm^{-1} which might be ν_6 (E_u). They found ν_3 and ν_4 unsplit in Rb_2GeF_6 and suggested that this might be due to the greater regularity of the Rb—F distances (all twelve are equal, whereas in K_2GeF_6 nine are 2·85 Å and three are 3·01 Å). Brown [190] studied the hexagonal forms of K_2GeF_6 and Rb_2GeF_6. These have the β-K_2MnF_6 structure with the space group C_{6v}^4 ($P6_3mc$) and anion site symmetry C_{3v}. This correlates with O_h according to $A_{1g} \rightarrow A_1, E_g \rightarrow E, F_{1u} \rightarrow A_1 + E, F_{2g} \rightarrow A_1 + E, F_{2u} \rightarrow A_2 + E$. Thus all bands become ir and Raman active, except for the inactive A_2 component of ν_6. Brown found bands at 608, 354 (broad), 335, 325 (both weak) in the ir spectrum of K_2GeF_6 and at 600, 351, 322 in that of Rb_2GeF_6.

Structures of lower symmetry still have been found for $CsAsF_6$ [X39] and $LiSbF_6$ [X27], which have space group S_6^2 and anion site symmetry S_6; Na_2SiF_6, space group D_3^2, site group D_3 or C_3 [X38]; and Na_2SnF_6, space group D_{2h}^{25} and site group D_{2h} [X25]. In none of these cases was any deviation from the free-ion selection rules detected. The assignment given for Na_2SnF_6 corrects that of Kriegsmann and Kessler [789] who gave for $(NH_4)_2SnF_6$, $\nu_1 = 593$, $\nu_2 = 620$, $\nu_3 = 564$, $\nu_5 = 342$. They found their ν_2 in the ir and not in the Raman spectrum. Kolditz and Preiss found ν_3 of K_2SnF_6 at 564 and of Na_2SnF_6 at 568 cm^{-1} [770]. Schwochau [1257] has studied $M_2^IM^{IV}F_6$ where M^I = K, Rb, M^{IV} = Re, Tc. These compounds have the K_2GeF_6 structure and show ν_3 at 550 cm^{-1} (M^{IV} = Re) and 574 cm^{-1} (M^{IV} = Tc). Meyers and Cotton [941] have studied a number of compounds containing the $[CoF_6]^{3-}$ ion. In K_2NaCoF_6, which is cubic, they found ν_3 at 510, ν_4 at 310 cm^{-1}. K_3CoF_6, which is tetragonally distorted, showed three bands in the ν_3 region (505, 480, 450), and ν_4 at 310 cm^{-1}. The Na, Li and Ba salts were also studied. Less work has been done on spectra-structure relations in chlorocomplexes. Adams and Morris [31–2] have studied the spectra of isostructural $M_2^IPtCl_6$ (M^I = K, Rb, Cs, Tl, Ag, NH_4) and found the adherence to the free-ion selection rules appropriate to the K_2PtCl_6 structure. They have also found a frequency in each spectrum, which they attributed to a lattice mode (106, 84, 73, 62, 82, 134 and 97) and have calculated ν_6, using an MUBFF (82, 87, 83, 79, 53, 84 cm^{-1}). These sets of values refer to the M^I in the order given in the text above: for the lattice frequencies the last two refer to NH_4^+. Creighton and Green [299] have noted a lattice mode in $(NH_4)_2TeCl_6$ at 107 cm^{-1}; since this compound has the factor group O_h, the F_{1u} lattice mode can mix with the neighbouring F_{1u} internal mode (ν_4, 141 cm^{-1}). The effect of this mixing should be to cause an increase in ν_4. Raman studies of $SnCl_6^=$ and $SeCl_6^=$ by Hiraishi [659] and $BiCl_6^{3-}$ [1087] by Oertel and Plane agree well with the data given in Table 9.4. Wendling and Rohmer [1501–2] have found bands at 345, 190, 80, 55

Table 9.4 Vibrational spectra of octahedral chloro-complexes.

Compound		Assignment						Ref.
		$\nu_1(A_{1g})$	$\nu_2(E_g)$	$\nu_3(F_{1u})$	$\nu_4(F_{1u})$	$\nu_5(F_{2g})$	$\nu_6(F_{1u})$	
$K_3TlCl_6, 2H_2O$	solid	280	262	—	—	155	—	1352
Cs_2GeCl_6	solid	309	211	310	213	198	—	102
Rb_2SnCl_6	solid	316	241	312	173	171	107	31
$(Et_4N)_2SnCl_6$	solid	309	232	306, 291	163	159	—	272
Rb_2PbCl_6	solid	293	220	281	139	147	83	31
$(NH_4)_2TiCl_6$	solid	331	284	330	193	194	(142)	34
$(Et_4N)_2TiCl_6$	solid	320	271	316	183	173	—	272
$(Et_4N)_2ZrCl_6$	solid	327	237	290	150	153	(90)	34
$(Et_4N)_2HfCl_6$	solid	333	237	288	145	157	(80)	34
$ThCl_6^{=}$ soln, CH_3CN		299	237	262	—	121	—	1544
$Py(HPCl_6)$	solid	355	273	444	285	237	—	102
$(Et_4N)AsCl_6$	solid	337	289	333	220	202	—	102
$(AsCl_4)^+(SbCl_6)^-$	solid	333	291	350	180	174	—	102
$(BuNH_3)^{3+}(BiCl_6)^{3-}$	solid	251	202	171	(120)	134	—	1485
Rb_2SeCl_6	solid	290	242	275	153	162	—	640
$(NH_4)_2TeCl_6$	R soln, ir solid	346	275	313	172	159	—	299
K_2OsCl_6		345	274	314	177	165	—	642
K_2IrCl_6		352	225	335	168	190	—	642
K_2PtCl_6		350	320	341	186	171	(82)	31
K_2PdCl_6		317	292	358	175	164	—	642
$WCl_6^{=}$		378	318	330	158	—	—	1622
$TaCl_6^-$		378	298	330	158	180	—	1622
$NbCl_6^-$		368	288	333	162	183	—	1622

cm^{-1} for Cs_2TiCl_6, but have not assigned them. For the chloro-complexes of Ti, Hf and Zr, Adams and Newton [34] have found the UBFF stretching force constant K to be inversely proportional to r^6, where r is the M—Cl distance calculated from the Pauling covalent radii. Hendra and Park [642] studied K_3RhCl_6, H_2O and found bands in its Raman spectrum at 350, 300, 285, 237, 175 cm^{-1}, and in its ir spectrum at 439, 335, 221, 195. 169, 105 cm^{-1}. They felt that the extra bands were probably not due to Jahn–Teller distortion, nor could they find evidence for the existence of $[RhCl_5.H_2O]^{2-}$. The symmetry may be reduced by the presence of the water molecule; however, the anhydrous compound also displays additional vibrations. Adams [28] has found ν_3, ν_4 at 324, 165 cm^{-1} in K_2WCl_6, and at 340, 174 in K_2MoCl_6. An additional band at 77, 74 cm^{-1} is attributed either to the lattice or to F_{2u} bending. The lack of cation mass dependence suggests the latter explanation, since for the Rb and Cs compounds the values are 66, 71 for the tungsten and 70, 70 cm^{-1} for the molybdenum complex.

For the $BiBr_6^{3-}$ ion, Oertel and Plane [1087] agree less exactly with Walton [1485] than for $BiCl_6^{3-}$. They give, for the Raman active fre-

Table 9.5 Vibrational spectra of octahedral bromo- and iodo-complexes.

Compound		Assignment						Ref.
		$\nu_1(A_{1g})$	$\nu_2(E_g)$	$\nu_3(F_{1u})$	$\nu_4(F_{1u})$	$\nu_5(F_{2g})$	$\nu_6(F_{2u})$	
$MgSnBr_6$	solid	183	144	—	—	69	—	658–9
$(Et_4N)_2SnBr_6$	solid	182	135	203	111	101	—	272
$SnBr_6^=$	soln H_2O	185	137	—	—	95	—	658–9
$(BuNH_3)_3BiBr_6$	solid	153	~120	126	—	92	—	1485
K_2SeBr_6	solid	179	157	—	122	105	—	640
$(NH_4)_2TeBr_6$	solid	—	—	198	117	99	—	299
K_2TeBr_6	solid	174	151	—	—	85 70	—	
K_2PdBr_6	solid	253	130	198	176	100	—	642
K_2PtBr_6	solid	207	190	240	90	97	(100)	642
		217	195	243	78	115		31
$ReBr_6^=$	ir solid, R soln H_2O	213	174	217	118	104	—	1543
$(Et_4N)_2SnI_6$	solid	—	—	156	90 79	—	—	272
$TeI_6^=$	soln CH_3CN* solid K salt	114*	105, 75, 60(?)	163	95	—	—	640

quencies of $(N_2H_5)_3BiBr_6$, 156, 133 and 63 cm^{-1}. For $PtBr_6^{2-}$, Hendra and Park's results represent the majority view, agreeing with those of Woodward and Creighton [1530] and Hiraishi [659]. Adams [28] has given ν_3, ν_4 for $M^I_2WBr_6$ (M^I = K, Rb, Cs) as 229, 74; 220, 78; 214, 60 cm^{-1} respectively. For $TeI_6^=$, the assignment to ν_2 is doubtful and one of the frequencies listed under ν_4 may be a lattice mode. Coupling between these modes has been suggested in $(NH_4)_2TeBr_6$ and K_2TeBr_6 by Creighton and Green [299], who also state that the C_i anion site symmetry in the latter compound is responsible for the splitting of ν_5. Hendra and Jovic [640] have assigned ν_1 and ν_2 as 113 cm^{-1} in a solution of $SeI_6^=$ in acetonitrile, and have given three components of ν_2 (104, 80, 59) for the pyridinium salt of this anion.

(b) Hydroxides, hydroxo- and oxo-complexes

The $Sn(OH)_6^=$ ion has been studied by Dupuis [400] and Kolditz and Preiss [770]. For the Zn, Cd, Co, Ni and Cu salts, Dupuis found bands in the regions 1300–1380, 1135–1190, 745–850 (usually very broad), 520–540 and 370–390 cm^{-1}. Kolditz and Preiss assigned bands at 492, 520 in $K_2Sn(OH)_6$ and 515 cm^{-1} in $Na_2Sn(OH)_6$ to Sn(OH) stretching. In a later study [881], the band near 530 cm^{-1} was interpreted as ν_3 (F_{1u}), and further bands found near 260 cm^{-1} (ν_4, F_{1u}) and 150 cm^{-1} (ν_6, F_{2u}). In an ir study of $NaSb(OH)_6$, Siebert [1300] gave ν(Sb—(OH)) as 528; 586,

600 (doublet); 628; 695; 735, 775 (doublet), while for $Te(OH)_6$ the corresponding frequencies were 605, 650, 658, 675, 708, 730 cm^{-1}. He considered the system Te—O—H to be nonlinear. $Te(OH)_6$ also showed a skeletal bending mode at 411 cm^{-1} in the ir and at 375 (ν_5, F_{2g}) in the Raman. He also studied the Raman spectrum of a 40 per cent solution of HIO_4 and assigned bands at 632, 594, 387 to ν_1, ν_2, ν_5 of the IO_6 skeleton [1294], comparing these with the values ascribed to these modes (647, 624, 357) by Venkateswaran [1455] for $Te(OH)_6$. Dupuis *et al.* [401, 403] have studied the ir spectra of some 'metastannates' $M^{II}SnO_3$. The Cd and Co compounds are perovskites; the structures of the Mg, Ni, Cu and Zn compounds are not known. They have interpreted the vibrations as being those of an SnO_6 octahedron but have not assigned them. Krishnan [804] studied the Raman spectrum of $Na_2Zn(SO_4)_2.4H_2O$, and identified bands at 370, 316, 242 cm^{-1} as characteristic of ZnO_6, and at 304, 256, 156 as characteristic of NaO_6.

(c) Aquo complexes

The information is fragmentary and the interpretations incomplete. Adams [1] has summarized the situation up to the end of 1964 and little of importance seems to have appeared since. A normal-coordinate analysis of the $M(H_2O)_6$ ion with T_h symmetry has been given by Nakagawa and Shimanouchi [1050]. This symmetry was taken as the result of a neutron diffraction analysis of $Fe(H_2O)_6SiF_6$. They calculated the frequencies of the F_u modes (under T_h symmetry, $2F_{1u}$ and F_{2u} both $\rightarrow$ F_u). In one paper [1049a] they gave all three values, but in the better-known paper [1050] only the stretching frequency is quoted. The results are in moderate agreement with those obtained by Lafont [818] and Kermarrec [742]. Raman data have been obtained for some hexahydrated salts of Mg and Al by Da Silveira [318–19], who has also attempted to relate ν_1 of the deuterated hexa-aquo complexes of Mg, Zn and Al to their lattice energies. It is possible to patch together an assignment of the skeletal vibrations for the Mg, Zn and Al complexes; for the other metals, we can say that ν_3 (F_{1u}, M—O stretching) has been given as 395 (Mn^{II}), 364 (Co^{II}), 405 (Ni^{II}), 490 (Cr^{III}); in all these Kermarrec [742] and Nakagawa and Shimanouchi [1050] agree. For Fe^{II}, Kermarrec gives 357 cm^{-1} while Nakagawa and Shimanouchi give 389 cm^{-1}. The distinction between the skeletal mode and the H_2O librational modes has been made on the basis of insensitivity of the skeletal mode to deuteration.

(d) Ammine complexes

The history of useful work on the skeletal vibrations of these complexes begins with the correct assignment by Powell and Sheppard [1163] of ν_3

Table 9.6 Vibrational assignments in hexa-aquo complexes.

Compound	ν_1	ν_2	ν_3	ν_4	ν_5	ν_6
$Mg(H_2O)_6^{2+}$	360	315	310	(153)	240	(148)
$Zn(H_2O)_6^{2+}$	394	239 or 200	364	(157)	—	(148)
$Al(H_2O)_6^{3+}$	525	447	540	305	340	—

References: ν_1, ν_2, ν_5 for Mg and Al complexes [318]; ν_3, ν_4 for Al complex [742]; ν_3, ν_4, ν_6 for Mg and Zn complexes [819]; ν_1, ν_2 for Zn complex [818].

(F_{1u}) to a weak band at 502 cm^{-1} for the $Co(NH_3)_6^{3+}$ ion. Shimanouchi and Nakagawa [1286] carried out a normal-coordinate analysis of this system, using a UBFF; they observed ν_3, ν_4 at 464 (w), 325 (s) in the bromide and at 464 (w), 318 (s) in the iodide. Their calculated values for the system were ν_1, 383; ν_2, 337; ν_3, 464; ν_4, 329; ν_5, 370; ν_6, 261 cm^{-1}. No complete study of a hexammino-cobalt(III) complex has been made; all attempts to obtain Raman spectra have been unsuccessful. Griffith [569] has given a complete assignment for the hexammino-Rh^{III}, Ru^{III} and Ir^{III} species. Infrared studies of $Cr(NH_3)_6Cl_3$ have been made by Nakagawa and Shimanouchi [1051], and by Terrasse *et al.* [1429]; they agree in finding ν_3 near 460 cm^{-1}, and neither observed ν_4. The hexammino-complexes of divalent metal ions have been studied by Terrasse *et al.* (Zn, Cd, Ni); by Nakagawa and Shimanouchi [1051] (Ni, Co); by Sacconi, Sabatini and Gans [1225] (Mn, Fe, Co, Ni, Zn, Cd) and by Viltange [1465–7] (Ni). Values for ν_3 [1225] of the chlorides, bromides and iodides respectively are Mn, 307, 299, 295; Fe, 321, 315, 306; Co, 327, 318, 312; Ni, 334, 327, 322; Zn, 300, 294, 282; Cd, 298, 291, 277 cm^{-1}. ν_4 of the chloride was found at 192 (Co) and 215 cm^{-1} (Ni). The only serious disagreement is with Terrasse *et al.* who gave ν_3 at 423 cm^{-1} for $Zn(NH_3)_6Cl_2$ and 370 cm^{-1} for the Cd complex; these would seem to be in error. Watt and Klett [1488] have noted that $Co(NH_3)_6Cl_3$ and $Co(NH_3)_6Br_3$ crystallize in the monoclinic system and should therefore show all three components of ν_3; this was found to be the case by Sacconi *et al.* [1225], who

Table 9.7 Vibrational assignments in hexammino complexes [569].

Compound	ν_1	ν_2	ν_3	ν_4	ν_5
$Ru(NH_3)_6Cl_3$	500	475	463	283, 263	248
$Ir(NH_3)_6Cl_3$	527	500	475	279, 264	262
$Rh(NH_3)_6Cl_3$	514	483	472	302, 287	240
$Ru(ND_3)_6Cl_3$	466	430	417	—	228
$Ir(ND_3)_6Cl_3$	498	471	440	255, 235	245
$Rh(ND_3)_6Cl_3$	489	455	433	278, 256	220

gave 499, 476, 449 for the chloride, and Shimanouchi and Nakagawa [1286], who gave 503, 492, 464 for the bromide. The iodide, however, crystallizes in the cubic system, with the complex ion on an O_h site, and should therefore give a single band. Shimanouchi and Nakagawa have found a band at 464 cm^{-1} for this substance. The splitting of ν_3 has also been observed for $Cr(NH_3)_6Cl_3$; components were given as 495, 470, 459 cm^{-1}.

(e) Nitro-complexes

The $[Co(NO_2)_6]^{3-}$ system has been studied in detail by Nakagawa and Shimanouchi [1052–3]; they recorded the spectra of the Na, K, Rb, Cs and $Co(NH_3)_6$ compounds. The K, Rb and Cs salts have a cubic structure, with the space group T_h^3, and T_h point symmetry for the anion. The Na salt has a somewhat different spectrum associated with S_6 symmetry of the anion. The only ir active skeletal modes for the K, Rb and Cs salts are of species F_u, and consist of one Co—N stretching and two skeletal bendings. In addition, two lattice translations of this class were expected, and observed. For the Rb salt, the observed values of these were respectively 415, 190 (?), 140 (?), 111 and 76 cm^{-1}, and, for the Cs salt, 412, not observed, 136, 92 and 63 cm^{-1}. Calculated values were 419, 179, 151, 112, 76 cm^{-1} for the Rb salt, and 417, 177, 147, 99, 63 cm^{-1} for the Cs salt.

9.3 Molecules with the formula XY_5Z

Two classes of compound have been studied. In the first, Y is a halogen and Z is another halogen, OH, oxygen or (O—O). In the second, Y is NH_3 and Z is a halogen. The only compound of the first class to have been studied more than once is SF_5Cl. The original assignment by Cross *et al.* [302] has been confirmed by Griffiths [576] except that, on the basis of polarization data, Cross's ν_6 (B_1, 396) and ν_{11} (E, 270) are reversed. Yeranos [1550] has calculated UBFF constants from Cross's assignment, based on transferability of constants from SF_6. There is a substantial disagreement for ν_1 (axial S—F stretching; observed, 854; calculated, 880 cm^{-1}). Smith and Begun [1341] have assigned IOF_5 on the basis of C_{4v} symmetry, a conclusion confirmed by NMR [X2], and microwave studies [X116] due to Pierce and co-workers. Keller and Chetham-Strode [736–7] have obtained the spectrum of K_2NbOF_5, H_2O, for which four structures are possible. These are (i) an octahedral $NbOF_5^{2-}$ ion of C_{4v} symmetry: (ii) $Nb(OH)_2F_5^{2-}$, (a) as a pentagonal bipyramid of D_{5h} symmetry, like $UO_2F_5^{2-}$, (b) with the NbF_7^{2-} structure of C_{2v} symmetry and (c) with the $NbOF_6^{3-}$ structure of C_{3v} symmetry, where the fluorine atoms form a distorted octahedron with the oxygen off one face. The spectrum was found to contain bands at 3643, 3572, 1626 cm^{-1} due to water; there was also one coincidence (932 ir, 935 R). If this band is an A_1 stretching mode of the C_{4v} structure, the coincidence is expected. The existence of the coincidence in any case eliminates structure ii(a) of D_{5h} symmetry, since its selection rules are A_1', E_2', E_1'' (R); A_2'', E_1' (ir). The spectrum is very dissimilar to that of K_2NbF_7, while the C_{3v} structure ii(c) is considered unstable with respect to ii(a) and ii(b). The C_{4v} structure has therefore been proposed as correct. For compounds of the type $M^I_2TaCl_5O_2$ (where the O_2 is a peroxo group), Dehand [340] has found bands at 565, 596, 864, 899 cm^{-1} for the NH_4 compound; the Rb and Cs compounds give almost identical spectra to this. Kolditz and Wendt [771] have found ν(As—F) at 675, ν(As—OH) at 562 in $KAsF_5OH$ and ν(Sb—F) at 640, ν(Sb—OH) at 560 cm^{-1} in $KSbF_5OH$, with similar values for the Na, Rb and Cs compounds. In the complete assignments of Table 9.8, the descriptions given are those of Smith and Begun for IOF_5, or (marked by a dagger) those of Griffiths for SF_5Cl.

Table 9.8 Vibrational assignments for molecules XY_5Z.

Class		Assignment and description	IF_5O [1341]	SF_5Cl [576]	WF_5Cl [26]	Cs_2WOCl_5* [1501–2]
A_1	ν_1	ν(X—Z)	927	402	407	960
	ν_2	$\nu_{sym}(XY_4)$	680	707	744 }	330
	ν_3	ν(XY axial)	640	855	703 }	
	ν_4	$\pi(XY_4)$	360	602	257	173
B_1	ν_5	$\nu(XY_4$ out of phase)	640	625	644	330
	ν_6	$\pi(XY_4)$	(275)	271	182	80
B_2	ν_7	$\delta(XY_2)$	305	505	377	—
E	ν_8	ν(XY axial)	710	909	661	313
	ν_9	$\delta(XY_2)(XY_4$ wag†)	369	597	307 }	
	ν_{10}	$\delta(XYZ)(XY_4$ rock†)	342	399	290 }	232, 80
	ν_{11}	$\delta(XY_2)(XY_4$ in plane bend†)	205	441	227 }	

* Wendling and Rohmer [1501–2] have also studied the corresponding compounds of V, Nb, Cr and Mo.
† Alternative descriptions.

A few substituted cobaltammine complexes have been studied. The first assignment, due to Shimanouchi and Nakagawa [1286] was criticized by Watt and Klett [1488] on the basis that it required the Co—X stretching force constant in $[Co(NH_3)_5X]^{2+}$ to vary in the order Co—Br > Co—I > Co—Cl, whereas the variation is expected (on simple electrostatic grounds) to be Co—Cl > Co—Br > Co—I. A subsequent assignment by Nakagawa and Shimanouchi [1051] contained substantial alterations, but did not quote Co—X stretching force constants; it was stated that they were estimated so as to give good agreement with the band assigned as ν_1 (ν(Co—X)), class A_1. They also observed bands in the spectrum of $[Cr(NH_3)_5Cl]Cl_2$ at 472, 459, 432 and 306 cm^{-1}.

Other substituted complexes to have been studied are

$$[Co(NH_3)_5NO_2]Cl_2$$

[164, 1053] and a number of complexes $[Co(NH_3)_5Z]B$, where Z is a monodentate carbonato- or sulphato-group, and B is Cl^-, Br^-, I^-, NO_3^- [549]. Fujita *et al.* [495] have also studied monodentate carbonato-complexes, and, on the basis of calculations made on the complexed $CO_3^=$ ion with C_s symmetry, have assigned ν(Co—O) as 369 cm^{-1}. Goldsmith *et al.* [547] gave 356 cm^{-1} in the carbonato chloride.

The original assignment by Shimanouchi and Nakagawa [1286] is not quoted in Table 9.9. The paper contains normal-coordinate analyses (using a UBFF) of $Co(NH_3)_5X$ and trans $Co(NH_3)_4X_2$. In the papers from which the tabulated assignments are taken, a normal-coordinate analysis is given for the halogen-substituted ammines [1051], and also for the nitropentammine [1053]. This has been carried out with regard to the geometry of the molecule as a whole; an X-ray study has shown the molecular point group to be C_{2v}. Under this symmetry the E vibrations of the skeleton go to $B_1 + B_2$, and some of the B_1 skeletal vibrations mix with the B_1 NO_2 rocking mode.

In the assignment to ν_{10}, the three observed bands correspond to B_1 and B_2 skeletal modes and B_1 rocking. For the calculated E frequencies, those in the upper line for each mode are the B_1 frequencies. In some of the pentammino-acido cobalt(III) complexes, though not those whose spectra are given in Table 9.9, Goldsmith *et al.* found additional bands. Two of the compounds were shown by them to be biaxial; this would imply a breakdown of the E modes, and activation of the B_1 and B_2 modes. A normal-coordinate analysis of the $[Co(NH_3)_5Z]$ system, using a GVFF, was given.

9.4 Molecules with the formula XY_4Z_2

Two disubstituted halides have been studied in some detail, $CsMoO_2Cl_4$ by Wendling and Rohmer [1501–2], and $K_2PtCl_4Br_2$ by Adams and Morris [31–2]. These both have the trans configuration, with D_{4h} symmetry. Clark [272] has studied some mixed halides of group IVA elements and has found them to have the cis-configuration, this being stabilized by π-donation into different F_{2g} orbitals.

The trans-tetrammino complexes $[Co(NH_3)_4Z_2]^+$ where Z = Cl, Br, NO_2, have been studied by Shimanouchi and Nakagawa [1051, 1286]; [1053]; the last reference also deals with the cis-dinitro complex. Gold-

Table 9.9 Vibrational assignments in monosubstituted cobaltammines. Substituent group Z.

Class	Assignment and description	F		Cl		Br		I		NO_2		CO_3	SO_4	
		Obs.	Calc.	Obs.	Calc.	Obs.	Calc.	Obs.	Calc.	Obs.	Calc.	Obs.	Obs.	Calc.*
A_1	ν_1 ν(X—Z)	449	480	457	476	462	475	428	473	482	469	356	346	352
	ν_2 ν(X—Y)	—	438	—	416	—	410	—	406	—	395	470	482	470
	ν_3 ν(X—Y)	328	343	285	285	319	319	319	271	335	329	496	509	490
	ν_4 δ	309	309	279	272	201	201	167	168	(290)	268	146	139	138
B_1	ν_5 ν(X—Y)	—	—	—	—	—	—	—	—	—	—	330	—	330
	ν_6 δ	—	—	—	—	—	—	—	—	—	—	(182)	—	173
B_2	ν_7 δ	—	—	—	—	—	—	—	—	—	—	—	—	—
E	ν_8 ν(X—Y)	504 493	498	487 479	498	501 487	497	496 473	498	493	504 501	434	438	450
	ν_9 δ	328	345	325	292	319	290	319	290	361	363 355	262	262	256
	ν_{10} δ	(309)	290	(285)	287	—	263	—	259	361 355 290	350 293 310	238	207	230
	ν_{11} δ	249	219	200 ?	188	—	146	—	132		183 173	182	196	182

* The same calculated values correspond to the last two sets of observed values.

smith *et al.* [547] have given an assignment of the skeletal vibrations of the carbonato- and phosphato-tetrammino cobalt(III) complexes, in which the acido group is chelate, leading to C_{2v} symmetry. Their assignment of the Co—O frequencies in the carbonato complex does not agree with that of Fujita *et al.* who gave 429, 370 cm^{-1} [495]. Beattie *et al.* have carried out a normal-coordinate analysis of both the cis- and trans-disubstituted systems, in connection with their work on compounds of the type MCl_4L_2 (where L is one of a number of ligands) [107]. They used a GVFF as did Goldsmith; Shimanouchi [1286] used a MUBFF. The assignment by Fujita *et al.* considers the complexed $CO_3^=$ group in isolation, and expressions for both GVFF and UBFF force constants are given. The latter constants are calculated.

Table 9.10 Vibrational assignments in molecules XY_4Z_2 with D_{4h} symmetry.

Class	Assignment and description		$Co(NH_3)_4Z_2$ Z = Cl* Obs.	Z = Cl* Calc.	Z = Br* Obs.	Z = Br* Calc.	$CsMoCl_4O_2$ Obs.	$K_2PtCl_4Br_2$ Obs.
A_{1g}	ν_1	ν(X—Z)					885	337
	ν_2	ν(X—Y)					384	232
A_{2u}	ν_3	ν(X—Z)	353	352	318	317	925	248
	ν_4	δ(YXZ)	290	287	228	227	185	183
B_{1g}	ν_5	ν(X—Y)					885	323
B_{2g}	ν_6	$\delta(XY_2)$						165
B_{2u}	ν_7	δ(YXZ)						
E_g	ν_8	δ(YXZ)						
E_u	ν_9	ν(X—Y)	501	499	486	497	323	337, 319
	ν_{10}	δ(YXZ)	290	289	295	280	240	234
	ν_{11}	$\delta(XY_2)$	178	184	140	120	85	92

* Taken from [1051]; the assignment in the earlier paper is different, and is based on the same force constants as that for $Co(NH_3)_5X$ criticized by Watt and Klett [1488].

The assignment for the dinitro complex was given in terms of the D_{2h} point symmetry of $[Co(NH_3)_4(NO_2)_2]^+$, the skeletal vibrations being assigned as B_{1u}, 371 (ν(X—Z)), 273 (δ(ZXY)); B_{2u}, 505 (ν(X—Y)), 345 and 282 ($\delta(XY_2)$ + NO_2 rocking), 157 (?) (δ(ZXY)); B_{3u}, 505 (ν(X—Y)), 314 ($\delta(XY_2)$), 157 (?) (δ(ZXY)). Viltange [1466–7] has studied trans-$Ni(NH_3)_4(H_2O)_2^{++}$ and assigns 363, 290 to stretching and 238, 218 to bending in the sulphate complex. The chlorate and bromate spectra are similar to this in that region.

Table 9.11 Vibrational assignments in molecules of the formula XY_4Z_2, with C_{2v} symmetry.

Class	Assignment and description		$[Co(NH_3)_4NO_2]Cl$ Obs.	$[Co(NH_3)_4NO_2]Cl$ Calc.	Assignment and description		$[Co(NH_3)_4CO_3]^+$ Obs.* I	$[Co(NH_3)_4CO_3]^+$ Obs.* II	$[Co(NH_3)_4CO_3]^+$ Calc.	$Co(NH_3)_4PO_4$ Obs.
A_1	ν_1	ν(X—Y)	483	473	ν_1	ν(X—Y)	528	516	527	490
	ν_2	ν(X—Y)	385	400	ν_2	ν(X—Y)	440	430	430	420
	ν_3	ν(X—Y) + δ	344	343	ν_3	δ	318	300	306	296
	ν_4	ν(X—Z) + δ	294	305	ν_4	δ	152	140	149	150
	ν_5	ν(X—Z) + δ	—	263	ν_5	δ	—	—	38	—
	ν_6	δ	140?	134	ν_6	ν(X—Z)	402	392	396	382
B_1	ν_9	ν(X—Y)	472	450	ν_9	ν(X—Y)	508	492	496	463
	ν_{10}	ν(X—Y) + ν(X—Z) + δ	315	322	ν_{10}	δ	282	270	273	246
	ν_{11}	ν(X—Z) + δ	—	268	ν_{11}	δ	204	198	201	228
B_2	ν_{12}	ν(X—Y)	518	516	ν_{12}	ν(X—Y)	488	458	459	430
	ν_{13}	δ + NO_2 rock	356	362	ν_{13}	δ	188	182	191	200
	ν_{14}	ν(X—Y) + δ	344	358	ν_{14}	ν(X—Z)	336	318	318	330
	ν_{15}	δ + NO_2 rock	180?	188	ν_{15}	δ	130	122	129	124

* I and II refer to the nitrate and chloride respectively.

Nakagawa [1053] also gives two other B_1 modes (δ + NO_2 wag) at 588 (mainly wag) and 210 (?) (mainly deformation). Calculated values corresponding to these are 602, 201 cm^{-1}.

Kharitonov and Kryazeva [755–6] have made some calculations on uranyl complexes of D_{4h}, D_{2h} and C_{2v} symmetry.

Wendling and Rohmer [1503] have studied compounds of the type M^I_2 ($M^{IV}O_2Cl_4$) where M^{IV} is Ti, Nb, Ta but were uncertain of the symmetry and have given a qualitative and rather fragmentary assignment.

9.5 Molecules with the formula X_2Y_5

The species to have been studied are $B_2O_5^{4-}$ [202, 614, 1500], N_2O_5 [453, 664], $S_2O_5^{2-}$, $Se_2O_5^{2-}$ [1208, 1306, 1313–17, 1323, 1327–31] and $Sn_2F_5^-$ [377]. The assignments are rendered difficult to make in many cases due to uncertainty about the molecular symmetry. Two of these species, $B_2O_5^{4-}$ in $Mg_2B_2O_5$ [X147] and $Sn_2F_5^-$ in $NaSn_2F_5$ [X104c] have been shown to have structure X with C_2 symmetry (see Table 9.1). For the latter molecule, a normal-coordinate analysis exists [377].

Table 9.12 Vibrational assignments for molecules X_2Y_5 with C_2 symmetry.

Type of vibration	Class	$B_2O_5^{4-}$	$Sn_2F_5^-$*
Terminal stretch	*A*	1496, 1285, 1270	504, 426, 345
	B	1436, 1025	432
Terminal bend	*A* }	713, 690	214, 195
	B }		216, 194, 179
Bridge stretch	*A*	834	268
	B	1168	460
Bridge bend	*B*	607	156

* Calculated values; the spectra of the Na, K, NH_4, Rb, Cs, Sr and Ba salts were obtained. A factor-group analysis of $NaSn_2F_5$ was carried out.

Although Teranishi and Decius [1427–8] have considered N_2O_5 to be NO_2^+ NO_3^- in the solid state, Hisatsune *et al.* [664] have found a richer spectrum which they interpret as that of a planar N_2O_5 molecule with C_{2v} symmetry, of structure V. They have classified the vibrations as $6A_1 + 2A_2 + 2B_1 + 5B_2$; Table 9.1 gives $5B_1 + 2B_2$ but this difference obviously arises from a different labelling of the symmetry planes. Fateley *et al.* [453] have studied the spectrum of N_2O_5 vapour and observed bands corresponding to ν_{11} (1720), ν_{12} (1240) and ν_3, ν_{14} (743 cm^{-1}) of Hisatsune.

Hisatsune did observe a weak band at 1033 cm^{-1} but preferred not to assign it to ν_{13}, though the UBFF calculation would suggest this. Simon *et al.* [1331] have discussed possible structures for the $S_2O_5^{2-}$ and $Se_2O_5^{2-}$

Table 9.13 Vibrational assignment of N_2O_5.

Class	Assignment and description	ir vapour	ir solid	Calc. (UBFF)
A_1	ν_1 ν_{as}(N—O)	1728	1742, 1702	1703
	ν_2 ν_{sym}(N—O)	1338	1335	1315
	ν_3 δNO_2	743	732, 712	803
	ν_4 NO_2 rock	~614	600	650
	ν_5 ν_{sym}(N—O—N)	~353	380, 337	369
	ν_6 δ(N—O—N)	(85)	—	104
B_1	ν_9 NO_2 wag	577	580	—
	ν_{10} NO_2 twist	—	—	—
B_2	ν_{11} ν_{as}(N—O)	1728	1742, 1702	1697
	ν_{12} ν_{sym}(N—O)	1247	1245	1316
	ν_{13} ν_{as}(N—O—N)	860	853	1061
	ν_{14} $\delta(NO_2)$	743	732, 712	679
	ν_{15} $\delta(NO_2)$ rock	353	380, 337	403

ions, and have concluded that the structure involves an angular bridge, and a molecular symmetry of C_{2v}. The structure chosen corresponds to structure VI of Table 9.1 and requires that 5 polarized Raman lines are observed, of which one will be in the region of 1000 cm^{-1} (symmetric in-phase terminal stretching) while the others will lie below about 650 cm^{-1} (terminal bending, bridge stretching and bending). Simon and Paetzold [1312, 1314–17] have made some assignments for alkali metal pyroselenites, though they have not classified them according to any symmetry considerations; while Rocchiciolli [1208] has studied some pyrosulphites and pyroselenites, without making any assignment. More recently, Herlinger and Long [1739] have studied aqueous solutions of sodium and potassium pyrosulphite, and have assigned the vibrational spectrum as that of a molecule of C_s symmetry, containing an S—S bond. The Raman spectrum of $K_2S_2O_5$ in aqueous solution shows 9 polarised lines (1085, 1052, 966, 655, 424, 309, 235, 200, 168 cm^{-1}) and 6 depolarised (1196, 1170, 637, 587, 558, 510 cm^{-1}), which is consistent with a structure of C_s symmetry. They attributed the five lines above 900 cm^{-1} to terminal S—O stretching, as the structure requires. They further point out the similarity between the spectra of the solution and the solid. The structure in the solid has been determined by Lindqvist and Mortsell [X96a] who have assigned the structure containing the S—S bridge to $K_2S_2O_5$.

9.6 Molecules with the formula $X(YZ)_3$

Boric acid, $B(OH)_3$, has been studied several times. Originally, the vibrations were assigned by treating the (OH) groups as point masses, and the

Table 9.14 The vibrational spectra of some pyrosulphites and pyroselenites (assuming C_{2v} symmetry).

Compound and band type	Stretching: Terminal		Stretching: Bridge	Bending: Terminal	Bending: Bridge	Ref.
$Na_2S_2O_5$	1266, 1233, 1187	B_1, B_2	999 B_2	629, 618,	457*	1208, 1331
(ir solid)	1146, 1123			551, 521, 512		
	1061	A_1	629 A_1			
$K_2S_2O_5$	1242, 1192, 1178, 1086	B_1, B_2	978 B_2	616, 562	443*	
(ir, R solid)	1061	A_1	629 A_1	518, 513		
$S_2O_5^{=}$	1193, 1089	B_1, B_2	966 B_2	620, 552	423*	
(R, soln. H_2O)	1052	A_1	655 A_1	506, 471		
$Na_2Se_2O_5$	901 A_1, 850		582 B_2	400, 369, 268	—	1208,
(ir, R solid)			491 A_1	(+ 463, 304)†		1314
$K_2Se_2O_5$	870 A_1, 845		557 B_2	320, 256	—	1314
(ir, R solid)			507 A_1			
$BaSe_2O_5$‡	863, 858, 830		574	417, 391, 381	—	1208
(ir solid)	816, 792, 784		538, 510	327, 312		
$CaSe_2O_5, 2H_2O$	866, 830, 808, 780		637	460, 417, 391	—	1208
(ir solid)				330, 324		

* May be a terminal bend.
† Observed by Rocchiciolli in addition to those found by Simon and Paetzold.
‡ The Ca salt has a very similar spectrum.

molecule as being of the type XY_3 with D_{3h} symmetry. For $B(OH)_3$, Bethell and Sheppard [141] give ν_1 (A_1'), 882; ν_2 (A''), 648; ν_3 (E'), 1450; ν_4 (E'), 540 cm^{-1}; on complete deuteration ν_1 fell to 850 cm^{-1}. Hornig and Plumb [679] later gave, for $B(OH)_3$, ν_1 885; ν_2 666 (^{10}B), 632 (^{11}B); ν_3 1490; ν_4 552 cm^{-1}. However, both sets of workers concluded that this was an inadequate basis for interpretation of the spectrum. Bethell and Sheppard suggested an interpretation in terms of a 'layer unit cell', and Durig [412] found this to provide a satisfactory basis for assignment of the internal modes, but not for the lattice modes. This 'layer unit cell' consists of a ring trimer with C_{6h} symmetry; this would require the B—O and O---H—O distances to be identical, the ring to be planar and the BOH angles equal.

Pistorius [1146, 1150] used the data obtained by Servoss and Clark [1273] and Bethell and Sheppard [141] in a normal-coordinate analysis of H_3BO_3 with C_{3h} symmetry—that is, considering an isolated molecule in which the B—O—H systems are nonlinear. His analysis, however, contained some errors which were later corrected [310–11]. A calculation of the A' and E' frequencies of $^{11}B(OH)_3$, $^{10}B(OH)_3$ and $^{11}B(OD)_3$ was made by Ottinger *et al.* [1096]. The selection rules for this system are $\Gamma_{vib} = 3A'$ (R) + $2A''$ (ir) + $4E'$ (ir, R) + E'' (R).

Table 9.15 Vibrational assignment for $B(OH)_3$ with C_{3h} symmetry [141, 1273].

Class	Assignment and description		$^{11}B(OH)_3$ Obs.	Calc.	$^{10}B(OH)_3$ Obs.	Calc.	$^{11}B(OD)_3$ Obs.	Calc.
A'	ν_1	$\nu(O{-}H)$	3250	3250	3250	3250	—	2370
	ν_2	$\nu(B{-}O)$	1060	1060	1060	1060	—	1016
	ν_3	$\delta(BOH)$	881	881	881	881	—	664
E'	ν_6	$\nu(O{-}H)$	3150	3171	3150	3171	2390	2317
	ν_7	$\nu(B{-}O)$	1428	1432	1490	1487	1424	1419
	ν_8	$\delta(OBO)$	544	546	545	547	504	499
	ν_9	$\delta(BOH)$	1183	1187	1195	1188	921	919

The observed values, given by Pistorius [1150], also included ν_4 (A''), BO_3 in-plane bending, at 668 in $^{10}B(OH)_3$ and 648 cm^{-1} in $^{11}B(OH)_3$, and ν_5 (A'') OH torsion at 824 cm^{-1} in $^{11}B(OH)_3$. A tentative assignment

Table 9.16 Vibrational assignments for the layer unit cell of $B(OH)_3$.

Class	Assignment and description		$B(OH)_3$ [141]	$B(OD)_3$ [141]	$B(OH)_3$ [412]	$B(OD)_3$ [412]	$B(OH)_3$ [679]*	$B(OH)_3$ [799]
A_g	ν_1	$\nu(O{-}H)$	3172	—	3168	2377	3171	3165
	ν_2	$\delta(O{-}H)$	1065	—	1077	835†	—	1085
	ν_3	$\nu(B{-}O)$	(882)	(850)	878	832	—	884
	ν_4	Lattice rotation	—	—	—	—	—	60
A_u	ν_5	$\pi(BO_3)$	824	709	808	560	750	798
	ν_6	$\pi(BOH)$	648	545†	625	665	—	647
E_{1g}	ν_{14}	$\pi(BOH)$	—	—	818	650	—	735
	ν_{15}	Lattice rotation	—	—	—	—	—	128
E_{1u}	ν_{16}	$\nu(O{-}H)$	3200	2400	3220	2400	3205	3210
	ν_{17}	$\nu(B{-}O)$	1450	1420	1440	1415	—	1460
	ν_{18}	$\delta(O{-}H)$	1197	929	1185	910	1196	1195
	ν_{19}	$\delta(OBO)$	540	485*	538	510	—	547
E_{2g}	ν_{20}	$\nu(O{-}H)$	3256	—	3245	2434	3253	3251
	ν_{21}	$\nu(B{-}O)$	1391	—	1365	1225	—	1384
	ν_{22}	$\delta(OH)$	1155	—	1163	1013	—	1172
	ν_{23}	$\delta(OBO)$	503	—	498	460	—	499
	ν_{24}	Lattice translation	—	—	—	—	—	210

* Hornig and Plumb label the irreducible representations of class E by single and double primes instead of numerical subscripts.

† Calculated from product rule.

of 209 cm^{-1} to ν_{10} (E''), OH torsion, was also suggested. For the layer unit cell containing the dimer of C_{6h} symmetry, there are 39 modes of vibration, the selection rules being $\Gamma = 4A_g + 2E_{1g} + 5E_{2g}$ (R) + $2A_u + 4E_{1u}$ (ir) + $3B_g + 4B_u + 2E_{2u}$ (inactive). The numbering and descriptions correspond to those of Bethell and Sheppard.

10. The vibrational spectra of molecules with eight or more atoms

10.1 Molecules with eight atoms

(a) Molecules with the formulae XY_7, XY_6Z, XY_5Z_2

The first such molecule to be studied was IF_7, by Lord *et al.* [877]. They gave selection rules for a large number of possible structures, and concluded that the observation of seven bands in the Raman spectrum and four in the infrared, with two of the Raman bands polarized, and no coincidences, was most nearly consistent with the pentagonal bipyramidal structure, whose point group is D_{5h}. The selection rules for this are $\Gamma = 2A_1' + E_1'' + 2E_2'$ (R) $+ 2A_2' + 3E_1'$ (ir) $+ E_2''$ (inactive). The only other structure for which no coincidences are predicted is the planar XY_7 with D_{7h} symmetry, but this would lead to only three Raman-active bands (one polarized) and three infrared. The two extra Raman bands observed by Lord were accounted for as difference bands, while one of the bending modes of class E_1' was not observed. Lord's original assignment was altered by Khanna [750], who carried out a normal-coordinate analysis using a GVFF, and this assignment was confirmed by Claassen *et al.* [257], in spite of their having discovered two (unspecified) mistakes in Khanna's G-matrix. The assignment of $UO_2F_5^{2-}$ by Nguyen-Quy-Dao [1072] reproduces Khanna's analysis. Claassen has given an assignment of ReF_7 corresponding to Khanna's for IF_7. Incomplete and unassigned spectra have been obtained for OsF_7 by Glemser [545], who found bands in the infrared spectrum of the solid at 715, 550, 483, 366, 336 (?) and 282 cm^{-1}, and for K_2NbF_7 [736], K_2TaF_7 [737] by Keller. These gave Raman lines at 782, 630, 388 and 640, 392, 275 cm^{-1} respectively.

The numbering and description of the modes listed in Table 10.1 is that derived from Khanna's internal coordinates.

Table 10.1 Assignments for 8-atomic molecules of D_{5h} symmetry.

Class	Assignment and description	IF_7 (ir, R) [877]	[750]	ReF_7 (ir, R) [257]	$UO_2F_5^{2-}$ (ir) [1073] obs.	[1073] calc.
A_1'	ν_1 ν_{eq}	678	678	736	—	668
	ν_2 $\nu_{sym, axial}$	635	635	645	816	816
A_2''	ν_3 $\nu_{as, axial}$	670	670	703	873	867
	ν_4 $\delta(\beta)$	368	368	299	380	354
E_1'	ν_5 ν_{eq}	547	670	703	740	738
	ν_6 $\delta(\alpha)$	(250)	(190)	353	425	451
	ν_7 $\delta(\beta)$	426	426	217	240	269
E_2'	ν_8 ν_{eq}	511	360	489	—	409
	ν_9 $\delta(\beta)$	313	313	352	—	660
E_1''	ν_{10} $\delta(\beta)$	360	511	597	—	526
E_2''	ν_{11} $\delta(\beta)$	(350)	340	—	—	304

$\delta(\alpha)$ and $\delta(\beta)$ refer to angle bending modes involving the angles of the F_5 planar pentagon, and the angles made by the axial bonds with the F_5 plane, respectively.

The main differences between the assignments of Khanna and of Lord are, firstly, in the reversal of ν_8 and ν_{10}, and, secondly, in the change of ν_5. Khanna considered that the value of 511 cm^{-1} for ν_{10} would be consistent with 368 cm^{-1} for ν_4 (calculated value, 363 cm^{-1}). He also found that a value of 547 cm^{-1} for ν_5 led to widely differing values for the equatorial and axial stretching force constants, whereas the closeness of ν_1 and ν_2 required them to have similar values; his assignment leads to f_r(eq.) = 3·01, f_R(axial) = 4·09 mdyn Å^{-1}. Claassen [257] has also pointed out that the strong bands above 1100 cm^{-1} can best be explained as binary combinations using Khanna's rather than Lord's assignment. He also mentions the possibility of some distortion of the D_{5h} structure; this has also been invoked by Nguyen-Quy-Dao to explain the appearance of ν_2 in the ir spectrum of $K_2UO_2F_5$ and $(NH_4)_2UO_2F_5$. The coincidence between ν_6 and ν_9 in the spectrum of ReF_7 was considered by Claassen to be accidental.

Keller and Chetham-Strode [737] have found bands attributed to M—O stretching at 862 cm^{-1} in $(NH_4)_3TaOF_6$, 910 cm^{-1} in $(NH_4)_3NbOF_6$, and 908 cm^{-1} in K_3NbOF_6. Other bands were found at 540, 305; 537, 288; 539, 419, 335, 295 cm^{-1} for the respective compounds.

(b) Molecules with the formula X_2Y_6

The two structural types observed are those resembling ethane (i.e., of the type $Y_3X—XY_3$) and those which are bridged,

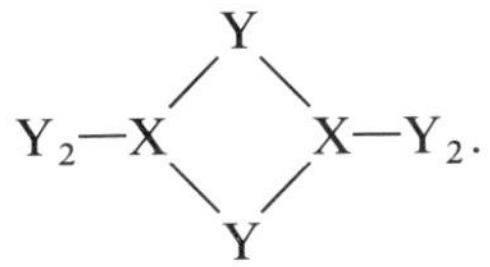

The first type may have symmetry D_{3d} (staggered XY_3 groups), D_{3h} (eclipsed groups) or D_3 (gauche); one may also consider the possibility of free rotation, giving the point group D'_{3h} which has the same selection rules as D_{3h}, but labels the irreducible representations differently. This arises because with free rotation the σ_h is destroyed; the single and double prime superscripts are not used, and $A'_1, A''_1 \rightarrow A_1, \bar{A}_1$, etc. The existence of freely rotating groups has been suggested in Si_2Cl_6, in the vapour, by Morino [997], and an assignment worked out on this basis by Katayama *et al.* [726]. Structural effects have been considered for $N_2H_6^{++}$ by Snyder and Decius [1347], and $S_2O_6^{=}$ by Buijs [205].

Although, as with bridged X_2Y_4, many possible arrangements exist for the second type, all assignments have been made on the basis that the molecule is planar and centrosymmetric, with the point group D_{2h}.

Table 10.2 Vibrational spectra of molecules X_2Y_6 with D_{3d} symmetry.

Class	Assignment and description	$N_2H_6^{++}$ [1347]	Si_2H_6 [142]	Si_2D_6 [142]	Ge_2H_6 [383] obs.	[383] calc.	$P_2O_6^{4-}$ [1106] obs.	$S_2O_6^{2-}$ [205] obs.
A_{1g}	ν_1 $\nu(XY)$	2650	2152	1548	(2070)	2090	1062	1102
	ν_2 $\delta(XY_3)$	1524	909	683	(765)	761	670	710
	ν_3 $\nu(X{-}X)$	1027	434	408	(229)	229	275	293
A_{1u}	ν_4 torsion	455	(115)	—	(144)	—	—	—
A_{2u}	ν_5 $\nu(XY)$	2600	2154	1549	2078	2095	942	1000
	ν_6 $\delta(XY_3)$	1485	844	625	755	753	562	577
E_u	ν_7 $\nu(XY)$	2739	2179	1585	2114	2095	1085	1240
	ν_8 $\delta(XY_3)$	1613	940	682	898	898	494	516
	ν_9 XY_3 rock	1096	379	277	407	399	200	204
E_g	ν_{10} $\nu(X{-}Y)$	2745	2155	1569	(2150)	2150	1168	1216
	ν_{11} $\delta(XY_3)$	1599	929	667	(875)	847	508	556
	ν_{12} XY_3 rock	1105	625	478	(417)	418	323	320

The $N_2H_6^{++}$ ion was studied in the form of the chloride (giving the tabulated results) and the fluoride. For Si_2H_6 and Si_2D_6, the complete study by Bethke and Wilson [142] agrees well with earlier partial studies by Stitt and Yost [1386] and Gutowsky and Stejskal [596]. Pfeiffer and Spangenberg [1134] calculated force constants based on these results and also determined the barrier to rotation as 1220 cal mol^{-1}, from which they estimated ν_4 (torsion) as 115 cm^{-1}. In the ir study by Dows and Hexter [383], the Raman active mode frequencies were all determined from combinations, as was the inactive A_{1u} frequency. Buijs [205] studied $K_2S_2O_6$; he considered the dithionate ion to have D_{3d} symmetry. Palmer [1106] has given an assignment of $S_2O_6^{2-}$ and $P_2O_6^{4-}$ on this basis. Stanley [X142] has shown that its space group is D_3^2, Z = 3, with one dithionate ion on D_3 and two on C_3 sites. Buijs has classified the observed vibrations in terms of C_3 symmetry and given as class A 1243, 1230; 1093 cm^{-1}, class E 1212, 1200; 998, 994 cm^{-1}. Katayama *et al.* [726] have assigned Si_2Cl_6 on the basis of D'_{3h} symmetry, giving ν(Si—Cl) as 354, 124 (A_1, $\bar{A}_1$) and 588 (E), ν(Si—Si) as 626 (A_1) and bending modes as 179, 132 (E), 211 ($\bar{E}$). Two recent studies of the vibrational spectrum of Si_2Cl_6 have led to differing conclusions about its structure. Griffiths [1723] has suggested a low or zero barrier to rotation, leading to D_{3h} or D'_{3h} symmetry; but Ozin [1828] finds the spectrum to be consistent with D_{3d} symmetry in all phases. Jolly [700, 701] has found fundamentals at 880 and 690 cm^{-1} in the ir spectrum of Sn_2H_6 at -196°K.

A greater number of bridged molecules has been studied, and B_2H_6 and its isotopically substituted derivatives have attracted much attention. The initial infrared and Raman study of $^{11}B_2H_6$ and $^{10}B_2D_6$ by Lord and Nielsen [878] led them to the conclusion that the molecule was centrosymmetric, of the asymmetric top type, while the product-rule ratios and the frequencies of the depolarized lines suggested D_{2h} rather than C_{2h} symmetry. The only assignment to be altered in subsequent work is that of ν_9, (B_{1u}), BH_2 rocking. Lord and Nielsen [878] suggested 950 cm^{-1} in $^{11}B_2H_6$ and 705 cm^{-1} in B_2D_6. Anderson and Barker [61] had suggested 369 cm^{-1} for this mode in B_2H_6, but this is assigned to ν_{10} (B_{1u}), BHB bridge bending. Brown and Longuet-Higgins [191] calculated ν_9 = 1860 cm^{-1}, and in the most recent study, Kaufman [731] gives ν_9 at 1858·6 cm^{-1} in $^{11}B_2H_6$, 1869 cm^{-1} in $^{10}B_2H_6$. In other respects his ir assignments agree with Lord and Nielsen's, as does the Raman study by Taylor and Emery [1422]. Lehmann [839] has also studied B_2H_5D and B_2D_5H, the unique atom being in the terminal position (giving C_s symmetry) or in the bridge position (giving C_{2v} symmetry).

The dimeric trihalides of Al, Ga, In, Fe and Au have been studied; there are some differences in assignment and observation in most cases. Adams [22] has obtained the ir spectra of tetraalkyl ammonium salts of $M_2X_6^{=}$ (M = Pd, Pt; X = Cl, Br, I). Stammreich and Kawano [1366] have observed the Raman spectrum of I_2Cl_6, which they consider is a dimer of square ICl_4 sharing an edge.

Holliday [672] has found three bands in the spectrum of $(Me_4N)_2B_2Cl_6$ at 694, 665, 600 cm^{-1}.

In considering the halides of the trivalent metals, we should note that the distribution of vibrations between symmetry classes is not the same for the dimers based on tetrahedral as on square planar monomer units.

Table 10.3 Vibrational assignments in B_2H_6 and its isotopically substituted derivatives.

Class	Assignment and description		$^{11}B_2H_6$ [878]	$^{10}B_2H_6$ [731, 1422]	$^{11}B_2D_6$ [839]	$^{10}B_2D_6$ [878]	C_{2v}: B_2H_5D [839]	C_{2v}: B_2D_5H [839]	C_s: B_2H_5D [839]	C_s: B_2D_5H [839]
A_g	ν_1	$\nu_s(XY_t, ip)$	2524	2537	1850	1860	2525	(1850)	2525	1910
	ν_2	$\nu_s(XY_b, ip)$	2104	2110	1500	1511	(2000)	(1980)	(1900)	(1510)
	ν_3	$\delta(XY_{2t}, ip)$	1180	1186	915	929	(1180)	(920)	1178	1073
	ν_4	Ring deformation	794	816	700	726	(780)	(720)	(785)	(715)
A_u	ν_5	XY_2 twist (anti-parallel)	829	—	592	592	(829)	(529)	761	(650)
B_{1g}	ν_6	$\nu_{as}(XY_t, oop)$	1768	1768	1270	1273	1710	1700	(1760 ?)	(1275)
	ν_7	XY_2 wag	(1035)	—	(860)	(870)	(1035)	(860)	1007	(900)
B_{1u}	ν_8	$\nu_{as}(XY_t, ip)$	2612	2625	1985	1999	2591	1985	2591	2565
	ν_9	XY_2 rock	1859*	1869	(730)	(705)	(950)	(740)	(940)	800
	ν_{10}	$\delta(XY_{2b})$	368	368	250	262	(310 ?)	(300)	(360)	(260)
B_{2g}	ν_{11}	$\nu_{as}(XY_t, oop)$	2591	2600	1975	1980	(2591)	(1975)	2550	1980
	ν_{12}	XY_2 rock	(920)	—	(725)	(740)	(920)	(725)	858 ?	(780)
B_{2u}	ν_{13}	$\nu_s(XY_b, oop)$	1915	1920	1460	1465	1493	1480	1900	1470
	ν_{14}	XY_2 wag	973	977	720	728	957	720	941	730
B_{3g}	ν_{15}	XY_2 twist (parallel)	1012	—	730	730	977	787	978	787
B_{3u}	ν_{16}	$\nu_s(XY_t, oop)$	2525	2528	1840	1845	2525	1850	1908	1850
	ν_{17}	$\nu_s(XY_b, ip)$	1602	1606	1199	1205	1252	1232	1606	1200
	ν_{18}	$\delta(XY_{2t}, oop)$	1177	1181	876	881	1178	1187	1079	898

* Assignment taken from Kaufmann. Lord gave 950 (calculated).

ip = in plane, oop = out of plane; subscripts t, b refer to terminal and bridging groups. The correlations between D_{2h} and the species of lower symmetry C_{2v}, C_s are given by Lehmann as $A_g \rightarrow A_1 \rightarrow A'$; $A_u \rightarrow A_2 \rightarrow A''$; $B_{1g} \rightarrow B_1 \rightarrow A''$; $B_{1u} \rightarrow B_2 \rightarrow A'$; $B_{2g} \rightarrow A_2 \rightarrow A'$; $B_{2u} \rightarrow A_1 \rightarrow A''$; $B_{3g} \rightarrow B_2 \rightarrow A''$; $B_{3u} \rightarrow B_1 \rightarrow A'$.

For the tetrahedral type, the distribution is as given in Tables 10.2, 10.3, but, where the monomer is square planar, we have $\Gamma = 4A_g + A_u + 3B_{1g} + 2B_{1u} + B_{2g} + 3B_{2u} + B_{3g} + 3B_{3u}$. The descriptions of the individual vibrations are the same in the two cases.

The original assignment for Al_2Br_6 given by Gerding and Smit [527] was altered by Miller [960], ν_1 and ν_{11} being interchanged; he considered their polarization data to be incorrect. Beattie has made a further change, altering Miller's ν_6 from 176 cm^{-1} to 354 cm^{-1}. The stronger Raman lines given by Perschina and Raskin [1125] agree with Adams' and Beattie's observations, and the data support the conclusion that both Al_2Cl_6 and Al_2Br_6 are dimers in the liquid as well as in the solid. The major disagreements concern Ga_2Cl_6 and Ga_2Br_6; these involve both observation and assignment. For Ga_2Br_6, Beattie [103] has observed no band at 220 cm^{-1}, while the intensity of the band at 313 cm^{-1} varies with time; this may be due to a hydrolysis product. In view of the disagreement, it is rather surprising that Beattie has given no detailed criticism of Balls' assignments, confining his comments to 'the agreement was disappointing'. ν_4, ν_6, ν_{14}, ν_{17} and ν_{18} are consistently assigned to much higher frequencies by Balls than by Beattie, and ν_9 to a much lower frequency, for Ga_2Cl_6 only. There seems to be no basis for decision as to the correctness of these assignments; Beattie's on the whole seem more likely, particularly with regard to the nodes involving the bridging atoms. Beattie [103] also disagrees to some extent with Adams [23] regarding the assignment of Au_2Cl_6. Beattie's calculations indicate that the bridge and terminal stretching frequencies should be roughly in the ratio 3:4 and there are no disagreements about these modes; they concern librational and ring deformation modes. Stammreich's assignment for I_2Cl_6 is given with alternatives.

Wilmshurst [1512] observed four bands in the ir spectrum of Fe_2Cl_6; they were 507, 468, 268, 399 cm^{-1}.

(c) Eight-membered rings

S_8 was first studied by Bernstein and Powling [131], who assigned to it the puckered octagonal structure with D_{4d} symmetry. The selection rules are $\Gamma = 2A_1 + 3E + 2E_3$ (R) $+ B_2 + 2E_1$ (ir) $+ B_1$ (ia). An assignment by Scott [1258] which led to satisfactory values of thermodynamic functions was subsequently revised by the same author [1259] after investigation of the far infrared. It was then noted that condensed-phase spectral data should not be applied to the calculation of thermodynamic properties in the vapour phase. The revised assignment is ν_1, 475, ν_2, 218 (A_1); ν_3, 411 (B_1); ν_4, 243 (B_2); ν_5, 471, ν_6, 191 (E_1); ν_7, 474, ν_8, 152, ν_9, 86 (E_2); ν_{10}, 437, ν_{11}, 248 (E_3). Force constants were calculated using a

Table 10.4 Vibrational assignments in bridged molecules X_2Y_6, based on tetrahedral monomer units.

Class	Assignment and description	Al_2Cl_6 [527, 766]	Al_2Br_6 [103]	Al_2I_6 [103]	Ga_2Cl_6 [101]	Ga_2Cl_6 [83]	Ga_2Br_6 [103]	Ga_2Br_6 [83]	Ga_2I_6 [103]	Ga_2I_6 [83]	In_2I_6 [103]
A_g	ν_1 νXY_t	506	417	348	413	412	291	287	229	—	187
	ν_2 νXY_b	340	210	148	318	342	204	201	143	—	134
	ν_3 $\delta XY_{2,t}$	217	140	95	167	167	119	118	85	—	69
	ν_4 Ring def	112	70	56	100	321	64	221	50	—	40
A_u	ν_5 XY_2 twist	(75)	—	—	—	—	—	—	—	—	—
B_{1g}	ν_6 νXY_b	438	354	—	231	(400)	241	(300)	195	—	114 or 165
	ν_7 XY_2 wag	164	85	82	128	107	85	83	64	—	55
B_{1u}	ν_8 νXY_t	(625)	502	415	477	420	347	346	273	270	226
	ν_9 XY_2 rock	(202)	—	—	202	123	102	93	(107)	(< 50)	(79)
	ν_{10} Ring def	—	—	—	(28)	73	(17)	(< 50)	(13)	(< 50)	(12)
B_{2g}	ν_{11} νXY_t	606	485	405	462	464	339	341	205	—	232, 226
	ν_{12} XY_2 rock	164	114	63	113	96	74	68	55	—	49
B_{2u}	ν_{13} νXY_b	420	345	291	310	365	232	227	189	185	157
	ν_{14} XY_2 wag	(108)	90	64	120	303	82	220	61	135	49
B_{3g}	ν_{15} XY_2 twist	(160)	79	54	215	228	158	142	68	—	44
B_{3u}	ν_{16} νXY_t	484	375	320	400	394	268	313	213	250	179
	ν_{17} νXY_b	(301)	198	140	280	466	188	264	134	210	125
	ν_{18} δXY_2	(177)	111	81	155	274	90	185	77	72	60

Table 10.5 Vibrational assignments in X_2Y_6 molecules, based on square-planar monomer units.

Class	Assignment and description		Au_2Cl_6 [23]	Au_2Cl_6 [104]	I_2Cl_6 [1366]	$PdCl_6^{2-}$	$PtCl_6^{2-}$ all	$PdBr_6^{2-}$ [22]	$PtBr_6^{2-}$	PdI_6^{2-}	PtI_6^{2-}
A_g	ν_1	$\nu(XY_t)$	380	350	312	—	—	—	—	—	—
	ν_2	$\nu(XY_b)$	331	329	83 or 76	—	—	—	—	—	—
	ν_3	$\delta(XY_{2t})$	168	105	197	—	—	—	—	—	—
	ν_4	Ring def	99	(64 or 55)	51	—	—	—	—	—	—
B_{1g}	ν_6	$\nu(XY_t)$	(376)	380	343	—	—	—	—	—	—
	ν_7	$\nu(XY_b)$	291	290	106	—	—	—	—	—	—
	ν_8	XY_2	124	97	76 or 83	—	—	—	—	—	—
B_{1u}	ν_9	XY_2 wag	137	136	—	155	159	88	85	85	—
	ν_{10}	Ring def	—	44	—	—	—	—	—	—	—
B_{2g}	ν_{11}	XY_2 wag	—	122	118 or 143	—	—	—	—	—	—
B_{2u}	ν_{12}	$\nu(XY_t)$	382	370	—	340	346	266	237	222	179
	ν_{13}	$\nu(XY_b)$	310	312	—	302	315	192	196	144	157
	ν_{14}	XY_2 rock	(148)	90	—	97	98	65	68	—	—
B_{3g}	ν_{15}	XY_2 wag	106	167	143 or 118	—	—	—	—	—	—
B_{3u}	ν_{16}	$\nu(XY_t)$	372	384	—	349	350	262	243	220	194
	ν_{17}	$\nu(XY_b)$	305	293	—	264	302	177	211	132	144
	ν_{18}	$\delta(XY_{2t})$	144	142	—	161	186	117	110	96	92

UBFF with additional terms for cis- non-bonded interactions. The activation of the B_1 mode is explicable in the solid since the molecule crystallizes in the orthorhombic system, space group D_{2h}^{24}, with C_2 site symmetry. As expected, it was not observed by Scott in CS_2 solution. The band originally attributed to it at 532 cm^{-1} by Bernstein and Powling is probably $\nu_9 + \nu_{10} = 523$ cm^{-1}.

N_4S_4 is also an eight-membered ring system, but diffraction measurements give conflicting interpretations. Lu and Donahue [X98] and Clark [X33] suggest a simple 8-membered ring of D_{2d} symmetry, but Hassel and Viervoll [X65] propose a tetracyclic 'cage' structure with the same point group in which the nitrogen atoms are bonded in pairs. Lippincott and Tobin [867] have attempted to resolve this question by means of ir and Raman spectroscopy. The selection rules in the vapour phase, for D_{2d} symmetry, are $\Gamma = 3A_1 + 2B_1$ (R) $+ 3B_2 + 4E$ (ir, R) $+ 2A_2$ (ia). These are relaxed in the solid, since the compound crystallizes with the monoclinic space group C_{2h}^5, and site symmetry C_1. This would account (e.g.) for the splitting of the band at 555 cm^{-1} (cyclohexane solution) into two components in the solid. Lippincott and Tobin unequivocally favour the cage model. This requires two N—N stretching modes ($A_1 + B_2$) which are assigned as 888 cm^{-1} (R; A_1) and 934 cm^{-1} (ir, R; B_2). The cage model also requires a total of seven stretching frequencies, while the ring model requires five. As well as the two N—N frequencies, N—S stretching modes are assigned as 615 cm^{-1} (A_1), 460 cm^{-1} (A_2, calc.), 519 (B_2) and 696, 555 cm^{-1} (E). The five bending modes are assigned as 177 (A_1 and B_2); 173 (A_2, calc.); 213 (B_1 and/or E). The absence of molecules containing N—N bonds from the reduction products of N_4S_4 had been considered to favour the simple ring; however, the compound $N_4S_4H_4$ has a spectrum very similar to that of N_4S_4 except for the absence of the bands near 890, 930 cm^{-1}, so it is presumed that the N—N bonds are preferentially reduced. Bands at 3320, 3285, 3220 (stretching) and 1302, 1296, 1262 (bending) in the ir spectrum of the reduction product confirm the existence of NH groups.

10.2 Molecules with nine atoms

(a) Molecules of the formula XY_8

The two structures found for such molecules are the Archimedean antiprism of D_{4d} symmetry, and the dodecahedron of D_{2d} symmetry. Na_3TaF_8 crystallizes in the monoclinic system, with space group C_{2h}^6 and site symmetry C_2, but according to Hartman and Miller [615] the observed spectrum suggests that the D_{4d} symmetry is retained in the solid. They found Ta—F stretching modes at 622, 426, 377 in the Raman and 500, 371 cm^{-1}

in the ir; the bands were at 331, 200 and 125 in the Raman and 325, 220, 180 cm^{-1} in the ir. Apart from one missing bending mode in the Raman, this is exactly correct for D_{4d} symmetry. An earlier Raman study by Keller and Chetham-Strode [737] gave very similar values for the stretching modes. Crystallographic studies of K_3CrO_8 have shown the CrO_8 group to be dodecahedral [X175–6, X145, X178]; Fergusson has shown K_3VO_8 to be isomorphous with K_3CrO_8 [X43], as has Bohm [X16] for the Ta and Nb compounds.

Griffith [568] has observed bands at 875, 670, 553, 424 and 294 cm^{-1} in the spectrum of K_3CrO_8; 807, 618, 552, 527, 440, 300 cm^{-1} for K_3TaO_8; and 813, 587, 530, 439, 300, 263 cm^{-1} for K_3NbO_8. Hartmann and Miller [615] have studied the complexes $K_4M(CN)_8$, $2H_2O$ (M = Mo, W), and concluded that the $M(CN)_8$ group has D_{4d} symmetry in solution, but D_{2d} in the solid. For the D_{4d} arrangement, there should be 15 Raman and 10 ir active bands, with no coincidences, while the D_{2d} arrangement requires 12 bands which are Raman active only, and 19 which are ir and Raman active. Clearly the complete spectrum was not observed, but the existence of nine coincidences in the solid favours the D_{2d} configuration.

(b) Molecules of the formulae X_2Y_7, $X_2Y_5Z_2$ and $X_2Y_3Z_4$

Most of the work carried out on molecules of these types involves approximate assignments; that is to say, the modes are simply classified as $\nu_{as}(XY_3)$, $\nu_{sym}(XY_3)$, $\nu_{as}(YXY)$, $\nu_{sym}(YXY)$, terminal bending and bridge bending. In a few cases, such as the study by Savoie and Giguere [1237] on Cl_2O_7, all 21 modes are described and assigned, but the assignment is not supported by a normal-coordinate analysis. Analyses have been carried out on pyrosilicates by Saksena [1227] (in Cartesian coordinates), and on pyrophosphates by Hezel and Ross [654] and Mooney and coworkers [985, 987] using a GVFF function. Two other normal-coordinate analyses have been carried out on systems geometrically similar to the pyro-anions; these are by Gutowsky on dimethyl mercury [594] and by Miles *et al.* on diammino-mercury and silver [942], and these used GVFF and UBFF potential functions respectively. In addition, a number of studies have been concerned largely or entirely with the frequencies of the bridging system; these constitute the subject of most of the disagreements. Since the bridge frequencies are strongly dependent on the bridge angle, and this is in many cases unknown, it is not always possible to resolve these disagreements. Possible geometries for the X_2Y_7 system have been listed by Simon and Wagner [1325].

There are no structures with any symmetry elements, for a system in which all the terminal bonds are equivalent, other than those listed in Table 10.6.

Table 10.6 Structures and selection rules for X_2Y_7 systems.

	Point group	Activity			
		R only	ir only	ir and R	Inactive
LINEAR BRIDGE					
XY_3 groups					
staggered	D_{3d}	$3A_{1g} + 3E_g$	$3A_{2u} + 4E_u$	—	$1A_{1u}$
eclipsed	D_{3h}	$3A_1' + 3E''$	$3A_2''$	$4E'$	$1A_1''$
gauche	D_3	$4A_1$	$3A_2$	$7E$	—
NON-LINEAR BRIDGE*					
XY_3 groups					
staggered	C_s	—	—	$13A' + 8A''$	—
eclipsed	C_{2v}	$4A_2$	—	$7A_1 + 4B_1 + 6B_2$	—
gauche	C_1	—	—	$21A$	—
NON-LINEAR BRIDGE†					
I	C_2	—	—	$11A + 10B$	—
II	C_s	—	—	$11A' + 10A''$	—

* One Y atom of each terminal group is in the plane of the bridge.
† Derived from the C_{2v} structure as follows:

The Y atoms which were in the plane of the bridge are now shifted so that the X—Y bonds are on opposite (I) or the same (II) side of the plane. They make equal angles with the plane.

These structures are ideal structures; in the solid state there are deviations from them owing to the differences in length between terminal bonds and the deviations of the terminal angles from the tetrahedral value. A good deal of the structural work has been carried out by Calvo [X29–30, X125]. There is also a pseudo-linearity of the bridge in a number of cases, such as β-$Mg_2P_2O_7$. Although X-ray diffraction measurements show the P—O—P bridge to be linear on the average, it has been found that the central oxygen atom has a large amplitude of vibration and that the linearity is due to thermal averaging. On the other hand, Cruickshank [X41] has expressly stated that the bond angle in $Sc_2Si_2O_7$ is 180° and that this is not an average value taken over random configurations. Steger and Leukroth [1375] have suggested that ZrP_2O_7 and its analogues do not contain discrete $P_2O_7^{4-}$ ions but are mixed oxides; but no crystallographic evidence contradicting Levi and Peyronel's formulation of the structure, which involves Zr^{4+} and $P_2O_7^{4-}$ in an NaCl-like arrangement [X95], has been adduced. Steger and Leukroth's main argument involves the existence of some low-frequency bands whose frequency varies from one cation to another. This, however, does not prove that the materials are mixed metal oxides. The situation is rather similar to that in spinels such as Na_2MoO_4, where the coordination polyhedron of the Mo involves much stronger bonds than that of the Na; in this case, the PO_4 tetrahedron involves much stronger bonds than the ZrO_6 polyhedron.

Table 10.7 Structural information about X_2Y_7 systems.

Compound	Space group	Site group	Bridge angle	Dihedral angle	Ref.
$Sc_2Si_2O_7$*	C^3_{2h}	C_{2h}	180°	—	X41
$Na_4P_2O_7$, $10H_2O$	C^4_{2h}	C_2	134°	13°	X40
MP_2O_7(M = Zr, Ti, Hf, Sn, Si)	T^6_h	C_{3i}	180°	—	X95
α-$Ca_2P_2O_7$	C^5_{2h}	C_1	130°	13° (e)	X30
β-$Ca_2P_2O_7$	—	—	131° 138°	6° (e) 11° (e)	X165
$Cd_2P_2O_7$	—	—	132°	28° (e)	X30
α$Cu_2P_2O_7$	—	—	157°	3° (s)	X125
α$Mg_2P_2O_7$	—	—	144°	15° (s)	X29
α$Co_2P_2O_7$	—	—	143°	14° (s)	X30
α$Ni_2P_2O_7$	—	—	137°	15° (s)	X99
Hemimorphite ($Zn_4(OH)_2Si_2O_7$, H_2O)	C^{20}_{2v}	C_{2v} or C_s	133°†	—	X6, X76
Ilvaite ($CaFe_2^{II}Fe^{III}O(OH)Si_2O_7$)	—	—	133°	—	830
Hardystonite, Ca_2ZnSiO_7	—	—	135°	—	830
Epidote, $CaH_2(Al, Fe)Al_2Si_3O_{13}$	C^2_{2h}	C_i or C_s	155°	—	X75
Rankinite $Ca_3Si_2O_7$	—	—	170°	—	830
Seidozerite	C^2_s or C^4_{2h}	C_1 C_i or C_2	180°	—	X138
$K_2Cr_2O_7$	C^1_i	C_i or C_1	124° 127·6°	5° (e) 10° (e)	X21
$(NH_4)_2Cr_2O_7$	C^6_{2h}	C_2	—	—	X31
$Na_2Mo_2O_7$‡	D^{18}_{2h}	C_{2h}, C_2, C_s, C_i, C_1	—	—	X96
$K_2S_2O_7$	C^6_{2h}	C_2	123°	—	X7

* This mineral, thortveitite, is isostructural with $Cd_2V_2O_7$, $Mg_2As_2O_7$, $Zn_2As_2O_7$, $Mn_2P_2O_7$ and the β forms of $M_2^{II}P_2O_7$ (M^{II} = Cu, Mg, Zn).

† Revised value due to Barclay *et al.* [X6]; Ito and West [X76] originally gave 180°.

‡ $Na_2W_2O_7$ has the same space group.

The symbols (e) and (s) denote departures from the eclipsed and staggered configurations, which appear to be favoured by compounds containing large and small cations respectively.

Before dealing with assignments of all the vibrations, the partial assignments concerned with the bridge frequencies alone are considered. There are three types of assignment:

(i) ν_{as} is about 50–100 cm^{-1} greater than ν_{sym}.

(ii) ν_{as} is about 200 cm^{-1} greater than ν_{sym}, and the bending frequency is about $\frac{1}{3}$ to $\frac{1}{2}$ of ν_{sym}.

(iii) ν_{as} is about twice as great as ν_{sym}, which is in turn about twice as great as the bending frequency.

Table 10.8 Type (i) assignments of bridge frequencies.

System	ν_{as}	ν_{sym}	δ	Ref.
$P_2O_7^{4-}$ (calc.)	940	909	201	654
$H_2S_2O_7$	806	732	—	534
$K_2S_2O_7$	790	730	320	1484
	800	743	321	1325
$K_2Se_2O_7$	680	555	236	1102
$S_2O_5Cl_2$	773	716	486	1309
$S_2O_5F_2$	818	731	488	1310
$P_2O_3Cl_4$	806	713	—	519

Table 10.9 Type (ii) assignments of bridge frequencies.

System	ν_{as}	ν_{sym}	δ	Ref.
Seidozerite	952	633	—	830
Epidote	951	661	—	830
$Na_4P_2O_7$	915	730	—	201
$Mn_2P_2O_7$	964	743	—	1374 (for $Mn_2P_2O_7$ and α-$Cu_2P_2O_7$)
α-$Cu_2P_2O_7$	972	735	—	
$P_2O_7^{4-}$ (calc.)	978	727	—	985
$Gd_4(P_2O_7)_3$	965, 900	800, 765	—	1129
SiP_2O_7	1012	715	—	684, 833–4
$Na_4As_2O_7$	735	550	245	201
$ZrAs_2O_7$	610 or 520	—	—	684
$Na_4V_2O_7$	780, 765	560	—	407
$K_2Cr_2O_7$	797, 764	566, 556	220	1359
$Na_2Mo_2O_7$	600	465	—	407 (for $Na_2Mo_2O_7$ and $Na_2W_2O_7$)
$Na_2W_2O_7$	625, 575	455	—	
Cl_2O_7	695	495	—	1237

Table 10.10 Type (iii) assignments of bridge frequencies.

System	ν_{as}	ν_{sym}	δ	Ref.
$Si_2O_7^{6-}$	1029	503	169	541
Si—O—Si	1110–1000	530–510	—	699 (for Si—O—Si to Sn^{IV}—O—Sn^{IV})
Ge—O—Ge	880–830	458	—	
Sn^{IV}—O—Sn^{IV}	775–735	490–465	—	
$H_2S_2O_7$	809	328	155	540, 541 (for $H_2S_2O_7$ to $P_2O_3Cl_4$)
$Na_2S_2O_7$	805	346	—	
$K_2S_2O_7$	800	321	—	
$S_2O_5F_2$	814	323	157	
$S_2O_5Cl_2$	760	298	147	
Cl_2O_7	695	280	—	
$P_2O_3F_4$	987	480	160	
$P_2O_3Cl_4$	962	425	159	

All these assignments are speculative. Type (i) and (ii) assignments have both been arrived at by normal-coordinate calculations. Lazarev [829–30] has noted the effect of the central angle size on the stretching frequencies, the separation between them increasing as the bridge becomes more nearly linear. His values for ν_{sym} in the series ilvaite (132°, 697 cm^{-1}), hardystonite (145°, 677 cm^{-1}), rankinite (165 ± 5°, 652 cm^{-1}) bear this out, as well as the values for seidozerite and epidote quoted in Table 10.8. This conclusion is borne out by approximate treatments in which the Y—X—Y bridge is treated as an isolated three-atom system. Gillespie and Robinson [541], who are almost the sole proponents of the type (iii) assignment, base their view on the following argument:

The asymmetric stretching frequency involves a large amplitude of vibration for the central Y atom and a small amplitude of vibration for the terminal groups: since the central atom is a light atom, the frequency of ν_{as} is expected to be high. Conversely, in the symmetric mode, the central atom has a small amplitude of vibration while the terminal groups have large amplitudes of vibration, and this should lead to a low frequency for ν_{sym}. They also draw attention to an increasing similarity of form between ν_{sym} and δ as the mass of X increases. The argument therefore depends on the assumption that the terminal XY_3 group moves as a whole in these vibrations, and also neglects the effect of varying the bridge angle.

Many compounds have been studied and the vibrations classified into the forms listed earlier; in some cases the classification has been further broken down, using the symmetry classes of the point groups C_{2v} or C_2. It would be impractical to list all such assignments; a representative selection is given. The two complete assignments given are those of Hezel and Ross [654] and Mooney [987] incorporating his later revisions [985].

The stretching force constants obtained by Mooney *et al.* [985, 987] are substantially higher than those given by Hezel and Ross [654]; this latter assignment has also been criticized by Steger and Kaessner [1374] due to the neglect of the terminal stretch–bridge stretch interaction force constants in obtaining the A_1' and A_2'' stretching frequencies. It is possible that inclusion of these would lead to a type (ii) assignment of the bridge stretching frequencies, as obtained by Mooney [985], rather than a type (i) assignment. A recent reassignment for Cl_2O_7 [1914] places the Cl—O—Cl symmetric stretching and bending frequencies at 699 and 161 cm^{-1} respectively; no value for ν_{as} (Cl—O—Cl) is given.

The bridge stretching frequencies for $S_2O_5F_2$ and $S_2O_5Cl_2$ have been quoted in Table 10.7; Simon and Lehmann have also given 1508, 1455 cm^{-1} for ν_{as} and 1265, 1225 for $\nu_{sym}(SO_3)$ in the fluoride and chloride respectively [1309, 1310]. Gerding [519] studied $P_2O_3Cl_4$; he considered the most likely structure to involve a planar OPOPO system with this plane bisecting the eclipsed terminal groups. He has found seven lines

Table 10.11 Complete assignments for the pyrophosphate ion.

Description	Class and assignment (D_{3h})	$\nu_{obs.}$ $Ca_2P_2O_7$	$\nu_{calc.}$ [654]	Class and assignment (D_{3d})	$\nu_{obs.}$ β-$Mg_2P_2O_7$	$\nu_{calc.}$ [985]
Terminal stretch	ν_1 A_1'	1212, 1190	1212	ν_1 A_{1g}	1094*	1047
Bridge stretch	ν_2	920	909	ν_2	727	713
Terminal bend	ν_3	494	477	ν_3	—	288
Torsion	ν_4 A_1''	—	—	ν_4 A_{1u}	—	69
Terminal stretch	ν_5 A_2''	1173, 1158	1165	ν_5 A_{2u}†	1170	1170
Bridge stretch	ν_6	973, 941	940	ν_6	978	978
Terminal bend	ν_7	559, 538	553	ν_7	558	558
Terminal stretch	ν_8 E''	1069, 1048 1030, 1003	999	ν_8 E_g	1094*	1133
Terminal bend	ν_9	584, 564	573	ν_9	—	445
Terminal bend	ν_{10}	450	432	ν_{10}	—	283
Terminal stretch	ν_{11} E'	1141, 1103	1124	ν_{11} E_u†	1118	1118
Terminal bend	ν_{12}	725	707	ν_{12}	586	586
Terminal bend	ν_{13}	620, 612	615	ν_{13}	440	440
Bridge bend	ν_{14}	229‡	201	ν_{14}	—	—

* Data on ZrP_2O_7 from Steger and Leukroth [1375].

† The A_{2u} and E_u mode assignments are from the revised version. The designations ν_5 and ν_6 have been interchanged to correspond with the D_{3h} assignment.

‡ In $Pb_2P_2O_7$.

which are polarized, consistent with this proposed structure, of C_{2v} symmetry. The disagreement with Robinson [1203] regarding the bridge stretching frequencies is noted in Tables 10.7 and 10.9.

(c) Nine-atomic systems involving rings

The trimeric metaborate, $(BO_2)_3^{3-}$, has been studied in a KBr matrix [666], and the spectrum of metaboric acid has also been recorded [1108a]. These assignments disagree as to the relative positions of the ring and terminal stretching frequencies. Grimm *et al.* [584] have studied boroxine, $H_3B_3O_3$. All these systems have been assigned according to D_{3h} symmetry, the six-membered rings being considered as planar. The selection rules are

$$3A_1' + 2E'' \text{ (R)} + 2A_2'' \text{ (ir)} + 5E' \text{ (ir, R)} + 2A_2' \text{ (inactive)}.$$

10.3 Molecules with ten or more atoms

The systems $(P_2O_8)^{4-}$, $(S_2O_8)^{2-}$ have been studied by Simon and Richter [1319–20]. They obtained the spectra of the NH_4 and K salts of the peroxodiphosphate, and of the Na, K, Rb and Cs salts of the peroxodisulphate. The spectra were assigned on the basis that the terminal XO_3 groups were

Table 10.12 Some vibrational assignments for systems X_2Y_7, where X is a Group IV or V element.

Type of vibration	System, Compound and ref.								
	$Si_2O_7^{6-}$		$P_2O_7^{4-}$					$As_2O_7^{4-}$	$V_2O_7^{4-}$
	Seidozerite [830]	Epidote [830]	Na [201]	Ba [654]	Mn^{II} [1374]	αCu [1374]	Si [684]	Na [201]	Na [407]
$\nu_{as}(XY_3)$	1115 (B_1)	1098 (B_1)	1150	1190	1152	1175	1186	900	895
	1015 (B_2)	1060 (B_2)	1135	1175	—	—	1134	890	880
	—	990 (A_1)	1122	1150	—	—	1103	875	860
	—	901 (A_2)	—	1120	—	—	—	—	842
	900 (A_2)	862 (B_1)	—	—	—	—	—	—	830
	870 (B_1)	823 (A_1)	—	—	—	—	—	—	—
$\nu_s(XY_3)$	—	—	1024	1030	1100	1110	1030	857	950
	—	—	986	1009	—	1095	—	—	940
	—	—	—	1001	—	—	—	—	922
	—	—	—	—	—	—	—	—	917
$\nu_{as}(YXY)$(bridge)	952 (B_1)	951 (B_1)	915	965, 942	964	972	753	735	780, 765
$\nu_{sym}(YXY)$(bridge)	633 (A_1)	661 (A_1)	730	908	743	735	—	550	560
$\delta(XY_3)$	568	575	559 as	735, 718	574 } as	658 }	741 }	420 as	464
	540	527	—	610, 600	558 }	619 } as	713 } as	—	408
	515	404	518 }	565	—	598 }	507 }	—	390
	480	—	488 } sym	550, 540	532 }	529 sym	460 sym	380 sym	371
	444	—	415 }	508, 490	493 } sym	—	—	—	350
	—	—	—	472	427 }	—	—	—	—
	—	—	—	430	—	—	—	—	—
$\delta(YXY)$(bridge)	—	—	—	vb; centre 193	—	—	—	245	—

Table 10.13 Some vibrational assignments for systems X_2Y_7, where X is a Group VI or VII element.

Type of vibration	System, compound and ref.							
	$Cr_2O_7^{2-}$ K [1359]	$Mo_2O_7^{2-}$ Na [407]	$W_2O_7^{2-}$ Na [407]	$S_2O_7^{2-}$ K [1325]	$Se_2O_7^{2-}$ K [1102]	Cl_2O_7 [1237]	Cl_2O_7 [1914]	$S_2O_7^{2-}$ K, melt [1484]
$\nu_{as}(XY_3)$	966, 956	880, 862	825	1302–1289	959	1294 (B_1)	1305	1370–1325 (B)
	950, 946	825, 780	775	1267	942	1260 (B_2)	1275	1295 (B)
	940, 934	—	—	—	—	1272 (A_1)	—	1250 (A)
	924	—	—	—	—	—	—	1200 (A)
$\nu_s(XY_3)$	908, 902	990, 950	962, 950	1108	895	1057 (B_2)	1048	1080 (A)
	891, 885	910	915, 885	1059	879	1034 (A_1)	1019	1060 (B)
$\nu_{as}(YXY)$(bridge)	796, 764	600	625, 575	800	680	695 (B_1)	—	790 (B)
$\nu_s(YXY)$(bridge)	566, 556	465	455	750	555	495 (A_1)	699	730 (A)
$\delta(XY_3)$	365	360	418	655	557	612, 595 (B_1)	591	646 (B)
	—	333	360	618	431	565 (B_2)	558	590 (B)
	—	315	—	589	412	554 (A_1)	—	555 (A)
	—	—	—	563	382	510 (B_2)	507	505 (A)
	—	—	—	526	359	501 (A_1)	—	455 (A)
	—	—	—	467	—	472, 464 (B_1)	435	420 (B)
						429 (A_1)	283	
	—	—	—	—	—	423 (?) (B_2)	272	—
$\delta(YXY)$(bridge)	220	—	—	321(?)	236	278 (A_1)	161	320 (A)

Table 10.14 Vibrational assignments in $P_2O_3Cl_4$ and $P_2O_3F_4$ [1203].

Type of vibration	Fluoride	Chloride
ν(P—O)	1390, 1370	1312
ν_{as}(POP)	987	962
ν_{sym}(POP)	480	425
$\nu P(Hal)_2$	951, 890, 855	611, 560, 531, 507
P=O wag	721, 700	713, 649
$P(Hal)_2$ bending	518, 440	290, 249
wag	395, 362	408, 364, 343 (wag, twist)
twist	352, 340	
rock	295, 273, 205	213, 192, 166
δPOP	160	159

Table 10.15 Assignments for metaborate, metaboric acid and boroxine.

Class	Assignment and description	$(^{11}BO_2)_3^{3-}$ [666]	$(^{10}BO_2)_3^{3-}$ [666]	$(BO_2)_3^{3-}$ [1108a]	$H_3{}^{11}B_3{}^{16}O_3$ [584] obs.	$H_3{}^{11}B_3{}^{16}O_3$ [584] calc.
A_1'	ν_1 Ring stretch	1138	1210	595	—	2616
	ν_2 Terminal stretch*	823	837	1239	—	906
	ν_3 Ring bend	616	633	†	—	811
A_2'	ν_4 Ring bend	690	708	1473	—	1448
	ν_5 Terminal bend	455	460	478	—	1183
E'	ν_6 Ring stretch	1400	1460	1360, 1342	1389	1403
	ν_7 Ring stretch	1222	1265	939	1213	1200
	ν_8 Terminal stretch	950	957	1397	2620	2613
	ν_9 Ring bend	473	478	462	530	530
	ν_{10} Terminal bend	(200)	(205)	280	990	998
A_2''	ν_{11} Ring bend (out of plane)	728	754	745	—	—
	ν_{12} Terminal bend (out of plane)	387	388	(200)	918	—
E''	ν_{13} Terminal bend (out of plane)	717	717	650	—	—
	ν_{14} Ring bend (out of plane)	(280)	(280)	(215)	—	—

* This expression refers to the stretching of the exocyclic B—O or B—H bonds. The descriptions and numbering are taken from [666]. The A_2' modes were activated when the D_{3h} symmetry was broken down in a trimer containing mixed isotopes. Grimm attributed the complexity of the spectrum in the 800 cm^{-1} of isotopically substituted boroxines to possible small deviations from non-planarity.

† Obscured by strong (OH) absorption.

trans, and the point group of the ion either C_{2h} or C_i. In both cases 12 ir and 12 Raman active vibrations are expected.

Nakamoto *et al.* [1056] have studied the bicarbonate dimer, proposing

Table 10.16 Assignments in X_2Y_8 systems ($Y_3X—Y—Y—XY_3$).

Type of vibration	$P_2O_8^{4-}$	$S_2O_8^{2-}$
$\nu_{as}(XY_3)$	1196, 1159, 1132	1303, 1295, 1267, 1262
$\nu_s(XY_3)$	1125, 1101, 1008	1088
$\nu_{as}(XYYX)$	988	1062
$\nu_s(XYYX)$	784	834
$\nu(YY)$	890	854
$\delta(XY_3)$	835, 732, 708	765, 700, 658
	683, 622, 585	590, 568, 560
	545, 528, 514, 425	491, 460, 425
$\delta(XYYX)$	397, 328	328, 236

a ring of C_{2h} symmetry ($\Gamma = 9A_g + 4A_u + 3B_g + 8B_u$). The ring occupies a C_i site. They have calculated the in plane ($A_g + B_u$) frequencies using a MUBFF. Their assignment shows some differences from that of Novak [1083]; the band at 1618 assigned by Novak to an overtone of the $CO_3^{=}$ out of plane bending mode is now ν_{as} (C=O), while the coupling suggested by Novak between the modes at 1405 (δ(OHO)) and 1367 (ν(C⩦O)) is not confirmed by the PED calculations; these show the 1405 cm^{-1} vibra-

Table 10.17 Vibrational assignment of the in-plane modes of $(HCO_3^-)_2$ and $(DCO_3^-)_2$.

Class	Assignment and description	$(HCO_3^-)_2$ obs.	$(HCO_3^-)_2$ calc.	$(DCO_3^-)_2$ obs.	$(DCO_3^-)_2$ calc.
A_g	ν_1 ν(O—H)	2590	2702	1890	1952
	ν_2 ν(C=O)	1682	1668	1668	1665
	ν_3 δ(OHO)	1448	1440	1054	1065
	ν_4 } ν(C⩦O) + ν(C=O)	1283	1289	1226	1289
				1220	
	ν_5 }	1029	1044	1021	1032
	ν_6 δ(OCO) + ν(C—O)	676	700	666	678
	ν_7 δ(OC=O)	635	626	616	620
	ν_8 ν(O---H) + δ(OCO)	—	258	250	257
	ν_9 Ring bend	134	139	—	—
B_u	ν_{10} ν(O—H)	2620	2703	2055	1954
	ν_{11} ν(C=O)	1618	1629	1615	1623
	ν_{12} δ(OHO)	1405	1460	1050	1026
	ν_{13} ν(C—O) + δ(OHO)	1367	1354	1392	1390
	ν_{14} ν(C—O) + ν(C⩦O)	1001	973	985	969
	ν_{15} δ(OC=O) + ν(O---H)	698	703	688	700
	ν_{16} δ(OCO)	655	640	630	636
	ν_{17} ν(O---H) + δ(OC=O)	248	246	248	244

tion to be pure. Out of plane bending modes were observed at 988, 832, 830 $(HCO_3^-)_2$ and 830, 825, 655 $(DCO_3^-)_2$.

Two types of twelve-atomic molecule have been studied; halides of the type X_2Y_{10}, and rings X_3Y_9. The halides S_2F_{10} and Te_2F_{10} have been studied, and the bands assigned on the basis of D_{4d} symmetry ($\Gamma = 4A_1 + 3E_2 + 4E_3$ (R) $+ 4B_2 + 4E_1$ (ir) $+ B_1$ (ia)). The assignments for S_2F_{10} due to Dodd *et al.* [371] and to Wilmshurst and Bernstein [1514] show some differences; Edelsen [425] had earlier recorded, but not assigned, some ir active fundamentals.

Table 10.18 Vibrational assignments in S_2F_{10} and Te_2F_{10}.

Class	Assignment and description		S_2F_{10} [371]	S_2F_{10} [1514]	Te_2F_{10}* [371]
A_1	ν_1	$\nu_{ax}(XY)$	913	908	726
	ν_2	$\nu_{eq}(XY_4)$	690	690	682
	ν_3	$\pi(XY_4)$	—	423	475
	ν_4	$\nu(XX)$	247	246	168
B_1	ν_5	Torsion	—	(89)	—
B_2	ν_6	$\nu_{ax}(XY)$	938	938	752
	ν_7	$\nu_{eq}(XY)$	684	683	734
	ν_8	$\pi(XY_4)$	571	412	468
E_1	ν_9	$\nu_{eq}(XY)$	826	728	714
	ν_{10}	Axial wag	544	504	734
	ν_{11}	$\pi(XY_4)$	410	283	—
	ν_{12}	$\nu_{ax}(XY)$	—	856	—
E_2	ν_{13}	$\nu_{eq}(XY)$	728(?)	634	670
	ν_{14}	$\pi(XY_4)$	634 or 624	624	See text.
	ν_{15}	$\delta_{eq}(XY_2)$	509	(57)	—
E_3	ν_{16}	$\nu_{eq}(XY)$	860	826	670
	ν_{17}	Axial wag	634 or 624	571	—
	ν_{18}	$\delta_{eq}(XY_2)$	425	541	—
	ν_{19}	XY_5 rock	188(?)	(37)	138

* Two bands at 319, 259 are also assigned to any or all of $\nu_{14,15,17,18}$.

Werder [1506] has given a full analysis of Nb_2Cl_{10} with D_{2h} symmetry ($\Gamma = 6A_g + 4B_{1g} + 2B_{2g} + 3B_{3g}$ (R) $+ 4B_{1u} + 5B_{2u} + 4B_{3u}$ (ir) $+ 2A_u$ (ia)). He studied the ir spectrum of the solid and of a solution in CCl_4, and the Raman spectrum of a single crystal. Observed frequencies were: A_g, 397, 357, 245, 168, 128; B_{1g}, 417, 262, 118; B_{2g}, 189; B_{3g}, 417; B_{1u}, 422, 171; B_{2u}, 400, 357, 250, 151, 105; B_{3u}, 412, 224, 105 cm^{-1}. $1A_g + 1B_{1g} + 2B_{1u} + 1B_{2g} + 2B_{3g} + 1B_{3u}$ were not found.

For rings of the type X_3Y_9 (X_3Y_9 skeleton) there are three possible structures. The planar (D_{3h}) structure has the selection rules $\Gamma = 4A_1' + 4E''$ (R) $+ 3A_2''$ (ir) $+ 6E'$ (ir, R) $+ 1A_1'' + 2A_2'$ (ia). Simon and Steger

[1322] have suggested this arrangement for $P_3O_9^{3-}$ in solution. The non-planar chair form, of C_{3v} symmetry ($\Gamma = 7A_1 + 10E$ (ir and R) $+ 3A_2$ (R)) has been proposed for $P_3O_9^{3-}$ in the solid state [204, 1324], for $Si_3O_9^{6-}$ [571], $As_3O_9^{3-}$ [571], and S_3O_9 [537, 1387]. No complete assignment of the planar ring exists; Simon and Steger [1322] divided some of the modes among symmetry classes and later [1324] assigned the stretching modes more exactly. For $Na_3P_3O_9$, Griffith and Rutt have given part of an exact assignment [573]; Bues and Gehrke [204] have allocated the observed bands to vibrational types. Griffith [571] has also given a full assignment for silicate and arsenate rings. The assignments for SO_3 by Stopperka [1387], Gillespie and Robinson [537], and Gerding and Nijveld [525] show some differences; in particular, Stopperka incorrectly tabulates the E ring modes to include 3 stretching and 1 bending, whereas 2 + 2 is correct. Ross [1217] has made a study of the (possibly hypothetical) B_3O_9 ring with C_{3v} symmetry; the assignments and descriptions are taken from that work.

Table 10.19 Vibrational assignments in X_3Y_9 rings.

Class	Assignment and description	'B_3O_9' [1217]	$BaTiSi_3O_9$ [571]	$Na_3P_3O_9$ [573]	β-$K_3As_3O_9$ [571]	S_3O_9 [1387]	S_3O_9 [537]	S_3O_9 [525]
A_1	ν_1 $\nu(XY_2)$	1190	1163	1260	952	1516	1515	1489
	ν_2 $\nu(XY_2)$	1057	962	1166	917	1274	1270	1271
	ν_3 ν ring	1397	570	695, 643	586, 566	698	368	664
	ν_4 δ ring	489	381	374	276	665	665	698
	ν_5 δ ring	344	—	—	—	369	657	369
	ν_6 $\delta(XY_2)$	389	441	545	410	545	534	657
	ν_7 Rock	92	481	530	—	654	698	775
A_2	ν_8 ν ring	1046	—	—	—	—	—	1515
	ν_9 Twist	262	—	—	—	—	—	597
	ν_{10} Wag	314	—	—	—	—	—	319
E	ν_{11} $\nu(XY_2)$	1190	1026	1290	962	1492	1489	1489
	ν_{12} $\nu(XY_2)$	1154	926	1120	899, 855	1232	1231	1271
	ν_{13} ν ring	897	} 746	780	800 }	865, 842	860	1231
	ν_{14} ν ring	786		730	741	772	840	840
	ν_{15} δ ring	442	259	220	—	289	290	289
	ν_{16} δ ring	419	—	—	—	—	407	408
	ν_{17} XY_2	471	441	—	410	545	480	657
	ν_{18} Rock	186	—	—	—	588	680	775
	ν_{19} Twist	328	353	300	—	317	322	597
	ν_{20} Wag	254	397	393	365	412	560	657

Appendices

Acknowledgments

Table A1.1, Appendix 2, Table A3.4, and Fig. A3.1 are reproduced from Wilson, Decius, and Cross: *Molecular Vibrations* © 1955 McGraw-Hill Book Co. Used by permission.

Appendix 6 is reproduced from Godfrey and Siddons: *Four-figure Tables* © 1947 Cambridge University Press. Used by permission.

Appendix 1. Character tables of crystal and molecular point groups

Table A1.1 Character tables of the crystal point groups and of the molecular point groups C_5, D_5, C_{5v}, C_{5h}, D_{5h}, D_{5d}, $C_{\infty v}$, $D_{\infty h}$.

C_s	I	σ_h		
A'	1	1	$T_x, T_y; R_z$	$\alpha_{xx}, \alpha_{yy}, \alpha_{zz}, \alpha_{xy}$
A''	1	-1	$T_z; R_x, R_y$	α_{yz}, α_{zx}

C_i	I	i		
A_g	1	1	R	α
A_u	1	-1	T	

C_2	I	C_2		
A	1	1	$T_z; R_z$	$\alpha_{xx}, \alpha_{yy}, \alpha_{zz}, \alpha_{xy}$
B	1	-1	$T_x, T_y; R_x, R_y$	α_{yz}, α_{zx}

C_3	I	C_3	C_3^2		$\varepsilon = e^{2\pi i/3}$
A	1	1	1	$T_z; R_z$	$\alpha_{xx} + \alpha_{yy}, \alpha_{zz}$
E	$\begin{cases}1\\1\end{cases}$	$\begin{matrix}\varepsilon\\\varepsilon^*\end{matrix}$	$\begin{matrix}\varepsilon^*\\\varepsilon\end{matrix}$	$(T_x, T_y); (R_x, R_y)$	$(\alpha_{xx} - \alpha_{yy}, \alpha_{xy}), (\alpha_{yz}, \alpha_{zx})$

C_4	I	C_4	C_2	C_4^3		
A	1	1	1	1	$T_z; R_z$	$\alpha_{xx} + \alpha_{yy}, \alpha_{zz}$
B	1	-1	1	-1		$\alpha_{xx} - \alpha_{yy}, \alpha_{xy}$
E	$\begin{cases}1\\1\end{cases}$	$\begin{matrix}i\\-i\end{matrix}$	$\begin{matrix}-1\\-1\end{matrix}$	$\begin{matrix}-i\\i\end{matrix}$	$(T_x, T_y); (R_x, R_y)$	$(\alpha_{yz}, \alpha_{zx})$

C_6	I	C_6	C_3	C_2	C_3^2	C_6^5		$\varepsilon = e^{2\pi i/6}$
A	1	1	1	1	1	1	T_z; R_z	$\alpha_{xx} + \alpha_{yy}$, α_{zz}
B	1	−1	1	−1	1	−1		
E_1	1	ε	$-\varepsilon^*$	−1	$-\varepsilon$	ε^*	(T_x, T_y); (R_x, R_y)	$(\alpha_{yz}, \alpha_{zx})$
	1	ε^*	$-\varepsilon$	−1	$-\varepsilon^*$	ε		
E_2	1	$-\varepsilon^*$	$-\varepsilon$	1	$-\varepsilon^*$	$-\varepsilon$		$(\alpha_{xx} - \alpha_{yy}, \alpha_{xy})$
	1	$-\varepsilon$	$-\varepsilon^*$	1	$-\varepsilon$	$-\varepsilon^*$		

C_{2v}	I	C_2	$\sigma_v(zx)$	$\sigma_v(yz)$		
A_1	1	1	1	1	T_z	α_{xx}, α_{yy}, α_{zz}
A_2	1	1	−1	−1	R_z	α_{xy}
B_1	1	−1	1	−1	T_x; R_y	α_{zx}
B_2	1	−1	−1	1	T_y; R_x	α_{yz}

C_{3v}	I	$2C_3$	$3\sigma_v$		
A_1	1	1	1	T_z	$\alpha_{xx} + \alpha_{yy}$, α_{zz}
A_2	1	1	−1	R_z	
E	2	−1	0	(T_x, T_y); (R_x, R_y)	$(\alpha_{xx} - \alpha_{yy}, \alpha_{xy})$; $(\alpha_{yz}, \alpha_{zx})$

C_{4v}	I	$2C_4$	C_2	$2\sigma_v$	$2\sigma_d$		
A_1	1	1	1	1	1	T_z	$\alpha_{xx} + \alpha_{yy}$, α_{zz}
A_2	1	1	1	−1	−1	R_z	
B_1	1	−1	1	1	−1		$\alpha_{xx} - \alpha_{yy}$
B_2	1	−1	1	−1	1		α_{xy}
E	2	0	−2	0	0	(T_x, T_y); (R_x, R_y)	$(\alpha_{yz}, \alpha_{zx})$

C_{6v}	I	$2C_6$	$2C_3$	C_2	$3\sigma_v$	$3\sigma_d$		
A_1	1	1	1	1	1	1	T_z	$\alpha_{xx} + \alpha_{yy}$, α_{zz}
A_2	1	1	1	1	−1	−1	R_z	
B_1	1	−1	1	−1	1	−1		
B_2	1	−1	1	−1	−1	1		
E_1	2	1	−1	−2	0	0	(T_x, T_y); (R_x, R_y)	$(\alpha_{yz}, \alpha_{zx})$
E_2	2	−1	−1	2	0	0		$(\alpha_{xx} - \alpha_{yy}, \alpha_{xy})$

C_{2h}	I	C_2	i	σ_h		
A_g	1	1	1	1	R_z	α_{xx}, α_{yy}, α_{zz}, α_{xy}
B_g	1	−1	1	−1	R_x, R_y	α_{yz}, α_{zx}
A_u	1	1	−1	−1	T_z	
B_u	1	−1	−1	1	T_x, T_y	

C_{3h}	I	C_3	C_3^2	σ_h	S_3	S_3^5		$\varepsilon = e^{2\pi i/3}$
A'	1	1	1	1	1	1	R_z	$\alpha_{xx} + \alpha_{yy}, \alpha_{zz}$
E'	1	ε	ε^*	1	ε	ε^*	(T_x, T_y)	$(\alpha_{xx} - \alpha_{yy}, \alpha_{xy})$
	1	ε^*	ε	1	ε^*	ε		
A''	1	1	1	−1	−1	−1	T_z	
E''	1	ε	ε^*	−1	$-\varepsilon$	$-\varepsilon^*$	(R_x, R_y)	$(\alpha_{yz}, \alpha_{zx})$
	1	ε^*	ε	−1	$-\varepsilon^*$	$-\varepsilon$		

C_{4h}	I	C_4	C_2	C_4^3	i	S_4^3	σ_h	S_4		
A_g	1	1	1	1	1	1	1	1	R_z	$\alpha_{xx} + \alpha_{yy}, \alpha_{zz}$
B_g	1	−1	1	−1	1	−1	1	−1		$\alpha_{xx} - \alpha_{yy}, \alpha_{xy}$
E_g	1	i	−1	$-i$	1	i	−1	$-i$	(R_x, R_y)	$(\alpha_{yz}, \alpha_{zx})$
	1	$-i$	−1	i	1	$-i$	−1	i		
A_u	1	1	1	1	−1	−1	−1	−1	T_z	
B_u	1	−1	1	−1	−1	1	−1	1		
E_u	1	i	−1	$-i$	−1	$-i$	1	i	(T_x, T_y)	
	1	$-i$	−1	i	−1	i	1	$-i$		

C_{6h}	I	C_6	C_3	C_2	C_3^2	C_6^5	i	S_3^5	S_6^5	σ_h	S_6	S_3		$\varepsilon = e^{2\pi i/6}$
A_g	1	1	1	1	1	1	1	1	1	1	1	1	R_z	$\alpha_{xx} + \alpha_{yy}, \alpha_{zz}$
B_g	1	−1	1	−1	1	−1	1	−1	1	−1	1	−1		
E_{1g}	1	ε	$-\varepsilon^*$	−1	$-\varepsilon$	ε^*	1	ε	$-\varepsilon^*$	−1	$-\varepsilon$	ε^*	(R_x, R_y)	$(\alpha_{yz}, \alpha_{zx})$
	1	ε^*	$-\varepsilon$	−1	$-\varepsilon^*$	ε	1	ε^*	$-\varepsilon$	−1	$-\varepsilon^*$	ε		
E_{2g}	1	$-\varepsilon^*$	$-\varepsilon$	1	$-\varepsilon^*$	$-\varepsilon$	1	$-\varepsilon^*$	$-\varepsilon$	1	$-\varepsilon^*$	$-\varepsilon$		$(\alpha_{xx} - \alpha_{yy}, \alpha_{xy})$
	1	$-\varepsilon$	$-\varepsilon^*$	1	$-\varepsilon$	$-\varepsilon^*$	1	$-\varepsilon$	$-\varepsilon^*$	1	$-\varepsilon$	$-\varepsilon^*$		
A_u	1	1	1	1	1	1	−1	−1	−1	−1	−1	−1	T_z	
B_u	1	−1	1	−1	1	−1	−1	1	−1	1	−1	1		
E_{1u}	1	ε	$-\varepsilon^*$	−1	$-\varepsilon$	ε^*	−1	$-\varepsilon$	ε^*	1	ε	$-\varepsilon^*$	(T_x, T_y)	
	1	ε^*	$-\varepsilon$	−1	$-\varepsilon^*$	ε	−1	$-\varepsilon^*$	ε	1	ε^*	$-\varepsilon$		
E_{2u}	1	$-\varepsilon^*$	$-\varepsilon$	1	$-\varepsilon^*$	$-\varepsilon$	−1	ε^*	ε	−1	ε^*	ε		
	1	$-\varepsilon$	$-\varepsilon^*$	1	$-\varepsilon$	$-\varepsilon^*$	−1	ε	ε^*	−1	ε	ε^*		

S_4	I	S_4	C_2	S_4^3		
A	1	1	1	1	R_z	$\alpha_{xx} + \alpha_{yy}, \alpha_{zz}$
B	1	−1	1	−1	T_z	$\alpha_{xx} - \alpha_{yy}, \alpha_{xy}$
E	1	i	−1	$-i$	(T_x, T_y); (R_x, R_y)	$(\alpha_{yz}, \alpha_{zx})$
	1	$-i$	−1	i		

S_6	I	C_3	C_3^2	i	S_6^5	S_6		$S_6 = C_3 \times i \quad \varepsilon = e^{2\pi i/3}$
A_g	1	1	1	1	1	1	R_z	$\alpha_{xx} + \alpha_{yy}, \alpha_{zz}$
E_g	1	ε	ε^*	1	ε	ε^*	(R_x, R_y)	$(\alpha_{xx} - \alpha_{yy}, \alpha_{xy})$; $(\alpha_{yz}, \alpha_{zx})$
	1	ε^*	ε	1	ε^*	ε		
A_u	1	1	1	−1	−1	−1	T_z	
E_u	1	ε	ε^*	−1	$-\varepsilon$	$-\varepsilon^*$	(T_x, T_y)	
	1	ε^*	ε	−1	$-\varepsilon^*$	$-\varepsilon$		

$D_2^* = V$	I	$C_2(z)$	$C_2(y)$	$C_2(x)$		
A	1	1	1	1		$\alpha_{xx}, \alpha_{yy}, \alpha_{zz}$
B_1	1	1	−1	−1	T_z, R_z	α_{xy}
B_2	1	−1	1	−1	T_y, R_y	α_{zx}
B_3	1	−1	−1	1	T_x, R_x	α_{yz}

D_3	I	$2C_3$	$3C_2$		
A_1	1	1	1		$\alpha_{xx} + \alpha_{yy}, \alpha_{zz}$
A_2	1	1	−1	T_z; R_z	
E	2	−1	0	(T_x, T_y); (R_x, R_y)	$(\alpha_{xx} - \alpha_{yy}, \alpha_{xy})$; $(\alpha_{yz}, \alpha_{zx})$

D_4	I	$2C_4$	$C_4^2 = C_2$	$2C_2'$	$2C_2''$		
A_1	1	1	1	1	1		$\alpha_{xx} + \alpha_{yy}, \alpha_{zz}$
A_2	1	1	1	−1	−1	T_z, R_z	
B_1	1	−1	1	1	−1		$\alpha_{xx} - \alpha_{yy}$
B_2	1	−1	1	−1	1		α_{xy}
E	2	0	−2	0	0	(T_x, T_y); (R_x, R_y)	$(\alpha_{yz}, \alpha_{zx})$

D_6	I	$2C_6$	$2C_3$	C_2	$3C_2'$	$3C_2''$		
A_1	1	1	1	1	1	1		$\alpha_{xx} + \alpha_{yy}, \alpha_{zz}$
A_2	1	1	1	1	−1	−1	T_z; R_z	
B_1	1	−1	1	−1	1	−1		
B_2	1	−1	1	−1	−1	1		
E_1	2	1	−1	−2	0	0	(T_x, T_y); (R_x, R_y)	$(\alpha_{yz}, \alpha_{zx})$
E_2	2	−1	−1	2	0	0		$(\alpha_{xx} - \alpha_{yy}, \alpha_{xy})$

* Because of the complete equivalence of the three twofold axes, the species notation is modified in this group as compared with the other D_n.

$D_{2h} = V_h$	I	$C_2(z)$	$C_2(y)$	$C_2(x)$	i	$\sigma(xy)$	$\sigma(zx)$	$\sigma(yz)$		
A_g	1	1	1	1	1	1	1	1		$\alpha_{xx}, \alpha_{yy}, \alpha_{zz}$
B_{1g}	1	1	−1	−1	1	1	−1	−1	R_z	α_{xy}
B_{2g}	1	−1	1	−1	1	−1	1	−1	R_y	α_{zx}
B_{3g}	1	−1	−1	1	1	−1	−1	1	R_x	α_{yz}
A_u	1	1	1	1	−1	−1	−1	−1		
B_{1u}	1	1	−1	−1	−1	−1	1	1	T_z	
B_{2u}	1	−1	1	−1	−1	1	−1	1	T_y	
B_{3u}	1	−1	−1	1	−1	1	1	−1	T_x	

D_{3h}	I	$2C_3$	$3C_2$	σ_h	$2S_3$	$3\sigma_v$		
A'_1	1	1	1	1	1	1		$\alpha_{xx} + \alpha_{yy}, \alpha_{zz}$
A'_2	1	1	−1	1	1	−1	R_z	
E'	2	−1	0	2	−1	0	(T_x, T_y)	$(\alpha_{xx} - \alpha_{yy}, \alpha_{xy})$
A''_1	1	1	1	−1	−1	−1		
A''_2	1	1	−1	−1	−1	1	T_z	
E''	2	−1	0	−2	1	0	(R_x, R_y)	$(\alpha_{yz}, \alpha_{zx})$

D_{4h}	I	$2C_4$	C_2	$2C'_2$	$2C''_2$	i	$2S_4$	σ_h	$2\sigma_v$	$2\sigma_d$		
A_{1g}	1	1	1	1	1	1	1	1	1	1		$\alpha_{xx} + \alpha_{yy}, \alpha_{zz}$
A_{2g}	1	1	1	−1	−1	1	1	1	−1	−1	R_z	
B_{1g}	1	−1	1	1	−1	1	−1	1	1	−1		$\alpha_{xx} - \alpha_{yy}$
B_{2g}	1	−1	1	−1	1	1	−1	1	−1	1		α_{xy}
E_g	2	0	−2	0	0	2	0	−2	0	0	(R_x, R_y)	$(\alpha_{yz}, \alpha_{zx})$
A_{1u}	1	1	1	1	1	−1	−1	−1	−1	−1		
A_{2u}	1	1	1	−1	−1	−1	−1	−1	1	1	T_z	
B_{1u}	1	−1	1	1	−1	−1	1	−1	−1	1		
B_{2u}	1	−1	1	−1	1	−1	1	−1	1	−1		
E_u	2	0	−2	0	0	−2	0	2	0	0	(T_x, T_y)	

D_{6h}	I	$2C_6$	$2C_3$	C_2	$3C'_2$	$3C''_2$	i	$2S_3$	$2S_6$	σ_h	$3\sigma_d$	$3\sigma_v$		
A_{1g}	1	1	1	1	1	1	1	1	1	1	1	1		$\alpha_{xx} + \alpha_{yy}, \alpha_{zz}$
A_{2g}	1	1	1	1	−1	−1	1	1	1	1	−1	−1	R_z	
B_{1g}	1	−1	1	−1	1	−1	1	−1	1	−1	1	−1		
B_{2g}	1	−1	1	−1	−1	1	1	−1	1	−1	−1	1		
E_{1g}	2	1	−1	−2	0	0	2	1	−1	−2	0	0	(R_x, R_y)	$(\alpha_{yz}, \alpha_{zx})$
E_{2g}	2	−1	−1	2	0	0	2	−1	−1	2	0	0		$(\alpha_{xx} - \alpha_{yy}, \alpha_{xy})$
A_{1u}	1	1	1	1	1	1	−1	−1	−1	−1	−1	−1		
A_{2u}	1	1	1	1	−1	−1	−1	−1	−1	−1	1	1	T_z	
B_{1u}	1	−1	1	−1	1	−1	−1	1	−1	1	−1	1		
B_{2u}	1	−1	1	−1	−1	1	−1	1	−1	1	1	−1		
E_{1u}	2	1	−1	−2	0	0	−2	−1	1	2	0	0	(T_x, T_y)	
E_{2u}	2	−1	−1	2	0	0	−2	1	1	−2	0	0		

$D_{2d} = V_d$	I	$2S_4$	C_2	$2C'_2$	$2\sigma_d$		
A_1	1	1	1	1	1		$\alpha_{xx} + \alpha_{yy}, \alpha_{zz}$
A_2	1	1	1	−1	−1	R_z	
B_1	1	−1	1	1	−1		$\alpha_{xx} - \alpha_{yy}$
B_2	1	−1	1	−1	1	T_z	α_{xy}
E	2	0	−2	0	0	(T_x, T_y); (R_x, R_y)	$(\alpha_{yz}, \alpha_{zx})$

D_{3d}	I	$2C_3$	$3C_2$	i	$2S_6$	$3\sigma_d$		
A_{1g}	1	1	1	1	1	1		$\alpha_{xx} + \alpha_{yy}, \alpha_{zz}$
A_{2g}	1	1	−1	1	1	−1	R_z	
E_g	2	−1	0	2	−1	0	(R_x, R_y)	$(\alpha_{xx} - \alpha_{yy}, \alpha_{xy}), (\alpha_{yz}, \alpha_{zx})$
A_{1u}	1	1	1	−1	−1	−1		
A_{2u}	1	1	−1	−1	−1	1	T_z	
E_u	2	−1	0	−2	1	0	(T_x, T_y)	

T	I	$4C_3$	$4C_3^2$	$3C_2$		$\varepsilon = e^{2\pi i/3}$
A	1	1	1	1		$\alpha_{xx} + \alpha_{yy} + \alpha_{zz}$
E	$\begin{cases}1\\1\end{cases}$	$\begin{matrix}\varepsilon\\\varepsilon^*\end{matrix}$	$\begin{matrix}\varepsilon^*\\\varepsilon\end{matrix}$	$\begin{matrix}1\\1\end{matrix}$		$(\alpha_{xx} + \alpha_{yy} - 2\alpha_{zz}, \alpha_{xx} - \alpha_{yy})$
F	3	0	0	−1	T, R	$(\alpha_{xy}, \alpha_{yz}, \alpha_{zx})$

$T_h = T \times i$; T in F_u; R in F_g; α in A_g, E_g, F_g

T_d	I	$8C_3$	$3C_2$	$6S_4$	$6\sigma_d$		
O	I	$8C_3$	$3C_2$	$6C_4$	$6C_2'$		
A_1	1	1	1	1	1		$\alpha_{xx} + \alpha_{yy} + \alpha_{zz}$
A_2	1	1	1	−1	−1		
E	2	−1	2	0	0		$(\alpha_{xx} + \alpha_{yy} - 2\alpha_{zz}, \alpha_{xx} - \alpha_{yy})$
F_1	3	0	−1	1	−1	R; T in O	
F_2	3	0	−1	−1	1	T in T_d	$(\alpha_{xy}, \alpha_{yz}, \alpha_{zx})$

T_h	I	$4C_3$	$4C_3^2$	$3C_2$	i	$4S_6$	$4S_6^5$	$3\sigma_h$		$\varepsilon = e^{2\pi i/3}$
A_g	1	1	1	1	1	1	1	1		$\alpha_{xx} + \alpha_{yy} + \alpha_{zz}$
E_g	$\begin{cases}1\\1\end{cases}$	$\begin{matrix}\varepsilon\\\varepsilon^2\end{matrix}$	$\begin{matrix}\varepsilon^2\\\varepsilon\end{matrix}$	$\begin{matrix}1\\1\end{matrix}$	$\begin{matrix}1\\1\end{matrix}$	$\begin{matrix}\varepsilon\\\varepsilon^2\end{matrix}$	$\begin{matrix}\varepsilon^2\\\varepsilon\end{matrix}$	$\begin{matrix}1\\1\end{matrix}$		$\begin{Bmatrix}\alpha_{xx} + \alpha_{yy} - 2\alpha_{zz},\\ \alpha_{xx} - \alpha_{yy}\end{Bmatrix}$
F_g	3	0	0	−1	3	0	0	−1	R	$\alpha_{xy}, \alpha_{xz}, \alpha_{yz}$
A_u	1	1	1	1	−1	−1	−1	−1		
E_u	$\begin{cases}1\\1\end{cases}$	$\begin{matrix}\varepsilon\\\varepsilon^2\end{matrix}$	$\begin{matrix}\varepsilon^2\\\varepsilon\end{matrix}$	$\begin{matrix}1\\1\end{matrix}$	$\begin{matrix}-1\\-1\end{matrix}$	$\begin{matrix}-\varepsilon\\-\varepsilon^2\end{matrix}$	$\begin{matrix}-\varepsilon^2\\-\varepsilon\end{matrix}$	$\begin{matrix}-1\\-1\end{matrix}$		
F_u	3	0	0	−1	−3	0	0	1	T	

O_h	I	$8C_3$	$3C_2$	$6C_4$	$6C_2'$	i	$8S_6$	$3\sigma_h$	$6S_4$	$6\sigma_d$		
A_{1g}	1	1	1	1	1	1	1	1	1	1		$\alpha_{xx}+\alpha_{yy}+\alpha_{zz}$
A_{2g}	1	1	1	−1	−1	1	1	1	−1	−1		
E_g	2	−1	2	0	0	2	−1	2	0	0		$(2\alpha_{zz}-\alpha_{xx}-\alpha_{yy}, \alpha_{xx}-\alpha_{yy})$
F_{1g}	3	0	−1	1	−1	3	0	−1	1	−1	R	
F_{2g}	3	0	−1	−1	1	3	0	−1	−1	1		$\alpha_{xy}, \alpha_{yz}, \alpha_{zx}$
A_{1u}	1	1	1	1	1	−1	−1	−1	−1	−1		
A_{2u}	1	1	1	−1	−1	−1	−1	−1	1	1		
E_u	2	−1	2	0	0	−2	1	−2	0	0		
F_{1u}	3	0	−1	1	−1	−3	0	1	−1	1	T	
F_{2u}	3	0	−1	−1	1	−3	0	1	1	−1		

C_5	I	C_5	C_5^2	C_5^3	C_5^4		$\varepsilon = e^{2\pi i/5}$
A	1	1	1	1	1	T_z, R_z	$\alpha_{xx} + \alpha_{yy}, \alpha_{zz}$
E_1	$\begin{cases}1\\1\end{cases}$	$\begin{matrix}\varepsilon\\\varepsilon^*\end{matrix}$	$\begin{matrix}\varepsilon^2\\\varepsilon^{2*}\end{matrix}$	$\begin{matrix}\varepsilon^{2*}\\\varepsilon^2\end{matrix}$	$\begin{matrix}\varepsilon^*\\\varepsilon\end{matrix}$	$(T_z, T_y); (R_x, R_y)$	$(\alpha_{yz}, \alpha_{zx})$
E_2	$\begin{cases}1\\1\end{cases}$	$\begin{matrix}\varepsilon^2\\\varepsilon^{2*}\end{matrix}$	$\begin{matrix}\varepsilon^*\\\varepsilon\end{matrix}$	$\begin{matrix}\varepsilon\\\varepsilon^*\end{matrix}$	$\begin{matrix}\varepsilon^{2*}\\\varepsilon^2\end{matrix}$		$(\alpha_{xx} - \alpha_{yy}, \alpha_{xy})$

D_5	I	$2C_5$	$2C_5^2$	$5C_2$		
A_1	1	1	1	1		$\alpha_{xx}+\alpha_{yy}, \alpha_{zz}$
A_2	1	1	1	-1	T_z; R_z	
E_1	2	2 cos 72°	2 cos 144°	0	(T_x, T_y); (R_x, R_y)	$(\alpha_{yz}, \alpha_{zx})$
E_2	2	2 cos 144°	2 cos 72°	0		$(\alpha_{xx}-\alpha_{yy}, \alpha_{xy})$

C_{5v}	I	$2C_5$	$2C_5^2$	$5\sigma_v$		
A_1	1	1	1	1	T_z	$\alpha_{xx}+\alpha_{yy}, \alpha_{zz}$
A_2	1	1	1	-1	R_z	
E_1	2	2 cos 72°	2 cos 144°	0	(T_x, T_y); (R_x, R_y)	$(\alpha_{yz}, \alpha_{zx})$
E_2	2	2 cos 144°	2 cos 72°	0		$(\alpha_{xx}-\alpha_{yy}, \alpha_{xy})$

C_{5h}	I	C_5	C_5^2	C_5^3	C_5^4	σ_h	S_5	S_5^7	S_5^3	S_5^9		$\varepsilon = e^{2\pi i/5}$
A'	1	1	1	1	1	1	1	1	1	1	R_z	$\alpha_{xx}+\alpha_{yy}, \alpha_{zz}$
E_1'	$\begin{cases}1\\1\end{cases}$	ε ε^*	ε^2 ε^{2*}	ε^{2*} ε^2	ε^* ε	1 1	ε ε^*	ε^2 ε^{2*}	ε^{2*} ε^2	ε^* ε	(T_x, T_y)	
E_2'	$\begin{cases}1\\1\end{cases}$	ε^2 ε^{2*}	ε^* ε	ε ε^*	ε^{2*} ε^2	1 1	ε^2 ε^{2*}	ε^* ε	ε ε^*	ε^{2*} ε^2		$(\alpha_{xx}-\alpha_{yy}, \alpha_{xy})$
A''	1	1	1	1	1	-1	-1	-1	-1	-1	T_z	
E_1''	$\begin{cases}1\\1\end{cases}$	ε ε^*	ε^2 ε^{2*}	ε^{2*} ε^2	ε^* ε	-1 -1	$-\varepsilon$ $-\varepsilon^*$	$-\varepsilon^2$ $-\varepsilon^{2*}$	$-\varepsilon^{2*}$ $-\varepsilon^2$	$-\varepsilon^*$ $-\varepsilon$	(R_x, R_y)	α_{yz}, α_{zx}
E_2''	$\begin{cases}1\\1\end{cases}$	ε^2 ε^{2*}	ε^* ε	ε ε^*	ε^{2*} ε^2	-1 -1	$-\varepsilon^2$ $-\varepsilon^{2*}$	$-\varepsilon^*$ $-\varepsilon$	$-\varepsilon$ $-\varepsilon^*$	$-\varepsilon^{2*}$ $-\varepsilon^2$		

D_{5h}	I	$2C_5$	$2C_5^2$	$5C_2$	σ_h	$2S_5$	$2S_5^3$	$5\sigma_v$		
A_1'	1	1	1	1	1	1	1	1		$\alpha_{xx}+\alpha_{yy}, \alpha_{zz}$
A_2'	1	1	1	-1	1	1	1	-1	R_z	
E_1'	2	2 cos 72°	2 cos 144°	0	2	2 cos 72°	2 cos 144°	0	(T_x, T_y)	
E_2'	2	2 cos 144°	2 cos 72°	0	2	2 cos 144°	2 cos 72°	0		$(\alpha_{xx}-\alpha_{yy}, \alpha_{xy})$
A_1''	1	1	1	1	-1	-1	-1	-1		
A_2''	1	1	1	-1	-1	-1	-1	1	T_z	
E_1''	2	2 cos 72°	2 cos 144°	0	-2	-2 cos 72°	-2 cos 144°	0	(R_x, R_y)	$(\alpha_{yz}, \alpha_{zx})$
E_2''	2	2 cos 144°	2 cos 72°	0	-2	-2 cos 144°	-2 cos 72°	0		

D_{5d}	I	$2C_5$	$2C_5^2$	$5C_2$	i	$2S_{10}^3$	$2S_{10}$	$5\sigma_d$		
A_{1g}	1	1	1	1	1	1	1	1		$\alpha_{xx}+\alpha_{yy}, \alpha_{zz}$
A_{2g}	1	1	1	-1	1	1	1	-1	R_z	
E_{1g}	2	2 cos 72°	2 cos 144°	0	2	2 cos 72°	2 cos 144°	0	(R_x, R_y)	$(\alpha_{yz}, \alpha_{zx})$
E_{2g}	2	2 cos 144°	2 cos 72°	0	2	2 cos 144°	2 cos 72°	0		$(\alpha_{xx}-\alpha_{yy}, \alpha_{xy})$
A_{1u}	1	1	1	1	-1	-1	-1	-1		
A_{2u}	1	1	1	-1	-1	-1	-1	1	T_z	
E_{1u}	2	2 cos 72°	2 cos 144°	0	-2	-2 cos 72°	-2 cos 144°	0	(T_x, T_y)	
E_{2u}	2	2 cos 72°	2 cos 72°	0	-2	-2 cos 144°	-2 cos 72°	0		

T_d	I	$8C_3$	$3C_2$	$6S_4$	$6\sigma_d$		
O	I	$8C_3$	$3C_2$	$6C_4$	$6C_2'$		
A_1	1	1	1	1	1		$\alpha_{xx}+\alpha_{yy}+\alpha_{zz}$
A_2	1	1	1	-1	-1		
E	2	-1	2	0	0		$(\alpha_{xx}+\alpha_{yy}-2\alpha_{zz}, \alpha_{xx}-\alpha_{yy})$
F_1	3	0	-1	1	-1	**R**; **T** in O	
F_2	3	0	-1	-1	1	**T** in T_d	$(\alpha_{xy}, \alpha_{yz}, \alpha_{zx})$

$C_{\infty v}$	I	$2C_\infty^\phi$	...	$\infty\sigma_v$		
$A_1 \equiv \Sigma^+$	1	1	...	1	T_z	$\alpha_{xx}+\alpha_{yy}, \alpha_{zz}$
$A_2 \equiv \Sigma^-$	1	1	...	-1	R_z	
$E_1 \equiv \Pi$	2	$2\cos\phi$	...	0	$(T_x, T_y); (R_x, R_y)$	$(\alpha_{yz}, \alpha_{zx})$
$E_2 \equiv \Delta$	2	$2\cos 2\phi$	...	0		$(\alpha_{xx}-\alpha_{yy}, \alpha_{xy})$
$E_3 \equiv \Phi$	2	$2\cos 3\phi$	...	0		
...	...	...				

$D_{\infty h}$	I	$2C_\infty^\phi$	...	$\infty\sigma_v$	i	$2S_\infty^\phi$	...	∞C_2		
Σ_g^+	1	1	...	1	1	1	...	1		$\alpha_{xx}+\alpha_{yy}, \alpha_{zz}$
Σ_g^-	1	1	...	-1	1	1	...	-1	R_z	
Π_g	2	$2\cos\phi$	...	0	2	$-2\cos\phi$	...	0	(R_x, R_y)	$(\alpha_{yz}, \alpha_{zx})$
Δ_g	2	$2\cos 2\phi$	...	0	2	$2\cos 2\phi$	...	0		$(\alpha_{xx}-\alpha_{yy}, \alpha_{zy})$
...	...	...	...	...	...	...	...	...		
Σ_u^+	1	1	...	1	-1	-1	...	-1	T_z	
Σ_u^-	1	1	...	-1	-1	-1	...	1		
Π_u	2	$2\cos\phi$	...	0	-2	$2\cos\phi$	...	0	(T_x, T_y)	
Δ_u	2	$2\cos 2\phi$	...	0	-2	$-2\cos 2\phi$	...	0		
...	...	...	...	...	...	...	...	...		

When a character table contains complex irreducible representations, it will be found that these occur in pairs, one member of the pair being the complex conjugate of the other. If the two representations are Γ, Γ^*, then the frequencies belonging to these two representations are accidentally degenerate, and we need calculate only the number belonging to Γ or Γ^*.

Alternatively, one can use the real representation formed by adding the complex conjugates, and divide a_j for this real representation by 2. Since this is a simpler process, the character tables for the point groups involved, containing the real reducible representations of the E species, are given.

Table A1.2 Character tables of point groups, complex representations being replaced by real representations.

C_3	I	$2C_3$
A	1	1
E	2	-1

C_5	I	$2C_5$	$2C_5^2$
A	1	1	1
E_1	2	2 cos 72°	2 cos 144°
E_2	2	2 cos 144°	2 cos 72°

C_4	I	$2C_4$	C_4^2
A	1	1	1
B	1	-1	1
E	2	0	2

C_6	I	C_6	$C_6^2 \equiv C_3$	$C_6^3 \equiv C_2''$
A	1	1	1	1
B	1	-1	1	-1
E_1	2	1	-1	-2
E_2	2	-1	-1	2

C_{3h}	I	C_3	σ_h	S_3
A'	1	1	1	1
A''	1	1	-1	-1
E'	2	-1	2	-1
E''	2	-1	-2	1

C_{4h}	I	C_4	$C_4^2 \equiv C_2''$	σ_h	S_4	i
A_g	1	1	1	1	1	1
A_u	1	1	1	-1	-1	-1
B_g	1	-1	1	1	-1	-1
B_u	1	-1	1	-1	1	-1
E_g	2	0	-2	-2	0	2
E_u	2	0	-2	2	0	-2

C_{5h}	I	$2C_5$	$2C_5^2$	σ_h	$2S_5^3$	$2S_5^2$
A'	1	1	1	1	1	1
A''	1	1	1	-1	-1	-1
E_1'	2	2 cos 72°	2 cos 144°	2	2 cos 72°	2 cos 144°
E_1''	2	2 cos 72°	2 cos 144°	-2	-2 cos 72°	-2 cos 144°
E_2'	2	2 cos 144°	2 cos 72°	2	2 cos 144°	2 cos 72°
E_2''	2	2 cos 144°	2 cos 72°	-2	-2 cos 144°	-2 cos 72°

C_{6h}	I	C_6	$C_6^2 \equiv C_3$	$C_6^3 \equiv C_2''$	σ_h	S_6	S_3	i
A_g	1	1	1	1	1	1	1	1
A_u	1	1	1	1	-1	-1	-1	-1
B_g	1	-1	1	-1	-1	1	-1	1
B_u	1	-1	1	-1	1	-1	1	-1
E_{1g}	2	1	-1	-2	-2	-1	1	2
E_{1u}	2	1	-1	-2	2	1	-1	-2
E_{2g}	2	-1	-1	2	2	-1	-1	2
E_{2u}	2	-1	-1	2	-2	1	1	-2

T	I	$8C_3$	$3C_2$
A	1	1	1
E	2	−1	2
F	3	0	−1

T_h	I	$8C_3$	$3C_2$	$8S_6$	$3\sigma_h$	i
A_g	1	1	1	1	1	1
A_u	1	1	1	−1	−1	−1
E_g	2	−1	2	−1	2	2
E_u	2	−1	2	1	−2	−2
F_g	3	0	−1	0	−1	3
F_u	3	0	−1	0	−1	−3

Appendix 2. Correlation tables for the species of a group and its subgroups

C_4	C_2
A	A
B	A
E	$2B$

C_6	C_3	C_2
A	A	A
B	A	B
E_1	E	$2B$
E_2	E	$2A$

D_2	C_2	C_2	C_2
A	A	A	A
B_1	A	B	B
B_2	B	A	B
B_3	B	B	A

D_3	C_3	C_2
A_1	A	A
A_2	A	B
E	E	$A + B$

D_4	C_4	C_2	C_2' C_2	C_2'' C_2
A_1	A	A	A	A
A_2	A	A	B	B
B_1	B	A	A	B
B_2	B	A	B	A
E	E	$2B$	$A + B$	$A + B$

D_5	C_5	C_2
A_1	A	A
A_2	A	B
E_1	E_1	$A + B$
E_2	E_2	$A + B$

D_6	C_6	C_2' D_3	C_2'' D_3	D_2	C_3	C_2	C_2' C_2	C_2'' C_2
A_1	A	A_1	A_1	A	A	A	A	A
A_2	A	A_2	A_2	B_1	A	A	B	B
B_1	B	A_1	A_2	B_2	A	B	A	B
B_2	B	A_2	A_1	B_3	A	B	B	A
E_1	E_1	E	E	$B_2 + B_3$	E	$2B$	$A + B$	$A + B$
E_2	E_2	E	E	$A + B_1$	E	$2A$	$A + B$	$A + B$

C_{2v}	C_2	$\sigma(zx)$ C_s	$\sigma(yz)$ C_s
A_1	A	A'	A'
A_2	A	A''	A''
B_1	B	A'	A''
B_2	B	A''	A'

C_{3v}	C_3	C_s
A_1	A	A'
A_2	A	A''
E	E	$A' + A''$

C_{4v}	C_4	σ_v C_{2v}	σ_d C_{2v}	C_2	σ_v C_s	σ_d C_s
A_1	A	A_1	A_1	A	A'	A'
A_2	A	A_2	A_2	A	A''	A''
B_1	B	A_1	A_2	A	A'	A''
B_2	B	A_2	A_1	A	A''	A'
E	E	$B_1 + B_2$	$B_1 + B_2$	$2B$	$A' + A''$	$A' + A''$

C_{5v}	C_5	C_s
A_1	A	A'
A_2	A	A''
E_1	E_1	$A + A''$
E_2	E_2	$A' + A''$

C_{6v}	C_6	σ_v C_{3v}	σ_d C_{3v}	$\sigma_v \rightarrow \sigma(zx)$ C_{2v}	C_3	C_2	σ_v C_s	σ_d C_s
A_1	A	A_1	A_1	A_1	A	A	A'	A'
A_2	A	A_2	A_2	A_2	A	A	A''	A''
B_1	B	A_1	A_2	B_1	A	B	A'	A''
B_2	B	A_2	A_1	B_2	A	B	A''	A'
E_1	E_1	E	E	$B_1 + B_2$	E	$2B$	$A' + A''$	$A' + A''$
E_2	E_2	E	E	$A_1 + A_2$	E	$2A$	$A' + A''$	$A' + A''$

C_{2h}	C_2	C_s	C_i
A_g	A	A'	A_g
B_g	B	A''	A_g
A_u	A	A''	A_u
B_u	B	A'	A_u

C_{3h}	C_3	C_s
A'	A	A'
E'	E	$2A'$
A''	A	A''
E''	E	$2A''$

C_{4h}	C_4	S_4	C_{2h}	C_2	C_s	C_i
A_g	A	A	A_g	A	A'	A_g
B_g	B	B	A_g	A	A'	A_g
E_g	E	E	$2B_g$	$2B$	$2A''$	$2A_g$
A_u	A	B	A_u	A	A''	A_u
B_u	B	A	A_u	A	A''	A_u
E_u	E	E	$2B_u$	$2B$	$2A'$	$2A_u$

C_{5h}	C_5	C_s
A'	A	A'
E_1'	E_1	$2A'$
E_2'	E_2	$2A'$
A''	A	A''
E_1''	E_1	$2A''$
E_2''	E_2	$2A''$

C_{6h}	C_6	C_{3h}	S_6	C_{2h}	C_3	C_2	C_s	C_i
A_g	A	A'	A_g	A_g	A	A	A'	A_g
B_g	B	A''	A_g	B_g	A	B	A''	A_g
E_{1g}	E_1	E''	E_g	$2B_g$	E	$2B$	$2A''$	$2A_g$
E_{2g}	E_2	E'	E_g	$2A_g$	E	$2A$	$2A'$	$2A_g$
A_u	A	A''	A_u	A_u	A	A	A''	A_u
B_u	B	A'	A_u	B_u	A	B	A'	A_u
E_{1u}	E_1	E'	E_u	$2B_u$	E	$2B$	$2A'$	$2A_u$
E_{2u}	E_2	E''	E_u	$2A_u$	E	$2A$	$2A''$	$2A_u$

D_{2h}	D_2	$C_2(z)$ C_{2v}	$C_2(y)$ C_{2v}	$C_2(x)$ C_{2v}	$C_2(z)$ C_{2h}	$C_2(y)$ C_{2h}	$C_2(x)$ C_{2h}	$C_2(z)$ C_2	$C_2(y)$ C_2	$C_2(x)$ C_2	$\sigma(xy)$ C_s	$\sigma(zx)$ C_s	$\sigma(yz)$ C_s
A_g	A	A_1	A_1	A_1	A_g	A_g	A_g	A	A	A	A'	A'	A'
B_{1g}	B_1	A_2	B_2	B_1	A_g	B_g	B_g	A	B	B	A'	A''	A''
B_{2g}	B_2	B_1	A_2	B_2	B_g	A_g	B_g	B	A	B	A''	A'	A''
B_{3g}	B_3	B_2	B_1	A_2	B_g	B_g	A_g	B	B	A	A''	A''	A'
A_u	A	A_2	A_2	A_2	A_u	A_u	A_u	A	A	A	A''	A''	A''
B_{1u}	B_1	A_1	B_1	B_2	A_u	B_u	B_u	A	B	B	A''	A'	A'
B_{2u}	B_2	B_2	A_1	B_1	B_u	A_u	B_u	B	A	B	A'	A''	A'
B_{3u}	B_3	B_1	B_2	A_1	B_u	B_u	A_u	B	B	A	A'	A'	A''

D_{3h}	C_{3h}	D_3	C_{3v}	$\sigma_h \to \sigma_v(zy)$ C_{2v}	C_3	C_2	σ_h C_s	σ_v C_s
A_1'	A'	A_1	A_1	A_1	A	A	A'	A'
A_2'	A'	A_2	A_2	B_2	A	B	A'	A''
E'	E'	E	E	$A_1 + B_2$	E	$A + B$	$2A'$	$A' + A''$
A_1''	A''	A_1	A_2	A_2	A	A	A''	A''
A_2''	A''	A_2	A_1	B_1	A	B	A''	A'
E''	E''	E	E	$A_2 + B_1$	E	$A + B$	$2A''$	$A' + A''$

D_{4h}	D_4	$C_2' \to C_2'$ D_{2d}	$C_2'' \to C_2'$ D_{2d}	C_{4v}	C_{4h}	C_2' D_{2h}	C_2'' D_{2h}	C_4	S_4
A_{1g}	A_1	A_1	A_1	A_1	A_g	A_g	A_g	A	A
A_{2g}	A_2	A_2	A_2	A_2	A_g	B_{1g}	B_{1g}	A	A
B_{1g}	B_1	B_1	B_2	B_1	B_g	A_g	B_{1g}	B	B
B_{2g}	B_2	B_2	B_1	B_2	B_g	B_{1g}	A_g	B	B
E_g	E	E	E	E	E_g	$B_{2g} + B_{3g}$	$B_{2g} + B_{3g}$	E	E
A_{1u}	A_1	B_1	B_1	A_2	A_u	A_u	A_u	A	B
A_{2u}	A_2	B_2	B_2	A_1	A_u	B_{1u}	B_{1u}	A	B
B_{1u}	B_1	A_1	A_2	B_2	B_u	A_u	B_{1u}	B	A
B_{2u}	B_2	A_2	A_1	B_1	B_u	B_{1u}	A_u	B	A
E_u	E	E	E	E	E_u	$B_{2u} + B_{3u}$	$B_{2u} + B_{3u}$	E	E

D_{4h} (cont.)	C_2' D_2	C_2'' D_2	C_2, σ_v C_{2v}	C_2, σ_d C_{2v}	C_2' C_{2v}	C_2' C_{2v}
A_{1g}	A	A	A_1	A_1	A_1	A_1
A_{2g}	B_1	B_1	A_2	A_2	B_1	B_1
B_{1g}	A	B_1	A_1	A_2	A_1	B_1
B_{2g}	B_1	A	A_2	A_1	B_1	A_1
E_g	$B_2 + B_3$	$B_2 + B_3$	$B_1 + B_2$	$B_1 + B_2$	$A_2 + B_2$	$A_2 + B_2$
A_{1u}	A	A	A_2	A_2	A_2	A_2
A_{2u}	B_1	B_1	A_1	A_1	B_2	B_2
B_{1u}	A	B_1	A_2	A_1	A_2	B_2
B_{2u}	B_1	A	A_1	A_2	B_2	A_2
E_u	$B_2 + B_3$	$B_2 + B_3$	$B_1 + B_2$	$B_1 + B_2$	$A_1 + B_1$	$A_1 + B_1$

D_{4h} (cont.)	C_2 C_{2h}	C_2' C_{2h}	C_2'' C_{2h}	C_2 C_2	C_2' C_2	C_2'' C_2	σ_h C_s	σ_v C_s	σ_d C_s	C_i
A_{1g}	A_g	A_g	A_g	A	A	A	A'	A'	A'	A_g
A_{2g}	A_g	B_g	B_g	A	B	B	A'	A''	A''	A_g
B_{1g}	A_g	A_g	B_g	A	A	B	A'	A'	A''	A_g
B_{2g}	A_g	B_g	A_g	A	B	A	A'	A''	A'	A_g
E_g	$2B_g$	$A_g + B_g$	$A_g + B_g$	$2B$	$A + B$	$A + B$	$2A''$	$A' + A''$	$A' + A''$	$2A_g$
A_{1u}	A_u	A_u	A_u	A	A	A	A''	A''	A''	A_u
A_{2u}	A_u	B_u	B_u	A	B	B	A''	A'	A'	A_u
B_{1u}	A_u	A_u	B_u	A	A	B	A''	A''	A'	A_u
B_{2u}	A_u	B_u	A_u	A	B	A	A''	A'	A''	A_u
E_u	$2B_u$	$A_u + B_u$	$A_u + B_u$	$2B$	$A + B$	$A + B$	$2A'$	$A' + A''$	$A' + A''$	$2A_u$

D_{5h}	D_5	C_{5v}	C_{5h}	C_5	$\sigma_h \rightarrow \sigma(zx)$ C_{2v}	C_2	σ_h C_s	σ_v C_s
A_1'	A_1	A_1	A'	A	A_1	A	A'	A'
A_2'	A_2	A_2	A'	A	B_1	B	A'	A''
E_1'	E_1	E_1	E_1'	E_1	$A_1 + B_1$	$A + B$	$2A'$	$A' + A''$
E_2'	E_2	E_2	E_2'	E_2	$A_1 + B_1$	$A + B$	$2A'$	$A' + A''$
A_1''	A_1	A_2	A''	A	A_2	A	A''	A''
A_2''	A_2	A_1	A''	A	B_2	B	A''	A'
E_1''	E_1	E_1	E_1''	E_1	$A_2 + B_2$	$A + B$	$2A''$	$A' + A''$
E_2''	E_2	E_2	E_2''	E_2	$A_2 + B_2$	$A + B$	$2A''$	$A' + A''$

D_{6h}	D_6	C_2' D_{3h}	C_2'' D_{3h}	C_{6v}	C_{6h}	C_2'' D_{3d}	C_2' D_{3d}	$\sigma_h \rightarrow \sigma(xy)$ $\sigma_v \rightarrow \sigma(yz)$ D_{2h}	C_6	C_{3h}	C_2' D_3	C_2'' D_3	σ_v C_{3v}	σ_d C_{3v}	S_6	D_2
A_{1g}	A_1	A_1'	A_1'	A_1	A_g	A_{1g}	A_{1g}	A_g	A	A'	A_1	A_1	A_1	A_1	A_g	A
A_{2g}	A_2	A_2'	A_2'	A_2	A_g	A_{2g}	A_{2g}	B_{1g}	A	A'	A_2	A_2	A_2	A_2	A_g	B_1
B_{1g}	B_1	A_1''	A_2''	B_2	B_g	A_{2g}	A_{1g}	B_{2g}	B	A''	A_1	A_2	A_2	A_1	A_g	B_1
B_{2g}	B_2	A_2''	A_1''	B_1	B_g	A_{1g}	A_{2g}	B_{3g}	B	A''	A_2	A_1	A_1	A_2	A_g	B_3
E_{1g}	E_1	E''	E''	E_1	E_{1g}	E_g	E_g	$B_{2g} + B_{3g}$	E_1	E''	E	E	E	E	E_g	$B_2 + B_3$
E_{2g}	E_2	E'	E'	E_2	E_{2g}	E_g	E_g	$A_g + B_{1g}$	E_2	E'	E	E	E	E	E_g	$A + B_1$
A_{1u}	A_1	A_1''	A_1''	A_2	A_u	A_{1u}	A_{1u}	A_u	A	A''	A_1	A_1	A_2	A_2	A_u	A
A_{2u}	A_2	A_2''	A_2''	A_1	A_u	A_{2u}	A_{2u}	B_{1u}	A	A''	A_2	A_2	A_1	A_1	A_u	B_1
B_{1u}	B_1	A_1'	A_2'	B_1	B_u	A_{2u}	A_{1u}	B_{2u}	B	A'	A_1	A_2	A_1	A_2	A_u	B_2
B_{2u}	B_2	A_2'	A_1'	B_2	B_u	A_{1u}	A_{2u}	B_{3u}	B	A'	A_2	A_1	A_2	A_1	A_u	B_3
E_{1u}	E_1	E'	E'	E_1	E_{1u}	E_u	E_u	$B_{2u} + B_{3u}$	E_1	E'	E	E	E	E	E_u	$B_2 + B_3$
E_{2u}	E_2	E''	E''	E_2	E_{2u}	E_u	E_u	$A_u + B_{1u}$	E_2	E''	E	E	E	E	E_u	$A + B_1$

D_{6h} (cont.)	C'_2 C_{2v}	C''_2 C_{2v}	C_2 C_{2h}	C'_2 C_{2h}	C''_2 C_{2h}	C_3	C_2 C_2
A_{1g}	A_1	A_1	A_g	A_g	A_g	A	A
A_{2g}	B_1	B_1	A_g	B_g	B_g	A	A
B_{1g}	A_2	B_2	B_g	A_g	B_g	A	B
B_{2g}	B_2	A_2	B_g	B_g	A_g	A	B
E_{1g}	$A_2 + B_2$	$A_2 + B_2$	$2B_g$	$A_g + B_g$	$A_g + B_g$	E	$2B$
E_{2g}	$A_1 + B_1$	$A_1 + B_1$	$2A_g$	$A_g + B_g$	$A_g + B_g$	E	$2A$
A_{1u}	A_2	A_2	A_u	A_u	A_u	A	A
A_{2u}	B_2	B_2	A_u	B_u	B_u	A	A
B_{1u}	A_1	B_1	B_u	A_u	B_u	A	B
B_{2u}	B_1	A_1	B_u	B_u	A_u	A	B
E_{1u}	$A_1 + B_1$	$A_1 + B_1$	$2B_u$	$A_u + B_u$	$A_u + B_u$	E	$2B$
E_{2u}	$A_2 + B_2$	$A_2 + B_2$	$2A_u$	$A_u + B_u$	$A_u + B_u$	E	$2A$

D_{6h} (cont.)	C'_2 C_2	C''_2 C_2	σ_h C_s	σ_d C_s	σ_v C_s	C_i
A_{1g}	A	A	A'	A'	A'	A_g
A_{2g}	B	B	A'	A''	A''	A_g
B_{1g}	A	B	A''	A'	A''	A_g
B_{2g}	B	A	A''	A''	A'	A_g
E_{1g}	$A + B$	$A + B$	$2A''$	$A' + A''$	$A' + A''$	$2A_g$
E_{2g}	$A + B$	$A + B$	$2A'$	$A' + A''$	$A' + A''$	$2A_g$
A_{1u}	A	A	A''	A''	A''	A_u
A_{2u}	B	B	A''	A'	A'	A_u
B_{1u}	A	B	A'	A''	A'	A_u
B_{2u}	B	A	A'	A'	A''	A_u
E_{1u}	$A + B$	$A + B$	$2A'$	$A' + A''$	$A' + A''$	$2A_u$
E_{2u}	$A + B$	$A + B$	$2A''$	$A' + A''$	$A' + A''$	$2A_u$

D_{2d}	S_4	$C_2 \rightarrow C_2(z)$ D_2	C_{2v}	C_2 C_2	C'_2 C_2	C_s
A_1	A	A	A_1	A	A	A'
A_2	A	B_1	A_2	A	B	A''
B_1	B	A	A_2	A	A	A''
B_2	B	B_1	A_1	A	B	A'
E	E	$B_2 + B_3$	$B_1 + B_2$	$2B$	$A + B$	$A' + A''$

D_{3d}	D_3	C_{3v}	S_6	C_3	C_{2h}	C_2	C_s	C_1
A_{1g}	A_1	A_1	A_g	A	A_g	A	A'	A_g
A_{2g}	A_2	A_2	A_g	A	B_g	B	A''	A_g
E_g	E	E	E_g	E	$A_g + B_g$	$A + B$	$A' + A''$	$2A_g$
A_{1u}	A_1	A_2	A_u	A	A_u	A	A''	A_u
A_{2u}	A_2	A_1	A_u	A	B_u	B	A'	A_u
E_u	E	E	E_u	E	$A_u + B_u$	$A + B$	$A' + A''$	$2A_u$

						C_2	C_2'	
D_{4d}	D_4	C_{4v}	S_8	C_4	C_{2v}	C_2	C_2	C_s
A_1	A_1	A_1	A	A	A_1	A	A	A'
A_2	A_2	A_2	A	A	A_2	A	B	A''
B_1	A_1	A_2	B	A	A_2	A	A	A''
B_2	A_2	A_1	B	A	A_1	A	B	A'
E_1	E	E	E_1	E	$B_1 + B_2$	$2B$	$A + B$	$A' + A''$
E_2	$B_1 + B_2$	$B_1 + B_2$	E_2	$2B$	$A_1 + A_2$	$2A$	$A + B$	$A' + A''$
E_3	E	E	E_3	E	$B_1 + B_2$	$2B$	$A + B$	$A' + A''$

D_{5d}	D_5	C_{5v}	C_5	C_2	C_s	C_i
A_{1g}	A_1	A_1	A	A	A'	A_g
A_{2g}	A_2	A_2	A	B	A''	A_g
E_{1g}	E_1	E_1	E_1	$A + B$	$A' + A''$	$2A_g$
E_{2g}	E_2	E_2	E_2	$A + B$	$A' + A''$	$2A_g$
A_{1u}	A_1	A_2	A	A	A''	A_u
A_{2u}	A_2	A_1	A	B	A'	A_u
E_{1u}	E_1	E_1	E_1	$A + B$	$A' + A''$	$2A_u$
E_{2u}	E_2	E_2	E_2	$A + B$	$A' + A''$	$2A_u$

D_{6d}	D_6	C_{6v}	C_6	D_{2d}	D_3	C_{3v}
A_1	A_1	A_1	A	A_1	A_1	A_1
A_2	A_2	A_2	A	A_2	A_2	A_2
B_1	A_1	A_2	A	B_1	A_1	A_2
B_2	A_2	A_1	A	B_2	A_2	A_1
E_1	E_1	E_1	E_1	E	E	E
E_2	E_2	E_2	E_2	$B_1 + B_2$	E	E
E_3	$B_1 + B_2$	$B_1 + B_2$	$2B$	E	$A_1 + A_2$	$A_1 + A_2$
E_4	E_2	E_2	E_2	$A_1 + A_2$	E	E
E_5	E_1	E_1	E_1	E	E	E

					C_2	C_2'	
D_{6d} (cont.)	D_2	C_{2v}	S_4	C_3	C_2	C_2	C_s
A_1	A	A_1	A	A	A	A	A'
A_2	B_1	A_2	A	A	A	B	A''
B_1	A	A_2	B	A	A	A	A''
B_2	B_1	A_1	B	A	A	B	A'
E_1	$B_2 + B_3$	$B_1 + B_2$	E	E	$2B$	$A + B$	$A' + A''$
E_2	$A + B_1$	$A_1 + A_2$	$2B$	E	$2A$	$A + B$	$A' + A''$
E_3	$B_2 + B_3$	$B_1 + B_2$	E	$2A$	$2B$	$A + B$	$A' + A''$
E_4	$A + B_1$	$A_1 + A_2$	$2A$	E	$2A$	$A + B$	$A' + A''$
E_5	$B_2 + B_3$	$B_1 + B_2$	E	E	$2B$	$A + B$	$A' + A''$

S_4	C_2
A	A
B	A
E	$2B$

S_6	C_3	C_i
A_g	A	A_g
E_g	E	$2A_g$
A_u	A	A_u
E_u	E	$2A_u$

S_8	C_4	C_2
A	A	A
B	A	A
E_1	E	$2B$
E_2	$2B$	$2A$
E_3	E	$2B$

T	D_2	C_3	C_2
A	A	A	A
E	$2A$	E	$2A$
F	$B_1 + B_2 + B_3$	$A + E$	$A + 2B$

T_h	T	D_{2h}	S_6	D_2
A_g	A	A_g	A_g	A
E_g	E	$2A_g$	E_g	$2A$
F_g	F	$B_{1g} + B_{2g} + B_{3g}$	$A_g + E_g$	$B_1 + B_2 + B_3$
A_u	A	A_u	A_u	A
E_u	E	$2A_u$	E_u	$2A$
F_u	F	$B_{1u} + B_{2u} + B_{3u}$	$A_u + E_u$	$B_1 + B_2 + B_2$

T_h (*cont.*)	C_{2v}	C_{2h}	C_3	C_2	C_s	C_i
A_g	A_1	A_g	A	A	A'	A_g
E_g	$2A_1$	$2A_g$	E	$2A$	$2A'$	$2A_g$
F_g	$A_2 + B_1 + B_2$	$A_g + 2B_g$	$A + E$	$A + 2B$	$A' + 2A''$	$3A_g$
A_u	A_2	A_u	A	A	A''	A_u
E_u	$2A_2$	$2A_u$	E	$2A$	$2A''$	$2A_u$
F_u	$A_1 + B_1 + B_2$	$A_u + 2B_u$	$A + E$	$A + 2B$	$2A' + A''$	$3A_u$

T_d	T	D_{2d}	C_{3v}	S_4	D_2	C_{2v}	C_3	C_2	C_s
A_1	A	A_1	A_1	A	A	A_1	A	A	A'
A_2	A	B_1	A_2	B	A	A_2	A	A	A''
E	E	$A_1 + B_1$	E	$A + B$	$2A$	$A_1 + A_2$	E	$2A$	$A' + A''$
F_1	F	$A_2 + E$	$A_2 + E$	$A + E$	$B_1 + B_2 + B_3$	$A_2 + B_1 + B_2$	$A + E$	$A + 2B$	$A' + 2A''$
F_2	F	$B_2 + E$	$A_1 + E$	$B + E$	$B_1 + B_2 + B_3$	$A_1 + B_1 + B_2$	$A + E$	$A + 2B$	$2A' + A''$

O	T	D_4	D_3	C_4	$3C_2$ D_2	$C_2, 2C_2'$ D_2	C_3	C_2	C_2
A_1	A	A_1	A_1	A	A	A	A	A	A
A_2	A	B_1	A_2	B	A	B_1	A	A	B
E	E	$A_1 + B_1$	E	$A + B$	$2A$	$A + B_1$	E	$2A$	$A + B$
F_1	F	$A_2 + E$	$A_2 + E$	$A + E$	$B_1 + B_2 + B_3$	$B_1 + B_2 + B_3$	$A + E$	$A + 2B$	$A + 2B$
F_2	F	$B_2 + E$	$A_1 + E$	$B + E$	$B_1 + B_2 + B_3$	$A + B_2 + B_3$	$A + E$	$A + 2B$	$2A + B$

O_h†	O	T_d	T_h	D_{4h}	D_{3d}
A_{1g}	A_1	A_1	A_g	A_{1g}	A_{1g}
A_{2g}	A_2	A_2	A_g	B_{1g}	A_{2g}
E_g	E	E	E_g	$A_{1g} + B_{1g}$	E_g
F_{1g}	F_1	F_1	F_g	$A_{2g} + E_g$	$A_{2g} + E_g$
F_{2g}	F_2	F_2	F_g	$B_{2g} + E_g$	$A_{1g} + E_g$
A_{1u}	A_1	A_2	A_u	A_{1u}	A_{1u}
A_{2u}	A_2	A_1	A_u	B_{1u}	A_{2u}
E_u	E	E	E_u	$A_{1u} + B_{1u}$	E_u
F_{1u}	F_1	F_2	F_u	$A_{2u} + E_u$	$A_{2u} + E_u$
F_{2u}	F_2	F_1	F_u	$B_{2u} + E_u$	$A_{1u} + E_u$

† To find correlations with smaller subgroups, carry out the correlation in two steps; for example, if the correlation of O_h with C_{2v} is desired, use the above table to pass from O_h to T_d and then employ the table for T_d to go on to C_{2v}.

Appendix 3. g matrix elements for bond stretching and valence-angle bending

The notation devised by Decius [331] involves three sets of designations.

(a) A double subscript consisting of two of the symbols r, ϕ, τ representing respectively stretching, bending and torsion. Note that no distinction is being made between in-plane and out-of-plane bending.
(b) A numerical superscript indicating the number of atoms common to the coordinates involved, where the element represents an interaction, or the total number of atoms in the coordinate, where the element represents a principal term in the inverse kinetic energy.
(c) A pair of numbers, in parentheses, denoting the numbers of non-common atoms on the upper left and lower left diagonals in the diagrams depicting the g elements. The Decius convention for numbering these diagrams is as follows:

Common atoms are drawn along a horizontal line and the numbers are enclosed in a double circle. The numbering always begins with the common atoms. The non-common atoms are then numbered, running outwards along the upper left, lower left, upper right and lower right diagonals in that order. It is important to realize that the Decius diagrams do not necessarily reproduce the geometry of the coordinates involved, and that a given atom in the molecule will not necessarily bear the same number in the diagrams for the g elements.

The Decius expressions for the g elements appropriate to r and ϕ-coordinates involve two other types of angle, designated ψ and τ. ψ is defined through the spherical-trigonometric relation

$$\cos \psi_{abc} = \frac{\cos \phi_{adc} - \cos \phi_{adb} \cos \phi_{bdc}}{\sin \phi_{adb} \sin \phi_{bdc}}.$$

Table A3.1 Descriptions and Decius symbols of g elements for bond stretching (r) and valence angle bending (ϕ).

Description	Decius symbol	Number of common atoms	Types of common atom
Principal stretching	g^2rr	2	
Principal bending	$g^3\phi\phi$	3	
Stretching interaction	g^1rr	1	Atom common to the two bonds
Stretch–bend interaction	$g^2r\phi$	2	The bond forms a side of the angle
	$g^1r\phi(\frac{1}{2})$	1	An atom of the bond is a terminal atom of the angle
	$g^1r\phi(\frac{1}{1})$	1	An atom of the bond is the central atom of the angle
Bend–bend interaction	$g^2\phi\phi(\frac{1}{1})$	2	The common atoms form the side common to the two angles; the non-common sides have an atom in common
	$g^2\phi\phi(\frac{1}{0})$	2	As for $g^2\phi\phi(\frac{1}{1})$ except that the non-common sides have no common atom (e.g., the O—O—H angles in H_2O_2)
	$g^1\phi\phi(\frac{2}{2})$	1	The common atom is a terminal atom of both angles
	$g^1\phi\phi(\frac{2}{1})$	1	The common atom is a terminal atom of one angle and the central atom of the other
	$g^1\phi\phi(\frac{1}{1})$	1	The common atom is the central atom of both angles

ψ is referred to in Wilson, Decius and Cross as a dihedral angle, but this term is very often used in structural chemistry to designate the angle τ. Wilson, Decius and Cross refer to τ as the torsional angle, and define it as follows:

Consider four atoms bonded in the sequence 1, 2, 3, 4. Then τ is the angle between the planes determined by atoms 1, 2, 3 and 2, 3, 4 respectively. If one considers the 'book' structure of H_2O_2, τ is the angle between the leaves of the book. Further, τ is conventionally restricted to the range $-\pi < \tau < \pi$, and is positive if, viewing the atoms along the bond 2, 3 from 2 to 3, the angle from the projection of 2, 1 to the projection of 3, 4 is traced in the clockwise direction.

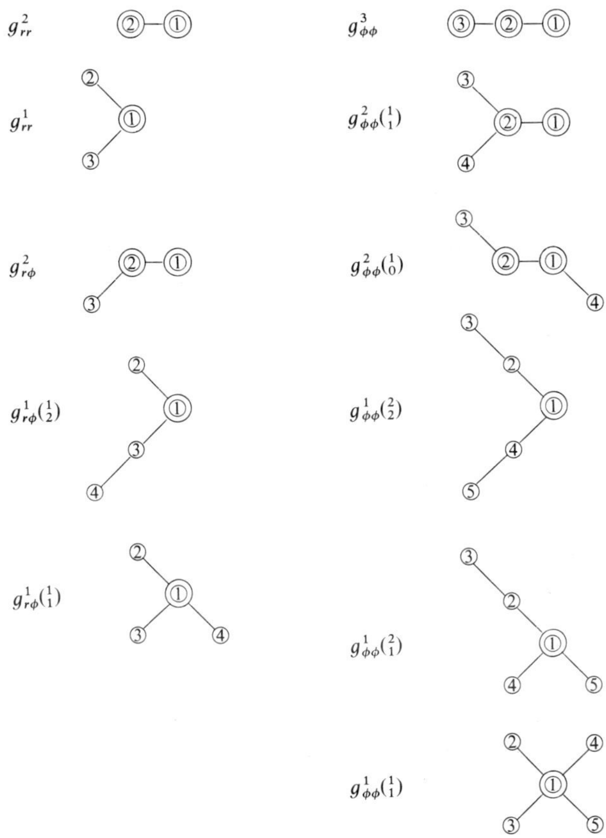

Fig. A3.1 Schematic representation of kinetic energy matrix elements.

Table A3.2 General formulae for g_{rr}, $g_{r\phi}$ and $g_{\phi\phi}$ elements ($c\phi$ = cos ϕ, $s\phi$ = sin ϕ and similarly for τ and ψ).

g_{rr}^2	$\mu_1 + \mu_2$
g_{rr}^1	$\mu_1 c\phi$
$g_{r\phi}^2$	$-\rho_{23}\mu_2 s\phi$
$g_{r\phi}^1\binom{1}{2}$	$\rho_{13}\mu_1 s\phi_1 c\tau$
$\binom{1}{1}$	$-(\rho_{13}s\phi_{213}c\psi_{234} + \rho_{14}s\phi_{214}c\psi_{243})\mu_1$
$g_{\phi\phi}^3$	$\rho_{12}^2\mu_1 + \rho_{23}^2\mu_3 + (\rho_{12}^2 + \rho_{23}^2 - 2\rho_{12}\rho_{23}c\phi)\mu_2$
$g_{\phi\phi}^2\binom{1}{1}$	$(\rho_{12}^2 c\psi_{314})\mu_1 + [(\rho_{12} - \rho_{23}c\phi_{123} - \rho_{24}c\phi_{124})\rho_{12}c\psi_{314} + (s\phi_{123}s\phi_{124}s^2\psi_{134} + c\phi_{324}c\psi_{314})\rho_{23}\rho_{24}]\mu_2$
$\binom{1}{0}$	$-\rho_{12}c\tau[(\rho_{12} - \rho_{14}c\phi_1)\mu_1 + (\rho_{12} - \rho_{23}c\phi_2)\mu_2]$
$g_{\phi\phi}^1\binom{2}{2}$	$-(s\tau_{25}s\tau_{34} + c\tau_{25}c\tau_{34}c\phi_1)\rho_{12}\rho_{14}\mu_1$
$\binom{2}{1}$	$[(s\phi_{214}c\phi_{415}c\tau_{34} - s\phi_{215}c\tau_{35})\rho_{14} + (s\phi_{215}c\phi_{415}c\tau_{35} - s\phi_{214}c\tau_{34})\rho_{15}]\dfrac{\rho_{12}\mu_1}{s\phi_{415}}$

$$\binom{1}{1}\ [(c\phi_{415} - c\phi_{314}c\phi_{315} - c\phi_{214}c\phi_{215} + c\phi_{213}c\phi_{214}c\phi_{315})\rho_{12}\rho_{13} + (c\phi_{413} - c\phi_{514}c\phi_{513} - c\phi_{214}c\phi_{213} + c\phi_{215}c\phi_{214}c\phi_{513})\rho_{12}\rho_{15} + (c\phi_{215} - c\phi_{312}c\phi_{315} - c\phi_{412}c\phi_{415} + c\phi_{413}c\phi_{412}c\phi_{315})\rho_{14}\rho_{13} + (c\phi_{213} - c\phi_{512}c\phi_{513} - c\phi_{412}c\phi_{413} + c\phi_{415}c\phi_{412}c\phi_{513})\rho_{14}\rho_{15}]\frac{\mu_1}{s\phi_{214}s\phi_{315}}.$$

Table A3.3 Special formulae, all $\phi = 109°\ 28'$.

g^2_{rr}	$\mu_1 + \mu_2$
g^1_{rr}	$-\frac{\mu_1}{3}$
$g^2_{r\phi}$	$-\frac{2^{3/2}}{3}\rho_{23}\mu_2$
$g^1_{r\phi}\binom{1}{2}$	$\frac{2^{3/2}}{3}\rho_{13}\mu_1 c\tau$
$\binom{1}{1}$	$\frac{2^{1/2}}{3}(\rho_{13} + \rho_{14})\mu_1$
$g^3_{\phi\phi}$	$\rho_{12}^2\mu_1 + \rho_{23}^2\mu_3 + \frac{1}{3}(3\rho_{12}^2 + 3\rho_{23}^2 + 2\rho_{12}\rho_{23})\mu_2$
$g^2_{\phi\phi}\binom{1}{1}$	$-\frac{1}{6}\{3\rho_{21}^2\mu_1 + [3\rho_{21}^2 + (\rho_{23} + \rho_{24})\rho_{21} - 5\rho_{23}\rho_{24}]\mu_2\}$
$\binom{1}{0}$	$-\frac{1}{3}\rho_{12}c\tau[(3\rho_{12} + \rho_{14})\mu_1 + (3\rho_{12} + \rho_{23})\mu_2]$
$g^1_{\phi\phi}\binom{2}{2}$	$-\frac{1}{3}(3s\tau_{25}s\tau_{34} - c\tau_{25}c\tau_{34})\rho_{12}\rho_{14}\mu_1$
$\binom{2}{1}$	$-\frac{1}{3}[(3c\tau_{35} + c\tau_{34})\rho_{14} + (3c\tau_{34} + c\tau_{35})\rho_{15}]\rho_{12}\mu_1$
$\binom{1}{1}$	$-\frac{2}{3}(\rho_{12} + \rho_{14})(\rho_{13} + \rho_{15})\mu_1.$

Table A3.4 Special formulae, all $\phi = 120°$.

g^2_{rr}	$\mu_1 + \mu_2$
g^1_{rr}	$-\frac{\mu_1}{2}$
$g^2_{r\phi}$	$-\frac{3^{1/2}}{2}\rho_{23}\mu_2$
$g^1_{r\phi}\binom{1}{2}$	$\frac{3^{1/2}}{2}\rho_{13}\mu_1 c\tau$
$\binom{1}{1}$	$\frac{3^{1/2}}{2}(\rho_{13} + \rho_{14})\mu_1$
$g^3_{\phi\phi}$	$\rho_{12}^2\mu_1 + \rho_{23}^2\mu_3 + (\rho_{12}^2 + \rho_{23}^2 + \rho_{12}\rho_{23})\mu_2$
$g^2_{\phi\phi}\binom{1}{1}$	$-\frac{1}{2}\{2\rho_{12}^2\mu_1 + [\rho_{12}(2\rho_{12} + \rho_{23} + \rho_{24}) - \rho_{23}\rho_{24}]\mu_2\}$
$\binom{1}{0}$	$-\frac{1}{2}\rho_{12}c\tau[(2\rho_{12} + \rho_{14})\mu_1 + (2\rho_{12} + \rho_{23})\mu_2]$
$g^1_{\phi\phi}\binom{2}{2}$	$\frac{1}{2}(c\tau_{25}c\tau_{34} - 2s\tau_{25}s\tau_{34})\rho_{12}\rho_{14}\mu_1$
$\binom{2}{1}$	$\frac{1}{2}c\tau_{34}(\rho_{14} - \rho_{15})\rho_{12}\mu_1$
$\binom{1}{1}$	Element of this type impossible with all $\phi = 120°$.

Table A3.5 Special formulae, all $\phi = 90°$.

g^2rr	$\mu_1 + \mu_2$
g^1rr	0
$g^2r\phi$	$-\rho_{23}\mu_2$
$g^1r\phi\binom{1}{2}$	$\rho_{13}\mu_1\, c\tau$
$g^1r\phi\binom{1}{1}$	0
$g^3\phi\phi$	$\rho_{12}^2\mu_1 + \rho_{23}^2\mu_3 + (\rho_{12}^2 + \rho_{23}^2)\mu_2$
$g^2\phi\phi\binom{1}{1}$	$\rho_{23}\rho_{24}\mu_2$
$g^2\phi\phi\binom{1}{0}$	$\rho_{12}^2(\mu_1 + \mu_2)\, c\tau$
$g^1\phi\phi\binom{2}{2}$	$-\rho_{12}\rho_{14}\mu_1\, s\tau_{25}s\tau_{34}$
$g^1\phi\phi\binom{2}{1}$	$-\rho_{12}\mu_1\,(\rho_{14}c\tau_{35} + \rho_{15}c\tau_{34})$
$g^1\phi\phi\binom{1}{1}$	$0.$

Table A3.6 Special formulae, all $\phi = 180°$.

g^2rr	$\mu_1 + \mu_2$
g^1rr	$-\mu_1$
$g^2r\phi$	0
$g^1r\phi\binom{1}{2}$	0
$g^3\phi\phi$	$(\rho_{12} + \rho_{13})^2\mu_2 + \rho_{12}^2\mu_1 + \rho_{23}^2\mu_3$
$g^2\phi\phi\binom{1}{0}$	$-[\rho_{12}(\rho_{12} + \rho_{14})\mu_1 + \rho_{12}(\rho_{12} + \rho_{23})\mu_2]$
$g^1\phi\phi\binom{2}{2}$	$\mu_1\rho_{12}\rho_{14}.$

No elements of the type $g^1r\phi\binom{1}{1}$, $g^2\phi\phi\binom{1}{1}$, $g^1\phi\phi\binom{2}{1}$ or $g^1\phi\phi\binom{1}{1}$ are possible with linear molecules.

Appendix 4. Table of site symmetries

The space groups are designated according to the number given in *International Tables for X-ray Crystallography*, Vol. I (ed. N. F. M. Henry and K. Lonsdale, Kynoch Press, 1965); the Herrmann–Mauguin and Schönflies symbols are also given. An entry of the type $m(n)$ in a column means that there are m sets of this kind, each set containing n sites. A single number entry is of type m and means that there are m sets each containing one site, as in space group 2. (8 sets of C_i sites, each containing 1 site.) The number G of general positions (or sites of C_1 symmetry) is not given, but can be found as follows:

G = number of symmetry operations of the factor group × 1, 2 or 4;

the multiplier is 1 for a P or R lattice, 2 for I, A, B or C and 4 for F. The number of symmetry operations of each factor group is given in Table 1.1.

Table A4.1 Space groups with general positions only.

Number	Symbol	
	HM	S
1	$P1$	C_1^1
4	$P2_1$	C_2^2
7	Pb	C_s^2
9	Bb	C_s^4
19	$P2_12_12_1$	D_2^4
29	$Pca2_1$	C_{2v}^5
33	$Pna2_1$	C_{2v}^9
76	$P4_1$	C_4^2
78	$P4_3$	C_4^4
144	$P3_1$	C_3^2
145	$P3_2$	C_3^3
169	$P6_1$	C_6^2
170	$P6_5$	C_6^3

For example, group 45 (*Iba*2, C_{2v}^{21}) has 8 general positions; the factor group C_{2v} contains 4 symmetry operations and the lattice is body-centred (*I*) orthorhombic. Thirteen of the space groups have general positions only.

Table A4.2 Site symmetries for the triclinic, monoclinic and orthorhombic groups.

Number	Symbol		Site						
	HM	S	C_i	C_2	C_s	C_{2h}	C_{2v}	D_2	D_{2h}
2	$P\bar{1}$	C_i^1	8						
3	$P2$	C_2^1		4					
5	$C2$	C_2^3		2(2)					
6	Pm	C_s^1			2				
8	Cm	C_s^3			1(2)				
10	$P2/m$	C_{2h}^1		4(2)	2(2)	8			
11	$P2_1/m$	C_{2h}^2	4(2)		1(2)				
12	$C2/m$	C_{2h}^3	2(4)	2(4)	1(4)	4(2)			
13	$P2/c$	C_{2h}^4	4(2)	2(2)					
14	$P2_1/c$	C_{2h}^5	4(2)						
15	$C2/c$	C_{2h}^6	4(4)	1(4)					
16	$P222$	D_2^1		12(2)				8	
17	$P222_1$	D_2^2		4(2)					
18	$P2_12_12$	D_2^3		2(2)					
20	$C222_1$	D_2^5		2(4)					
21	$C222$	D_2^6		7(4)				4(2)	
22	$F222$	D_2^7		6(8)				4(4)	
23	$I222$	D_2^8		6(4)				4(2)	
24	$I2_12_12_1$	D_2^9		3(4)					
25	$Pmm2$	C_{2v}^1			4(2)		4		
26	$Pmc2_1$	C_{2v}^2			2(2)				
27	$Pcc2$	C_{2v}^3		4(2)					
28	$Pma2$	C_{2v}^4		2(2)	1(2)				
30	$Pnc2$	C_{2v}^6		2(2)					
31	$Pmn2_1$	C_{2v}^7			1(2)				
32	$Pba2$	C_{2v}^8		2(2)					
34	$Pnn2$	C_{2v}^{10}	2(2)	2(2)					
35	$Cmm2$	C_{2v}^{11}		1(4)	2(4)		2(2)		
36	$Cmc2_1$	C_{2v}^{12}			1(4)				
37	$Ccc2$	C_{2v}^{13}		3(4)					
38	$Amm2$	C_{2v}^{14}			3(4)		2(2)		
39	$Abm2$	C_{2v}^{15}		2(4)	1(4)				
40	$Ama2$	C_{2v}^{16}		1(4)	1(4)				
41	$Aba2$	C_{2v}^{17}		1(4)					
42	$Fmm2$	C_{2v}^{18}		1(8)	2(8)		1(4)		
43	$Fdd2$	C_{2v}^{19}		1(8)					
44	$Imm2$	C_{2v}^{20}			2(4)		2(2)		
45	$Iba2$	C_{2v}^{21}		2(4)					
46	$Ima2$	C_{2v}^{22}		1(4)	1(4)				
47	$Pmmm$	D_{2h}^1			6(4)		12(2)		8
48	$Pnnn$	D_{2h}^2	2(4)	6(4)					4(2)
49	$Pccm$	D_{2h}^3		8(4)	1(4)	4(2)		4(2)	
50	$Pban$	D_{2h}^4	2(4)	6(4)				4(2)	

Table A4.2 (*cont.*).

Number	Symbol		Site						
	HM	S	C_i	C_2	C_s	C_{2h}	C_{2v}	D_2	D_{2h}
51	*Pmma*	D_{2h}^{5}		2(4)	3(4)	4(2)	2(2)		
52	*Pnna*	D_{2h}^{6}	2(4)	2(4)					
53	*Pmna*	D_{2h}^{7}		3(4)	1(4)	4(2)			
54	*Pcca*	D_{2h}^{8}	2(4)	3(4)					
55	*Pbam*	D_{2h}^{9}		2(4)	2(4)	4(2)			
56	*Pccn*	D_{2h}^{10}	2(4)	2(4)					
57	*Pbcm*	D_{2h}^{11}	2(4)	1(4)	1(4)				
58	*Pnnm*	D_{2h}^{12}		2(4)	1(4)	4(2)			
59	*Pmmm*	D_{2h}^{13}	2(4)		2(4)		2(2)		
60	*Pbcn*	D_{2h}^{14}	2(4)	1(4)					
61	*Pbca*	D_{2h}^{15}	2(4)						
62	*Pnma*	D_{2h}^{16}	2(4)		1(4)				
63	*Cmcm*	D_{2h}^{17}	1(8)	1(8)	2(8)	2(4)	1(4)		
64	*Cmca*	D_{2h}^{18}	1(8)	2(8)	1(8)		2(4)		
65	*Cmmm*	D_{2h}^{19}		1(8)	4(8)	2(4)	6(4)		4(2)
66	*Cccm*	D_{2h}^{20}		5(8)	1(8)	4(4)		2(4)	
67	*Cmma*	D_{2h}^{21}		5(8)	2(8)	4(4)	1(4)	2(4)	
68	*Ccca*	D_{2h}^{22}	2(8)	4(8)				2(4)	
69	*Fmmm*	D_{2h}^{23}		3(16)	3(16)	3(8)	3(8)	1(8)	2(4)
70	*Fddd*	D_{2h}^{24}	2(16)	3(16)				2(8)	
71	*Immm*	D_{2h}^{25}	1(8)		3(8)		6(4)		4(2)
72	*Ibam*	D_{2h}^{26}	1(8)	4(8)	1(8)	2(4)		2(4)	
73	*Ibca*	D_{2h}^{27}	2(8)	3(8)					
74	*Imma*	D_{2h}^{28}		2(8)	2(8)	4(4)	1(4)		

Table A4.3 Site symmetries for the tetragonal groups.

Number	Symbol		Site													
	HM	S	C_i	C_2	C_s	C_{2h}	C_{2v}	D_2	D_{2h}	C_4	S_4	D_4	C_{4v}	C_{4h}	D_{2d}	D_{4h}
75	$P4$	C_4^1		1(2)						2						
77	$P4_2$	C_4^3		3(2)												
79	$I4$	C_4^5		1(4)						1(2)						
80	$I4_1$	C_4^6		1(4)												
81	$P\bar{4}$	S_4^1		3(2)							4					
82	$I\bar{4}$	S_4^2		2(4)							4					
83	$P4/m$	C_{4h}^1		1(4)	2(4)	2(2)				2(2)				4		
84	$P4_2/m$	C_{4h}^2		3(4)	1(4)	4(2)					2(2)					
85	$P4/n$	C_{4h}^3	2(4)	1(4)						1(2)	2(2)					
86	$P4_2/n$	C_{4h}^4	2(4)	2(4)							2(2)					
87	$I4/m$	C_{4h}^5	1(8)	1(8)	1(8)	1(4)				1(4)	1(4)			2(2)		
88	$I4_1/a$	C_{4h}^6	2(8)	1(8)							2(4)					
89	$P422$	D_4^1		7(4)				2(2)		2(2)		4				
90	$P42_12$	D_4^2		3(4)				2(2)		1(2)						
91	$P4_122$	D_4^3		3(4)												
92	$P4_12_12$	D_4^4		1(4)												
93	$P4_22_12$	D_4^5		9(4)				6(2)								
94	$P4_22_12$	D_4^6		4(4)				2(2)								
95	$P4_32_12$	D_4^7		3(4)												
96	$P4_32_12$	D_4^8		1(4)												
97	$I422$	D_4^9		5(8)				2(4)		1(4)		2(2)				
98	$I4_122$	D_4^{10}		4(8)				2(4)								
99	$P4mm$	C_{4v}^1			3(4)		1(2)						2			
100	$P4bm$	C_{4v}^2			1(4)		1(2)			1(2)						
101	$P4_2cm$	C_{4v}^3		1(4)	1(4)		2(2)									
102	$P4_2nm$	C_{4v}^4		1(4)	1(4)		1(2)									
103	$P4cc$	C_{4v}^5		1(4)						2(2)						
104	$P4nc$	C_{4v}^6		1(4)						1(2)						
105	$P4_2mc$	C_{4v}^7			2(4)		3(2)									
106	$P4_2bc$	C_{4v}^8		2(4)												
107	$I4mm$	C_{4v}^9			2(8)		1(4)						1(2)			
108	$I4cm$	C_{4v}^{10}			1(8)		1(4)			1(4)						
109	$I4_1md$	C_{4v}^{11}			1(8)		1(4)									

Table A4.3 (*cont.*).

Number	Symbol		Site													
	HM	S	C_i	C_2	C_s	C_{2h}	C_{2v}	D_2	D_{2h}	C_4	S_4	D_4	C_{4v}	C_{4h}	D_{2d}	D_{2h}
110	$I4_1cd$	C_{4v}^{12}		1(8)												
111	$P\bar{4}2m$	D_{2d}^{1}		5(4)	1(4)		2(2)	2(2)							4	
112	$P\bar{4}2c$	D_{2d}^{2}		7(4)				4(2)			2(2)					
113	$P\bar{4}2_1m$	D_{2d}^{3}		1(4)	1(4)		1(2)				2(2)					
114	$P\bar{4}2_1c$	D_{2d}^{4}		2(4)							2(2)					
115	$P\bar{4}m2$	D_{2d}^{5}		2(4)	2(4)		3(2)								4	
116	$P\bar{4}c2$	D_{2d}^{6}		5(4)				2(2)			2(2)					
117	$P\bar{4}b2$	D_{2d}^{7}		4(4)				2(2)			2(2)					
118	$P\bar{4}n2$	D_{2d}^{8}		4(4)				2(2)			2(2)					
119	$I\bar{4}m2$	D_{2d}^{9}		2(8)	1(8)		2(4)								4(2)	
120	$I\bar{4}c2$	D_{2d}^{10}		4(8)				2(4)			2(4)					
121	$I\bar{4}2m$	D_{2d}^{11}		3(8)	1(8)		1(4)	1(4)			1(4)				2(2)	
122	$I\bar{4}2d$	D_{2d}^{12}		2(8)							2(4)					
123	$P4/mmm$	D_{4h}^{1}			5(8)		7(4)		2(2)				2(2)			4
124	$P4/mcc$	D_{4h}^{2}		4(8)	1(8)	1(4)		1(4)		2(4)		2(2)		2(2)		
125	$P4/nbm$	D_{4h}^{3}		4(8)	1(8)	2(2)	1(4)			1(4)		2(2)			2(2)	
126	$P4/nnc$	D_{4h}^{4}	1(8)	4(8)				1(4)		1(4)	1(4)	2(2)				
127	$P4/mbm$	D_{4h}^{5}			3(8)		3(4)		2(2)	1(4)				2(2)		
128	$P4/mnc$	D_{4h}^{6}		2(8)	1(8)	1(4)		1(4)		1(4)				2(2)		
129	$P4/nmm$	D_{4h}^{7}		2(8)	2(8)	2(4)	1(4)						1(2)		2(2)	
130	$P4/ncc$	D_{4h}^{8}	1(8)	2(8)				1(4)		1(4)	1(4)					
131	$P4_2/mmc$	D_{4h}^{9}		1(8)	3(8)		7(4)		4(2)						2(2)	
132	$P4_2/mcm$	D_{4h}^{10}		3(8)	2(8)	1(4)	4(4)	1(4)	2(2)						2(2)	
133	$P4_2/nbc$	D_{4h}^{11}	1(8)	5(8)				3(4)			1(4)					
134	$P4_2/nnm$	D_{4h}^{12}		5(8)	1(8)	2(4)	1(4)	2(4)							2(2)	
135	$P4_2/mbc$	D_{4h}^{13}		3(8)	1(8)	2(4)		1(4)			1(4)					
136	$P4_2/mnm$	D_{4h}^{14}		1(8)	2(8)	1(4)	3(4)		2(2)		1(4)					
137	$P4_2/nmc$	D_{4h}^{15}	1(8)	1(8)		2(4)								2(2)		
138	$P4_2/ncm$	D_{4h}^{16}		3(8)	1(8)	2(4)	1(4)	1(4)			1(4)					
139	$I4/mmm$	D_{4h}^{17}		1(16)	3(16)	1(8)	4(8)		1(4)				1(4)		1(4)	2(2)
140	$I4/mcm$	D_{4h}^{18}		2(16)	2(16)	1(8)	2(8)		1(4)	1(8)		1(4)		1(4)	1(4)	
141	$I4_1/amd$	D_{4h}^{19}		2(16)	2(16)	2(8)	1(8)								2(4)	
142	$I4_1/acd$	D_{4h}^{20}	1(16)	2(16)				1(8)			1(8)					

Table A4.4 Site symmetries for trigonal groups.

Number	Symbol		Site								
	HM	S	C_i	C_2	C_s	C_{2h}	C_3	C_{3i}≡S_6	D_3	C_{3v}	D_{3d}
143	$P3$	C_3^1					3				
146	$R3$	C_3^4					1				
147	$P\bar{3}$	C_{3i}^1	2(3)				2(2)	2			
148	$R\bar{3}$	C_{3i}^2	2(3)				1(2)	2			
149	$P312$	D_3^1		2(3)			3(2)		6		
150	$P321$	D_3^2		2(3)			2(2)		2		
151	$P3_112$	D_3^3		2(3)							
152	$P3_121$	D_3^4		2(3)							
153	$P3_112$	D_3^5		2(3)							
154	$P3_221$	D_3^6		2(3)							
155	$R32$	D_3^7		2(3)			1(2)		2		
156	$P3m1$	C_{3v}^1			1(3)					3	
157	$P31m$	C_{3v}^2			1(3)		1(2)			1	
158	$P3c1$	C_{3v}^3					3(2)				
159	$P31c$	C_{3v}^4					2(2)				
160	$R3m$	C_{3v}^5			1(3)					1	
161	$R3c$	C_{3v}^6					1(2)				
162	$P\bar{3}1m$	D_{3d}^1		2(6)	1(6)	2(3)	1(4)		2(2)	1(2)	2
163	$P\bar{3}1c$	D_{3d}^2	1(6)	1(6)			2(4)	1(2)	3(2)		
164	$P\bar{3}m1$	D_{3d}^3	1(6)	2(6)		2(3)				2(2)	2
165	$P\bar{3}c1$	D_{3d}^4	1(6)	1(6)			2(4)	1(2)	1(2)		
166	$R\bar{3}m$	D_{3d}^5		2(6)	1(6)	2(3)				1(2)	2
167	$R\bar{3}c$	D_{3d}^6	1(6)	1(6)			1(4)	1(2)	1(2)		

There are no sites of symmetry C_{2v}, D_2 and D_{2h} in trigonal groups.

Table A4.5 Site symmetries for hexagonal groups.

Number	Symbol		Site																		
	HM	S	C_i	C_2	C_s	C_{2h}	C_{2v}	D_2	D_{2h}	C_3	C_{3i}	D_3	C_{3v}	D_{3d}	C_6	C_{3h}	D_6	D_{3h}	C_{6h}	C_{6v}	D_{6h}
168	$P6$	C_6^1		1(3)						1(2)											
171	$P6_2$	C_6^4		2(3)																	
172	$P6_4$	C_6^5		2(3)																	
173	$P6_3$	C_6^6								2(2)											
174	$P\bar{6}$	C_{3h}^1			2(3)					3(2)						6					
175	$P6/m$	C_{6h}^1		1(6)	2(6)	2(3)				1(4)					1(2)	2(2)			2		
176	$P6_3/m$	C_{6h}^2	1(6)		1(6)					2(4)	1(2)					3(2)					
177	$P622$	D_6^1		5(6)				2(3)		1(4)		2(2)			1(2)		2				
178	$P6_122$	D_6^2		2(6)																	
179	$P6_522$	D_6^3		2(6)																	
180	$P6_222$	D_6^4		6(6)				4(3)													
181	$P6_422$	D_6^5		6(6)				4(3)													
182	$P6_322$	D_6^6		2(6)						2(4)		4(2)									
183	$P6mm$	C_{6v}^1			2(6)		1(3)						1(2)							1	
184	$P6cc$	C_{6v}^2		1(6)						1(4)					1(2)						
185	$P6_3cm$	C_{6v}^3			1(6)					1(4)			1(2)								
186	$P6_3mc$	C_{6v}^4			1(6)								2(2)								
187	$P\bar{6}m2$	D_{3h}^1			3(6)		2(3)						3(2)					6			
188	$P\bar{6}c2$	D_{3h}^2		1(6)	1(6)	3(2)				3(4)		3(2)									
189	$P\bar{6}2m$	D_{3h}^3			3(6)		2(3)			1(4)			1(2)			2(2)		2			
190	$P\bar{6}2c$	D_{3h}^4		1(6)	1(6)					2(4)		1(2)				3(2)					
191	$P6/mmm$	D_{6h}^1			4(12)		5(6)		2(3)				1(4)					2(2)		1(2)	2
192	$P6/mcc$	D_{6h}^2		3(12)	1(12)	1(6)		1(6)		1(8)		1(4)			1(4)	1(4)	1(2)		1(2)		
193	$P6_3/mcm$	D_{6h}^3		1(12)	2(12)	1(6)	1(6)			1(8)		1(4)	1(4)	1(2)		1(4)		1(2)			
194	$P6_3/mmc$	D_{6h}^4		1(12)	2(12)	1(6)	1(6)						2(4)	1(2)				3(2)			

Table A4.6 Site symmetries for cubic groups.

Number	Symbol		Site									
	HM	S	C_2	C_s	C_{2h}	C_{2v}	D_2	D_{2h}	C_4	S_4	D_4	C_{4v}
195	$P23$	T^1	4(6)				2(3)					
196	$F23$	T^2	2(24)									
197	$I23$	T^3	2(12)				2(6)					
198	$P2_13$	T^4										
199	$I2_13$	T^5	1(12)									
200	$Pm3$	T_h^1		2(12)		4(6)		2(3)				
201	$Pn3$	T_h^2	2(12)				1(6)					
202	$Fm3$	T_h^3	1(48)	1(48)	1(24)	1(24)						
203	$Fd3$	T_h^4	1(48)									
204	$Tm3$	T_h^5		1(24)		2(12)		1(6)				
205	$Pa3$	T_h^6										
206	$Ia3$	T_h^7	1(24)									
207	$P432$	O^1	3(12)						2(6)		2(3)	
208	$P4_232$	O^2	5(12)				3(6)					
209	$F432$	O^3	3(48)				1(24)		1(24)			
210	$F4_132$	O^4	2(48)									
211	$I432$	O^5	3(24)				1(12)		1(12)		1(6)	
212	$P4_332$	O^6	1(12)									
213	$P4_132$	O^7	1(12)									
214	$I4_132$	O^8	3(24)				2(12)					
215	$P\bar{4}3m$	T_d^1	1(12)	1(12)		2(6)						
216	$F\bar{4}3m$	T_d^2		1(48)		2(24)						
217	$I\bar{4}3m$	T_d^3	1(24)	1(24)		1(12)				1(12)		
218	$P\bar{4}3n$	T_d^4	3(12)				1(6)			2(6)		
219	$F\bar{4}3c$	T_d^5	2(48)							2(24)		
220	$I\bar{4}3d$	T_d^6	1(24)							2(12)		
221	$Pn3m$	O_h^1		3(24)		3(12)						2(6)
222	$Pn3n$	O_h^2	2(24)						1(12)	1(12)	1(6)	
223	$Pm3n$	O_h^3	1(24)	1(24)		3(12)		1(6)				
224	$Pn3m$	O_h^4	3(24)	1(24)		1(12)	1(12)					
225	$Fm3m$	O_h^5		2(96)		3(48)		1(24)				1(24)
226	$Fm3c$	O_h^6	1(96)	1(96)		1(48)						
227	$Fd3m$	O_h^7	1(96)	1(96)		1(48)						
228	$Fd3c$	O_h^8	2(96)							1(48)		
229	$Im3m$	O_h^9	1(48)	2(48)		2(24)						1(12)
230	$Ia3d$	O_h^{10}	2(48)				1(24)			1(24)		

Table A4.6 (*cont.*).

C_{4h}	D_{2d}	D_{4h}	C_3	C_{3i}	D_3	C_{3v}	D_{3d}	T	T_h	T_d	O	O_h
			1(4)					2				
			1(16)					4(4)				
			1(8)					1(2)				
			1(4)									
			1(8)									
			1(8)						2			
			1(8)	2(6)				1(2)				
			1(32)					1(8)	2(4)			
			1(32)	2(16)				2(8)				
			1(16)	1(8)					1(2)			
			1(8)	2(4)								
			1(16)	2(8)								
			1(8)								2	
			1(8)		2(4)			1(2)				
			1(32)					1(8)			2(4)	
			1(32)		2(16)			2(8)				
			1(16)		1(8)						1(2)	
			1(8)		2(4)							
			1(8)		2(4)							
			1(16)		2(8)							
	2(3)					1(4)				2		
						1(16)				4(4)		
	1(6)					1(8)				1(2)		
			1(8)					1(2)				
			1(32)					2(8)				
			1(16)									
		2(3)				1(8)						2
			1(16)	1(8)							1(2)	
	2(6)		1(16)		1(8)				1(2)			
	1(6)					1(8)	2(4)			1(2)		
						1(32)				1(8)		2(4
1(24)	1(24)		1(64)						1(8)		1(8)	
						1(32)	2(16)			2(8)		
			1(64)	1(32)	1(32)			1(16)				
	1(12)	1(6)				1(16)	1(8)					1(2
			1(32)	1(16)	1(16)							

Appendix 5. Tables of λ values

$$\lambda = 4\pi^2 c^2 \nu^2 \text{ where } \nu \text{ is the frequency in cm}^{-1}$$
$$= 0{\cdot}588851\left(\frac{\nu}{1000}\right)^2$$

Conversely, $\nu = 1303{\cdot}16\,\lambda^{1/2}$.

Values of ν are given corresponding to the range of λ between 0·100 and 10·00.

Values of ν corresponding to values of λ outside this range are obtained by multiplying or dividing the appropriate λ value in the table by $(100)^n$, and the ν value in the table by $(10)^n$.

Table A5.1 λ from 0·100 to 0·999, in intervals of 0·001.

	0·000	0·001	0·002	0·003	0·004	0·005	0·006	0·007	0·008	0·009
0·10	412·10	414·14	416·19	418·24	420·26	422·28	424·28	426·28	428·26	430·24
0·11	432·21	434·17	436·12	438·06	440·00	441·93	443·84	445·75	447·65	449·54
0·12	451·43	453·30	455·17	457·03	458·90	460·73	462·57	464·41	466·23	468·06
0·13	469·86	471·67	473·46	475·25	477·04	478·81	480·58	482·35	484·10	485·86
0·14	487·60	489·34	491·07	492·79	494·51	496·23	497·94	499·65	501·34	503·03
0·15	504·71	506·40	508·06	509·73	511·40	513·05	514·71	516·35	517·99	519·64
0·16	521·26	522·89	524·51	526·12	527·74	529·34	530·95	532·55	534·14	535·13
0·17	537·31	538·88	540·46	542·02	543·59	545·15	546·70	548·25	549·80	551·34
0·18	552·88	554·42	555·94	557·47	558·99	560·52	562·03	563·53	565·04	566·54
0·19	568·03	569·53	571·02	572·50	573·98	575·46	576·94	578·41	579·87	581·33
0·20	582·79	584·25	585·69	587·15	588·59	590·03	591·47	592·90	594·33	595·77
0·21	597·18	598·61	600·01	601·43	602·84	604·25	605·66	607·05	608·45	609·84
0·22	611·24	612·63	614·01	615·39	616·77	618·14	619·51	620·89	622·25	623·61
0·23	624·97	626·33	627·68	629·06	630·39	631·73	633·08	634·42	635·75	637·09
0·24	638·42	639·75	641·06	642·39	643·71	645·03	646·34	647·66	648·97	650·28
0·25	651·58	652·88	654·19	655·48	656·77	658·07	659·35	660·64	661·93	663·20
0·26	664·48	665·76	667·04	668·31	669·58	670·84	672·11	673·37	674·63	675·88
0·27	677·15	678·40	679·65	680·89	682·14	683·38	684·63	685·87	687·10	688·33
0·28	689·57	690·79	692·03	693·26	694·48	695·69	696·92	698·13	699·35	700·57
0·29	701·78	702·98	704·19	705·39	706·60	707·80	709·00	710·20	711·38	712·58
0·30	713·77	714·95	716·15	717·32	718·51	719·70	720·87	722·06	723·23	724·40
0·31	725·57	726·73	727·91	729·07	730·24	731·40	732·56	733·72	734·87	736·03
0·32	737·19	738·33	739·48	740·63	741·77	742·92	744·05	745·20	746·33	747·48
0·33	748·61	749·75	750·87	752·00	753·14	754·26	755·39	756·51	757·63	758·75
0·34	759·87	760·98	762·10	763·21	764·33	765·44	766·54	767·65	768·76	769·86
0·35	770·96	772·06	773·17	774·26	775·35	776·45	777·54	778·63	779·72	780·81
0·36	781·90	782·98	784·06	785·14	786·22	787·30	788·38	789·47	790·54	791·60
0·37	792·69	793·76	794·82	795·89	796·96	798·01	799·09	800·27	801·21	802·26
0·38	803·32	804·38	805·43	806·49	807·54	808·59	809·64	810·68	811·74	812·78
0·39	813·82	814·87	815·91	816·95	817·98	819·02	820·07	821·10	822·13	823·15
0·40	824·20	825·23	826·24	827·27	828·30	829·33	830·35	831·38	832·52	833·41

Table A5.1 (*cont.*).

	0·000	0·001	0·002	0·003	0·004	0·005	0·006	0·007	0·008	0·009
0·41	834·43	835·44	836·46	837·48	838·49	839·50	840·51	841·53	842·53	843·54
0·42	844·54	845·56	846·56	847·55	848·55	849·56	850·56	851·55	852·55	853·54
0·43	854·53	855·54	856·53	857·52	858·51	859·50	860·48	861·47	862·46	863·44
0·44	864·41	865·40	866·38	867·36	868·36	869·31	870·29	871·27	872·24	873·21
0·45	874·16	875·16	876·13	877·09	878·07	879·04	880·00	880·96	881·93	882·89
0·46	883·84	884·81	885·77	886·72	887·69	888·64	889·59	890·54	891·51	892·46
0·47	893·41	894·36	895·31	896·26	897·20	898·15	899·09	900·04	900·98	901·91
0·48	902·86	903·79	904·73	905·67	906·60	907·54	908·48	909·41	910·34	911·28
0·49	912·21	913·14	914·08	915·00	915·93	916·85	917·78	918·70	919·63	920·55
0·50	921·48	922·39	923·31	924·23	925·15	926·06	926·99	927·90	928·81	929·73
0·51	930·64	931·55	932·46	933·37	934·29	935·20	936·10	937·01	937·91	938·82
0·52	939·72	940·62	941·53	942·43	943·33	944·24	945·13	946·03	946·93	947·81
0·53	948·71	949·61	950·50	951·40	952·28	953·18	954·07	954·96	955·84	956·74
0·54	957·63	958·51	959·40	960·29	961·16	962·04	962·93	963·80	964·79	965·66
0·55	966·45	967·32	968·21	969·08	969·96	970·83	971·70	972·57	973·45	974·32
0·56	975·20	976·07	976·94	977·80	978·67	979·53	980·41	981·27	982·14	983·00
0·57	983·36	984·73	985·59	986·45	987·31	988·17	989·03	989·89	990·74	991·61
0·58	992·47	993·31	994·17	995·02	995·88	996·73	997·59	998·43	999·28	1000·12
0·59	1000·97	1001·83	1002·68	1003·54	1004·38	1005·23	1006·08	1006·93	1007·76	1008·61
0·60	1009·43	1010·26	1011·11	1011·94	1012·78	1013·62	1014·46	1015·29	1016·13	1016·96
0·61	1017·79	1018·63	1019·46	1020·30	1021·13	1021·96	1022·80	1023·52	1024·35	1025·28
0·62	1026·11	1026·94	1027·76	1028·58	1029·42	1030·24	1031·06	1031·88	1032·70	1033·54
0·63	1034·36	1035·18	1035·99	1036·81	1037·63	1038·45	1039·27	1040·08	1040·90	1041·71
0·64	1042·53	1043·34	1044·16	1044·97	1045·79	1046·59	1047·40	1048·21	1049·02	1049·84
0·65	1050·65	1051·46	1052·26	1053·06	1053·87	1054·67	1055·48	1056·29	1057·09	1057·89
0·66	1058·69	1059·50	1060·29	1061·10	1061·89	1062·70	1063·50	1064·29	1065·09	1065·88
0·67	1066·69	1067·48	1068·28	1069·07	1069·87	1070·65	1071·45	1072·24	1073·04	1073·82
0·68	1074·61	1075·41	1076·19	1076·99	1077·77	1078·56	1079·34	1080·13	1080·92	1081·70
0·69	1082·49	1083·27	1084·06	1084·84	1085·63	1086·41	1087·19	1087·97	1088·74	1089·52
0·70	1090·31	1091·09	1091·86	1092·64	1093·42	1094·19	1094·97	1095·74	1096·52	1097·29
0·71	1098·06	1098·84	1099·61	1100·38	1101·16	1101·93	1102·70	1103·47	1104·24	1105·01
0·72	1105·78	1106·55	1107·32	1108·07	1108·84	1109·61	1110·38	1111·13	1111·90	1112·61
0·73	1113·43	1114·20	1114·95	1115·71	1116·46	1117·22	1117·97	1118·74	1119·50	1120·25
0·74	1121·02	1121·78	1122·53	1123·29	1124·05	1124·80	1125·56	1126·31	1127·07	1127·83
0·75	1128·58	1129·32	1130·08	1130·83	1131·57	1132·33	1133·07	1133·83	1134·57	1135·33
0·76	1136·07	1136·81	1137·57	1138·31	1139·06	1139·80	1140·54	1141·30	1142·04	1142·78
0·77	1143·52	1144·27	1145·01	1145·74	1146·48	1147·22	1147·97	1148·71	1149·44	1150·18
0·78	1150·93	1151·66	1152·40	1153·13	1153·87	1154·60	1155·34	1156·07	1156·80	1157·55
0·79	1158·28	1159·01	1159·75	1160·48	1161·21	1161·94	1162·67	1163·40	1164·13	1164·86
0·80	1165·59	1166·31	1167·03	1167·76	1168·49	1169·22	1169·95	1170·67	1171·40	1172·11
0·81	1172·84	1173·57	1174·29	1175·02	1175·74	1176·45	1177·18	1177·90	1178·61	1179·34
0·82	1180·06	1180·78	1181·49	1182·21	1182·93	1183·66	1184·73	1185·09	1185·81	1186·51
0·83	1187·23	1187·95	1188·67	1189·38	1190·10	1190·80	1191·52	1192·24	1192·94	1193·66
0·84	1194·37	1195·08	1195·79	1196·50	1197·20	1197·92	1198·62	1199·34	1200·04	1200·75
0·85	1201·45	1202·17	1202·87	1203·57	1204·28	1204·98	1205·68	1206·39	1207·09	1207·80
0·86	1208·50	1209·20	1209·91	1210·61	1211·32	1212·01	1212·71	1213·41	1214·12	1214·81
0·87	1215·51	1216·20	1216·91	1217·60	1218·30	1218·99	1219·69	1220·38	1221·09	1221·78
0·88	1222·47	1223·17	1223·86	1224·55	1225·24	1225·94	1226·64	1227·33	1228·02	1228·71
0·89	1229·40	1230·09	1230·78	1231·47	1232·16	1232·84	1233·53	1234·22	1234·91	1235·61
0·90	1236·28	1236·97	1237·66	1238·34	1239·03	1239·71	1240·40	1241·09	1241·77	1242·45
0·91	1243·14	1243·81	1244·51	1245·18	1245·86	1246·55	1247·23	1247·91	1248·58	1249·26
0·92	1249·95	1250·63	1251·31	1251·98	1252·66	1253·34	1254·02	1254·69	1255·37	1256·05
0·93	1256·73	1257·39	1258·06	1258·74	1259·42	1260·09	1260·76	1261·44	1262·10	1262·78
0·94	1263·46	1264·13	1264·80	1265·47	1266·14	1266·80	1267·47	1268·13	1268·80	1269·48
0·95	1270·16	1270·83	1271·49	1272·17	1272·84	1273·50	1274·16	1274·83	1275·49	1276·17
0·96	1276·84	1277·50	1278·17	1278·83	1279·50	1280·15	1280·82	1281·48	1282·15	1282·81
0·97	1283·47	1284·12	1284·79	1285·45	1286·10	1286·77	1287·43	1288·08	1288·75	1289·40
0·98	1290·06	1290·71	1291·38	1292·03	1292·70	1293·35	1294·01	1294·66	1295·32	1295·97
0·99	1296·63	1297·28	1297·93	1298·59	1299·24	1299·90	1300·55	1301·21	1301·86	1302·51

Table A5.2 λ from 1·00 to 4·99 in intervals of 0·01.

	0·00	0·01	0·02	0·03	0·04	0·05	0·06	0·07	0·08	0·09
1·0	1303·16	1309·66	1316·13	1322·56	1328·97	1335·34	1341·69	1348·00	1354·28	1360·54
1·1	1366·77	1372·96	1379·13	1385·28	1391·31	1397·48	1403·55	1409·58	1415·59	1421·58
1·2	1427·54	1433·48	1439·39	1445·27	1451·14	1456·98	1462·79	1468·59	1474·36	1480·10
1·3	1485·83	1491·53	1497·22	1502·88	1508·52	1514·13	1519·73	1525·31	1530·87	1536·40
1·4	1541·92	1547·42	1552·89	1558·35	1563·79	1569·21	1574·61	1580·00	1585·36	1590·71
1·5	1596·04	1601·35	1606·64	1611·92	1617·18	1622·42	1627·65	1632·85	1638·05	1643·22
1·6	1648·38	1653·52	1658·65	1663·76	1668·86	1673·94	1679·00	1684·05	1689·09	1694·11
1·7	1699·11	1704·10	1709·08	1714·04	1718·99	1723·92	1728·84	1733·74	1738·63	1743·51
1·8	1748·37	1753·22	1758·06	1762·88	1767·69	1772·49	1777·27	1782·04	1786·80	1791·55
1·9	1796·28	1801·00	1805·71	1810·41	1815·09	1819·76	1824·42	1829·07	1833·71	1838·33
2·0	1842·95	1847·55	1852·14	1856·72	1861·28	1865·84	1870·39	1874·92	1879·44	1883·96
2·1	1888·46	1892·95	1897·43	1901·90	1906·36	1910·81	1915·25	1919·67	1924·09	1928·50
2·2	1932·90	1937·29	1941·66	1946·03	1950·39	1954·74	1959·08	1963·41	1967·73	1972·04
2·3	1976·34	1980·63	1984·91	1989·19	1993·45	1997·71	2001·95	2006·19	2010·42	2014·64
2·4	2018·85	2023·05	2027·24	2031·42	2035·60	2039·77	2043·93	2048·08	2052·22	2056·35
2·5	2060·48	2064·59	2068·70	2072·80	2076·89	2080·98	2085·06	2089·12	2093·18	2097·24
2·6	2101·28	2105·32	2109·35	2113·37	2117·38	2121·39	2125·39	2129·38	2133·36	2137·34
2·7	2141·31	2145·27	2149·23	2153·17	2157·11	2161·05	2164·97	2168·89	2172·80	2176·71
2·8	2180·60	2184·49	2188·38	2192·25	2196·12	2199·99	2203·84	2207·49	2211·54	2215·37
2·9	2219·20	2223·02	2226·84	2230·65	2234·45	2238·25	2242·04	2245·82	2249·60	2253·37
3·0	2257·14	2260·90	2264·65	2268·40	2272·14	2275·87	2279·60	2283·32	2287·04	2290·75
3·1	2294·45	2298·15	2301·84	2305·52	2309·20	2312·88	2316·55	2320·21	2323·87	2327·52
3·2	2331·16	2334·80	2338·44	2342·06	2345·69	2349·30	2352·92	2356·52	2360·12	2363·72
3·3	2367·31	2370·89	2374·47	2378·04	2381·61	2385·17	2388·73	2392·28	2395·83	2399·37
3·4	2402·91	2406·44	2409·96	2413·49	2417·00	2420·51	2424·02	2427·52	2431·01	2434·50
3·5	2437·99	2441·47	2444·94	2448·41	2451·88	2455·34	2458·80	2462·25	2465·69	2469·13
3·6	2472·57	2476·00	2479·43	2482·85	2486·27	2489·68	2493·09	2496·49	2499·89	2503·29
3·7	2506·68	2510·06	2513·44	2516·82	2520·19	2523·56	2526·92	2530·28	2533·63	2536·98
3·8	2540·33	2543·67	2547·00	2550·33	2553·66	2556·98	2560·30	2563·62	2566·93	2570·23
3·9	2573·53	2576·83	2580·12	2583·41	2586·70	2589·98	2593·25	2596·53	2599·79	2603·06
4·0	2606·32	2609·57	2612·83	2616·07	2619·32	2622·56	2625·79	2629·03	2632·25	2635·48
4·1	2638·70	2641·91	2645·12	2648·33	2651·54	2654·74	2657·93	2661·13	2668·32	2667·50
4·2	2670·68	2673·86	2677·03	2680·20	2683·37	2686·53	2689·69	2692·85	2696·00	2699·14
4·3	2702·29	2705·43	2708·57	2711·70	2714·83	2717·95	2721·08	2724·20	2727·31	2730·42
4·4	2733·53	2730·63	2739·74	2742·83	2745·93	2749·02	2752·10	2755·19	2758·27	2761·35
4·5	2764·42	2767·49	2770·55	2773·62	2776·68	2779·73	2782·79	2785·84	2788·88	2791·93
4·6	2794·97	2798·00	2801·04	2804·06	2807·09	2810·11	2813·13	2816·15	2819·16	2822·18
4·7	2825·18	2828·19	2831·19	2834·18	2837·18	2840·17	2843·16	2846·14	2849·12	2852·10
4·8	2855·08	2858·05	2861·02	2863·99	2866·95	2869·91	2872·87	2875·82	2878·77	2881·72
4·9	2884·67	2887·61	2890·55	2893·48	2896·42	2899·35	2902·27	2905·20	2908·12	2911·04

Table A5.3 λ from 5·00 to 10·00, in intervals of 0·05.

	0·00	0·05	0·10	0·15	0·20	0·25	0·30	0·35	0·40	0·45
5·00	2913·95	2928·49	2942·95	2957·34	2971·66	2985·91	3000·10	3014·22	3028·27	3042·26
5·50	3056·18	3070·04	3083·84	3097·57	3111·25	3124·87	3138·82	3151·92	3165·36	3178·75
6·00	3192·08	3205·35	3218·57	3231·73	3244·84	3257·90	3270·90	3283·86	3296·76	3309·61
6·50	3322·42	3335·17	3347·88	3360·53	3373·14	3385·71	3398·22	3410·69	3423·12	3435·50
7·00	3447·84	3460·13	3472·38	3484·58	3496·74	3508·86	3520·94	3532·98	3544·98	3556·93
7·50	3568·85	3580·73	3592·56	3604·36	3616·12	3627·84	3639·53	3651·17	3662·78	3674·35
8·00	3685·89	3697·39	3708·86	3720·29	3731·68	3743·04	3754·37	3765·66	3776·91	3788·14
8·50	3799·33	3810·49	3821·61	3832·71	3843·77	3854·80	3865·80	3876·76	3887·70	3898·60
9·00	3909·48	3920·32	3931·14	3941·92	3952·68	3963·40	3974·10	3984·77	3995·41	4006·02
9·50	4016·61	4027·16	4037·69	4048·19	4058·67	4069·11	4079·53	4089·93	4100·30	4110·64
10·00	4120·95									

Appendix 6. Table of reciprocals

	0	1	2	3	4	5	6	7	8	9	SUBTRACT 1	2	3	4	5	6	7	8	9
1·0	1·0000	·9901	·9804	·9709	·9615	·9524	·9434	·9346	·9259	·9174	9	18	27	36	45	55	64	73	82
1·1	·9091	·9009	·8929	·8850	·8772	·8696	·8621	·8547	·8475	·8403	8	15	23	30	38	45	53	61	68
1·2	·8333	·8264	·8197	·8130	·8065	·8000	·7937	·7874	·7813	·7752	6	13	19	26	32	38	45	51	58
1·3	·7692	·7634	·7576	·7519	·7463	·7407	·7353	·7299	·7246	·7194	5	11	16	22	27	33	38	44	49
1·4	·7143	·7092	·7042	·6993	·6944	·6897	·6849	·6803	·6757	·6711	5	10	14	19	24	29	33	38	43
1·5	·6667	·6623	·6579	·6536	·6494	·6452	·6410	·6369	·6329	·6289	4	8	13	17	21	25	29	33	38
1·6	·6250	·6211	·6173	·6135	·6098	·6061	·6024	·5988	·5952	·5917	4	7	11	15	18	22	26	29	33
1·7	·5882	·5848	·5814	·5780	·5747	·5714	·5682	·5650	·5618	·5587	3	7	10	13	16	20	23	26	30
1·8	·5556	·5525	·5495	·5464	·5435	·5405	·5376	·5348	·5319	·5291	3	6	9	12	15	18	20	23	26
1·9	·5263	·5236	·5208	·5181	·5155	·5128	·5102	·5076	·5051	·5025	3	5	8	11	13	16	18	21	24
2·0	·5000	·4975	·4950	·4926	·4902	·4878	·4854	·4831	·4808	·4785	2	5	7	10	12	14	17	19	21
2·1	·4762	·4739	·4717	·4695	·4673	·4651	·4630	·4608	·4587	·4566	2	4	7	9	11	13	15	17	20
2·2	·4545	·4525	·4505	·4484	·4464	·4444	·4425	·4405	·4386	·4367	2	4	6	8	10	12	14	16	18
2·3	·4348	·4329	·4310	·4292	·4274	·4255	·4237	·4219	·4202	·4184	2	4	5	7	9	11	13	14	16
2·4	·4167	·4149	·4132	·4115	·4098	·4082	·4065	·4049	·4032	·4016	2	3	5	7	8	10	12	13	15
2·5	·4000	·3984	·3968	·3953	·3937	·3922	·3906	·3891	·3876	·3861	2	3	5	6	8	9	11	12	14
2·6	·3846	·3831	·3817	·3802	·3788	·3774	·3759	·3745	·3731	·3717	1	3	4	6	7	8	10	11	13
2·7	·3704	·3690	·3676	·3663	·3650	·3636	·3623	·3610	·3597	·3584	1	3	4	5	7	8	9	11	12
2·8	·3571	·3559	·3546	·3534	·3521	·3509	·3497	·3484	·3472	·3460	1	2	4	5	6	7	9	10	11
2·9	·3448	·3436	·3425	·3413	·3401	·3390	·3378	·3367	·3356	·3344	1	2	3	5	6	7	8	9	10
3·0	·3333	·3322	·3311	·3300	·3289	·3279	·3268	·3257	·3247	·3236	1	2	3	4	5	6	7	9	10
3·1	·3226	·3215	·3205	·3195	·3185	·3175	·3165	·3155	·3145	·3135	1	2	3	4	5	6	7	8	9
3·2	·3125	·3115	·3106	·3096	·3086	·3077	·3067	·3058	·3049	·3040	1	2	3	4	5	6	7	8	9
3·3	·3030	·3021	·3012	·3003	·2994	·2985	·2976	·2967	·2959	·2950	1	2	3	4	4	5	6	7	8
3·4	·2941	·2933	·2924	·2915	·2907	·2899	·2890	·2882	·2874	·2865	1	2	3	3	4	5	6	7	8
3·5	·2857	·2849	·2841	·2833	·2825	·2817	·2809	·2801	·2793	·2786	1	2	2	3	4	5	6	6	7
3·6	·2778	·2770	·2762	·2755	·2747	·2740	·2732	·2725	·2717	·2710	1	2	2	3	4	5	5	6	7
3·7	·2703	·2695	·2688	·2681	·2674	·2667	·2660	·2653	·2646	·2639	1	1	2	3	4	4	5	6	6
3·8	·2632	·2625	·2618	·2611	·2604	·2597	·2591	·2584	·2577	·2571	1	1	2	3	3	4	5	5	6
3·9	·2564	·2558	·2551	·2545	·2538	·2532	·2525	·2519	·2513	·2506	1	1	2	3	3	4	4	5	6
4·0	·2500	·2494	·2488	·2481	·2475	·2469	·2463	·2457	·2451	·2445	1	1	2	2	3	4	4	5	5
4·1	·2439	·2433	·2427	·2421	·2415	·2410	·2404	·2398	·2392	·2387	1	1	2	2	3	3	4	5	5
4·2	·2381	·2375	·2370	·2364	·2358	·2353	·2347	·2342	·2336	·2331	1	1	2	2	3	3	4	4	5
4·3	·2326	·2320	·2315	·2309	·2304	·2299	·2294	·2288	·2283	·2278	1	1	2	2	3	3	4	4	5
4·4	·2273	·2268	·2262	·2257	·2252	·2247	·2242	·2237	·2232	·2227	1	1	2	2	3	3	4	4	5
4·5	·2222	·2217	·2212	·2208	·2203	·2198	·2193	·2188	·2183	·2179	0	1	1	2	2	3	3	4	4
4·6	·2174	·2169	·2165	·2160	·2155	·2151	·2146	·2141	·2137	·2132	0	1	1	2	2	3	3	4	4
4·7	·2128	·2123	·2119	·2114	·2110	·2105	·2101	·2096	·2092	·2088	0	1	1	2	2	3	3	4	4
4·8	·2083	·2079	·2075	·2070	·2066	·2062	·2058	·2053	·2049	·2045	0	1	1	2	2	3	3	3	4
4·9	·2041	·2037	·2033	·2028	·2024	·2020	·2016	·2012	·2008	·2004	0	1	1	2	2	2	3	3	4
5·0	·2000	·1996	·1992	·1988	·1984	·1980	·1976	·1972	·1969	·1965	0	1	1	2	2	2	3	3	4
5·1	·1961	·1957	·1953	·1949	·1946	·1942	·1938	·1934	·1931	·1927	0	1	1	2	2	2	3	3	3
5·2	·1923	·1919	·1916	·1912	·1908	·1905	·1901	·1898	·1894	·1890	0	1	1	1	2	2	3	3	3
5·3	·1887	·1883	·1880	·1876	·1873	·1869	·1866	·1862	·1859	·1855	0	1	1	1	2	2	3	3	3
5·4	·1852	·1848	·1845	·1842	·1838	·1835	·1832	·1828	·1825	·1821	0	1	1	1	2	2	2	3	3
	0	**1**	**2**	**3**	**4**	**5**	**6**	**7**	**8**	**9**	**1**	**2**	**3**	**4**	**5**	**6**	**7**	**8**	**9**

	0	1	2	3	4	5	6	7	8	9	SUBTRACT 1	2	3	4	5	6	7	8	9
5·5	·1818	·1815	·1812	·1808	·1805	·1802	·1799	·1795	·1792	·1789	0	1	1	1	2	2	2	3	3
5·6	·1786	·1783	·1779	·1776	·1773	·1770	·1767	·1764	·1761	·1757	0	1	1	1	2	2	2	3	3
5·7	·1754	·1751	·1748	·1745	·1742	·1739	·1736	·1733	·1730	·1727	0	1	1	1	2	2	2	2	3
5·8	·1724	·1721	·1718	·1715	·1712	·1709	·1706	·1704	·1701	·1698	0	1	1	1	1	2	2	2	3
5·9	·1695	·1692	·1689	·1686	·1684	·1681	·1678	·1675	·1672	·1669	0	1	1	1	1	2	2	2	3
6·0	·1667	·1664	·1661	·1658	·1656	·1653	·1650	·1647	·1645	·1642	0	1	1	1	1	2	2	2	3
6·1	·1639	·1637	·1634	·1631	·1629	·1626	·1623	·1621	·1618	·1616	0	1	1	1	1	2	2	2	2
6·2	·1613	·1610	·1608	·1605	·1603	·1600	·1597	·1595	·1592	·1590	0	1	1	1	1	2	2	2	2
6·3	·1587	·1585	·1582	·1580	·1577	·1575	·1572	·1570	·1567	·1565	0	0	1	1	1	1	2	2	2
6·4	·1563	·1560	·1558	·1555	·1553	·1550	·1548	·1546	·1543	·1541	0	0	1	1	1	1	2	2	2
6·5	·1538	·1536	·1534	·1531	·1529	·1527	·1524	·1522	·1520	·1517	0	0	1	1	1	1	2	2	2
6·6	·1515	·1513	·1511	·1508	·1506	·1504	·1502	·1499	·1497	·1495	0	0	1	1	1	1	2	2	2
6·7	·1493	·1490	·1488	·1486	·1484	·1481	·1479	·1477	·1475	·1473	0	0	1	1	1	1	2	2	2
6·8	·1471	·1468	·1466	·1464	·1462	·1460	·1458	·1456	·1453	·1451	0	0	1	1	1	1	2	2	2
6·9	·1449	·1447	·1445	·1443	·1441	·1439	·1437	·1435	·1433	·1431	0	0	1	1	1	1	1	2	2
7·0	·1429	·1427	·1425	·1422	·1420	·1418	·1416	·1414	·1412	·1410	0	0	1	1	1	1	1	2	2
7·1	·1408	·1406	·1404	·1403	·1401	·1399	·1397	·1395	·1393	·1391	0	0	1	1	1	1	1	2	2
7·2	·1389	·1387	·1385	·1383	·1381	·1379	·1377	·1376	·1374	·1372	0	0	1	1	1	1	1	2	2
7·3	·1370	·1368	·1366	·1364	·1362	·1361	·1359	·1357	·1355	·1353	0	0	1	1	1	1	1	2	2
7·4	·1351	·1350	·1348	·1346	·1344	·1342	·1340	·1339	·1337	·1335	0	0	1	1	1	1	1	1	2
7·5	·1333	·1332	·1330	·1328	·1326	·1325	·1323	·1321	·1319	·1318	0	0	1	1	1	1	1	1	2
7·6	·1316	·1314	·1312	·1311	·1309	·1307	·1305	·1304	·1302	·1300	0	0	1	1	1	1	1	1	2
7·7	·1299	·1297	·1295	·1294	·1292	·1290	·1289	·1287	·1285	·1284	0	0	0	1	1	1	1	1	1
7·8	·1282	·1280	·1279	·1277	·1276	·1274	·1272	·1271	·1269	·1267	0	0	0	1	1	1	1	1	1
7·9	·1266	·1264	·1263	·1261	·1259	·1258	·1256	·1255	·1253	·1252	0	0	0	1	1	1	1	1	1
8·0	·1250	·1248	·1247	·1245	·1244	·1242	·1241	·1239	·1238	·1236	0	0	0	1	1	1	1	1	1
8·1	·1235	·1233	·1232	·1230	·1229	·1227	·1225	·1224	·1222	·1221	0	0	0	1	1	1	1	1	1
8·2	·1220	·1218	·1217	·1215	·1214	·1212	·1211	·1209	·1208	·1206	0	0	0	1	1	1	1	1	1
8·3	·1205	·1203	·1202	·1200	·1199	·1198	·1196	·1195	·1193	·1192	0	0	0	1	1	1	1	1	1
8·4	·1190	·1189	·1188	·1186	·1185	·1183	·1182	·1181	·1179	·1178	0	0	0	1	1	1	1	1	1
8·5	·1176	·1175	·1174	·1172	·1171	·1170	·1168	·1167	·1166	·1164	0	0	0	1	1	1	1	1	1
8·6	·1163	·1161	·1160	·1159	·1157	·1156	·1155	·1153	·1152	·1151	0	0	0	1	1	1	1	1	1
8·7	·1149	·1148	·1147	·1145	·1144	·1143	·1142	·1140	·1139	·1138	0	0	0	1	1	1	1	1	1
8·8	·1136	·1135	·1134	·1133	·1131	·1130	·1129	·1127	·1126	·1125	0	0	0	1	1	1	1	1	1
8·9	·1124	·1122	·1121	·1120	·1119	·1117	·1116	·1115	·1114	·1112	0	0	0	1	1	1	1	1	1
9·0	·1111	·1110	·1109	·1107	·1106	·1105	·1104	·1103	·1101	·1100	0	0	0	1	1	1	1	1	1
9·1	·1099	·1098	·1096	·1095	·1094	·1093	·1092	·1091	·1089	·1088	0	0	0	0	1	1	1	1	1
9·2	·1087	·1086	·1085	·1083	·1082	·1081	·1080	·1079	·1078	·1076	0	0	0	0	1	1	1	1	1
9·3	·1075	·1074	·1073	·1072	·1071	·1070	·1068	·1067	·1066	·1065	0	0	0	0	1	1	1	1	1
9·4	·1064	·1063	·1062	·1060	·1059	·1058	·1057	·1056	·1055	·1054	0	0	0	0	1	1	1	1	1
9·5	·1053	·1052	·1050	·1049	·1048	·1047	·1046	·1045	·1044	·1043	0	0	0	0	1	1	1	1	1
9·6	·1042	·1041	·1040	·1038	·1037	·1036	·1035	·1034	·1033	·1032	0	0	0	0	1	1	1	1	1
9·7	·1031	·1030	·1029	·1028	·1027	·1026	·1025	·1024	·1022	·1021	0	0	0	0	1	1	1	1	1
9·8	·1020	·1019	·1018	·1017	·1016	·1015	·1014	·1013	·1012	·1011	0	0	0	0	1	1	1	1	1
9·9	·1010	·1009	·1008	·1007	·1006	·1005	·1004	·1003	·1002	·1001	0	0	0	0	0	1	1	1	1
	0	**1**	**2**	**3**	**4**	**5**	**6**	**7**	**8**	**9**	**1**	**2**	**3**	**4**	**5**	**6**	**7**	**8**	**9**

Bibliography

This is divided into two parts. The larger part (Part I) contains references dealing with vibrational–spectroscopic studies. All such references are designated by a number. All other references are collected in Part II and in every case the reference number is preceded by a letter or letters. The significance of these literal prefixes is explained at the head of Part II.

Note: Where a reference to a Russian journal is given, the page numbers apply to the English version if the title of the journal is given in English; otherwise they apply to the Russian version.

Part I

1. ADAMS, D. M., *Metal-Ligand and Related Vibrations*, Edward Arnold (London), 1st edition 1967.

1a. ADAMS, D. M. and D. C. NEWTON, *Tables for factor-group and point-group analysis*. Beckmann–RIIC, 1970.

2. BHAGAVANTAM, S. and T. VENKATARAYUDU, *Theory of Groups and its Application to Physical Problems*, Andhra University (Waltair), 3rd edition 1962.

3. BORN, M. and K. HUANG, *Dynamical Theory of Crystal Lattices*, Clarendon Press (Oxford), 1st edition 1954.

4. BRILLOUIN, L., *Wave Propagation in Periodic Structures*, Dover, 2nd edition 1953.

5. COLTHUP, N. B., L. H. DALY and S. E. WIBERLEY, *Introduction to Infra-red and Raman Spectroscopy*, Academic Press (New York), 1st edition 1964.

6. COTTON, F. A., *Chemical Applications of Group Theory*, Interscience (New York), 1st edition 1963.

7. FERRARO, J. R. and J. S. ZIOMEK, *Introductory Group Theory and its Application to Molecular Structure*, Plenum Press (New York), 1st edition 1969.

7a. GANS, P., *Vibrating Molecules*. Chapman and Hall, 1st edition (1971).

8. HADNI, A., *Essentials of Modern Physics Applied to the Study of the Infra-red*, Pergamon Press (Oxford), 1st edition 1967.

9. HERZBERG, G., *Molecular Spectra and Molecular Structure*, D. Van Nostrand (Princeton), 1st edition 1945.

10. JAFFE, H. H. and M. ORCHIN, *Symmetry in Chemistry*, John Wiley (New York), 1st edition 1965.

10a. LEECH, J. W. and D. J. NEWMAN, *How to use Groups*. Methuen, 1st edition 1969.

11. MARADUDIN, A. A., E. W. MONTROLL and G. H. WEISS, *Lattice Dynamics in the Harmonic Approximation*, Academic Press (New York), 1963.
12. MATOSSI, F., *Gruppentheorie der Eigenschwingungen von Punktsystemen*, Springer Verlag (Berlin), 1st edition 1961.
13. NAKAMOTO, K., *Infra-red Spectra of Inorganic and Co-ordination Compounds*, John Wiley (New York), 1st edition 1963.
14. SCHONLAND, D., *Molecular Symmetry*, D. Van Nostrand (London), 1st edition 1965.
15. SIEBERT, H., *Anwendungen des Schwingungsspektroscopie in der Anorganischen Chemie*, Springer Verlag (Berlin), 1st edition 1966.
16. SMITH, R. A., *Wave Mechanics of Crystalline Solids*, Chapman and Hall (London), 1st edition 1961.
17. TINKHAM, M., *Group Theory and Quantum Mechanics*, McGraw-Hill (New York), 1st edition 1964.
18. WILSON, E. B., J. C. DECIUS and P. C. CROSS, *Molecular Vibrations*. McGraw-Hill (New York), 1st edition 1955.
19. ZBINDEN, R., *Infra-red Spectra of High Polymers*, Academic Press (New York), 1st edition 1964.
20. ABRAMOWITZ, S., N. ACQUISTA and I. W. LEVIN, *Proceedings of 22nd Symposium on Molecular Structure and Spectroscopy*, Columbus, Ohio, 1962.
21. ADAMS, D. M., *Proc. Chem. Soc.,* 335 (1961).
22. ADAMS, D. M., P. J. CHANDLER and R. G. CHURCHILL, *J. Chem. Soc.* **A**. 1272 (1967).
23. ADAMS, D. M. and R. G. CHURCHILL, *J. Chem. Soc.* **A**. 2141 (1968).
24. ADAMS, D. M. and R. G. CHURCHILL, *J. Chem. Soc.* **A**. 2310 (1968).
25. ADAMS, D. M., J. M. DAVIDSON and J. GERRATT, *J. Chem. Soc.* 2189 (1963).
26. ADAMS, D. M., G. W. FRASER, D. M. MORRIS and R. D. PEACOCK, *J. Chem. Soc.* **A**. 1131 (1968).
27. ADAMS, D. M. and H. A. GEBBIE, *Spectrochim. Acta* **19**, 925 (1963).
28. ADAMS, D. M., H. A. GEBBIE and R. D. PEACOCK, *Nature* **199**, 278 (1963).
29. ADAMS, D. M., M. GOLDSTEIN and E. F. MOONEY, *Trans. Far. Soc.* **59**, 2228 (1963).
30. ADAMS, D. M. and P. J. LOCK, *J. Chem. Soc.* **A**. 145 (1967).
31. ADAMS, D. M. and D. M. MORRIS, *J. Chem. Soc.* **A**. 1666 (1967).
32. ADAMS, D. M. and D. M. MORRIS, *J. Chem. Soc.* **A**. 1669 (1967).
33. ADAMS, D. M. and D. M. MORRIS, *J. Chem. Soc.* **A**. 2067 (1967).
34. ADAMS, D. M. and D. C. NEWTON, *J. Chem. Soc.* **A**. 2262 (1968).
35. ADDISON, C. C., D. W. AMOS, D. SUTTON and W. H. H. HOYLE, *J. Chem. Soc.* **A**. 808 (1967).
36. ADDISON, C. C. and B. M. GATEHOUSE, *Chem. and Ind.* 464 (1958).
37. ADDISON, C. C. and B. M. GATEHOUSE, *J. Chem. Soc.* 613 (1960).
38. ADDISON, C. C. and W. B. SIMPSON, *J. Chem. Soc.* 598 (1965).
39. ADDISON, C. C. and A. WALKER, *J. Chem. Soc.* 1220 (1963).
40. ADLER, H. H., *Am. Mineralogist* **50**, 1553 (1965).
41. ADLER, H. H. and P. F. KERR, *Am. Mineralogist* **47**, 700 (1962).
42. ADLER, H. H. and P. F. KERR, *Am. Mineralogist* **48**, 124 (1963).
43. ADLER, H. H. and P. F. KERR, *Am. Mineralogist* **48**, 839 (1963).
44. ADLER, H. H. and P. F. KERR, *Am. Mineralogist* **50**, 132 (1965).
45. AGAHIGIAN, H., A. P. GRAY and G. D. VICKERS, *Can. J. Chem.* **40**, 157 (1952).
46. ALDOUS, J. and I. M. MILLS, *Spectrochim. Acta* **18**, 1073 (1962).

47. Aldous, J. and I. M. Mills, *Spectrochim. Acta* **19**, 1567 (1963).
48. Aleksandrovskaya, A. M. and I. N. Godnev, *Optika i Spektroskopiya* **9**, 273 (1960).
49. Alers, G. A. and J. R. Neighbours, *Rev. Mod. Phys.* **31**, 675 (1959).
50. Allen, G. and C. A. McDowell, *J. Chem. Phys.* **23**, 209 (1955).
50a. Allen, H. C. and E. K. Plyler, *J. Chem. Phys.* **25**, 1132 (1956).
50b. Allen, H. C., E. D. Tidwell and E. K. Plyler, *J. Chem. Phys.* **25**, 302 (1956).
51. Allpress, J. G. and A. N. Hambly, *Australian J. Chem.* **12**, 569 (1959).
52. Ananthanarayanan, V., *Z. Physik* **163**, 144 (1961).
53. Ananthanarayanan, V., *Z. Physik* **166**, 318 (1962).
54. Ananthanarayanan, V., *Proc. Ind. Acad. Sci.* **56A**, 188 (1962).
55. Ananthanarayanan, V., *Z. phys. Chem. (Leipzig)* **222**, 102 (1963).
56. Ananthanarayanan, V., *Ind. J. Pure Appl. Phys.* **1**, 58 (1963).
57. Ananthanarayanan, V., *J. Chem. Phys.* **48**, 573 (1968).
58. Ananthanarayanan, V. and A. Danti, *J. Mol. Spec.* **20**, 88 (1966).
59. Anbar, M., M. Halmann and S. Pinchas, *J. Chem. Soc.* 1242 (1960).
60. Anderson, T. F., E. N. Lassettre and D. M. Yost, *J. Chem. Phys.* **4**, 703 (1936).
61. Anderson, W. E. and E. F. Barker, *J. Chem. Phys.* **18**, 698 (1950).
62. Angino, E. E., *Am. Mineralogist* **52**, 137 (1967).
63. Arvia, A. J., L. F. R. Cafferata and J. Schumacher. *Chem. Ber.* **96**, 1187 (1963).
64. Asano, S. and Y. Timishima, *J. Phys. Soc. Japan* **12**, 900 (1957).
65. Axe, J. D., *Phys. Rev.* **157**, 429 (1967).
66. Axe, J. D. and G. D. Petit, *Phys. Rev.* **157**, 435 (1967).
67. Aynsley, E. E., R. E. Dodd and R. Little, *Spectrochim. Acta* **18**, 1005 (1962).
68. Azman, A., A. Ocvirk, D. Hadzi, P. A. Giguere and M. Schneider, *Can. J. Chem.* **45**, 1347 (1967).
69. Babaeva, A. V., Yu. Ya. Kharitonov and Z. M. Novozhenyuk, *Russ. J. Inorg. Chem.* **6**, 1151 (1961).
70. Babaeva, A. V., Yu. Ya. Kharitonov and Z. M. Novozhenyuk, *Russ. J. Inorg. Chem.* **6**, 1159 (1961).
71. Bader, R. F. W. and K. P. Huang, *J. Chem. Phys.* **43**, 3760 (1965).
72. Baglin, F. G., S. F. Bush and J. R. Durig, *J. Chem. Phys.* **47**, 2104 (1967).
73. Bailey, C. R. and J. W. Thompson, *Nature* **135**, 913 (1935).
74. Bain, O. and P. A. Giguere, *Can. J. Chem.* **33**, 527 (1955).
75. Balasubramanyam, K. and L. Nanis, *J. Chem. Phys.* **40**, 2657 (1964).
76. Balasubramanyam, K. and L. Nanis, *J. Chem. Phys.* **42**, 676 (1965).
77. Baldwin, M. E., *J. Chem. Soc.* 3123 (1963).
78. Balkanski, M., P. Moch and G. Parisot, *J. Chem. Phys.* **44**, 940 (1966).
79. Balkanski, M., P. Moch and M. K. Teng, *Compt. rend.* **261**, 1492 (1965).
80. Balkanski, M., P. Moch and M. K. Teng, *J. Chem. Phys.* **46**, 1621 (1967).
81. Balkanski, M., M. Nusimovici and R. Le Toullec, *J. Phys. (Paris)* **25**, 305 (1964).
82. Ballantyne, J. M., *Phys. Rev.* **136**, 429 (1964).
83. Balls, A. A., A. J. Downs, N. N. Greenwood and B. P. Straughan, *Trans. Far. Soc.* **62**, 521 (1966).
83a. Bannister, E. and F. A. Cotton, *J. Chem. Soc.* 2276 (1960).

83b. Barakat, T. M., N. Legge and A. D. E. Pullin, *Trans. Far. Soc.* **59**, 1764 (1963).
84. Baran, E. J. and P. J. Aymonino, *Z. anorg. allgem. Chem.* **354**, 85 (1967).
85. Baran, E. J. and P. J. Aymonino, *Monatsh.* **99**, 1584 (1968).
86. Baran, E. J. and P. J. Aymonino, *Spectrochim. Acta* **24A**, 288 (1968).
87. Baran, E. J. and P. J. Aymonino, *Spectrochim. Acta* **24A**, 291 (1968).
88. Baran, E. J. and P. J. Aymonino, *Z. Naturforsch.* **23B**, 107 (1968).
89. Barcelo, J. R. and C. Otero, *An. Real Soc. españ Fisica Quim, Madrid* **51**, 223 (1955).
90. Barker, A. S., jr., *Phys. Rev.* **132**, 1474 (1963).
91. Barker, A. S., jr., *Phys. Rev.* **135**, 742 (1964).
92. Barker, A. S., jr., *Phys. Rev.* **136**, 1290 (1964).
92a. Barker, A. S., jr., and M. Tinkham, *Phys. Rev.* **125**, 1527 (1962).
93. Barrowcliffe, T., I. R. Beattie, P. Day and K. Livingston, *J. Chem. Soc.* **A**. 1810 (1967).
94. Bartlett, N. and L. E. Levchuk, *Proc. Chem. Soc.* 342 (1963).
95. Bass, C. D., L. Lynds, T. Wolfram and R. E. Dewames, *J. Chem. Phys.* **40**, 3611 (1964).
96. Baudler, M. and G. Fricke, *Z. anorg. allgem. Chem.* 345, 129 (1966).
97. Baudler, M. and L. Schmidt, *Naturwiss.* **44**, 488 (1957).
98. Baudler, M. and L. Schmidt, *Z. anorg. allgem. Chem.* **289**, 219 (1957).
99. Baun, W. L. and N. T. McDevitt, *J. Am. Ceram. Soc.* **46**, 294 (1963).
100. Beattie, I. R. and H. Chudzynska, *J. Chem. Soc.* **A**. 984 (1967).
101. Beattie, I. R., T, Gilson and P. Cocking, *J. Chem. Soc.* **A**. 702 (1967).
102. Beattie, I. R., T. Gilson, K. Livingston, V. Fawcett and G. A. Ozin, *J. Chem. Soc.* **A**. 712 (1967).
103. Beattie, I. R., T. Gilson and G. A. Ozin, *J. Chem. Soc.* **A**. 813 (1968).
104. Beattie, I. R., T. R. Gilson and G. A. Ozin, *J. Chem. Soc.* **A**. 2765 (1968).
105. Beattie, I. R. and M. Webster, *J. Chem. Soc.* 38 (1963).
106. Beattie, I. R. and M. Webster, *J. Chem. Soc.* 3507 (1964).
107. Beattie, I. R., M. Webster and G. W. Chantry, *J. Chem. Soc.* (Suppl.) 6172 (1967).
108. Becarud, N., *Comm. Energie At. France Rappt.* 2895.
109. Becarud, N. and C. Duval, *Compt. rend.* **257**, 1930 (1963).
110. Becher, H. J. and R. Mattes, *Spectrochim. Acta* **23A**, 2449 (1967).
111. Beck, W., *Chem. Ber.* **95**, 341 (1962).
112. Beck, W., *Z. Naturforsch.* **17B**, 130 (1962).
113. Beck, W., *Chem. Ber.* **98**, 298 (1965).
114. Beckmann, L., L. Gutjahr and R. Mecke, *Spectrochim. Acta* **20**, 1295 (1964).
115. Beckmann, L., L. Gutjahr and R. Mecke, *Spectrochim. Acta* **21**, 141 (1965).
116. Beckmann, L., L. Gutjahr and R. Mecke, *Spectrochim. Acta* **21**, 307 (1965).
117. Begun, G. M. and W. H. Fletcher, *J. Mol. Spect.* **4**, 388 (1960).
118. Begun, G. M. and W. H. Fletcher, *Spectrochim. Acta* **19**, 1343 (1963).
119. Begun, G. M., W. H. Fletcher and D. F. Smith, *J. Chem. Phys.* **42**, 2236 (1965).
120. Begun, G. M. and A. C. Rutenberg, *Inorg. Chem.* **6**, 2212 (1967).
121. Bell, R. J., N. F. Bird and P. Dean, *J. Phys.* (C) **1**, 299 (1968).
122. Bell, R. P and H. C. Longuet-Higgins, *Proc. Roy. Soc.* **183A**, 357 (1945).

123. Bender, P. and J. M. Wood, *J. Chem. Phys.* **23**, 1316 (1955).
123a. Benedict, W. S., N. Gailar and E. K. Plyler, *J. Chem. Phys.* **24**, 1139 (1956).
123b. Benedict, W. S., E. K. Plyler and E. D. Tidwell, *Can. J. Phys.* **35**, 1235 (1957).
123c. Benedict, W. S., E. K. Plyler and E. D. Tidwell, *J. Chem. Phys.* **29**, 829 (1958).
123d. Benedict, W. S., E. K. Plyler and E. D. Tidwell, *J. Chem. Phys.* **32**, 32 (1960).
124. Benoit, C. and A. M. Vergnoux, *J. Phys.* (Paris) Supp. **5–6**, 35 (1966).
125. Berkowitz, J. *J. Chem. Phys.* **29**, 1386 (1958).
126. Berkowitz, J., *J. Chem. Phys.* **32**, 1519 (1960).
127. Berkowitz, J., W. A. Chupka, E. Bromels and R. L. Belford, *J. Chem. Phys.* **47**, 4320 (1967).
128. Bernitt, D. L., K. O. Hartman and I. C. Hisatsune, *J. Chem. Phys.* **42**, 3553 (1965).
129. Bernitt, D. L., R. H. Miller and I. C. Hisatsune, *Spectrochim. Acta* **23A**, 237 (1967).
130. Bernstein, H. J. and J. Powling, *J. Chem. Phys.* **18**, 685 (1950).
131. Bernstein, H. J. and J. Powling, *J. Chem. Phys.* **18**, 1018 (1950).
132. Bernstein, H. J. and J. Powling, *J. Chem. Phys.* **19**, 139 (1951).
133. Berreman, D. W., *Bull. Amer. Phys. Soc.* **7**, 545 (1962).
134. Berreman, D. W., *Bull. Amer. Phys. Soc.* **8**, 194 (1963).
135. Berreman, D. W., *Phys. Rev.* **130**, 2193 (1963).
136. Berthold, G., *Z. Physik* **185**, 400 (1965).
137. Bertie, J. E., H. J. Labbé and E. Whalley, *J. Chem. Phys.* **49**, 775 (1968).
138. Bertie, J. E., H. J. Labbé and E. Whalley, *J. Chem. Phys.* **49**, 2141 (1968).
139. Bertie, J. E. and E. Whalley, *J. Chem. Phys.* **40**, 1637 (1964).
140. Bethell, D. E. and N. Sheppard, *J. Chem. Phys.* **21**, 1421 (1953).
141. Bethell, D. E. and N. Sheppard, *Trans. Far. Soc.* **51**, 9 (1955).
142. Bethke, G. W. and M. K. Wilson, *J. Chem. Phys.* **26**, 1107 (1957).
143. Bhagavantam, S., *Proc. Ind. Acad. Sci.* **13A**, 543 (1941).
144. Bhagavantam, S. and K. Venkatarayudu, *Proc. Ind. Acad. Sci.* **9A**, 224 (1939).
145. Bigeleisen, J., M. Goppert-Mayer, P. C. Stevenson and J. Turkevich, *J. Chem. Phys.* **16**, 442 (1948).
146. Billes, F., *Acta Chim. Acad. Sci. Hung.* **45**, 285 (1965).
147. Billes, F., *Acta Chim. Acad. Sci. Hung.* **49**, 97 (1966).
148. Birchall, T. and R. J. Gillespie, *Spectrochim. Acta* **22**, 681 (1966).
149. Birman, J. L., *Phys. Rev.* **131**, 1489 (1963).
150. Blackman, M., *Proc. Roy. Soc.* **148A**, 365 (1935).
151. Blackman, M., *Proc. Roy. Soc.* **148A**, 384 (1935).
152. Blackman, M., *Proc. Roy. Soc.* **149A**, 177 (1935).
153. Blackman, M., *Phil. Trans. Roy. Soc.* **236A**, 103 (1936).
154. Blackman, M. *Proc. Roy. Soc.* **159A**, 416 (1937).
155. Blanchard, S., *J. Chim. Phys.* **61**, 747 (1964).
156. Blanchard, S., *J. Chim. Phys.* **62**, 919 (1965).
157. Blanchard, S., *Commis. Energ. At.* (Fr.) Rapp. No: CEA-R 3194.
158. Blau, E. J. and B. F. Hochheimer, *J. Chem. Phys.* **41**, 1174 (1964).
159. Blomgren, G. E., *Ph.D. Thesis* Washington (1956).
160. Bloor, D., *Spectrochim. Acta* **21**, 133 (1965).

161. BLUE, R. W., *J. Chem. Phys.* **22**, 280 (1954).
162. BLYHOLDER, G. and N. FORD, *Proc. Int. Symposium Mol. Struct. Spectroscopy* **A321**, Tokyo (1962).
163. BLYHOLDER, G. and N. FORD, *J. Phys. Chem.* **68**, 1496 (1964).
164. BLYHOLDER, G. and A. KITTILA, *J. Phys. Chem.* **67**, 2147 (1963).
165. BLYHOLDER, G. and S. VERGEZ, *J. Phys. Chem.* **67**, 2149 (1963).
166. BOBOVICH, YA. S., *Optika i Spektroskopiya* **11**, 161 (1961).
167. BOBOVICH, YA. S., *Optika i Spektroskopiya* **13**, 459 (1962).
168. BOBOVICH, YA. S. and E. V. BURSIAN, *Optika i Spektroskopiya* **11**, 131 (1961).
169. BODE, H. and E. VOSS, *Z. anorg. allgem. Chem.* **290**, 1 (1957).
170. BONADEO, H., E. D'ALESSIO and E. SILBERMAN, *J. Mol. Spec.* **22**, 402 (1967).
171. BONHAM, R. A., *J. Mol. Spec.* **14**, 404 (1964).
172. BONN, R., R. METSELAAR and J. VAN DER ELSKEN, *J. Chem. Phys.* **46**, 1988 (1967).
173. BOTTGER, G. L. and A. L. GEDDES, *J. Chem. Phys.* **47**, 4858 (1967).
174. BOUHET, C. and R. LAFONT *J. Chim. Phys.*, **46**, 66 (1949).
175. BOUTIN, H., G. J. SAFFORD and H. R. DANNER, *J. Chem. Phys.* **40**, 2670 (1964).
176. BOVEY, L. F. H., *J. Chem. Phys.* **18**, 1684 (1950).
177. BOVEY, L. F. H., *J. Opt. Soc. Am.* **41**, 836 (1951).
178. BRADLEY, E. B., C. R. BENNETT and E. A. JONES, *Spectrochim. Acta* **21**, 1505 (1965).
179. BRADLEY, E. B., M. S. MATHUR and C. A. FRENZEL, *J. Chem. Phys.* **47**, 4325 (1967).
180. BRAHMS, S. and J. P. MATHIEU, *Compt. rend.* **251**, 938 (1960).
181. BRANDLE, K., M. SCHMEISSER and W. LÜTTKE, *Chem. Ber.* **93**, 2300 (1960).
182. BREARLEY, N. and S. SUNDARAM, *Z. Physik. Chem.* (Frankfurt), **32**, 219 (1962).
183. BREWER, L., G. R. SOMAYASULU and E. BRACKETT, *Chem. Rev.* **63**, 111 (1963).
184. BRIEUX DE MANDIROLA, O., *J. Mol. Structure* **1**, 203 (1968).
185. BRIEUX DE MANDIROLA, O. and J. F. WESTERKAMP, *Spectrochim. Acta* **20**, 1633 (1964).
186. BRIEUX DE MANDIROLA, O. and J. F. WESTERKAMP, *Spectrochim. Acta* **21**, 1101 (1965).
187. BRINTZINGER, H. and R. E. HESTER, *Inorg. Chem.* **5**, 980 (1966).
187a. BRODERSEN, K., *Chem. Ber.* **90**, 2703 (1957).
188. BROWN, D. A., *Nature* **190**, 804 (1961).
189. BROWN, D. A., D. CUNNINGHAM and W. K. GLASS, *Spectrochim. Acta* **24A**, 965 (1968).
190. BROWN, D. A., K. R. DIXON, C. M. LIVINGSTON, R. H. NUTTALL and D. W. A. SHARP, *J. Chem. Soc.* **A.**, 100 (1967).
191. BROWN, D. A. and H. C. LONGUET-HIGGINS, *J. Inorg. Nucl. Chem.* **1**, 352 (1955).
192. BROWN, R. D., G. P. PEZ and M. F. O'DWYER, *Australian J. Chem.* **18**, 627 (1965).
192a. BROWN, R. G., J. DENNING, A. HALLETT and S. D. ROSS, *Spectrochim. Acta* **26A**, 963 (1970).
193. BRYANT, J. I., *J. Chem. Phys.* **38**, 2845 (1963).
194. BRYANT, J. I., *J. Chem. Phys.* **40**, 3195 (1964).
195. BRYANT, J. I., *J. Chem. Phys.* **45**, 689 (1966).

196. BUECHLER, A. and W. KLEMPERER, *J. Chem. Phys.* **29**, 121 (1958).
197. BUECHLER, A., W. KLEMPERER and A. G. EMSLIE, *J. Chem. Phys.* **36**, 2499 (1962).
198. BUECHLER, A., E. P. MARRAM and J. L. STAUFFER, *J. Chem. Phys.* **71**, 4139 (1967).
199. BUEHLER, K. and W. BUES, *Z. anorg. allgem. Chem.* **308**, 62 (1961).
200. BUES, W., *Z. Physik Chem.* **10**, 1 (1957).
201. BUES, W., K. BUEHLER and P. KUHNLE, *Z. anorg. allgem. Chem.* **325**, 8 (1963).
202. BUES, W., G. FOERSTER and R. SCHMITT, *Z. anorg. allgem. Chem.* **344**, 148 (1966).
203. BUES, W. and H. GEHRKE, *Z. anorg. allgem. Chem.* **288**, 291 (1956).
204. BUES, W. and H. GEHRKE, *Z. anorg. allgem. Chem.* **288**, 307 (1956).
205. BUIJS, K. *J. Chem. Phys.* **36**, 861 (1962).
206. BUIJS, K. and C. J. H. SCHUTTE, *Spectrochim. Acta* **17**, 917 (1961).
207. BUIJS, K. and C. J. H. SCHUTTE, *Spectrochim. Acta* **17**, 921 (1961).
208. BUIJS, K. and C. J. H. SCHUTTE, *Spectrochim. Acta* **17**, 926 (1961).
209. BUIJS, K. and C. J. H. SCHUTTE, *Spectrochim. Acta* **18**, 307 (1962).
210. BULANIN, M. O. and N. D. ORLOVA, *Optics and Spectroscopy* **15**, 112 (1963).
211. BURKE, T. G., *J. Chem. Phys.* **25**, 791 (1956).
212. BURKE, T. G. and E. A. JONES, *J. Chem. Phys.* **19**, 1611 (1951).
213. BURKE, T. G., D. F. SMITH and A. H. NIELSEN, *J. Chem. Phys.* **20**, 447 (1952).
214. BURNS, W. G. and H. J. BERNSTEIN, *J. Chem. Phys.* **18**, 1669 (1950).
215. BUSEY, H. and O. KELLER, jr., *J. Chem. Phys.* **41**, 225 (1964).
216. BUSLAEV, YU. A., YU. YA. KHARITONOV and R. L. DAVIDOVITCH, *Izv. Akad. Nauk. SSSR. Neorg. Mater.* **3**, 589 (1967).
217. BUTCHER, F. K., W. GERRARD, M. HOWARTH, E. F. MOONEY and H. A. WILLIS, *Spectrochim. Acta* **19**, 905 (1963).
218. CABANNES, J., *Rev. Scient.* **80**, 407 (1942).
218aa. CALIFANO, S. and B. CRAWFORD, jr., *Z. Elektrochem.* **64**, 571 (1960).
218a. CAMERON, D. M., W. C. SEARS and H. H. NIELSEN, *J. Chem. Phys.* **7**, 994 (1939).
219. CAMPBELL, J. A., *Spectrochim. Acta* **21**, 851 (1965).
220. CARDONA, M., *J. Appl. Phys.* **36**, 2181 (1965).
221. CARLSON, G., *Spectrochim. Acta* **19**, 1291 (1963).
222. CARREIRA, L. A., V. A. MARONI, J. W. SWAINE and R. C. PLUMB, *J. Chem. Phys.* **45**, 2216 (1966).
223. CARROLL, R. L. and A. H. COWLEY, *Chem. Comm.* 872 (1966).
224. CARTER, R. P. and R. R. HOLMES, *Inorg. Chem.* **4**, 738 (1965).
225. CATALANO, E. and R. H. SANBORN, *J. Chem. Phys.* **38**, 2273 (1963).
226. CATALANO, E., R. H. SANBORN and J. W. FRAZER, *J. Chem. Phys.* **38**, 2265 (1963).
227. CAUNT, A. D., L. N. SHORT and L. A. WOODWARD, *Trans. Far. Soc.* **48**, 873 (1952).
228. CAVELL, R. G., *Spectrochim. Acta* **23A**, 249 (1967).
229. CAVELL, R. G. and H. C. CLARK, *Inorg. Chem.* **4**, 738 (1965).
230. CHALMERS, A. A. and D. C. MCKEAN, *Spectrochim. Acta* **22**, 251 (1966).
231. CHANDRASEKHARAN, V., *Current Sci.* (India) **19**, 277 (1950).
232. CHANTRY, G. W. and V. C. EWING, *Mol. Phys.* **5**, 209 (1962).
233. CHANTRY, G. W. and L. A. WOODWARD, *Trans. Far. Soc.* **56**, 1110 (1960).
234. CHAPELLE, J., *Bull. soc. franc. min. et crist.* **73**, 511 (1950).

235. Chapman, A. E. and L. E. Thirlwell, *Spectrochim. Acta* **20**, 937 (1964).
236. Charlton, T. L. and K. B. Harvey, *Can. J. Chem.* **44**, 2717 (1966).
237. Chen, S. H. and V. Dvorak, *J. Chem. Phys.* **48**, 4060 (1968).
238. Chernick, C. L., H. H. Claassen and B. Weinstock, *J. Amer. Chem. Soc.* **83**, 3165 (1961).
239. Chiba, T., *J. Chem. Phys.* **39**, 947 (1963).
240. Chin, D. and P. Giguere, *J. Chem. Phys.* **34**, 690 (1961).
241. Christe, K. O. and J. P. Guertin, *Inorg. Chem.* **4**, 905 (1965).
242. Christe, K. O. and J. P. Guertin, *Inorg. Chem.* **4**, 1785 (1965).
243. Christe, K. O., J. P. Guertin, A. E. Pavlath and W. Sawodny, *Inorg. Chem.* **6**, 533 (1967).
244. Christe, K. O., J. P. Guertin and W. Sawodny, *Inorg. Chem.* **7**, 626 (1968).
245. Christe, K. O. and W. Sawodny, *Inorg. Chem.* **6**, 313 (1967).
246. Christe, K. O. and W. Sawodny, *Inorg. Chem.* **6**, 1783 (1967).
247. Christe, K. O. and W. Sawodny, *Inorg. Chem.* **7**, 1685 (1968).
248. Christe, K. O. and W. Sawodny, *Z. anorg. allgem. chem.* **357**, 125 (1968).
249. Christe, K. O., W. Sawodny and J. P. Guertin, *Inorg. Chem.* **6**, 1159 (1967).
250. Christensen, D. H. and O. F. Nielsen, *J. Mol. Spec.* **27**, 489 (1968).
251. Cilento, G., D. A. Ramsay and R. N. Jones, *J. Amer. Chem. Soc.* **71**, 2753 (1949).
252. Ciruna, J. A. and A. E. Robinson, *Can. J. Chem.* **46**, 1715 (1968).
253. Cismaru, D., Z. Simon and C. Volanschi, *Rev. Roumaine. Chim.* **9**, 681 (1964).
254. Claassen, H. H., C. L. Chernick and J. G. Malm, *J. Amer. Chem. Soc.* **85**, 1927 (1963).
255. Claassen, H. H., C. L. Chernick and J. G. Malm, *Noble Gas Cpds.* 287 (1963).
256. Claassen, H. H., E. L. Gasner, H. Kim and J. L. Huston, *J. Chem. Phys.* **49**, 253 (1968).
257. Claassen, H. H., E. L. Gasner and H. Selig, *J. Chem. Phys.* **49**, 1803 (1968).
258. Claassen, H. H., G. L. Goodman, J. G. Malm and F. Schreiner, *J. Chem. Phys.* **42**, 1229 (1965).
259. Claassen, H. H. and G. Knapp, *J. Amer. Chem. Soc.* **86**, 2341 (1964).
260. Claassen, H. H., J. G. Malm and H. Selig, *J. Chem. Phys.* **36**, 2890 (1962).
261. Claassen, H. H. and H. Selig, *J. Chem. Phys.* **43**, 103 (1965).
262. Claassen, H. H. and H. Selig, *J. Chem. Phys.* **44**, 4039 (1965).
263. Claassen, H. H., H. Selig and J. G. Malm, *J. Chem. Phys.* **36**, 2888 (1962).
264. Claassen, H. H. and B. Weinstock, *J. Chem. Phys.* **33**, 436 (1960).
265. Claassen, H. H., B. Weinstock and J. G. Malm, *J. Chem. Phys.* **25**, 426 (1956).
266. Claassen, H. H., B. Weinstock and J. G. Malm, *J. Chem. Phys.* **28**, 285 (1958).
267. Claassen, H. H. and A. J. Zielen, *J. Chem. Phys.* **22**, 707 (1954).
268. Claeys, E. J. and G. P. van der Kelen, *Spectrochim. Acta* **22**, 2095 (1966).
269. Claeys, E. J. and G. P. van der Kelen, *Spectrochim. Acta* **22**, 2103 (1966).
270. Clark, G. M. and W. P. Doyle, *J. Inorg. Nucl. Chem.* **28**, 381 (1966).
271. Clark, G. M. and W. P. Doyle, *Spectrochim. Acta* **22**, 1441 (1966).
272. Clark, R. J. H., L. Maresca and J. Puddephatt, *Inorg. Chem.* **7**, 1603 (1968).
273. Clarke, J. H. R. and C. Solomons, *J. Chem. Phys.* **47**, 1823 (1967).

274. Clifford, A. F. and P. A. Faeth. *Proc. 2nd Conf. Rare Earth Res.* (Glenwood) 1961, p. 35.
275. Cochran, W., *Rep. Progr. Physics*, **26**, 1 (1963).
276. Cochran, W. and R. A. Cowley, *J. Phys. Chem. Solids* **23**, 447 (1962).
277. Cohen, B., A. J. Edwards, M. Mercer and R. D. Peacock, *Chem. Comm.* 322 (1965).
278. Cohn, H., *J. Chem. Soc.* 4282 (1952).
278a. Cohn, H., C. K. Ingold and H. G. Poole, *J. Chem. Soc.* 4272 (1952).
279. Comeford, J. J., *J. Chem. Phys.* **45**, 3463 (1966).
280. Comeford, J. J., D. E. Mann, L. J. Schoen and D. R. Lide, *J. Chem. Phys.* **38**, 461 (1963).
281. Condrate, R. A. and K. Nakamoto, *Bull. Chem. Soc. Japan* **39**, 1108 (1966).
282. Conn, G. K. T. and C. K. Wu, *Trans. Far. Soc.* **34**, 1483 (1938).
283. Cook, D., S. J. Kuhn and G. A. Olah, *J. Chem. Phys.* **33**, 1669 (1960).
284. Coomber, R. and W. P. Griffith, *J. Chem. Soc.* **A**. 1128 (1968).
285. Cooney, R. P. J., J. R. Hall and M. A. Hooper, *Austral. J. Chem.* **21**, 2145 (1968).
286. Costa, G., E. Pauluzzi and A. Puxeddu, *Gazz. chim. ital.* **87**, 885 (1957).
287. Coté, G. L. and H. W. Thompson, *Proc. Roy. Soc.* **210A**, 206 (1951).
288. Coté, G. L. and H. W. Thompson, *Proc. Roy. Soc.* **210A**, 217 (1951).
288a. Cotton, F. A., D. M. L. Goodgame and R. H. Soderberg, *Inorg. Chem.* **2**, 1163 (1963).
289. Cotton, F. A. and W. D. Horrocks, *Spectrochim. Acta* **16**, 358 (1960).
290. Coulson, C. A., *J. Chem. Phys.* **44**, 468 (1966).
291. Couture, J., *Ann. Phys.* **2**, 5 (1944).
292. Couture, L. and J.-P. Mathieu, *Compt. rend.* **226**, 1261 (1948).
293. Couture, L. and J.-P. Mathieu, *Compt. rend.* **230**, 1054 (1950).
294. Couture-Mathieu, L. and J.-P. Mathieu, *J. Chim. Phys.* **49**, 226 (1952).
295. Couture-Mathieu, L. and J.-P. Mathieu, *Ann. Phys.* (Paris), **9**, 255 (1954).
296. Cowan, R. D., *J. Chem. Phys.* **18**, 1101 (1950).
297. Coyle, T. D., J. J. Ritter and T. C. Farrar, *Proc. Chem. Soc.* 25 (1964).
297a. Craine, G. D. and H. W. Thompson, *Trans. Far. Soc.* **49**, 1273 (1953).
298. Creighton, J. A., *J. Chem. Soc.* 6589 (1965).
299. Creighton, J. A. and J. H. S. Green, J. *Chem. Soc.* **A**. 808 (1968).
300. Creighton, J. A., J. H. S. Green and W. Kynaston, *J. Chem. Soc.* 208 (1966).
301. Creighton, J. A. and L. A. Woodward, *Trans. Far. Soc.* **58**, 1077 (1962).
302. Cross, L. H., H. L. Roberts, P. Goggin and L. A. Woodward, *Trans. Far. Soc.* **56**, 945 (1960).
303. Crow, T. T. and R. T. Lagemann, *Spectrochim. Acta* **12**, 143 (1958).
304. Curtis, E. C., D. Pilipovitch and W. H. Moberly, *J. Chem. Phys.* **46**, 2904 (1967).
305. Curtis, E. E., *J. Mol. Spect.* **17**, 108 (1965).
306. Curtis, N. F. and Y. M. Curtis, *Inorg. Chem.* **4**, 804 (1965).
307. Cyvin, S. J., *Acta Chem. Scand.* **20**, 1389 (1966).
308. Cyvin, S. J., *Acta Chem. Scand.* **20**, 2616 (1966).
309. Cyvin, S. J., I. Elvebredd, G. Hagen and J. Brunvoll, *J. Chem. Phys.* **49**, 3561 (1968).
310. Cyvin, S. J., R. W. Mooney, J. Brunvoll and L. A. Kristiansen, *Acta Chem. Scand.* **19**, 983 (1965).

311. Cyvin, S. J., R. W. Mooney, J. Brunvoll and L. A. Kristiansen, *Acta Chem. Scand.* **19**, 1031 (1965).
312. Dachille, F. and R. Roy, *Nature* **183**, 1257 (1959).
313. Damen, T. C., S. P. S. Porto and B. Tell, *Phys. Rev.* **142**, 570 (1966).
314. Dasent, W. and T. C. Waddington, *J. Chem. Soc.* 2429 (1960).
315. Dasent, W. and T. C. Waddington, *J. Chem. Soc.* 3350 (1960).
316. Dasent, W. E. and T. C. Waddington, *J. Inorg. Nucl. Chem.* **25**, 132 (1963).
317. Dasgupta, S. and A. P. B. Sinha, *Trans. Far. Soc.* **53**, 909 (1957).
318. Silveira, A. Da, M. A. Marques and N. M. Marques, *Compt. rend.* **252**, 3983 (1961).
319. Silveira, A. Da., M. A. Marques and N. M. Marques, *Mol. Phys.* **9**, 271 (1965).
320. Dautel, R. and W. Zeil, *Z. Elektrochem.* **64**, 1234 (1960).
321. Davidson, G., N. Logan and A. Morris, *Chem. Comm.* 1044 (1968).
322. Davies, J. E. D. and D. A. Long, *J. Chem. Soc.* **A**. 2050 (1968).
323. Davies, J. E. D. and D. A. Long, *J. Chem. Soc.* **A**. 2055 (1968).
324. Davis, A. R. and D. E. Irish, *Inorg. Chem.* **7**, 1699 (1968).
325. Davis, P. W. and R. A. Oetjen, *J. Mol. Spec.* **2**, 253 (1958).
326. Deabriges, J. *Bull. Soc. chim. France* 4005 (1968).
327. Deacon, G. B., J. H. S. Green and W. Kynaston, *Austral. J. Chem.* **19**, 1603 (1966).
328. Deb, S. K. and A. D. Yoffe, *Trans. Far. Soc.* **55**, 106 (1959).
329. Deb, S. K. and A. D. Yoffe, *Proc. Roy. Soc.* **254A**, 528 (1960).
330. Décamps, E. and A. Hadni, *J. Chim. Phys.* **65**, 1030 (1968).
331. Decius, J. C., *J. Chem. Phys.* **16**, 1025 (1948).
332. Decius, J. C., *J. Chem. Phys.* **17**, 1315 (1949).
333. Decius, J. C., *J. Chem. Phys.* **22**, 1941 (1954).
334. Decius, J. C., *J. Chem. Phys.* **22**, 1946 (1954).
335. Decius, J. C., *J. Chem. Phys.* **23**, 1290 (1955).
336. Decius, J. C., *J. Chem. Phys.* **38**, 241 (1963).
337. Decius, J. C., *Spectrochim. Acta* **21**, 15 (1965).
338. Decius, J. C., E. H. Coker and G. L. Brenna, *Spectrochim. Acta* **19**, 1281 (1963).
339. Decius, J. C., O. G. Malan and H. W. Thompson, *Proc. Roy. Soc.* **275A**, 295 (1963).
340. Dehand, J., J. E. Guerchais and R. Rohmer, *Bull. Soc. chim. France* 346 (1966).
341. Dehnicke, K. and J. Weidlein, *Z. anorg. allgem. chem.* **323**, 267 (1963).
342. Dehnicke, K. and J. Weidlein, *Chem. Ber.* **98**, 1087 (1965).
343. Dehnicke, K. and J. Weidlein, *Angew. Chem.* (Int. Eng. Ed.) **5**, 1041 (1966).
344. Deiters, R. M. and R. R. Holmes, *J. Chem. Phys.* **48**, 4796 (1968).
345. Delhaye, M., N. Durrieu-Mercier and M. Migeon, *Compt. rend.* **267B**, 135 (1968).
346. Dellepiane, G. and G. Zerbi, *J. Mol. Spec.* **24**, 62 (1967).
347. De Loach, B. C. and W. H. Shaffer, *J. Mol. Spec.* **6**, 229 (1961).
348. Delwaulle, M. L., *J. Phys. Chem.* **56**, 355 (1952).
349. Delwaulle, M. L., *Compt. rend.* **238**, 84 (1954).
350. Delwaulle, M. L., *Compt. rend.* **238**, 2522 (1954).
351. Delwaulle, M. L., *Bull. Soc. chim. France* 1294 (1955).
352. Delwaulle, M. L., *Compt. rend.* **240**, 2132 (1955).

353. Delwaulle, M. L., M. B. Buisset and M. Delhaye, *J. Amer. Chem. Soc.* **74**, 5768 (1952).
354. Delwaulle, M. L. and F. Francois, *Compt. rend.* **220**, 173 (1945).
355. Delwaulle, M. L. and F. Francois, *Compt. rend.* **220**, 817 (1945).
356. Delwaulle, M. L. and F. Francois, *Bull. Soc. chim. France* **13**, 205 (1946).
356a. Delwaulle, M. L. and F. Francois, *J. Phys. Radium* **7**, 15 (1946).
356b. Delwaulle, M. L. and F. Francois, *J. Phys. Radium* **7**, 53 (1946).
357. Delwaulle, M. L. and F. Francois, *Compt. rend.* **226**, 896 (1948).
358. Delwaulle, M. L. and F. Francois, *J. Chim. Phys.* **45**, 50 (1948).
359. Delwaulle, M. L. and F. Francois, *J. Chim. Phys.* **46**, 87 (1949).
360. Delwaulle, M. L. and F. Francois, *Compt. rend.* **228**, 1585 (1949).
361. Delwaulle, M. L., F. Francois, M. B. Delhaye-Buisset and M. Delhaye, *J. Phys. Radium* **15**, 206 (1954).
362. Delwaulle, M. L. and G. Schilling, *Compt. rend* **244**, 70 (1957).
363. Detoni, S. and D. Hadzi, *J. Chim. Phys.* **53**, 760 (1956).
364. Devlin, J. P., *J. Chem. Phys.* **41**, 2951 (1964).
365. Devlin, J. P. and I. C. Hisatsune, *Spectrochim. Acta* **17**, 206 (1961).
366. Devlin, J. P. and I. C. Hisatsune, *Spectrochim. Acta* **17**, 218 (1961).
367. Diallo, A. O. *Can. J. Chem.* **46**, 2641 (1968).
368. Djega-Mariadassou, G., R. Kircher, R. Diament, J. Breiss, F. Gans and G. Pannetier, *Bull. Soc. chim. France* 2726 (1968).
369. Dodd, R. E., *Trans. Far. Soc.* **55**, 1480 (1959).
370. Dodd, R. E., J. A. Rolfe and L. A. Woodward, *Trans. Far. Soc.* **52**, 145 (1956).
370a. Dodd, R. E., L. A. Woodward and H. L. Roberts, *Trans. Far. Soc.* **52**, 1052 (1956).
371. Dodd, R. E., L. A. Woodward and H. L. Roberts, *Trans. Far. Soc.* **53**, 1545 (1957).
372. Donaldson, J. D., J. F. Knifton, J. O'Donoghue and S. D. Ross, *Spectrochim. Acta* **22**, 1173 (1966).
373. Donaldson, J. D., J. F. Knifton and S. D. Ross, *Spectrochim. Acta* **20**, 847 (1964).
374. Donaldson, J. D., J. F. Knifton and S. D. Ross, *Spectrochim. Acta* **21**, 275 (1965).
375. Donaldson, J. D., J. F. Knifton and S. D. Ross, *Spectrochim. Acta* **21**, 1043 (1965).
376. Donaldson, J. D., R. Oteng and B. J. Senior, *Chem. Comm.* 618 (1965).
377. Donaldson, J. D., S. D. Ross and B. J. Senior, *Spectrochim. Acta* **24A**, 1899 (1968).
378. D'Or, L. and J. Fuger, *Bull. Soc. roy. Sci. Liège* **25**, 14 (1956).
379. D'Or, L. and P. Tarte, *J. Chem. Phys.* **19**, 1064 (1951).
379a. D'Or, L. and P. Tarte, *Bull. Soc. roy. Sci. Liège* **22**, 276 (1953).
380. Dove, M. F. A., J. A. Creighton and L. A. Woodward, *Spectrochim. Acta* **18**, 267 (1962).
381. Dows, D. A., *J. Chem. Phys.* **31**, 1637 (1959).
382. Dows, D. A. and G. Bottger, *J. Chem. Phys.* **34**, 689 (1961).
383. Dows, D. A. and R. M. Hexter, *J. Chem. Phys.* **24**, 1029 (1956).
384. Dows, D. A., E. Whittle and G. C. Pimentel, *J. Chem. Phys.* **23**, 1475 (1955).
385. Dows, D. A., E. Whittle and G. C. Pimentel, *J. Chem. Phys.* **23**, 1606 (1955).

386. DOYLE, W. P. and P. EDDY, *Spectrochim. Acta* **23A**, 1903 (1967).
387. DOYLE, W. P. and I. KIRKPATRICK, *Spectrochim. Acta* **24A**, 1495 (1968).
388. DRESKA, N., K. N. RAO and L. H. JONES, *J. Mol. Spec.* **18**, 404 (1965).
389. DREYFUS, R. W. and P. W. LEVY, *Proc. Roy. Soc.* **246A**, 233 (1958).
390. DROUIN, M. J.-P. MATHIEU and H. POULET, *J. Chim. Phys.* **58**, 322 (1961).
391. DUDLEY, F. B., G. H. CADY and D. F. EGGERS, *J. Amer. Chem. Soc.* **78**, 1553 (1956).
392. DUNCAN, J. L., *J. Mol. Spec.* **13**, 338 (1964).
393. DUNCAN, J. L., *Spectrochim. Acta* **20**, 1807 (1964).
394. DUNCAN, J. L. and I. M. MILLS, *Spectrochim. Acta* **20**, 523 (1964).
395. DUNCAN, J. L. and I. M. MILLS, *Spectrochim. Acta* **20**, 1807 (1964).
396. DUNCAN, J. L., *J. Mol. Spec.* **18**, 62 (1965).
397. DUNLAP, J. L. and E. A. JONES, *Spectroscopia Mol.* **9**, 32 (1960).
398. DUPUIS, T., *Compt. rend.* **246**, 3332 (1958).
399. DUPUIS, T., *Compt. rend.* **250**, 1237 (1960).
400. DUPUIS, T., C. DUVAL and J. LECOMTE, *Compt. rend* **257**, 3080 (1963).
401. DUPUIS, T. and V. LORENZELLI, *Compt. rend* **259**, 4585 (1964).
402. DUPUIS, T. and V. LORENZELLI, *Compt. rend* **262B**, 489 (1966).
403. DUPUIS, T. and V. LORENZELLI, *Compt. rend.* **264B**, 1019 (1967).
404. DUPUIS, T. and V. LORENZELLI, *Ann. Chim.* (Rome) **57**, 391 (1967).
405. DUPUIS, T. and C. ROCCHICIOLLI, *Compt. rend.* **266B**, 397 (1968).
406. DUPUIS, T. and M. VILTANGE, *Compt. rend* **255**, 2582 (1962).
407. DUPUIS, T. and M. VILTANGE, *Mikrochim. Ichnoanal. Acta* 232 (1963).
408. DURIG, J. R., D. J. ANTION and F. G. BAGLIN, *J. Chem. Phys.* **49**, 666 (1968).
409. DURIG, J. R., S. F. BUSH and E. C. MERCER, *J. Chem. Phys.* **44**, 4235 (1966).
410. DURIG, J. R. and J. W. CLARK, *J. Chem. Phys.* **46**, 3057 (1967).
411. DURIG, J. R. and J. W. CLARK, *J. Chem. Phys.* **48**, 3216 (1968).
412. DURIG, J. R., W. H. GREEN and A. L. MARSTON, *J. Mol. Structure* **2**, 19 (1968).
413. DURIG, J. R. and R. C. LORD, *Spectrochim. Acta* **19**, 1877 (1963).
414. DUVAL, C. and J. LECOMTE, *Bull. Soc. Chim. France* **11**, 376 (1944).
415. DUVAL, C. and J. LECOMTE, *Compt. rend* **227**, 1153 (1948).
416. DUVAL, C. and J. LECOMTE, *Bull. Soc. chim. France* 101 (1952).
417. DUVAL, C. and J. LECOMTE, *Compt. rend* **239**, 249 (1954).
418. DUVAL, C. and J. LECOMTE, *J. Opt. Soc. Amer.* **44**, 261 (1954).
419. DUVAL, C. and J. LECOMTE, *Compt. rend* **240**, 66 (1955).
420. DUVAL, C. and J. LECOMTE, *Z. Elektrochem.* **64**, 582 (1960).
421. DUVAL, C. and J. LECOMTE, *Compt. rend* **262B**, 1000 (1966).
421a. DUVAL, C., J. LECOMTE and J. MORANDAT, *Bull. Soc. Chim. France* 745 (1951).
422. EAGELS, D. M., *J. Phys. Chem. Solids* **25**, 1243 (1964).
423. EARLEY, J. E., D. FORTNUM, A. WOJCIKI and J. O. EDWARDS, *J. Amer. Chem. Soc.* **81**, 1295 (1959).
424. EBERHARDT, W. H. and T. G. BURKE, *J. Chem. Phys.* **20**, 529 (1952).
425. EDELSON, D., *J. Amer. Chem. Soc.* **74**, 262 (1952).
426. EDELSON, D. and K. B. MCAFEE, *J. Chem. Phys.* **19**, 1311 (1951).
427. EDWARDS, A. J., M. A. MOUTY, R. D. PEACOCK and A. J. SUDDEN, *J. Chem. Soc.* 4087 (1964).
428. EDWARDS, D. A. and R. N. HAYWARD, *Can. J. Chem.* **46**, 3443 (1968).
429. EDWARDS, H. G. M., M. J. WARE and L. A. WOODWARD, *Chem. Comm.* 540 (1968).

430. EDWARDS, J. O., G. C. MORRISON, V. F. ROSS and J. W. SCHULTZ, *J. Amer. Chem. Soc.* **77**, 266 (1955).
431. EICHHOFF, H. J. and F. WEIGEL, *Z. anorg. allgem. chem.* **275**, 267 (1954).
432. ELCOMBE, M. M., *Proc. Phys. Soc.* **91**, 947 (1967).
433. ELLIOTT, H. and B. J. HATHAWAY, *Spectrochim. Acta* **21**, 1047 (1965).
434. EMERY, A. R. and R. C. TAYLOR, *J. Chem. Phys.* **28**, 1029 (1958).
435. EMERY, A. R. and R. C. TAYLOR, *Spectrochim. Acta* **16**, 1455 (1960).
436. ESKENAZI, R., J. RASKOVAN and R. LEVITUS, *J. Inorg. Nucl. Chem.* **28**, 521 (1966).
437. ETTINGER, R., *J. Chem. Phys.* **38**, 2427 (1963).
438. EUCKEN, A. and F. SAUTER, *Z. Physik. Chem.* (Leipzig) **26**, 463 (1934).
439. EVANS, J. A. and D. A. LONG, *J. Chem. Soc.* **A**. 1688 (1968).
440. EVANS, J. C., *J. Mol. Spec.* **4**, 435 (1960).
441. EVANS, J. C. and H. BERNSTEIN, *Can. J. Chem.* **33**, 1270 (1955).
442. EVANS, J. C. and G. Y-S. LO, *J. Chem. Phys.* **44**, 3638 (1966).
443. EVANS, J. C. and G. Y-S. LO, *J. Chem. Phys.* **45**, 1069 (1966).
443a. EVANS, J. C. and G. Y-S. LO, *J. Phys. Chem.* **70**, 11 (1966).
444. EVANS, J. C. and G. Y-S. LO, *J. Phys. Chem.* **71**, 3942 (1967).
445. EVANS, J. C. and G. Y-S. LO, *J. Mol. Spec.* **26**, 147 (1968).
446. EVSTAF'EVA, O. N., N. N. CHALISOVA and S. I. GINZBURG, *Zhur. Neorg. Khim.* **11**, 2587 (1966).
447. EVSTAF'EVA, O. N., A. K. MOLODKIN, G. G. DVORYANTSOVA, O. M. IVANOVA and M. I. STRUCHKOVA, *Zhur. Neorg. Khim.* **11**, 1306 (1966).
448. FADINI, A., *Z. Naturforsch* **21A**, 484 (1966).
449. FADINI, A., B. KREBS and A. MUELLER, *Z. Naturforsch.* **20A**, 1124 (1965).
450. FADINI, A. and A. MUELLER, *Mol. Phys.* **12**, 145 (1967).
451. FADINI, A., A. MUELLER and B. KREBS, *Z. Naturforsch.* **20A**, 1241 (1965).
452. FALK, M. and P. A. GIGUERE, *Can. J. Chem.* **35**, 1680 (1958).
453. FATELEY, W. G., H. A. BENT and B. CRAWFORD, jr., *J. Chem. Phys.* **31**, 204 (1959).
454. FATELEY, W. G. and F. A. MILLER, *Spectrochim. Acta* **17**, 857 (1961).
455. FATELEY, W. G. and F. A. MILLER, *Spectrochim. Acta* **18**, 977 (1962).
456. FATELEY, W. G. and F. A. MILLER, *Spectrochim. Acta* **19**, 611 (1963).
457. FATELEY, W. G., R. K. HARRIS, F. A. MILLER and R. E. WITKOWSKI, *Spectrochim. Acta* **21**, 231 (1965).
457a. FAWCETT, F. S. and R. D. LIPSCOMB, *J. Amer. Chem. Soc.* **86**, 2576 (1964).
458. FEHER, F. and M. BAUDLER, *Z. Elektrochem.* **47**, 844 (1941).
459. FERGUSON, J. E., C. J. WILKINS and J. F. YOUNG, *J. Chem. Soc.* 2136 (1962).
460. FERIGLE, S. M. and A. G. MEISTER, *Am. J. Phys.* **20**, 421 (1952).
461. FERRARO, J. R., *J. Mol. Spec.* **4**, 99 (1960).
462. FERRARO, J. R. and A. WALKER, *J. Chem. Phys.* **42**, 1273 (1965).
463. FERRARO, J. R. and A. WALKER, *J. Chem. Phys.* **42**, 1278 (1965).
464. FERRARO, J. R. and A. WALKER, *J. Chem. Phys.* **45**, 550 (1966).
465. FERRARO, J. R., A. WALKER and C. CRISTALLINI, *Inorg. Nucl. Chem. Letters* **1**, 25 (1965).
466. FERRISO, C. C. and D. F. HORNIG, *J. Am. Chem. Soc.* **75**, 4113 (1953).
467. FERRISO, C. C. and D. F. HORNIG, *J. Chem. Phys.* **23**, 1464 (1955).
468. FERTEL, J. H. and C. PERRY, *J. Phys. Chem. Solids* **26**, 1773 (1965).
469. FIELD, B. O. and C. J. HARDY, *Proc. Chem. Soc.* 76 (1962).
470. FIELD, B. O. and C. J. HARDY, *Proc. Chem. Soc.* 11 (1963).
471. FIELD, B. O. and C. J. HARDY, *J. Chem. Soc.* 5278 (1963).

472. FIELD, B. O. and C. J. HARDY, *J. Chem. Soc.* 4428 (1964).
473. FINCH, A., I. J. HYAMS and D. STEELE, *Spectrochim. Acta* **21**, 1423 (1965).
474. FINCH, A., I. J. HYAMS and D. STEELE, *Trans. Far. Soc.* **61**, 398 (1965).
475. FINHOLT, J. E., R. W. ANDERSON, J. A. FYFE and K. C. CAULTON, *Inorg. Chem.* **4**, 43 (1965).
475a. FONTEYNE, R., *Natuurw. Tijdschr.* (Ghent) **25**, 173 (1943).
476. FORD, T. A. and W. J. ORVILLE-THOMAS, *Spectrochim. Acta* **23A**, 579 (1967).
477. FORDYCE, J. S. and R. L. BAUM, *J. Chem. Phys.* **43**, 843 (1965).
478. FORSTER, D. and D. M. L. GOODGAME, *J. Chem. Soc.* 262 (1965).
479. FORSTER, D. and D. M. L. GOODGAME, *J. Chem. Soc.* 1286 (1965).
480. FORTNUM, D. and J. O. EDWARDS, *J. Inorg. Nucl. Chem.* **2**, 264 (1956).
481. FOURNIER, R. P., R. SAVOIE, F. BESSETTE and A. CABANA, *J. Chem. Phys.* **49**, 1159 (1968).
482. FOX, J. J. and A. E. MARTIN, *Proc. Roy. Soc.* **174A**, 234 (1940).
483. FRANKISS, S. G., *J. Mol. Structure* **2**, 271 (1968).
484. FRANKISS, S. G. and F. A. MILLER, *Spectrochim. Acta* **21**, 1235 (1965).
485. FRANKISS, S. G., F. A. MILLER, H. STAMMREICH and TH. TEXEIRA SANS, *Chem. Comm.* 318 (1966).
486. FRANKISS, S. G., F. A. MILLER, H. STAMMREICH and TH. TEXEIRA SANS, *Spectrochim. Acta* **23A**, 543 (1967).
487. FREEMAN, D. E., *J. Mol. Spec.* **20**, 463 (1966).
488. FREEMAN, D. E., K. H. RHEE and M. K. WILSON, *J. Chem. Phys.* **39**, 2908 (1963).
489. FREIBERG, M., A. RON and O. SCHNEPP, *J. Phys. Chem.* **72**, 3526 (1968).
489a. FREITAG, W. O. and E. R. NIXON, *J. Chem. Phys.* **24**, 109 (1956).
490. FREUND, I. and R. S. HALFORD, *J. Chem. Phys.* **43**, 3795 (1965).
491. FREYMANN, M., R. FREYMANN, R. ROHMER and P. HAMON, *Compt. rend.* **230**, 1524 (1950).
492. FRITZ, H. P. and H. KELLER, *Chem. Ber.* **94**, 1524 (1961).
493. FRLEC, B. and H. H. CLAASSEN, *J. Chem. Phys.* **46**, 4603 (1967).
494. FUJITA, J., A. E. MARTELL and K. NAKAMOTO, *J. Chem. Phys.* **36**, 324 (1962).
495. FUJITA, J., A. E. MARTELL and K. NAKAMOTO, *J. Chem. Phys.* **36**, 339 (1962).
496. FUJITA, J., K. NAKAMOTO and M. KOBAYASHI, *J. Amer. Chem. Soc.* **78**, 3293 (1956).
497. FUNCK, E., *Ber. Bunsenges. Phys. Chem.* **68**, 617 (1964).
498. FUNCK, E., *Ber. Bunsenges. Phys. Chem.* **71**, 170 (1967).
499. GALLEZOT, P., M. V. MATHIEU, D. WEIGEL and B. IMELIK, *Bull. soc. chim. France* 882 (1968).
500. GAMO, I., *Compt. rend.* **239**, 1478 (1954).
501. GAMO. I., *Bull. Chem. Soc. Japan* **34**, 760 (1961).
502. GAMO, I., *Bull. Chem. Soc. Japan* **34**, 764 (1961).
503. GAMO, I., *Bull. Chem. Soc. Japan* **34**, 1420 (1961).
504. GAMO, I., *Bull. Chem. Soc. Japan* **34**, 1433 (1961).
505. GAMO, I., *Bull. Chem. Soc. Japan* **35**, 1055 (1962).
506. GAMO, I., *Bull. Chem. Soc. Japan* **35**, 1058 (1962).
507. GATEHOUSE, B. M. and A. E. COMYNS, *J. Chem. Soc.* 3965 (1958).
507a. GATEHOUSE, B. M., S. E. LIVINGSTONE and R. S. NYHOLM, *J. Chem. Soc.* 4222 (1957).
508. GATEHOUSE, B. M., S. E. LIVINGSTONE and R. S. NYHOLM, *J. Chem. Soc.* 3137 (1958).

509. GATEHOUSE, B. M., S. E. LIVINGSTONE and R. S. NYHOLM, *J. Inorg. Nucl. Chem.* **8**, 75 (1959).
510. GATTI, R., R. L. KREIGER, J. E. SICRE and H. J. SCHUMACHER, *J. Inorg. Nucl. Chem.* **28**, 655 (1966).
511. GATTOW, G., A. FRANKE and A. MUELLER, *Naturwiss.* **52**, 428 (1965).
512. GAUNT, J., *Trans. Far. Soc.* **49**, 1122 (1953).
513. GAUNT, J. *Trans. Far. Soc.* **50**, 209 (1954).
514. GAUNT, J. and J. B. AINSCOUGH, *Spectrochim. Acta* **10**, 52 (1957).
515. GAUNT, J. and J. B. AINSCOUGH, *Spectrochim. Acta* **10**, 57 (1957).
516. GAYLES, J. N. and J. SELF, *J. Chem. Phys.* **40**, 3530 (1964).
517. GEICK, R., C. H. PERRY and G. RUPPRECHT, *Phys. Rev.* **146**, 543 (1966).
518. GERBAUX, X. and A. HADNI, *J. Phys. Radium* **23**, 877 (1962).
519. GERDING, H., H. GISBEN, B. NIEWENKUISSE and J. G. v. RAAPHORST, *Rec. trav. chim.* **79**, 41 (1960).
520. GERDING, H. and H. HOUTGRAAF, *Rec. trav. chim.* **73**, 737 (1954).
521. GERDING, H. and H. HOUTGRAAF, *Rec. trav. chim.* **73**, 759 (1954).
522. GERDING, H. and H. HOUTGRAAF, *Rec. trav. chim.* **75**, 589 (1956).
523. GERDING, H. and H. HOUTGRAAF, *Rec. trav. chim.* **74**, 5 (1955).
524. GERDING, H. and J. A. KONIGSTEIN, *Rec. trav. chim.* **79**, 46 (1960).
525. GERDING, H. and W. J. NIJVELD, *Rec. trav. chim.* **59**, 1209 (1940).
526. GERDING, H. and P. C. NOBEL, *Rec. trav. chim.* **77**, 472 (1958).
527. GERDING, H. and E. SMIT, *Z. Phys. Chem.* **B50**, 171 (1941).
528. GESMUNDO, F. and V. LORENZELLI, *Atti. Accad. Naz. Lincei, Rend. Classe, Sci. fis. mat. nat.* **38**, 88 (1965).
529. GIGUERE, P. A., *J. Chem. Phys.* **18**, 88 (1950).
530. GIGUERE, P. A. and J. P. ARRADEAU, *Compt. rend.* **257**, 1692 (1963).
531. GIGUERE, P. A. and O. BAIN, *J. Phys. Chem.* **56**, 340 (1952).
532. GIGUERE, P. A. and C. CHAPADOS, *Spectrochim. Acta* **22**, 1131 (1966).
532a. GIGUERE, P. A. and K. B. HARVEY, *J. Mol. Spec.* **3**, 36 (1959).
533. GIGUERE, P. A. and I. D. LIU, *J. Chem. Phys.* **20**, 136 (1952).
534. GIGUERE, P. A. and R. SAVOIE, *Can. J. Chem.* **38**, 2467 (1960).
534a. GILLARD, R. D. and G. WILKINSON, *J. Chem. Soc.* 1640 (1964).
535. GILLESPIE, R. J. and H. J. CLASE, *J. Chem. Phys.* **47**, 1071 (1967).
536. GILLESPIE, R. J. and E. A. ROBINSON, *Can. J. Chem.* **39**, 2171 (1961).
537. GILLESPIE, R. J. and E. A. ROBINSON, *Can. J. Chem.* **39**, 2189 (1961).
538. GILLESPIE, R. J. and E. A. ROBINSON, *Spectrochim. Acta* **18**, 1473 (1962).
539. GILLESPIE, R. J. and E. A. ROBINSON, *Can. J. Chem.* **40**, 644 (1962).
540. GILLESPIE, R. J. and E. A. ROBINSON, *Spectrochim. Acta* **19**, 741 (1963).
541. GILLESPIE, R. J. and E. A. ROBINSON, *Can. J. Chem.* **42**, 2496 (1964).
542. GIULOTTO, L. *J. Chem. Phys.* **19**, 1316 (1957).
543. GIULOTTO, L. and G. OLIVELLI, *J. Chem. Phys.* **16**, 556 (1948).
544. GIULOTTO, L. and G. OLIVELLI, *J. Chem. Phys.* **16**, 668 (1948).
545. GLEMSER, O., H. W. ROESKY, K. H. HELLBERG and H. V. WERTHER, *Chem. Ber.* **99**, 2652 (1966).
546. GOGGIN, P. L., H. L. ROBERTS and L. A. WOODWARD, *Trans. Far. Soc.* **57**, 1877 (1961).
547. GOLDSMITH, J. A., A. HEZEL and S. D. ROSS, *Spectrochim. Acta* **24A**, 1139 (1968).
548. GOLDSMITH, J. A. and S. D. ROSS, *Spectrochim. Acta* **22**, 1069 (1966).
549. GOLDSMITH, J. A. and S. D. ROSS, *Spectrochim. Acta* **23A**, 1909 (1967).
550. GOLDSMITH, J. A. and S. D. ROSS, *Spectrochim. Acta* **24A**, 993 (1968).

551. GOLDSTEIN, M., *Spectrochim. Acta* **22**, 1389 (1966).
551a. GOMPERTZ, G. and W. J. ORVILLE-THOMAS, *J. Phys. Chem.* **63**, 1331 (1959).
552. GOUBEAU, J., R. BAUMGAERTNER and H. WEISS, *Z. anorg. allgem. chem.* **348**, 286 (1966).
553. GOUBEAU, J. and W. BUES, *Z. anorg. allgem. chem.* **268**, 221 (1952).
554. GOUBEAU, J. and H. KALLFASS, *Z. anorg. allgem. chem.* **299**, 160 (1959).
555. GOUBEAU, J. and K. LAITENBERGER, *Z. anorg. allgem. chem.* **320**, 78 (1963).
556. GOUBEAU, J., D. E. RICHTER and H. J. BECHER, *Z. anorg. allgem. chem.* **278**, 12 (1955).
557. GOULDEN, J. D. S., A. MACCOLL and D. J. MILLEN, *J. Chem. Soc.* 1635 (1950).
558. GOULDEN, J. D. S. and D. J. MILLEN, *J. Chem. Soc.* 2620 (1950).
559. GRANDJEAN, D. and A. HADNI, *J. Chim. Phys.* **62**, 981 (1965).
560. GRAY, P. and T. C. WADDINGTON, *Trans. Far. Soc.* **53**, 901 (1957).
561. GRAY, P. and A. D. YOFFE, *Chem. Revs.* **55**, 1069 (1955).
562. GREENBERG, J. and L. J. HALLGREN, *J. Chem. Phys.* **33**, 900 (1960).
563. GREENWOOD, N. N., *J. Chem. Soc.* 3811 (1954).
564. GREENWOOD, N. N., N. R. LITTLE and M. J. SPRAGUE, *J. Chem. Soc.* 1292 (1964).
565. GREENWOOD, N. N., D. J. PRINCE and B. P. STRAUGHAN, *J. Chem. Soc.* **A**. 1694 (1968).
566. GREENWOOD, N. N. and B. P. STRAUGHAN, *J. Chem. Soc.* **A**. 962 (1966).
567. GREENWOOD, N. N., B. P. STRAUGHAN and A. E. WILSON, *J. Chem. Soc.* **A**. 1479 (1966).
568. GRIFFITH, W. P., *J. Chem. Soc.* 5248 (1964).
569. GRIFFITH, W. P., *J. Chem. Soc.* **A**. 899 (1966).
570. GRIFFITH, W. P., *J. Chem. Soc.* **A**. 1467 (1966).
571. GRIFFITH, W. P., *J. Chem. Soc.* **A**. 905 (1967).
572. GRIFFITH, W. P., *J. Chem. Soc.* **A**. 1663 (1968).
573. GRIFFITH, W. P. and K. J. RUTT, *J. Chem. Soc.* **A**. 2331 (1968).
574. GRIFFITHS, J. E., *J. Chem. Phys.* **42**, 2632 (1965).
575. GRIFFITHS, J. E., *Can. J. Chem.* **45**, 2639 (1967).
576. GRIFFITHS, J. E., *Spectrochim. Acta* **23A**, 2145 (1967).
577. GRIFFITHS, J. E., *J. Chem. Phys.* **49**, 642 (1968).
578. GRIFFITHS, J. E., R. P. CARTER, jr. and R. R. HOLMES, *J. Chem. Phys.* **41**, 863 (1964).
579. GRIFFITHS, J. E. and D. E. IRISH, *Inorg. Chem.* **3**, 1134 (1964).
580. GRIFFITHS, J. E., T. N. SRIVASTAVA and M. ONYSZCHUK, *Can. J. Chem.* **40**, 579 (1962).
581. GRIGOR'EV, A. I., *Russ. J. Inorg. Chem.* **8**, 409 (1963).
582. GRIGOR'EV, A. I. and V. N. MAKSIMOV, *Zhur. neorg. khim.* **9**, 1060 (1964).
583. GRIGOR'EV, A. I., IU-V. ORLOVA, V. A. SPIACHEV and A. V. NOVOSELOVA, *Doklady. Acad. Nauk SSSR.* **152**, 134 (1963).
584. GRIMM, F. A., L. BARTON and R. F. PORTER, *Inorg. Chem.* **7**, 1309 (1968).
585. GRUBB, E. L. and R. L. BELFORD, *J. Chem. Phys.* **39**, 244 (1963).
586. GUERCHAIS, J. E., *Compt. rend.* **259**, 2850 (1964).
587. GUERCHAIS, J. E., A. HAUMESSER and R. ROHMER, *Compt. rend.* **260**, 5571 (1965).
588. GUERCHAIS, J. E. and M. J. LEROY, *Compt. rend.* **260**, 6915 (1965).
589. GUERCHAIS, J. E., M. J. LEROY and R. ROHMER, *Compt. rend.* **261**, 3628 (1965).

590. GULLIKSON, C. W., J. R. NIELSEN and A. T. STAIR, *J. Mol. Spec.* **1**, 151 (1957).
591. GUSSONI, M. and G. ZERBI, *Atti. Accad. Naz. Lincei, Rend. Cl. Sci. Fis. Mat. Nat.* **40**, 1032 (1966).
592. GUSSONI, M. and G. ZERBI, *Chem. Phys. Letters* **2**, 145 (1968).
593. GUSSONI, M. and G. ZERBI, *J. Mol. Spec.* **26**, 485 (1968).
594. GUTOWSKY, H. S., J. *Chem. Phys.* **17**, 128 (1949).
595. GUTOWSKY, H. J. and A. D. LIEHR, *J. Chem. Phys.* **20**, 1652 (1952).
596. GUTOWSKY, H. J. and E. O. STEJKSAL, *J. Chem. Phys.* **22**, 939 (1954).
597. HAARHOFF, P. C. and C. W. F. T. PISTORIUS, *Z. Naturforsch.* **14A**, 972 (1959).
598. HAAS, C. and D. F. HORNIG, *J. Chem. Phys.* **32**, 1763 (1960).
599. HADNI, A., E. DÉCAMPS and J.-P. HERBEUVAL, *J. Chim. Phys.* **65**, 959 (1968).
600. HADNI, A., C. HENRY, J.-P. MATHIEU and H. POULET, *Compt. rend.* **252**, 1585 (1961).
601. HADNI, A. and R. L. MICHEL, *Compt. rend.* **244**, 2602 (1957).
602. HAENDLER, H. M., S. W. BUKATA, B. MILLARD, E. I. GOODMAN and J. LITTMAN, *J. Chem. Phys.* **22**, 1939 (1954).
603. HAFELE, H. G., *Z. Naturforsch.* **18A**, 331 (1963).
604. HAFNER, S. and F. LAVES, *Z. Krist.* **115**, 321 (1961).
605. HAGENMULLER, P. and B. TANGUY, *Compt. rend.* **260**, 3974 (1965).
606. HALFORD, R. S., *J. Chem. Phys.* **14**, 8 (1946).
606a. HALL, R. T. and G. C. PIMENTEL, *J. Chem. Phys.* **38**, 1889 (1963).
607. HAMMAKER, R. M. and J. P. WALTERS, *Spectrochim. Acta* **20**, 1311 (1964).
608. HANWICK, T. J. and P. O. HOFFMANN, *J. Chem. Phys.* **17**, 1166 (1949).
609. HANWICK, T. J. and P. O. HOFFMANN, *J. Chem. Phys.* **19**, 708 (1951).
610. HARADA, J. and G. HONJO, *J. Phys. Soc. Japan* **22**, 45 (1967).
611. HARDWICK, J. L. and G. E. LEROI, *Inorg. Chem.* **7**, 1683 (1968).
612. HARMELIN, M., *Compt. rend.* **262B**, 620 (1966).
613. HARMONY, M. D. and R. J. MYERS, *J. Chem. Phys.* **37**, 636 (1962).
614. HART, P. B. and S. E. F. SMALLWOOD, *J. Inorg. Nucl. Chem.* **24**, 1047 (1962).
615. HARTMAN, K. O. and F. A. MILLER, *Spectrochim. Acta* **24A**, 669 (1968).
616. HARVEY, K. B., B. A. MORROW and H. F. SHURVELL, *Can. J. Chem.* **41**, 1181 (1963).
617. HASS, M., *J. Phys. and Chem. Solids* **24**, 1159 (1963).
618. HASS, M. and G. B. B. M. SUTHERLAND, *Proc. Roy. Soc.* **236A**, 427 (1956).
619. HATHAWAY, B. J., D. G. HOLAH and M. HUDSON, *J. Chem. Soc.* 4586 (1963).
620. HATHAWAY, B. J. and A. E. UNDERHILL, *J. Chem. Soc.* 3709 (1960).
621. HATHAWAY, B. J. and D. E. WEBSTER, *Proc. Chem. Soc.* 14 (1963).
622. HAWKINS, N. J. and D. R. CARPENTER, *J. Chem. Phys.* **23**, 1700 (1955).
623. HAWKINS, N. J., H. C. MATTRAW and W. W. SABOL, *J. Chem. Phys.* **23**, 2191 (1955).
624. HAWKINS, N. J. and W. W. SABOL, *J. Chem. Phys.* **25**, 775 (1956).
625. HAYNIE, W. H. and H. H. NIELSEN, *J. Chem. Phys.* **21**, 1839 (1953).
626. HEATH, D. F. and J. W. LINNETT, *Nature* **161**, 314 (1948).
627. HEATH, D. F. and J. W. LINNETT, *Trans. Far. Soc.* **44**, 873 (1948).
628. HEATH, D. F. and J. W. LINNETT, *Trans. Far. Soc.* **44**, 878 (1948).
629. HEATH, D. F. and J. W. LINNETT, *Trans. Far. Soc.* **44**, 884 (1948).
630. HEDBERG, K., *J. Chem. Phys.* **19**, 509 (1951).
631. HEDBERG, K. and R. M. BADGER, *J. Chem. Phys.* **19**, 508 (1951).
632. HEICKLEN, J. and V. KNIGHT, *Spectrochim. Acta* **20**, 295 (1964).
633. HEINEMANN, A. *Naturwiss.* **48**, 568 (1961).

634. Heinemann, A., *Ber. Bunsenges* **68**, 280 (1964).
635. Heinemann, A., *Ber. Bunsenges* **68**, 287 (1964).
636. Hendra, P. J., *Nature* **212**, 179 (1966).
637. Hendra, P. J., *J. Chem. Soc.* **A**. 1298 (1967).
638. Hendra, P. J., *Spectrochim. Acta* **23A**, 2871 (1967).
639. Hendra, P. J., *Spectrochim. Acta* **24A**, 125 (1968).
640. Hendra, P. J. and Z. Jovic, *J. Chem. Soc.* **A**. 600 (1968).
641. Hendra, P. J. and J. R. Mackenzie, *Chem. Comm.* 760 (1968).
642. Hendra, P. J. and P. J. D. Park, *Spectrochim. Acta* **23A**, 1635 (1967).
643. Hendra, P. J. and P. J. D. Park, *J. Chem. Soc.* **A**. 908 (1968).
644. Herschbach, D. R., *J. Chem. Phys.* **27**, 975 (1957).
645. Herschbach, D. R., *J. Chem. Phys.* **31**, 91 (1959).
645a. Herzberg, G. and C. Reid, *Diss. Far. Soc.* **9**, 92 (1950).
646. Heslop, W. R. and J. W. Linnett, *Trans. Far. Soc.* **49**, 1262 (1953).
647. Hester, R. G. and W. E. L. Grossman, *Inorg. Chem.* **5**, 1308 (1966).
648. Hester, R. E. and R. A. Plane, *Inorg. Chem.* **3**, 513 (1964).
649. Hester, R. E., R. A. Plane and G. E. Walrafen, *J. Chem. Phys.* **38**, 249 (1963).
650. Hewkin, D. J. and W. P. Griffith, *J. Chem. Soc.* **A**. 472 (1966).
651. Hexter, R. M., *Spectrochim. Acta* **10**, 281 (1958).
652. Hexter, R. M., *Spectrochim. Acta* **10**, 291 (1958).
653. Hezel, A. and S. D. Ross, *Spectrochim. Acta* **22**, 1949 (1966).
654. Hezel, A. and S. D. Ross, *Spectrochim. Acta* **23A**, 1583 (1967).
655. Hezel, A. and S. D. Ross, *Spectrochim. Acta* **24A**, 131 (1968).
656. Hezel, A. and S. D. Ross, *Spectrochim. Acta* **24A**, 985 (1968).
657. Hildebrand, J. H., *J. Chem. Phys.* **15**, 727 (1947).
658. Hiraishi, J., I. Nakagawa and T. Shimanouchi, *Proc. Int. Symp. Mol. Struct. Spectry.* **A** 313, Tokyo (1962).
659. Hiraishi, J., I. Nakagawa and T. Shimanouchi, *Spectrochim. Acta* **20**, 819 (1964).
660. Hirota, E., *Bull. Chem. Soc. Japan* **31**, 130 (1958).
661. Hirschmann, R. P., L. R. Anderson, D. F. Harnish and W. B. Fox, *Spectrochim. Acta* **24A**, 1267 (1968).
662. Hisatsune, I. C. and J. P. Devlin, *Spectrochim. Acta* **16**, 401 (1960).
663. Hisatsune, I. C., J. P. Devlin and S. Califano, *Spectrochim. Acta* **16**, 450 (1960).
664. Hisatsune, I. C., J. P. Devlin and Y. Wada, *Spectrochim. Acta* **18**, 1641 (1962).
665. Hisatsune, I. and R. V. Fitzsimmons, *Spectrochim. Acta* **15**, 206 (1959).
666. Hisatsune, I. C. and N. H. Suarez, *Inorg. Chem.* **3**, 168 (1964).
667. Hobbs, W. E., *J. Chem. Phys.* **28**, 1220 (1958).
668. Hoekstra, H. R., *J. Inorg. Nucl. Chem.* **27**, 801 (1965).
669. Hoffman, P. and T. J. Hanwick, *J. Chem. Phys.* **16**, 1163 (1948).
670. Hollenberg, J. L., *J. Chem. Phys.* **46**, 3271 (1967).
671. Hollenberg, J. L. and D. A. Dows, *Spectrochim. Acta* **16**, 1155 (1960).
672. Holliday, A. K., M. E. Peach and T. C. Waddington, *Proc. Chem. Soc.* 220 (1961).
673. Holmes, R. R., *J. Chem. Phys.* **46**, 3718 (1967).
674. Holmes, R. R., *J. Chem. Phys.* **46**, 3724 (1967).
675. Holmes, R. R., *J. Chem. Phys.* **46**, 3730 (1967).
676. Hooge, F. N. and J. A. A. Ketelaar, *Rec. trav. chim.* **77**, 902 (1958).

676a. Hooper, M. A. and D. W. Jakes, *Austral. J. Chem.* **21**, 2379 (1968).
677. Horn, H. G., A. Mueller and O. Glemser, *Z. Naturforsch.* **20A**, 765 (1966).
678. Hornig, D. F., *J. Chem. Phys.* **16**, 1063 (1948).
679. Hornig, D. F. and R. C. Plumb, *J. Chem. Phys.* **26**, 637 (1957).
680. Hornig, D. F., H. F. White and F. P. Reding, *Spectrochim. Acta* **12**, 338 (1958).
681. Hoskins, L. C., *J. Chem. Phys.* **42**, 2631 (1965).
682. Hoskins, L. C. and R. C. Lord, *J. Chem. Phys.* **43**, 155 (1965).
683. Hougen, J. T., *Pure and Appl. Chem.* **11**, 481 (1965).
684. Hubin, R. and P. Tarte, *Spectrochim. Acta* **23A**, 1815 (1967).
685. Hunt, G. R., C. H. Perry and J. Ferguson, *Phys. Rev.* **134**, 688 (1964).
686. Hunt, G. R. and M. K. Wilson, *Spectrochim. Acta* **16**, 570 (1960).
687. Hunt, G. R. and M. K. Wilson, *Spectrochim. Acta* **18**, 959 (1962).
688. Hunt, R. H., R. A. Leacock, C. W. Peters and K. T. Hecht, *J. Chem. Phys.* **42**, 1931 (1965).
689. Hyman, H. H., *Science* **145**, 773 (1964).
690. Ikegami, S., *J. Phys. Soc. Japan* **19**, 46 (1964).
691. Ikrami, D. D., E. D. Ruchkin and N. S. Nikolaev, *Zh. Strukt. Khim.* **8**, 354 (1967).
692. Ito, K. and H. J. Bernstein, *Can. J. Chem.* **34**, 170 (1956).
693. Ivanov, V. I., M. B. Varfolomeev, K. I. Petrov, V. G. Pervykh and V. E. Plyuschev, *Sov. Phys.—Crystallography* **10**, 425 (1965).
694. Jacobson, J. L. and E. R. Nixon, *J. Phys. and Chem. Solids* **29**, 967 (1968).
694a. Jacox, M. E. and D. E. Milligan, *Spectrochim. Acta* **17**, 1196 (1961).
694b. Jacox, M. E. and D. E. Milligan, *J. Chem. Phys.* **40**, 2457 (1964).
695. Janz, G. J. and D. W. James, *J. Chem. Phys.* **38**, 902 (1963).
696. Janz, G. J. and D. W. James, *J. Chem. Phys.* **38**, 905 (1963).
697. Janz, G. J. and J. D. E. McIntyre, *J. Electrochem. Soc.* **109**, 845 (1962).
698. Janz, G. J. and Y. Mikawa, *J. Mol. Spec.* **5**, 92 (1960).
699. Jezowska-Trzebiatowska, B., J. Hanuza and W. Wojciechowski, *Spectrochim. Acta* **23A**, 2631 (1967).
700. Jolly, W. J., *Angew. Chem.* **72**, 268 (1960).
701. Jolly, W. J., *J. Amer. Chem. Soc.* **83**, 335 (1961).
702. Jones, E. A., J. S. Kirby-Smith, P. J. Woltz and A. H. Nielsen, *J. Chem. Phys.* **19**, 242 (1951).
703. Jones, E. A., J. S. Kirby-Smith, P. J. H. Woltz and A. H. Nielsen, *J. Chem. Phys.* **19**, 337 (1951).
704. Jones, E. A., T. F. Parkinson and R. B. Murray, *J. Chem. Phys.* **17**, 501 (1949).
705. Jones, G. O., D. H. Martin, P. A. Mawer and C. H. Perry, *Proc. Roy. Soc.* **261A**, 10 (1961).
706. Jones, L. H., *J. Chem. Phys.* **23**, 2105 (1955).
707. Jones, L. H., *J. Chem. Phys.* **25**, 1069 (1956).
708. Jones, L. H., *J. Chem. Phys.* **28**, 1234 (1958).
709. Jones, L. H., *Spectrochim. Acta* **15**, 409 (1959).
710. Jones, L. H., *J. Mol. Spec.* **5**, 133 (1960).
711. Jones, L. H., *J. Mol. Spec.* **8**, 105 (1962).
712. Jones, L. H., *Coord. Chem. Rev.* **1**, 351 (1966).
713. Jones, L. H., L. B. Asprey and R. R. Ryan, *J. Chem. Phys.* **47**, 3371 (1967).
713a. Jones, L. H., R. M. Badger and G. E. Moore, *J. Chem. Phys.* **19**, 1599 (1951).

714. JONES, L. H. and E. MCLAREN, *J. Chem. Phys.* **22**, 1796 (1954).
715. JONES, L. H. and R. A. PENNEMAN, *J. Chem. Phys.* **21**, 542 (1953).
716. JONES, L. H. and R. A. PENNEMAN, *J. Chem. Phys.* **22**, 781 (1954).
717. JONES, M. M., E. A. JONES, D. F. HARMON and R. T. SEMMES, *J. Amer. Chem. Soc.* **83**, 2038 (1961).
718. JONES, W. E. and T. W. SWADDLE, *Can. J. Chem.* **45**, 2647 (1967).
719. KAHOVEC, L. and K. W. F. KOHLRAUSCH, *Monatsh.* **77**, 180 (1947).
720. KAISER, W., W. G. SPITZER, R. H. KAISER and L. E. HOWARTH, *Phys. Rev.* 127, 1950 (1962).
721. KARO, A. M., *J. Chem. Phys.* **31**, 1489 (1959).
722. KARO, A. M. and J. R. HARDY, *Phys. Rev.* **129**, 2024 (1963).
723. KARPOV, S. V. and A. A. SHULTIN, *J. Phys. Chem. Solids* **29**, 475 (1968).
724. KASTLER, A., *Compt. rend.* **227**, 1024 (1948).
725. KASTLER, A., *Proc. Ind. Acad. Sci.* **28A**, 349 (1948).
726. KATAYAMA, M., T. SHIMANOUCHI, Y. MORINO and S. MIZUSHIMA, *J. Chem. Phys.* **18**, 506 (1950).
727. KATIYAR, R. S. and R. S. KRISHNAN, *Can. J. Phys.* **45**, 3079 (1967).
728. KATO, R. and J. ROLFE, *J. Chem. Phys.* **47**, 1901 (1967).
729. KATSAROS, N. and J. W. GEORGE, *Chem. Comm.* 613 (1965).
730. KATZ, B., A. RON and O. SCHNEPP, *J. Chem. Phys.* **47**, 5303 (1967).
730a. KATZIN, L., *J. Inorg. Nucl. Chem.* **24**, 245 (1962).
731. KAUFMAN, J. J., W. S. KOSKI and R. ANACREON, *J. Mol. Spec.* **11**, 1, (1963).
732. KEANE, A., *J. Proc. Roy. Soc., N. S. Wales* **89**, 173 (1955).
733. KECKI, Z., *Spectrochim. Acta* **18**, 1165 (1962).
734. KECKI, Z. and J. MANKOWSKI, *Roczniki Chem.* **36**, 375 (1962).
735. KEEN, N. and M. C. R. SYMONS, *Proc. Chem. Soc.* **383** (1960).
736. KELLER, O. L., jr., *Inorg. Chem.* **2**, 783 (1963).
737. KELLER, O. L., jr. and A. CHETHAM-STRODE, jr., *Inorg. Chem.* **5**, 367 (1966).
738. KELLER, W. E. and R. S. HALFORD, *J. Chem. Phys.* **17**, 26 (1949).
739. KELLERMANN, E. W., *Phil. Trans. Roy. Soc.* **238**, 513 (1940).
740. KELLERMANN, E. W., *Proc. Roy. Soc.* **178A**, 17 (1941).
741. KENNEDY, T. and D. S. PAYNE, *J. Chem. Soc.* 1229 (1959).
742. KERMARREC, Y., *Compt. rend.* **258**, 5836 (1964).
743. KESSLER, G. and H. A. LEHMANN, *Z. anorg. allgem. chem.* **339**, 87 (1965).
744. KETELAAR, J. A. A. and R. L. FULTON, *Z. Elektrochem.* **64**, 641 (1960).
745. KETELAAR, J. A. A., C. HAAS and J. VAN DER ELSKEN, *J. Chem. Phys.* **24**, 624 (1956).
746. KETELAAR, J. A. A., F. N. HOOGE and G. BLASSE, *Rec. trav. chim.* **75**, 220 (1956).
747. KETELAAR, J. A. A. and C. J. H. SCHUTTE, *Spectrochim. Acta* **17**, 1240 (1961).
748. KETELAAR, J. A. A. and W. VEDDER, *J. Chem. Phys.* **19**, 654 (1951).
749. KHANNA, R. K., *J. Sci. Ind. Res., India* **20B**, 483 (1961).
750. KHANNA, R. K., *J. Mol. Spec.* **8**, 134 (1962).
751. KHANNA, R. K., *J. Sci. Ind. Res., India* **21B**, 352 (1962).
752. KHANNA, R. K., J. LINGSCHEID and J. C. DECIUS, *Spectrochim. Acta* **20**, 1109 (1964).
753. KHANNA, R. K. and E. R. LIPPINCOTT, *Spectrochim. Acta* **24A**, 905 (1968).
754. KHARITONOV, YU. YA. and E. N. DEICHMANN, *Zhur. neorg. khim.* **10**, 853 (1965).
755. KHARITONOV, YU. YA. and N. A. KRYAZEVA, *Zhur. neorg. khim.* **11**, 2700 (1966).

756. KHARITONOV, YU. YA. and N. A. KRYAZEVA, *Zhur. neorg. khim.* **11**, 2713 (1966).
757. KHARITONOV, YU. YA. and N. A. KRYAZEVA, *Russ. J. Inorg. Chem.* **12**, 837 (1967).
758. KIM, H., P. A. SOUDER and H. H. CLAASSEN, *J. Mol. Spec.* **26**, 46 (1968).
759. KING, C. M. and E. R. NIXON, *J. Chem. Phys.* **48**, 1685 (1968).
760. KING, S. T. and J. OVEREND, *Spectrochim. Acta* **22**, 689 (1966).
761. KING, S. T. and J. OVEREND, *Spectrochim. Acta* **23A**, 61 (1967).
762. KINUMAKI, S. and K. AIDA, *Sci. Rep. Res. Inst. Tohoku Univ.* **6A**, 186 (1954).
763. KISHIDA, S. and K. NAKAMOTO, *J. Chem. Phys.* **41**, 1554 (1964).
764. KISHIDA, S. and K. NAKAMOTO, *J. Chem. Phys.* **41**, 1558 (1964).
765. KLEE, W. E., *Z. anorg. allgem. chem.* **343**, 58 (1966).
766. KLEMPERER, W., *J. Chem. Phys.* **24**, 353 (1956).
767. KLEMPERER, W., *J. Electrochem. Soc.* **110**, 1023 (1963).
768. KLEMPERER, W. and L. LINDEMANN, *J. Chem. Phys.* **25**, 397 (1956).
769. KLEMPERER, W. and W. G. NORRIS, *J. Chem. Phys.* **34**, 1071 (1961).
770. KOLDITZ, L. and H. PREISS, *Z. anorg. allgem. chem.* **325**, 252 (1963).
771. KOLDITZ, L. and J. WENDT, *Z. Chem.* **3**, 312 (1963).
772. KOMISSAROVA, L. N., P. P. MELNIKOV, R. P. KRAVTSOVA and V. I. SPITSYN, *Z. Chem.* **5**, 429 (1965).
773. KOMISSAROVA, L. N., G. YA. PUSHKINA and E. G. TETERIN, *Zhur. neorg. khim.* **12**, 2586 (1967).
774. KOSTER, D. F. and F. A. MILLER, *Spectrochim. Acta* **24A**, 1487 (1968).
775. KOTOV, YU. I. and V. M. TATEVSKII, *Optika i Spėktroskopiya* **14**, 443 (1963).
776. KOVALEV, I. F., *Optika i Spektroskopiya* **2**, 310 (1957).
777. KRAUZMANN, N., *Compt. rend.* **263B**, 922 (1966).
778. KREBS, B., G. GATTOW and A. MUELLER, *Z. anorg. allgem. chem.* **337**, 275 (1965).
779. KREBS, B. and A. MUELLER, *Z. Naturforsch.* **20A**, 1124 (1965).
780. KREBS, B. and A. MUELLER, *Z. Naturforsch.* **20A**, 1664 (1965).
781. KREBS, B. and A. MUELLER, *Spectrochim. Acta* **22**, 1532 (1966).
782. KREBS, B. and A. MUELLER, *Spectrochim. Acta* **22**, 1536 (1966).
783. KREBS, B. and A. MUELLER, *J. Mol. Spec.* **22**, 290 (1967).
784. KREBS, B. and A. MUELLER, *Z. Chem.* **7**, 243 (1967).
785. KREBS, B. and A. MUELLER, *J. Inorg. Nucl. Chem.* **30**, 463 (1968).
786. KREBS, B., A. MUELLER and A. FADINI, *J. Mol. Spec.* **24**, 198 (1967).
787. KREBS, B., A. MUELLER and G. GATTOW, *Z. Naturforsch.* **20B**, 1017 (1965).
788. KREBS, B., A. MUELLER and H. W. ROESKY, *Mol. Phys.* **12**, 469 (1967).
789. KREIGSMANN, H. and G. KESSLER, *Z. anorg. allgem. chem.* **318**, 277 (1962).
790. KRISHNAMURTHY, N., *Proc. Ind. Acad. Sci.* **61A**, 118 (1965).
791. KRISHNAMURTHY, V. N. and S. SOUNDARARAJAN, *Z. anorg. allgem. chem.* **349**, 220 (167).
792. KRISHNAMURTI, D., *Proc. Ind. Acad. Sci.* **42A**, 77 (1955).
793. KRISHNAMURTI, D., *Proc. Ind. Acad. Sci.* **43A**, 210 (1956).
794. KRISHNAMURTI, D., *Proc. Ind. Acad. Sci.* **46A**, 183 (1957).
795. KRISHNAMURTI, D., *Proc. Ind. Acad. Sci.* **47A**, 276 (1958).
796. KRISHNAMURTI, D., *Proc. Ind. Acad. Sci.* **48A**, 355 (1959).
797. KRISHNAMURTI, D., *Proc. Ind. Acad. Sci.* **51A**, 285 (1960).
798. KRISHNAMURTI, D., *Proc. Ind. Acad. Sci.* **55A**, 291 (1962).
799. KRISHNAN, K., *Proc. Ind. Acad. Sci.* **57A**, 103 (1962).
800. KRISHNAN, R. S., *Nature* **155**, 452 (1945).

801. KRISHNAN, R. S., *Proc. Ind. Acad. Sci.* **31A**, 435 (1950).
802. KRISHNAN, R. S. and R. S. KATIYAR, *J. Phys.* (Paris) **26**, 627 (1965).
803. KRISHNAN, R. S. and P. S. NARAYANAN, *Proc. Ind. Acad. Sci.* **41A**, 121 (1955).
804. KRISHNAN, R. S. and P. NEELAKANTAN, *Ind. J. Pure Appl. Phys.* **3**, 4 (1965).
805. KRISHNAN, R. S. and C. SHANTAKUMARI, *Proc. Ind. Acad. Sci.* **32A**, 105 (1950).
806. KRISHNAN, T. S., *Proc. Ind. Acad. Sci.* **44A**, 96 (1956).
807. KRISHNAPILLAI, M. G., K. RAMASWAMY and A. PERUMAL, *Ind. J. Pure Appl. Phys.* **3**, 180 (1965).
808. KRYNAUW, G. N., *Z. Phys. Chem.* (Frankurt) **43**, 191 (1964).
809. KRYNAUW, G. N., C. W. F. T. PISTORIUS and M. C. PISTORIUS, *Z. Phys. Chem.* (Frankfurt) **43**, 217 (1964).
810. KRYNAUW, G. N. and C. J. H. SCHUTTE, *Spectrochim. Acta* **21**, 1947 (1965).
811. KUCHER, W., H. ECKE and H. G. BECKER, *Z. anorg. allgem. chem.* **313**, 138 (1961).
812. KUHN, L. and E. R. LIPPINCOTT, *J. Amer. Chem. Soc.* **78**, 1820 (1956).
813. KUROSAWA, T., *J. Phys. Soc. Japan* **16**, 1298 (1961).
814. KYNASTON, W., B. E. LARCOMBE and H. S. TURNER, *J. Chem. Soc.* 1772 (1960).
815. LAFFERTY, W. J., A. G. MAKI and E. K. PLYLER, *J. Chem. Phys.* **40**, 224 (1964).
816. LAFONT, R., *Compt. rend.* **236**, 678 (1953).
817. LAFONT, R., *Compt. rend.* **242**, 1154 (1956).
818. LAFONT, R., *Compt. rend.* **244**, 1481 (1957).
819. LAFONT, R., *Ann. Phys.* **4**, 905 (1959).
820. LAFONT, R. and C. BOUHET, *J. Chim. Phys.* **50**, C91 (1953).
821. LAGEMANN, R. T. and E. A. JONES, *J. Chem. Phys.* **19**, 534 (1951).
822. LANDAU, L., *J. Mol. Spec.* **4**, 276 (1960).
823. LANDER, J. J., *J. Chem. Phys.* **17**, 892 (1949).
824. LANGSETH, A. and C. K. MOELLER, *Acta Chem. Scand.* **4**, 725 (1950).
825. LAPERCHES, J. P. and P. TARTE, *Spectrochim. Acta* **22**, 1201 (1966).
826. LASCOMBE, J., *J. Chim. Phys.* **56**, 79 (1959).
827. LAST, J. T., *Phys. Rev.* **105**, 1740 (1957).
828. LAUNER, P. J., *Am. Mineralogist* **37**, 764 (1952).
829. LAZAREV, A. N., *Optika i Spektroskopiya* **9**, 195 (1960).
830. LAZAREV, A. N., *Kristallografiya* **6**, 125 (1961).
831. LAZAREV, A. N., *Optics and Spectroscopy* **12**, 28 (1962).
832. LAZAREV, A. N. and V. S. AKSELROD, *Optika i Spektroskopiya* **9**, 326 (1960).
833. LECOMTE, J., A. BOULLE, C. DOREMIEUX-MORIN and B. BELONG, *Compt. rend.* **258**, 131 (1964).
834. LECOMTE, J., A. BOULLE, C. DOREMIEUX-MORIN and B. BELONG, *Compt. rend.* **258**, 1447 (1964).
835. LECOMTE, J., C. DUVAL and C. WADIER, *Compt. rend.* **249**, 1991 (1959).
836. LECOMTE, J. and R. FREYMANN, *Bull. Soc. Chim. France* 622 (1941).
837. LEE, S. M. and W. H. SHAFFER, *J. Mol. Spec.* **21**, 183 (1966).
838. LEECH, R. C., D. B. POWELL and N. SHEPPARD, *Spectrochim. Acta* **22**, 1931 (1966).
839. LEHMANN, W. J., J. F. DITTER and I. SHAPIRO, *J. Chem. Phys.* **29**, 1248 (1958).
840. LEHMANN, W. L., L. BECKMANN and L. GUTJAHR, *J. Chem. Phys.* **44**, 1654 (1966).

841. LE NARVOR, A., P. SAUMAGNE and A. NOVAK, *J. Chim. Phys.* **64**, 1643 (1967).
842. LE POSTOLLEC, M., *Compt. rend.* **254**, 2988 (1962).
843. LE POSTOLLEC, M. and J.-P. MATHIEU, *Compt. rend.* **254**, 1800 (1962).
844. LEROI, G. E., T. C. JAMES, J. T. HOUGEN and W. KLEMPERER, *J. Chem. Phys.* **36**, 2879 (1962).
845. LEROY, M. J. F. and G. KAUFMANN, *Bull. Soc. Chim. France* 3090 (1966).
846. LEROY, M. J. F. and G. KAUFMANN, *Bull. Soc. Chim. France* 1001 (1967).
847. LEROY, M. J. F. and G. KAUFMANN, *Compt. rend.* **265C**, 1322 (1967).
848. LEROY, M. J. F. and G. KAUFMANN, *Bull. Soc. Chim. France* 3586 (1968).
849. LEROY, M. J. F. and G. KAUFMANN, *Bull. Soc. Chim. France* 4028 (1968).
850. LEROY, M. J. F., G. KAUFMANN, R. CHARLIONET and R. ROHMER, *Compt. rend.* **263C**, 601 (1966).
851. LEROY, M. J. F., G. KAUFMANN, A. MUELLER and H. W. ROESKY, *Compt. rend.* **267C**, 563 (1968).
852. LEROY, M. J. F., G. KAUFMANN and R. ROHMER, *Bull. Soc. Chim. France* 3640 (1967).
853. LEVIN, I. W., *J. Chem. Phys.* **47**, 4685 (1967).
854. LEVIN, I. W. and S. ABRAMOWITZ, *Inorg. Chem.* **5**, 2024 (1966).
855. LEVIN, I. W., and BERNEY, C. V., *J. Chem. Phys.* **44**, 2557 (1966).
856. LEVIN, I. W. and H. ZIFFER, *J. Chem. Phys.* **43**, 4023 (1965).
857. LEWIS, J. and G. WILKINSON, *J. Inorg. Nucl. Chem.* **6**, 12 (1958).
858. LIDE, D. R. and D. E. MANN, *J. Chem. Phys.* **25**, 1128 (1956).
859. LIDE, D. R., D. E. MANN and J. J. COMEFORD, *Spectrochim. Acta* **21**, 497 (1965).
860. LINDEMANN, L. P. and M. K. WILSON, *J. Chem. Phys.* **24**, 242 (1956).
861. LINEVSKY, M. J., *J. Chem. Phys.* **38**, 658 (1963).
862. LINEVSKY, M. J., *Sci. Tech. Aerospace Rept.* **3**, 720 (1965).
863. LINNETT, J. W. and D. F. HEATH, *Trans. Far. Soc.* **48**, 592 (1952).
864. LINNETT, J. W. and C. J. S. M. SIMPSON, *Trans. Far. Soc.* **55**, 857 (1959).
865. LIPPINCOTT, E. R., *J. Chem. Phys.* **17**, 1351 (1949).
866. LIPPINCOTT, E. R., J. A. PSELLOS and M. C. TOBIN, *J. Chem. Phys.* **20**, 536 (1952).
867. LIPPINCOTT, E. R. and M. C. TOBIN, *J. Chem. Phys.* **21**, 1559 (1953).
868. LOEHR, T. M. and R. A. PLANE, *Inorg. Chem.* **7**, 1708 (1968).
869. LOEWENSCHUSS, A., A. RON and O. SCHNEPP, *J. Chem. Phys.* **49**, 272 (1968).
870. LOHMAN, J. B., F. P. REDING and D. F. HORNIG, *J. Chem. Phys.* **19**, 252 (1951).
871. LONG, D. A. and R. T. BAILEY, *Trans. Far. Soc.* **59**, 792 (1963).
872. LONG, D. A. and R. T. BAILEY, *Spectrochim. Acta* **19**, 1607 (1963).
873. LONG, D. A. and J. Y. H. CHAU, *Trans. Far. Soc.* **58**, 2325 (1962).
874. LONG, D. A. and D. T. L. JONES, *Trans. Far. Soc.* **59**, 273 (1963).
875. LONG, D. A., F. S. MURFIN and R. L. WILLIAMS, *Proc. Roy. Soc.* **A223**, 251 (1954).
876. LONG, D. A., T. V. SPENCER, D. N. WATERS and L. A. WOODWARD, *Proc. Roy. Soc.* **A240**, 499 (1957).
877. LORD, R. C., M. A. LYNCH, W. C. SCHUMB and E. J. SLOWINSKI, *J. Am. Chem. Soc.* **72**, 522 (1950).
878. LORD, R. C. and E. NIELSEN, *J. Chem. Phys.* **19**, 1 (1951).
879. LORENZELLI, V., *Atti. Accad. naz. Lincei, Rend. Classe Sci. fis. mat. nat.* **31**, 416 (1961).
880. LORENZELLI, V., *Compt. rend.* **252**, 3219 (1961).

881. LORENZELLI, V., T. DUPUIS and J. LECOMTE, *Compt. rend.* **259**, 1057 (1964).
882. LORENZELLI, V. and F. GESMUNDO, *Atti. Accad. naz. Lincei, Rend. Classe Sci. fis. mat. nat.* **36**, 485 (1964).
883. LORENZELLI, V. and F. GESMUNDO, *Atti. Accad. naz. Lincei, Rend. Classe Sci. fis. mat. nat.* **37**, 70 (1964).
884. LORENZELLI, V. and K. D. MOELLER, *Compt. rend* **248**, 1980 (1959).
885. LOUISFERT, J., *Compt. rend* **233**, 381 (1951).
886. LOUISFERT, J., *Compt. rend* **235**, 287 (1952).
887. LOUISFERT, J., *Compt. rend* **241**, 940 (1955).
888. LOUISFERT, J., *Compt. rend* **248**, 1150 (1959).
889. LOUISFERT, J. *Compt. rend* **248**, 1497 (1959).
890. LOUISFERT, J., *Compt. rend* **252**, 3565 (1961).
891. LOVEJOY, R. W. and E. L. WAGNER, *J. Phys. Chem.* **68**, 544 (1964).
892. LYNDS, L., *J. Chem. Phys.* **42**, 1124 (1965).
893. LYNDS, L., *J. Chem. Phys.* **44**, 1721 (1966).
894. LYNDS, L., *Spectrochim. Acta* **22**, 2123 (1966).
895. LYNDS, L. and C. D. BASS, *J. Chem. Phys.* **41**, 3165 (1964).
896. LYNDS, L., T. WOLFRAM and C. D. BASS, *J. Chem. Phys.* **43**, 3775 (1965).
897. MACCOLL, A., *J. Proc. Roy. Soc., New South Wales*, **77**, 130 (1943).
898. MADDEN, R. P. and W. S. BENEDICT, *J. Chem. Phys.* **25**, 594 (1956).
899. MAGNASCO, V., *Nuovo cimento* **24**, 425 (1962).
900. MAKI, A., *J. Chem. Phys.* **38**, 1261 (1963).
901. MAKI, A. and J. C. DECIUS, *J. Chem. Phys.* **31**, 772 (1959).
902. MAKI, A. and R. FORNERIS, *Spectrochim. Acta* **23A**, 867 (1967).
903. MAKI, A., E. K. PLYLER and E. D. TIDWELL, *J. Res. Nat. Bur. Stds.* **66A**, 163 (1962).
904. MALM, J. G., H. SELIG and S. FRIED, *J. Amer. Chem. Soc.* **82**, 1510 (1960).
905. MALM, J. G., B. WEINSTOCK and H. H. CLAASSEN, *J. Chem. Phys.* **23**, 2192 (1955).
906. MANLEY, T. R. and D. A. WILLIAMS, *Spectrochim. Acta* **21**, 1773 (1965).
907. MANN, D. E., G. V. CALDER, K. S. SESHADRI, D. WHITE and M. J. LINEVSKY, *J. Chem. Phys.* **46**, 1138 (1967).
908. MANN, D. E. and L. FANO, *J. Chem. Phys.* **26**, 1665 (1957).
909. MARA, R. T., G. B. B. M. SUTHERLAND and H. J. V. TYRRELL, *Phys. Rev.* **96**, 801 (1954).
910. MARSULT-HERAIL, F. and N. TARTAR, *Compt. rend* **267B**, 270 (1968).
911. MARTIN, K. and E. STEGER, *Z. anorg. allgem. chem.* **345**, 306 (1966).
912. MARTINEZ, J. V. and E. L. WAGNER, *J. Chem. Phys.* **27**, 1110 (1957).
913. MARTZ, D. E. and R. T. LAGEMANN, *J. Chem. Phys.* **22**, 1193 (1954).
914. MASON, E. A. and M. M. KREEVOY, *J. Amer. Chem. Soc.* **77**, 5808 (1965).
915. MATHIEU, J.-P., *Compt. rend* **229**, 1068 (1949).
916. MATHIEU, J.-P., *Compt. rend* **234**, 2272 (1952).
917. MATHIEU, J.-P., *Compt. rend* **240**, 2508 (1955).
918. MATHIEU, J.-P. and L. COUTURE-MATHIEU, *J. Phys. Radium* **13**, 271 (1952).
919. MATHIEU, J.-P. and L. COUTURE-MATHIEU, *J. Chim. Phys.* **50**, 513 (1953).
920. MATHIEU, J.-P. and J. JACQUES, *Compt. rend* **215**, 346 (1946).
921. MATHIEU, J.-P. and M. LOUNSBURY, *Diss. Far. Soc.* **9**, 196 (1960).
922. MATHIEU, J.-P. and H. POULET, *Spectrochim. Acta* **16**, 696 (1960).
923. MATOSSI, F., *Ann. Physik* **11**, 22 (1963).
924. MATTRAW, H. C., N. J. HAWKINS, D. R. CARPENTER and W. W. SABOL, *J. Chem. Phys.* **23**, 985 (1955).

925. Matutano, J. R. B. and C. Otero, *Anales real soc. españ. fis y quim* (Madrid) **51B**, 223 (1955).
926. McConaghie, V. M. and H. H. Nielsen, *Proc. Natl. Acad. Sci. U.S.* **34**, 455 (1948).
927. McCory, L. D., R. C. Paule and J. L. Margrave, *J. Phys. Chem.* **67**, 1086 (1963).
928. McDaniel, D. H. and R. E. Vallee, *Inorg. Chem.* **2**, 996 (1963).
929. McDevitt, N. T. and W. L. Baun, *Spectrochim. Acta* **20**, 779 (1964).
930. McDevitt, N. T. and A. D. Davison, *J. Opt. Soc. Amer.* **56**, 636 (1966).
931. McDowell, C. A., *Trans. Far. Soc.* **49**, 371 (1953).
932. McDowell, R. S., *Inorg. Chem.* **6**, 1759 (1967).
933. McDowell, R. S. and L. B. Asprey, *J. Chem. Phys.* **37**, 165 (1962).
934. McGraw, G. E., D. L. Bernitt and I. C. Hisatsune, *Spectrochim. Acta* **23A**, 25 (1967).
935. McWhinnie, W. R., *J. Inorg. Nuclear Chem.* **26**, 21 (1964).
936. Meal, J. H. and M. K. Wilson, *J. Chem. Phys.* **24**, 385 (1956).
937. Mehra, A. and P. Venkateswarlu, *J. Chem. Phys.* **47**, 2334 (1967).
938. Meister, A. G. and F. F. Cleveland, *Amer. J. Phys.* **14**, 13 (1946).
939. Meister, A. G., F. F. Cleveland and M. J. Murray, *Amer. J. Phys.* **11**, 239 (1943).
940. Meisingseth, E., *Acta Chem. Scand.* **16**, 1607 (1962).
941. Meyers, M. D. and F. A. Cotton, *J. Amer. Chem. Soc.* **82**, 5027 (1960).
942. Miles, M. G., J. H. Patterson, C. W. Hobbs, M. J. Hopper, J. Overend and R. S. Tobias, *Inorg. Chem.* **7**, 1721 (1968).
943. Millen, D. J., *J. Chem. Soc.* 2589 (1950).
944. Millen, D. J., *J. Chem. Soc.* 2606 (1950).
945. Millen, D. J., C. N. Polydoropoulos and D. Watson, *J. Chem. Soc.* 687 (1960).
946. Miller, F. A. and W. K. Baer, *Spectrochim. Acta* **17**, 114 (1961).
947. Miller, F. A. and W. K. Baer, *Spectrochim. Acta* **18**, 1311 (1962).
948. Miller, F. A. and G. L. Carlson, *Spectrochim. Acta* **16**, 6 (1960).
949. Miller, F. A. and G. L. Carlson, *Spectrochim. Acta* **16**, 1148 (1960).
950. Miller, F. A. and G. L. Carlson, *Spectrochim. Acta* **17**, 977 (1961).
951. Miller, F. A., G. L. Carlson, F. F. Bentley and W. H. Jones, *Spectrochim. Acta* **16**, 135 (1960).
952. Miller, F. A., G. L. Carlson and W. B. White, *Spectrochim. Acta* **15**, 709 (1959).
953. Miller, F. A. and L. R. Cousins, *J. Chem. Phys.* **26**, 329 (1957).
954. Miller, F. A. and W. G. Fateley, *Spectrochim. Acta* **20**, 253 (1964).
955. Miller, F. A., W. G. Fateley and R. E. Witkowski, *Spectrochim. Acta* **23A**, 891 (1967).
956. Miller, F. A., D. H. Lemmon and R. E. Witkowski, *Spectrochim. Acta* **21**, 1709 (1965).
957. Miller, F. A. and C. H. Wilkins, *Anal. Chem.* **25**, 1253 (1952).
958. Miller, R. E. and D. F. Eggers, *J. Chem. Phys.* **45**, 3028 (1966).
959. Miller, R. E. and G. E. Leroi, *J. Chem. Phys.* **49**, 2789 (1968).
960. Miller, R. H., *Diss. Abs.* **26**, 5744 (1966).
961. Miller, R. H., D. L. Bernitt and I. C. Hisatsune, *Spectrochim. Acta* **23A**, 223 (1967).
962. Miller, R. L. and D. F. Hornig, *J. Chem. Phys.* **34**, 265 (1961).
963. Miller, S. A., H. H. Caspers and H. E. Rast, *Phys. Rev.* **168**, 964 (1968).

964. MILLIGAN, D. E. and M. E. JACOX, *J. Chem. Phys.* **40**, 2461 (1964).
964a. MILLIGAN, D. E. and M. E. JACOX, *J. Chem. Phys.* **41**, 2838 (1964).
965. MILLIGAN, D. E., M. E. JACOX and J. D. MCKINLEY, *J. Chem. Phys.* **42**, 902 (1965).
966. MILLIKAN, R. C. and K. S. PITZER, *J. Chem. Phys.* **27**, 1305 (1957).
967. MILLS, I. M., *Spectrochim. Acta* **16**, 35 (1960).
968. MILLS, I. M., *Spectrochim. Acta* **19**, 1585 (1963).
969. MIRRI, A. M., *J. Chem. Phys.* **47**, 2823 (1967).
970. MITCHELL, A., *Trans. Far. Soc.* **62**, 530 (1966).
971. MITRA, G., *J. Amer. Chem. Soc.* **80**, 5639 (1958).
972. MITRA, S. S., *Z. physik. chem.* (Frankfurt) **23**, 190 (1960).
973. MITRA, S. S., *Solid State Phys.* **13**, 1 (1962).
974. MITRA, S. S. and J. I. BRYANT, *Bull. Amer. Phys. Soc.* **10**, 333 (1965).
975. MITRA, S. S. and P. GIELISSE, *Progress in Infra-Red Spectroscopy* **2**, 47–126 (1964).
976. MITRA, S. S. and S. K. JOSHI, *Z. physik. chem.* (Frankfurt) **26**, 2141 (1960).
977. MITSUISHI, A., H. YOSHINAGA and S. FUJITA, *J. Phys. Soc. Japan* **13**, 1235 (1958).
978. MITSUISHI, A., H. YOSHINAGA and S. FUJITA, *J. Phys. Soc. Japan* **13**, 1236 (1958).
979. MITSUISHI, A., H. YOSHINAGA, S. FUJITA and Y. SUEMOTO, *Japan J. Appl. Phys.* **1**, 1 (1962).
980. MIYAZAWA, T. and K. FUKUSHIMA, *J. Mol. Spec.* **15**, 308 (1963).
981. MOEHNKE, H., *Infrared Phys.* **2**, 111 (1962).
982. MONOSTORI, B. and A. WEBER, *J. Chem. Phys.* **33**, 1867 (1960).
983. MOONEY, R. W., S. J. CYVIN and J. BRUNVOLL, *Acta Chem. Scand.* **19**, 983 (1965).
984. MOONEY, R. W., S. J. CYVIN, J. BRUNVOLL and L. A. KRISTIANSEN, *J. Chem. Phys.* **42**, 3741 (1965).
985. MOONEY, R. W. and R. L. GOLDSMITH, *J. Inorg. Nucl. Chem.* **31**, 933 (1969).
986. MOONEY, R. W. and S. Z. TOMA, *J. Chem. Phys.* **46**, 3364 (1967).
987. MOONEY, R. W., S. Z. TOMA and J. BRUNVOLL, *Spectrochim. Acta* **23A**, 1541 (1967).
988. MOONEY, R. W., S. Z. TOMA, R. L. GOLDSMITH and K. H. BUTLER, *J. Inorg. Nucl. Chem.* **30**, 1669 (1968).
989. MOORE, G. E., *J. Opt. Soc. Amer.* **43**, 1045 (1953).
990. MOORE, G. E. and R. M. BADGER, *J. Amer. Chem. Soc.* **74**, 6076 (1952).
991. MORANDAT, J. V. LORENZELLI and J. LECOMTE, *J. Phys.* (Paris) **28**, 156 (1967).
992. MORANDAT, J., V. LORENZELLI and H. PAQUET, *Compt. rend* **263B**, 697 (1966).
993. MORGAN, H. W., *J. Inorg. Nucl. Chem.* **16**, 367 (1960–61).
994. MORGAN, H. W. and P. A. STAATS, *J. Appl. Phys.* **33**, 364 (1962).
995. MORGAN, H. W., P. A. STAATS and J. H. GOLDSTEIN, *J. Chem. Phys.* **27**, 1212 (1957).
996. MORGAN, H. W., P. A. STAATS and J. H. GOLDSTEIN, *J. Chem. Phys.* **27**, 1213 (1957).
997. MORINO, Y., *J. Chem. Phys.* **24**, 164 (1956).
997a. MORINO, Y. and K. KUCHITSU, *J. Chem. Phys.* **20**, 1809 (1952).
998. MORINO, Y. and Y. MURATA, *Bull. Chem. Soc. Japan* **38**, 104 (1965).

999. Morino, Y., Y. Murata, T. Ito and J. Nakamura, *J. Phys. Soc. Japan* **17** Supp. B-II, 37 (1962).
1000. Morris, D. F. C., E. L. Short and D. N. Waters, *J. Inorg. Nucl. Chem.* **25**, 975 (1963).
1000a. Mould, H. M., W. C. Price and G. R. Wilkinson, *Spectrochim. Acta* **15**, 313 (1959).
1001. Moureau, H., M. Magat and G. Wetroff, *Compt. rend* **205**, 276 (1937).
1002. Moureau, H., M. Magat and G. Wetroff, *Compt. rend* **205**, 545 (1937).
1003. Mueller, A., *Z. Naturforsch* **20A**, 745 (1965).
1004. Mueller, A., *Z. Naturforsch* **21A**, 433 (1966).
1005. Mueller, A., *Z. Chem.* **7**, 35 (1967).
1006. Mueller, A., R. Ahlrichs and B. Krebs, *Z. Naturforsch* **21B**, 720 (1966).
1007. Mueller, A. and A. Fadini, *Spectrochim. Acta* **22**, 1523 (1966).
1008. Mueller, A. and A. Fadini, *Z. Naturforsch* **21B**, 585 (1966).
1009. Mueller, A. and A. Fadini, *Z. anorg. allgem. chem.* **349**, 164 (1967).
1010. Mueller, A. and A. Fadini, *Z. Phys. Chem.* (Frankfurt) **54**, 129 (1967).
1011. Mueller, A. and G. Gattow, *Z. anorg. allgem. chem.* **348**, 71 (1966).
1012. Mueller, A., G. Gattow and B. Krebs, *Z. anorg. allgem. chem.* **347**, 24 (1966).
1013. Mueller, A., G. Gattow and H. Seidel, *Z. anorg. allgem. chem.* **347**, 24 (1966).
1014. Mueller, A., O. Glemser and E. Niecke, *Z. anorg. allgem. chem.* **347**, 275 (1966).
1015. Mueller, A., O. Glemser and E. Niecke, *Z. Naturforsch* **21B**, 732 (1966).
1016. Mueller, A., R. Kebabcioglu, M. J. F. Leroy and G. Kaufmann, *Z. Naturforsch* **23B**, 740 (1968).
1017. Mueller, A. and B. Krebs, *Naturwiss* **52**, 448 (1965).
1018. Mueller, A. and B. Krebs, *Z. Naturforsch* **20A**, 967 (1965).
1019. Mueller, A. and B. Krebs, *Spectrochim. Acta* **22**, 1535 (1966).
1020. Mueller, A. and B. Krebs, *Z. Naturforsch* **21B**, 3 (1966).
1021. Mueller, A. and B. Krebs, *Z. anorg. allgem. chem.* **344**, 56 (1966).
1022. Mueller, A. and B. Krebs, *Z. anorg. allgem. chem.* **347**, 261 (1966).
1023. Mueller, A. and B. Krebs, *J. Mol. Spec.* **24**, 180 (1967).
1024. Mueller, A., B. Krebs, R. Kebabcioglu, M. Stockburger and O. Glemser, *Spectrochim. Acta* **24A**, 1831 (1968).
1025. Mueller, A., B. Krebs, E. Niecke and A. Ruoff, *Ber. Bunsenges. Phys. Chem.* **71**, 571 (1967).
1026. Mueller, A., B. Krebs and C. Peacock, *Spectrochim. Acta* **24A**, 1662 (1968).
1027. Mueller, A., B. Krebs, W. Rittner and M. Stockburger, *Ber. Bunsenges. Phys. Chem.* **71**, 182 (1967).
1028. Mueller, A., E. Niecke and O. Glemser, *Z. anorg. allgem. chem.* **350**, 246 (1967).
1029. Mueller, A., E. Niecke and O. Glemser, *Z. anorg. allgem. chem.* **350**, 256 (1967).
1030. Mueller, A., E. Niecke, B. Krebs and O. Glemser, *Z. Naturforsch* **23B**, 588 (1968).
1031. Mueller, A. and C. J. Peacock, *Z. Chem.* **8**, 69 (1968).
1032. Mueller, A. and W. Rittner, *Spectrochim. Acta* **23A**, 1831 (1967).
1033. Mueller, A., H. W. Roesky and B. Krebs, *Z. Chem.* **7**, 159 (1967).
1034. Mueller, A. and M. Stockburger, *Z. Naturforsch* **20A**, 1242 (1965).

1035. Mueller, H. and K. Dehnicke, *Z. anorg. allgem. chem.* **350**, 231 (1967).
1036. Mullhaupt, J. T. and D. F. Hornig, *J. Chem. Phys.* **24**, 169 (1956).
1037. Murata, H. and K. Kawai, *J. Chem. Phys.* **25**, 589 (1956).
1038. Murata, H. and K. Kawai, *J. Chem. Phys.* **25**, 796 (1956).
1039. Murata, H. and K. Kawai, *J. Chem. Phys.* **27**, 605 (1957).
1040. Murata, H. and K. Kawai, *J. Chem. Phys.* **28**, 516 (1958).
1041. Murchison, C., S. Reichman, D. Anderson, J. Overend and F. Schreiner, *J. Amer. Chem. Soc.* **90**, 5690 (1968).
1042. Mutschin, A. and K. Maenuchen, *Z. anal. chem.* **160**, 81 (1958).
1043. Nachbaur, E., *Monatsh* **93**, 135 (1962).
1044. Nagarajan, G., *Bull. Soc. chim. belges* **71**, 82 (1962).
1045. Nagarajan, G., *J. Sci. Res. Ind.* **21B**, 356 (1962).
1046. Nagarajan, G., *Bull. Soc. chim. belges* **71**, 276 (1963).
1047. Nagarajan, G., *Bull. Soc. chim. belges* **72**, 563 (1963).
1048. Nagarajan, G., *Ind. J. Pure Appl. Phys.* **2**, 278 (1964).
1049. Nagarajan, G., A. Mueller and H. G. Horn, *Z. Chem.* **6**, 319 (1966).
1049a. Nakagawa, I. and T. Shimanouchi, *Proc. Int. Symp. Mol. Struct. Spect.* **A314**, Tokyo (1962).
1050. Nakagawa, I. and T. Shimanouchi, *Spectrochim. Acta* **20**, 429 (1964).
1051. Nakagawa, I. and T. Shimanouchi, *Spectrochim. Acta* **22**, 759 (1966).
1052. Nakagawa, I. and T. Shimanouchi, *Spectrochim. Acta* **22**, 1707 (1966).
1053. Nakagawa, I. and T. Shimanouchi, *Spectrochim. Acta* **23A**, 2099 (1967).
1054. Nakagawa, I., A. Tsuchida and T. Shimanouchi, *J. Chem. Phys.* **47**, 982 (1967).
1054a. Nakamoto, K., J. Fujita and H. Murata, *J. Amer. Chem. Soc.* **80**, 4817 (1958).
1055. Nakamoto, K., J. Fujita, S. Tanaka and M. Kobayashi, *J. Amer. Chem. Soc.* **79**, 4904 (1957).
1056. Nakamoto, K., Y. A. Sarma and H. Ogoshi, *J. Chem. Phys.* **43**, 1177 (1965).
1057. Nakamura, K., *J. Chem. Soc. Japan (Pure Chem.)* **79**, 1411 (1958).
1058. Narayanan, P. S. and K. Vedam, *Z. Physik* **163**, 158 (1961).
1059. Narayanaswamy, P. K., *Proc. Ind. Acad. Sci.* **28A**, 417 (1948).
1060. Nedungadi, T. M. K., *Proc. Ind. Acad. Sci.* **10A**, 197 (1939).
1061. Nelson, L. Y. and G. C. Pimentel, *Inorg. Chem.* **6**, 1758 (1967).
1062. Nelson, L. Y. and G. C. Pimentel, *J. Chem. Phys.* **47**, 3671 (1967).
1063. Nelson, L. Y. and G. C. Pimentel, *Inorg. Chem.* **7**, 1695 (1968).
1064. Neu, J. T. and W. D. Gwinn, *J. Amer. Chem. Soc.* **70**, 3463 (1948).
1065. Newman, G. A., *Chem. and Ind.* 514 (1968).
1066. Newman, G. A. and D. B. Powell, *Spectrochim. Acta* **19**, 213 (1963).
1067. Newman, R., *J. Chem. Phys.* **20**, 444 (1952).
1068. Newman, R., *J. Chem. Phys.* **20**, 1663 (1952).
1069. Newman, R. and R. M. Badger, *J. Chem. Phys.* **19**, 1207 (1951).
1070. Newman, R. and R. S. Halford, *J. Chem. Phys.* **18**, 1276 (1950).
1071. Newman, R. and R. S. Halford, *J. Chem. Phys.* **18**, 1291 (1950).
1072. Nguyen-Quy-Dao, *Bull. Soc. chim. France* 3976 (1968).
1073. Nibler, J. W. and G. C. Pimentel, *J. Chem. Phys.* **47**, 710 (1967).
1074. Nielsen, A. H. and H. H. Nielsen, *J. Chem. Phys.* **5**, 277 (1957).
1075. Nielsen, A. H. and P. J. H. Woltz, *J. Chem. Phys.* **20**, 1878 (1952).
1075a. Nielsen, J. R. and L. H. Berryman, *J. Chem. Phys.* **17**, 659 (1949).
1076. Nilsen, W. G. and J. G. Skinner, *J. Chem. Phys.* **48**, 2240 (1968).

1077. Nimon, L. A. and V. D. Neff, *J. Mol. Spec.* **26**, 175 (1968).
1078. Nimon, L. A., V. D. Neff, R. E. Cantley and R. O. Buttlar, *J. Mol. Spec.* **22**, 105 (1967).
1079. Nixon, E. R., *J. Phys. Chem.* **60**, 1054 (1956).
1080. Nixon, J. and R. A. Plane, *J. Amer. Chem. Soc.* **84**, 4445 (1962).
1081. Noble, P. N. and G. C. Pimentel, *J. Chem. Phys.* **49**, 3165 (1968).
1082. Noble, P. N. and G. C. Pimentel, *Spectrochim. Acta* **24A**, 797 (1968).
1083. Novak, A., P. Saumagne and L. D. C. Bok, *J. Chem. Phys.* **60**, 1385 (1963).
1084. Nyquist, R., *Spectrochim. Acta* **23A**, 845 (1967).
1085. Ockman, N., *Advan. Phys.* **7**, 199 (1958).
1086. Ockman, N. and G. B. B. M. Sutherland, *Proc. Roy. Soc.* **247A**, 434 (1958).
1087. Oertel, R. P. and R. A. Plane, *Inorg. Chem.* **6**, 1960 (1967).
1088. Olie, K., C. C. Smitskamp and H. Gerding, *Inorg. Nucl. Chem. Letters* **4**, 129 (1968).
1089. Olivelli, G., *Nuovo cimento* **10**, 343 (1953).
1090. O'Loane, J. K. and M. K. Wilson, *J. Chem. Phys.* **23**, 1313 (1955).
1091. Omori, K. and P. F. Kerr, *Bull. Geol. Soc. Amer.* **74**, 709 (1963).
1092. Omori, K. and P. F. Kerr, *Sci. Rept. Tohoku. Univ., Ser. 3,* **9**, 1 (1964).
1093. Ortner, M. H., *J. Chem. Phys.* **34**, 556 (1961).
1093a. Osberg, W. E. and D. F. Hornig, *J. Chem. Phys.* **20**, 1345 (1952).
1094. O'Shea, D. C., R. V. Kolluri and H. Z. Cummins, *Solid State Comm.* **5**, 241 (1967).
1095. Otero, C. and J. R. B. Matutano, *Anales real. soc. españ. fis y quim.* (Madrid) **52B**, 291 (1956).
1096. Ottinger, R., S. J. Cyvin, R. W. Mooney, L. A. Kristiansen and J. Brunvoll, *Acta Chem. Scand.* **20**, 1389 (1966).
1097. Overend, J. and J. C. Evans, *Trans. Far. Soc.* **55**, 1817 (1959).
1098. Pace, E. L., *J. Chem. Phys.* **18**, 881 (1950).
1099. Pace, E. L. and L. Pierce, *J. Chem. Phys.* **23**, 1248 (1955).
1100. Paetzold, R., *Z. Chem.* **4**, 272 (1964).
1101. Paetzold, R. and H. Amoulong, *Z. anorg. allgem. chem.* **317**, 166 (1962).
1102. Paetzold, R., H. Amoulong and A. Ruzicka, *Z. anorg. allgem. chem.* **336**, 278 (1965).
1103. Paetzold, R. and K. Aurich, *Z. anorg. allgem. chem.* **335**, 281 (1965).
1104. Paetzold, R. and K. H. Ziegenbalg, *Z. Chem.* **4**, 461 (1964).
1105. Pailette, P., A. Lankoandé, A. Hadni, J. Masson, C. Duc-Maugé and H. Guerin, *Bull. Soc. chim. France* 432 (1960).
1106. Palmer, W. G., *J. Chem. Soc.* 1552 (1961).
1107. Pannetier, G. and J. M. Bregeault, *Bull. Soc. Chim. France* 609 (1965).
1108. Papazian, H. A., *J. Chem. Phys.* **34**, 1614 (1961).
1108a. Parsons, J. L., *J. Chem. Phys.* **33**, 1860 (1960).
1109. Parsons, J. L. and L. Rimai, *Solid State Comm.* **5**, 423 (1967).
1110. Patil, K. C., R. K. Gosavi and C. N. R. Rao, *Inorg. Chim. Acta* **1**, 155 (1967).
1111. Patil, K. C. and C. N. R. Rao, *Inorg. Nucl. Chem. Lett.* **2**, 329 (1966).
1112. Peacock, C. J. and A. Mueller, *J. Mol. Spec.* **26**, 454 (1968).
1113. Peacock, C. J., A. Mueller and R. Kebabcioglu, *J. Mol. Spec.* **27**, 351 (1968).
1114. Peacock, R. D. and D. W. A. Sharp, *J. Chem. Soc.* 2762 (1959).
1115. Pemsler, J. P. and W. G. Planet, *J. Chem. Phys.* **24**, 920 (1956).
1116. Penland, R. B., T. J. Lane and J. V. Quagliano, *J. Amer. Chem. Soc.* **78**, 887 (1956).

1117. Penot, D., K. Poulet and J.-P. Mathieu, *Compt. rend* **243**, 1303 (1956).
1118. Perkins, W. D. and M. K. Wilson, *J. Chem. Phys.* **20**, 1792 (1952).
1119. Perry, C., D. P. Athens, E. Young, R. Durig and B. R. Mitchel, *Spectrochim. Acta* **23A**, 1137 (1967).
1120. Perry, C. H., *Japan J. Appl. Phys., Suppl.* **4**, 564 (1964).
1121. Perry, C. H., J. H. Fertel and T. F. McNelly, *J. Chem. Phys.* **47**, 1619 (1967).
1122. Perry, C. H. and D. B. Hall, *Phys. Rev. Letters* **15**, 700 (1965).
1123. Perry, C. H., B. N. Khanna and G. Rupprecht, *Phys. Rev.* **135**, 408 (1964).
1124. Perry, C. H. and E. F. Young, *J. Appl. Phys.* **38**, 4616 (1967).
1125. Perschina, E. W. and S. S. Raskin, *Optika i Spektroskopiya* **13**, 488 (1962).
1126. Person, W. B., G. R. Anderson, J. N. Fordemwalt, H. Stammreich and R. Forneris, *J. Chem. Phys.* **35**, 908 (1961).
1127. Petreanu, E., S. Pinchas and D. Samuel, *J. Inorg. Nucl. Chem.* **27**, 2519 (1965).
1128. Petrov, K. I., V. I. Ivanov, V. G. Pervykh, *Zh. Strukt. Khim.* **8**, 356 (1967).
1129. Petrov, K. I., I. V. Tananaev, V. G. Pervykh and S. M. Petushkova, *Russ. J. Inorg. Chem.* **12**, 1395 (1967).
1130. Petrov, K. I., V. P. Vasil'eva and V. G. Pervykh, *Zhur. neorg. khim.* **11**, 1837 (1966).
1131. Petru, F. and A. Muck, *Z. Chem.* **6**, 386 (1966).
1132. Petru, F. and A. Muck, *Z. Chem.* **7**, 27 (1967).
1133. Petru, F. and A. Muck, *Z. Chem.* **7**, 159 (1967).
1134. Pfeiffer, M. and H.-J. Spangenberg, *Z. Phys. Chem.* (Leipzig) **232**, 47 (1966).
1135. Philippot, E. and M. Maurin, *Compt. rend.* **267C**, 243 (1968).
1136. Pichai, R., M. G. K. Pillai and K. Ramaswamy, *Austral. J. Chem.* **20**, 1055 (1967).
1137. Pierce, L., *J. Chem. Phys.* **34**, 498 (1961).
1138. Pillai, M. G. K. and P. P. Pillai, *Can. J. Chem.* **46**, 2393 (1968).
1138a. Pinchas, S. and M. Halmann, *J. Chem. Phys.* **31**, 1692 (1959).
1139. Pinchas, S., D. Samuel and E. Petreanu, *J. Inorg. Nucl. Chem.* **29**, 335 (1967).
1140. Pinczuk, A., W. Taylor, E. Burstein and I. Lefkowitz, *Solid State Comm.* **5**, 429 (1967).
1141. Piseri, L. and G. Zerbi, *J. Mol. Spec.* **26**, 254 (1968).
1142. Pistorius, C. W. F. T., *J. Chem. Phys.* **27**, 965 (1957).
1143. Pistorius, C. W. F. T., *J. Chem. Phys.* **28**, 514 (1958).
1144. Pistorius, C. W. F. T., *J. Chem. Phys.* **29**, 1328 (1958).
1145. Pistorius, C. W. F. T., *J. Mol. Spec.* **2**, 287 (1958).
1146. Pistorius, C. W. F. T., *Bull. Soc. chim. belges* **67**, 566 (1958).
1147. Pistorius, C. W. F. T., *Z. Physik. Chem.* (Frankfurt) **17**, 292 (1958).
1148. Pistorius, C. W. F. T., *Bull. Soc. chim. belges* **68**, 630 (1959).
1149. Pistorius, C. W. F. T., *J. Chem. Phys.* **30**, 332 (1959).
1150. Pistorius, C. W. F. T., *J. Chem. Phys.* **31**, 1454 (1959).
1151. Pistorius, C. W. F.T. and P. C. Haarhoff, *J. Mol. Spec.* **3**, 621 (1959).
1152. Pitts, J. J., S. Konopischa and A. W. Jache, *Inorg. Chem.* **4**, 257 (1965).
1153. Plumb, R. C. and D. F. Hornig, *J. Chem. Phys.* **23**, 947 (1955).
1154. Polo, S. R. and M. K. Wilson, *J. Chem. Phys.* **22**, 900 (1954).
1155. Porto, S. P. S., P. A. Fleury and T. C. Damen, *Phys. Rev.* **154**, 522 (1967).

1156. PORTO, S. P. S., J. A. GIORDMAINE and T. C. DAMEN, *Phys. Rev.* **147**, 608 (1966).
1157. PORTO, S. P. S. and R. S. KRISHNAN, *J. Chem. Phys.* **47**, 1009 (1967).
1158. PORTO, S. P. S. and J. F. SCOTT, *Phys. Rev.* **157**, 716 (1967).
1159. POSTMUS, C. and J. R. FERRARO, *J. Chem. Phys.* **48**, 3605 (1968).
1160. POULET, H. and J.-P. MATHIEU, *J. Chim. Phys.* 442 (1963).
1161. POULET, H. and J.-P. MATHIEU, *Ann. Phys.* (Paris) **9**, 549 (1964).
1162. POULET, H. and J.-P. MATHIEU, *Compt. rend.* **258**, 2043 (1964).
1163. POWELL, D. B. and N. SHEPPARD, *J. Chem. Soc.* 3108 (1956).
1164. POWELL, F. X. and E. R. LIPPINCOTT, *J. Chem. Phys.* **32**, 1883 (1960).
1165. PRASK, H. J. and H. BOUTIN, *J. Chem. Phys.* **45**, 699 (1966).
1166. PRASK, H. J. and H. BOUTIN, *J. Chem. Phys.* **45**, 3284 (1966).
1167. PREUDHOMME, J., *Spectrochim. Acta* **20**, 275 (1964).
1168. PRICE, W. C., *J. Chem. Phys.* **17**, 1044 (1949).
1169. PRICE, W. C., H. C. LONGUET-HIGGINS, B. RICE and T. F. YOUNG, *J. Chem. Phys.* **17**, 217 (1949).
1170. PRIGENT, J., *Compt. rend* **247**, 1739 (1958).
1171. PURCELL, K. F. and R. S. DRAGO, *J. Chem. Phys.* **48**, 3837 (1968).
1172. QUICKSALL, C. O. and T. G. SPIRO, *Inorg. Chem.* **5**, 2232 (1966).
1173. RADHAKRISHNAN, M., *Z. phys. Chem.* (Frankfurt) **36**, 227 (1963).
1174. RADAKRISHNAN, M., *J. Mol. Spec.* **10**, 111 (1963).
1175. RAJAGOPAL, A. K. and R. SRINIVASA, *J. Phys. Chem. Solids* **23**, 633 (1962).
1176. RAJESWARA RAO, N. and S. SAKKU, *Ind. J. Pure Appl. Phys.* **6**, 4 (1968).
1177. RAKOV, *Res. Mol. Spectroscopy* (Proceedings of the P.N. Lebedev Physics Institute) **27**, 109 (1965).
1178. RAMDAS, A. K., *Proc. Ind. Acad. Sci.* **35A**, 249 (1952).
1179. RAMDAS, A. K., *Proc. Ind. Acad. Sci.* **36A**, 55 (1952).
1180. RAMDAS, A. K., *Proc. Ind. Acad. Sci.* **37A**, 441 (1953).
1181. RAMDAS, A. K., *Proc. Ind. Acad. Sci.* **39A**, 81 (1954).
1182. RANDI, G., *Atti. Accad. naz. Lincei, Rend. Classe Sci. fis. mat. nat.* **41**, 197 (1966).
1183. RANDI, G. and F. GESMUNDO, *Atti. Accad. naz. Lincei, Rend. Classe Sci. fis. mat. nat.* **41**, 200 (1966).
1184. RAO, K. V. K., *Proc. Ind. Acad. Sci.* **14A**, 521 (1941).
1185. RAO, P. B. and K. S. MURTI, *Current Sci.* (India) **29**, 14 (1960).
1186. RAO, P. K. and P. B. RAO, *Z. Physik. Chem.* (Frankfurt) **42**, 166 (1964).
1187. RASKIN, SH. SH., *Dokl. Akad. Nauk SSSR* **141**, 900 (1961).
1188. RAUCH, J. E. and J. C. DECIUS, *Spectrochim. Acta* **22**, 1963 (1966).
1189. REDING, F. P. and D. F. HORNIG, *J. Chem. Phys.* **22**, 1926 (1954).
1190. REDING, F. P. and D. F. HORNIG, *J. Chem. Phys.* **23**, 1053 (1955).
1191. REDING, F. P. and D. F. HORNIG, *J. Chem. Phys.* **27**, 1024 (1957).
1192. REDINGTON, R. L., W. B. OLSON and P. C. CROSS, *J. Chem. Phys.* **36**, 1311 (1962).
1192a. REDLICH, O. and L. E. NIELSEN, *J. Amer. Chem. Soc.* **65**, 654 (1943).
1193. REED, P. R., jr., and R. W. LOVEJOY, *Spectrochim. Acta* **24A**, 1795 (1968).
1194. REICH, P. and H. PREISS, *Z. Chem.* **7**, 115 (1967).
1195. REISFELD, M. J. and G. A. CROSBY, *Inorg. Chem.* **4**, 65 (1965).
1196. RENNEKE, D. R. and D. W. LYNCH, *Phys. Rev.* **138**, 530 (1965).
1197. REY, T., *Z. Krist.* **123**, 263 (1966).
1198. RICE, O. K. *Statistical mechanics, thermodynamics and kinetics*, Chapter 9, W. H. Freeman (San Francisco), 1st edition, 1967.

1198a. RICHERT, H., *Z. anorg. allgem. chem.* **309**, 171 (1961).
1198b. RICHERT, H. and O. GLEMSER, *Z. anorg. allgem. chem.* **307**, 328 (1961).
1199. RICHMAN, I., *J. Opt. Soc. Amer.* **56**, 1589 (1966).
1200. RIMAI, L. and J. L. PARSONS, *Solid State Comm.* **5**, 387 (1967).
1201. RITCHIE, R. K. and H. LEW, *J. Phys.* **42**, 43 (1964).
1202. ROBIN-KANDARE, S. and J. ROBIN, *Compt. rend* **262B**, 1211 (1966).
1203. ROBINSON, E. A., *Can. J. Chem.* **40**, 1725 (1962).
1204. ROCCHICIOLLI, C., *Compt. rend.* **242**, 2922 (1956).
1205. ROCCHICIOLLI, C., *Compt. rend.* **244**, 2704 (1957).
1206. ROCCHICIOLLI, C., *Compt. rend.* **247**, 1108 (1958).
1207. ROCCHICIOLLI, C., *Compt. rend.* **249**, 236 (1959).
1208. ROCCHICIOLLI, C., *Compt. rend.* **250**, 2347 (1960).
1209. ROCCHICIOLLI, C., *Compt. rend.* **256**, 1707 (1963).
1210. ROCCHICIOLLI, C., *Compt. rend.* **257**, 3851 (1963).
1211. ROCHKIND, M. M. and G. C. PIMENTEL, *J. Chem. Phys.* **42**, 1361 (1965).
1212. ROLFE, J. A., D. E. SHEPPARD and L. A. WOODWARD, *Trans. Far. Soc.* **50**, 1275 (1954).
1213. ROLFE, J. A. and L. A. WOODWARD, *Trans. Far. Soc.* **51**, 779 (1955).
1214. ROLFE, J. A., L. A. WOODWARD and D. A. LONG, *Trans. Far. Soc.* **49**, 1388 (1953).
1215. ROSS, S. D., *Spectrochim. Acta* **18**, 225 (1962).
1216. ROSS, S. D., *Spectrochim. Acta* **18**, 1575 (1962).
1217. ROSS, S. D., *J. Mol. Spec.* **27**, 131 (1969).
1218. ROSS, S. D. and J. A. GOLDSMITH, *Spectrochim. Acta* **20**, 781 (1964).
1219. ROSZINSKI, H., R. DAUTEL and W. ZEIL, *Z. Physik. Chem.* (Frankfurt) 36, 26 (1963).
1219a. ROTHSCHILD, W. G., *J. Opt. Soc. Amer.* **54**, 20 (1964).
1220. ROUSSEAU, D. L., R. E. MILLER and G. E. LEROI, *J. Chem. Phys.* **48**, 3409 (1968).
1221. RUSSELL, J. P. and R. LOUDON, *Proc. Phys. Soc.* **85**, 1029 (1965).
1222. RYASON, R. and M. K. WILSON, *J. Chem. Phys.* **22**, 2000 (1954).
1222a. RYSKIN, YA. Y., *Optika i Spektroskopiya* **6**, 113 (1959).
1223. SABATINI, A. and L. SACCONI, *J. Amer. Chem. Soc.* **86**, 17 (1964).
1224. SABATINI, A., L. SACCONI and V. SCHETTINO, *Inorg. Chem.* **3**, 1775 (1964).
1225. SACCONI, L., A. SABATINI and P. GANS, *Inorg. Chem.* **3**, 1772 (1964).
1226. SAHINI, V. E. and M. DAMASCHIN, *Rev. Chem. Acad. R.P.R.* **8**, 193 (1963).
1227. SAKSENA, B. D., *Trans. Far. Soc.* **57**, 242 (1961).
1228. SALTHOUSE, J. A. and T. C. WADDINGTON, *Spectrochim. Acta* **23A**, 1069 (1967).
1229. SANBORN, R. H., *J. Chem. Phys.* **33**, 1855 (1960).
1230. SARALIDZE, O. D., L. P. SHKLOVER, K. I. PETROV and V. I. PLYUSHCHEV, *Zh. Strukt. Khim.* **8**, 57 (1967).
1231. SATHIANADAN, K., L. D. MCCORY and J. L. MARGRAVE, *Spectrochim. Acta* **20**, 957 (1964).
1232. SATHIANADAN, K., K. RAMASWAMY, S. SUNDARAM and F. F. CLEVELAND, *J. Mol. Spec.* **13**, 214 (1964).
1233. SATO, Y., *J. Phys. Soc. Japan* **20**, 275 (1965).
1234. SATO, Y., K. GESI and Y. TAKAGI, *J. Phys. Soc. Japan* **19**, 449 (1964).
1235. SATYANARAYANA, B. S., *Proc. Ind. Acad. Sci.* **15A**, 414 (1942).
1236. SAUER, P., *Z. Physik* **199**, 280 (1967).
1237. SAVOIE, R. and P. A. GIGUERE, *Can. J. Chem.* **40**, 991 (1962).

1238. SAVOIE, R. and P. A. GIGUERE, *J. Chem. Phys.* **41**, 2698 (1964).
1239. SAWODNY, W. and K. DEHNICKE, *Z. anorg. allgem. chem.* **349**, 169 (1967).
1240. SAWODNY, W., A. FADINI and K. BALLEIN, *Spectrochim. Acta* **21**, 995 (1965).
1241. SAWODNY, W., A. RUOFF, C. J. PEACOCK and A. MUELLER, *Mol. Phys.* **14**, 433 (1968).
1242. SCHAEFER, K. and E. WICKE, *Z. Elektrochem.* **52**, 205 (1948).
1243. SCHATZ, P. N., *J. Chem. Phys.* **29**, 481 (1958).
1244. SCHAUFELE, R. F. and M. J. WEBER, *J. Chem. Phys.* **46**, 2859 (1967).
1245. SCHMELZ, M. J., T. MIYAZAWA, S. MIZUSHIMA, T. J. LANE and J. V. QUAGLIANO, *Spectrochim. Acta* **9**, 51 (1957).
1246. SCHROEDER, R. A., E. R. LIPPINCOTT and C. E. WEIR, *J. Inorg. Nucl. Chem.* **28**, 1397 (1966).
1247. SCHROEDER, R. A. and L. L. LYONS, *J. Inorg. Nucl. Chem.* **28**, 115 (1966).
1248. SCHROEDER, R. A., C. E. WEIR and E. R LIPPINCOTT, *J. Res. Natl. Bur. Stds.* **66A**, 407 (1962).
1249. SCHULTZ, J. W. and D. F. HORNIG, *J. Phys. Chem.* **65**, 2131 (1961).
1250. SCHUTTE, C. J. H., *Spectrochim. Acta* **16**, 1054 (1960).
1251. SCHUTTE, C. J. H., *Nature* **189**, 745 (1961).
1252. SCHUTTE, C. J. H., *Nature* **190**, 805 (1961).
1253. SCHUTTE, C. J. H., *Z. phys. chem.* (Frankfurt) **39**, 241 (1963).
1254. SCHUTTE, C. J. H., *Z. Krist.* **126**, 397 (1968).
1255. SCHUTTE, C. J. H. and K. BUIJS, *Nature* **192**, 351 (1961).
1256. SCHUTTE, C. J. H. and K. BUIJS, *Spectrochim. Acta* **20**, 187 (1964).
1257. SCHWOCHAU, K., *Angew. Chem.* **76**, 608 (1964).
1258. SCOTT, D. W. and J. P. MCCULLOUGH, *J. Mol. Spec.* **6**, 372 (1961).
1259. SCOTT, D. W., J. P. MCCULLOUGH and F. H. KRUSE, *J. Mol. Spec.* **13**, 313 (1964).
1260. SCOTT, J. F., *J. Chem. Phys.* **48**, 874 (1968).
1261. SCOTT, J. F., *J. Chem. Phys.* **49**, 2766 (1968).
1262. SCOTT, R. A. and H. SCHERAGA, *J. Chem. Phys.* **42**, 2209 (1965).
1263. SCRUBY, R. E., J. R. LACHER and J. D. PARK, *J. Chem. Phys.* **19**, 386 (1951).
1264. SEEL, F. and J. BOUDIER, *Z. anorg. allgem. chem.* **342**, 173 (1966).
1265. SEEL, F. and R. BUDENZ, *Chem. Ber.* **98**, 251 (1965).
1266. SEEL, F. and D. GOELITZ, *Z. anorg. allgem. chem.* **327**, 32 (1964).
1267. SEIDEL, H., *Naturwiss* **52**, 257 (1965).
1268. SEIDEL, H., *Naturwiss* **52**, 539 (1965).
1269. SELIG, H., *Science* **144**, 537 (1964).
1270. SELIG, H., H. H. CLAASSEN, C. L. CHERNICK, J. G. MALM and J. L. HUSTON, *Science* **143**, 1322 (1964).
1271. SELIG, H., A. REIS and E. L. GASNER, *J. Inorg. Nucl. Chem.* **30**, 2087 (1968).
1272. SENIOR, W. A. and W. K. THOMPSON, *Nature* **205**, 170 (1965).
1273. SERVOSS, R. R. and H. M. CLARK, *J. Chem. Phys.* **26**, 1175 (1957).
1274. SEVCHENKO, A. I. and B. I. STEPANOV, *Zh. Ekspt. Teor. Fiz.* **19**, 1113 (1949).
1275. SHAMIR, J. and H. H. HYMAN, *Spectrochim. Acta* **23A**, 1191 (1967).
1276. SHAMIR, J. and H. H. HYMAN, *Spectrochim. Acta* **23A**, 1899 (1967).
1277. SHANTAKUMARI, C., *Proc. Ind. Acad. Sci.* **28A**, 500 (1948).
1278. SHANTAKUMARI, C., *Current Sci.* (India) **19**, 204 (1950).
1279. SHANTAKUMARI, C., *Proc. Ind. Acad. Sci.* **32A**, 177 (1950).
1280. SHANTAKUMARI, C., *Proc. Ind. Acad. Sci.* **37A**, 393 (1953).
1281. SHAPIRO, S. M., D. S. O'SHEA and H. Z. CUMMINS, *Phys. Rev. Letters* **19**, 361 (1967).

1282. SHARP, D. W. A., *J. Chem. Soc.* 3761 (1957).
1283. SHELDRICK, G. M., *J. Mol. Spec.* **20**, 295 (1966).
1284. SHIMANOUCHI, T., *J. Chem. Phys.* **17**, 245 (1949).
1285. SHIMANOUCHI, T., *J. Chem. Phys.* **25**, 660 (1956).
1286. SHIMANOUCHI, T. and I. NAKAGAWA, *Spectrochim. Acta* **18**, 89 (1962).
1287. SHIMANOUCHI, T. and I. NAKAGAWA, *Inorg. Chem.* **3**, 1805 (1964).
1288. SHIMANOUCHI, T. and I. SUZUKI, *J. Mol. Spec.* **6**, 277 (1961).
1289. SHIMANOUCHI, T. and I. SUZUKI, *J. Mol. Spec.* **8**, 222 (1962).
1290. SHIMANOUCHI, T. and I. SUZUKI, *J. Chem. Phys.* **42**, 296 (1965).
1291. SHIMANOUCHI, T., M. TSUBOI and T. MIYAZAWA, *J. Chem. Phys.* **35**, 1597 (1961).
1292. SHIMIZU, K. and H. SHINGU, *Spectrochim. Acta* **22**, 1999 (1966).
1292a. SHURVELL, H. F., *Spectrochim. Acta* **25A**, 973 (1969).
1293. SIEBERT, H., *Z. anorg. allgem. chem.* **265**, 303 (1951).
1294. SIEBERT, H., *Z. anorg. allgem. chem.* **273**, 21 (1953).
1295. SIEBERT, H., *Z. anorg. allgem. chem.* **275**, 210 (1955).
1296. SIEBERT, H., *Z. anorg. allgem. chem.* **275**, 225 (1955).
1297. SIEBERT, H., *Z. anorg. allgem. chem.* **289**, 15 (1957).
1298. SIEBERT, H., *Z. anorg. allgem. chem.* **296**, 280 (1958).
1299. SIEBERT, H., *Z. anorg. allgem. chem.* **298**, 51 (1959).
1300. SIEBERT, H., *Z. anorg. allgem. chem.* **301**, 161 (1959).
1301. SIEBERT, H., *Z. anorg. allgem. chem.* **303**, 162 (1960).
1302. SIEBERT, H., *Z. anorg. allgem. chem.* **304**, 266 (1960).
1303. SILVERMAN, B. D. and G. F. KOSTER, *Z. Physik* **165**, 334 (1961).
1304. SILVIDI, A. A., J. W. MCGRATH and D. F. HOLCOMB, *J. Chem. Phys.* **41**, 105 (1964).
1305. SIMON, A. and H. KREIGSMANN, *Naturwiss* **42**, 12 (1955).
1306. SIMON, A. and H. KREIGSMANN, *Chem. Ber.* **89**, 2443 (1956).
1307. SIMON, A., H. KREIGSMANN and H. DUTZ, *Chem. Ber.* **89**, 2378 (1956).
1308. SIMON, A. and H. KUECHLER, *Z. anorg. allgem. chem.* **260**, 161 (1949).
1309. SIMON, A. and R. LEHMANN, *Z. anorg. allgem. chem.* **311**, 212 (1961).
1310. SIMON, A. and R. LEHMANN, *Z. anorg. allgem. chem.* **311**, 224 (1961).
1311. SIMON, A. and R. PAETZOLD, *Naturwiss* **44**, 108 (1957).
1312. SIMON, A. and R. PAETZOLD, *Z. Elektrochem.* **64**, 209 (1959).
1313. SIMON, A. and R. PAETZOLD, *Z. anorg. allgem. chem.* **301**, 246 (1959).
1314. SIMON, A. and R. PAETZOLD, *Z. anorg. allgem. chem.* **303**, 39 (1960).
1315. SIMON, A. and R. PAETZOLD, *Z. anorg. allgem. chem.* **303**, 46 (1960).
1316. SIMON, A. and R. PAETZOLD, *Z. anorg. allgem. chem.* **303**, 53 (1960).
1317. SIMON, A. and R. PAETZOLD, *Z. anorg. allgem. chem.* **303**, 79 (1960).
1318. SIMON, A. and H. RICHTER, *Z. anorg. allgem. chem.* **301**, 154 (1959).
1319. SIMON, A. and H. RICHTER, *Z. anorg. allgem. chem.* **304**, 1 (1960).
1320. SIMON, A. and H. RICHTER, *Z. anorg. allgem. chem.* **315**, 196 (1962).
1321. SIMON, A. and W. SCHMIDT, *Z. Elektrochem.* **64**, 737 (1960).
1322. SIMON, A. and E. STEGER, *Z. anorg. allgem. chem.* **277**, 209 (1954).
1323. SIMON, A. and E. STEGER, *Z. anorg. allgem. chem.* **288**, 131 (1957).
1324. SIMON, A. and E. STEGER, *Z. anorg. allgem. chem.* **296**, 305 (1958).
1325. SIMON, A. and H. WAGNER, *Z. anorg. allgem. chem.* **311**, 102 (1961).
1326. SIMON, A. and K. WALDMANN, *Z. Physik. Chem.* **204**, 235 (1955).
1327. SIMON, A. and K. WALDMANN, *Z. anorg. allgem. chem.* **281**, 113 (1955).
1328. SIMON, A. and K. WALDMANN, *Z. anorg. allgem. chem.* **281**, 135 (1955).
1329. SIMON, A. and K. WALDMANN, *Z. anorg. allgem. chem.* **283**, 359 (1956).

1330. Simon, A. and K. Waldmann, *Naturwiss* **45**, 128 (1958).
1331. Simon, A., K. Waldmann and E. Steger, *Z. anorg. allgem. chem.* **288**, 131 (1956).
1332. Siratori, K. and Y. Aiyama, *J. Phys. Soc. Japan* **20**, 1962 (1965).
1333. Siratori, K., A. Tsuchida and Y. Tomono, *J. Appl. Phys.* **36**, 1050 (1965).
1334. Skiens, W. E. and G. H. Cady, *J. Amer. Chem. Soc.* **80**, 5640 (1958).
1335. Smith, A. L., W. E. Keller and H. L. Johnston, *J. Chem. Phys.* **19**, 189 (1951).
1336. Smith, D. C., *J. Chem. Phys.* **19**, 384 (1951).
1337. Smith, D. F., *J. Amer. Chem. Soc.* **85**, 816 (1963).
1338. Smith, D. F., *Science* **141**, 61 (1963).
1339. Smith, D. F., *Science* **141**, 1039 (1963).
1340. Smith, D. F., *J. Chem. Phys.* **48**, 1429 (1968).
1341. Smith, D. F. and G. M. Begun, *J. Chem. Phys.* **43**, 2001 (1965).
1342. Smith, D. F., G. M. Begun and W. H. Fletcher, *Spectrochim. Acta* **20**, 1763 (1964).
1343. Smith, H. M. J., *Proc. Roy. Soc.* **241A**, 105 (1948).
1344. Snelson, A., *J. Chem. Phys.* **46**, 3652 (1967).
1345. Snelson, A., *J. Phys. Chem.* **71**, 3202 (1967).
1346. Snelson, A., *J. Phys. Chem.* **72**, 250 (1968).
1347. Snyder, R. G. and J. C. Decius, *Spectrochim. Acta* **13**, 280 (1959).
1348. Snyder, R. G. and I. C. Hisatsune, *J. Mol. Spec.* **1**, 139 (1957).
1349. Solomons, C., J. H. R. Clarke and J. O'M. Bockris, *J. Chem. Phys.* **49**, 445 (1968).
1350. Soulen, J. R. and W. F. Schwartz, *J. Phys. Chem.* **66**, 2066 (1962).
1351. Spinner, E., *J. Chem. Soc.* **B**, 879 (1967).
1352. Spiro, T. G., *Inorg. Chem.* **4**, 1290 (1965).
1353. Spiro, T. G., *Inorg. Chem.* **6**, 569 (1967).
1354. Spitzer, W. G. and D. A. Kleinman, *Phys. Rev.* **121**, 1234 (1961).
1355. Spitzer, W. G., R. C. Miller, D. A. Kleinman and L. E. Howarth, *Phys. Rev.* **126**, 1710 (1962).
1356. Staats, P. A., H. W. Morgan and J. H. Goldstein, *J. Chem. Phys.* **24**, 916 (1956).
1356a. Staats, P. A., H. W. Morgan and J. H. Goldstein, *J. Chem. Phys.* **25**, 582 (1956).
1357. Stafsudd, O. M., F. A. Haak and K. Radisavljevic, *J. Opt. Soc. Amer.* **57**, 1475 (1967).
1358. Stammreich, H., D. Bassi and O. Sala, *Spectrochim. Acta* **12**, 403 (1958).
1359. Stammreich, H., D. Bassi, O. Sala and H. Siebert, *Spectrochim. Acta* **13**, 192 (1959).
1360. Stammreich, H. and R. Forneris, *Spectrochim. Acta* **8**, 46 (1956).
1361. Stammreich, H. and R. Forneris, *Spectrochim. Acta* **16**, 363 (1960).
1362. Stammreich, H., R. Forneris and K. Sone, *J. Chem. Phys.* **23**, 972 (1955).
1363. Stammreich, H., R. Forneris and Y. Tavares, *J. Chem. Phys.* **25**, 580 (1956).
1364. Stammreich, H., R. Forneris and Y. Tavares, *J. Chem. Phys.* **25**, 1278 (1956).
1365. Stammreich, H., K. Kawai and Y. Tavares, *Spectrochim. Acta* **15**, 438 (1959).
1366. Stammreich, H. and Y. Kawano, *Spectrochim. Acta* **24A**, 899 (1968).
1367. Stammreich, H., O. Sala and D. Bassi, *Spectrochim. Acta* **19**, 593 (1963).

1368. STAMMREICH, H., O. SALA and K. KAWAI, *Spectrochim. Acta* **17**, 226 (1961).
1369. STAMMREICH, H., Y. TAVARES and D. BASSI, *Spectrochim. Acta* **17**, 661 (1961).
1370. STEELE, W. C. and J. C. DECIUS, *J. Chem. Phys.* **25**, 1184 (1956).
1371. STEGER, E., *Z. anorg. allgem. chem.* **296**, 305 (1958).
1372. STEGER, E., I. C. CIUREA and A. FADINI, *Z. anorg. allgem. chem.* **350**, 225 (1967).
1373. STEGER, E. and K. HERZOG, *Z. anorg. allgem. chem.* **331**, 169 (1964).
1374. STEGER, E. and B. KAESSNER, *Spectrochim. Acta* **24A**, 447 (1968).
1375. STEGER, E. and G. LEUKROTH, *Z. anorg. allgem. chem.* **303**, 169 (1960).
1376. STEGER, E. and K. MARTIN, *Z. anorg. allgem. chem.* **308**, 330 (1961).
1377. STEGER, E. and K. MARTIN, *Z. anorg. allgem. chem.* **322**, 215 (1963).
1378. STEGER, E. and K. MARTIN, *Z. anorg. allgem. chem.* **323**, 108 (1963).
1379. STEGER, E. and W. SCHMIDT, *Ber. Bunsenges. Phys. Chem.* **68**, 102 (1964).
1380. STEIN, L., *J. Amer. Chem. Soc.* **81**, 1273 (1959).
1381. STEINHARDT, R. G., jr., G. E. S. FETSCH and M. W. JORDAN, *J. Chem. Phys.* **43**, 4528 (1965).
1382. STEKHANOV, A. I., *Optika i Spektroskopiya* **3**, 143 (1957).
1383. STEPHENSON, C. V. and E. A. JONES, *J. Chem. Phys.* **20**, 1830 (1952).
1384. STERZEL, W., *Naturwiss* **51**, 505 (1964).
1385. STERZEL, W. and E. CHORINSKY, *Spectrochim. Acta* **24A**, 353 (1968).
1386. STITT, F. and D. M. YOST, *J. Chem. Phys.* **5**, 90 (1937).
1387. STOPPERKA, K., *Z. anorg. allgem. chem.* **345**, 277 (1966).
1388. STORCHAI, N. T. and V. P. MOROZOV, *Dopov. Akad. Nauk Ukr. RSR* 1596 (1966).
1389. STRAUB, D. K., R. S. DRAGO and J. T. DONOGHUE, *Inorg. Chem.* **1**, 848 (1962).
1390. STRAUCH, B. and L. M. KOMISSAROVA, *Z. Chem.* **6**, 474 (1966).
1391. STREY, G., *J. Mol. Spec.* **17**, 265 (1965).
1392. SULLIVAN, J. J., *J. Phys. and Chem. Solids* **25**, 1039 (1964).
1393. SUMMITT, R., *J. Appl. Phys.* **39**, 3762 (1968).
1394. SUNDARAM, S., *J. Chem. Phys.* **33**, 708 (1960).
1395. SUNDARAM, S., *Can. J. Phys.* **39**, 370 (1961).
1396. SUNDARAM, S. and F. F. CLEVELAND, *J. Mol. Spec.* **5**, 61 (1960).
1397. SUTHERLAND, G. B. B. M., *Proc. Roy. Soc.* **A141**, 342 (1933).
1398. SUTHERLAND, G. B. B. M., *Proc. Roy. Soc.* **A141**, 535 (1933).
1399. SUTTON, L. E., *Nature* **169**, 235 (1952).
1400. SWALEN, J. D. and C. C. COSTAIN, *J. Chem. Phys.* **31**, 1562 (1959).
1401. SZYMANSKI, H. A., R. YELIN and L. MARABELLA, *J. Chem. Phys.* **47**, 1877 (1967).
1402. TANAKA, M., K. BALASUBRAMANYAM and J. O'M. BOCKRIS, *Electrochim. Acta* **8**, 621 (1963).
1403. TANAKA, N., H. SUGI and J. FUJITA, *Bull. Chem. Soc. Japan* **37**, 640 (1964).
1404. TANDON, S. P., *Proc. Rajasthan Acad. Sci* **9**, 3 (1962).
1405. TANNER, K. N. and A. B. F. DUNCAN, *J. Amer. Chem. Soc.* **73**, 1164 (1951).
1406. TARTE, P., *Anal. soc. sci. Bruxelles* (Ser. 1) **70**, 244 (1956).
1407. TARTE, P., *Nature* **191**, 1002 (1961).
1408. TARTE, P., *Spectrochim. Acta* **18**, 467 (1962).
1409. TARTE, P., *Spectrochim. Acta* **19**, 25 (1963).
1410. TARTE, P., *Spectrochim. Acta* **19**, 49 (1963).
1411. TARTE, P., *Acta Cryst.* **16**, 228 (1963).
1412. TARTE, P., *Spectrochim. Acta* **20**, 238 (1964).

1413. Tarte, P., *Spectrochim. Acta* **21**, 313 (1965).
1414. Tarte, P., *J. Inorg. Nucl. Chem.* **27**, 1933 (1965).
1415. Tarte, P., *Spectrochim. Acta* **23A**, 2127 (1967).
1416. Tarte, P. and G. Nizet, *Spectrochim. Acta* **20**, 503 (1964).
1417. Tarte, P. and J. Preudhomme, *Acta Cryst.* **16**, 227 (1963).
1418. Tarte, P. and A. E. Ringwood, *Nature* **193**, 971 (1962).
1418a. Taylor, J. H., W. S. Benedict and J. Strong, *J. Chem. Phys.* **20**, 1884 (1952).
1419. Taylor, M. J. and L. A. Woodward, *J. Chem. Soc.* 4670 (1963).
1420. Taylor, R. C., *J. Chem. Phys.* **18**, 898 (1950).
1421. Taylor, R. C. and P. C. Cross, *J. Chem. Phys.* **24**, 41 (1956).
1422. Taylor, R. C. and A. R. Emery, *Spectrochim. Acta* **10**, 419 (1958).
1423. Taylor, R. C. and G. L. Vidale, *J. Amer. Chem. Soc.* **78**, 5999 (1956).
1424. Taylor, W., *Phys. Lett.* **24A**, 556 (1967).
1425. Taylor, W. J., *J. Chem. Phys.* **18**, 1301 (1950).
1426. Tell, B., T. C. Damen and S. P. S. Porto, *Phys. Rev.* **144**, 771 (1966).
1427. Teranishi, R. and J. C Decius, *J. Chem. Phys.* **21**, 1116 (1953).
1428. Teranishi, R. and J. C. Decius, *J. Chem. Phys.* **22**, 896 (1954).
1429. Terrasse, J. M., H. Poulet and J.-P. Mathieu, *Spectrochim. Acta* **20**, 305 (1964).
1430. Theoret, A. and C. Sandorfy, *Can. J. Chem.* **42**, 57 (1964).
1431. Thompson, W. K., *J. Chem. Soc.* 998 (1963).
1432. Thompson, W. T. and W. H. Fletcher, *Spectrochim. Acta* **22**, 1907 (1966).
1433. Topping, G., *Spectrochim. Acta* **21**, 1743 (1965).
1434. Torkington, P., *J. Chem. Phys.* **17**, 1026 (1947).
1435. Torkington, P., *Trans. Far. Soc.* **47**, 105 (1951).
1435a. Toyuki, H., H. Horiuchi and K. Shimuzu, *Kagaka To Kogyo* (Osaka), **40**, 35 (1966).
1436. Toyuki, H. and K. Shimizu, *Bull. Chem. Soc. Japan* **39**, 2364 (1966).
1437. Toyuki, H., K. Shimizu and H. Hiyama, *Bull. Chem. Soc. Japan* **36**, 783 (1963).
1438. Tramer, A., *Bull. Acad. polon. Sci., Ser. Sci. Math, astron. et phys.* **6**, 659 (1958).
1439. Tramer, A., *Compt. rend.* **249**, 2755 (1959).
1440. Tramer, A. and J.-P. Mathieu, *Compt. rend.* **249**, 392 (1959).
1441. Tsuboi, M., *J. Amer. Chem. Soc.* **79**, 1351 (1957).
1442. Tsuboi, M., *Proc. Intern. Symp. Mol. Struct. Spectry. Tokyo* **A305**, 1/4 (1962).
1443. Tsuboi, M., *J. Chem. Phys.* **40**, 1326 (1964).
1444. Tsuboi, M. and A. Wada, *J. Chem. Phys.* **48**, 2615 (1968).
1445. Tsuchida, A. and I. Nakagawa, *J. Phys. Soc. Japan* **20**, 1726 (1965).
1446. Ulbricht, K. and H. Kreigsmann, *Z. Chem.* **5**, 266 (1965).
1447. Ulbricht, K. and H. Kreigsmann, *Z. Chem.* **6**, 232 (1966).
1448. Ulbricht, K. and H. Kreigsmann, *Z. Chem.* **7**, 244 (1967).
1449. Ulbricht, K. and H. Kreigsmann, *Z. anorg. allgem. chem.* **358**, 193 (1968).
1449a. Urey, H. C. and C. A. Bradley, *Phys. Rev.* **38**, 1969 (1931).
1450. Van Der Elsken, J. and D. W. Robinson, *Spectrochim. Acta* **17**, 1249 (1961).
1451. Vanderryn, J., *J. Chem. Phys.* **30**, 331 (1959).
1452. Van Der Voorn, P. C., K. F. Purcell and R. S. Drago, *J. Chem. Phys.* **43**, 3457 (1965).

1453. VDOVENKO, V. M., L. G. MASHIROV and D. N. SUGLOVOV, *Radiokhimiya* **6**, 299 (1964).
1454. VEDDER, W. and D. F. HORNIG, *J. Chem. Phys.* **35**, 1560 (1961).
1454a. VENKATESWARAN, C. S., *Proc. Ind. Acad. Sci.* **1A**, 850 (1935).
1455. VENKATESWARAN, C. S., *Proc. Ind. Acad. Sci.* **7A**, 144 (1938).
1456. VENKATESWARLU, K. and K. V. RAJALAKSHMI, *J. Sci. Ind. Res.* (India) **21B**, 349 (1962).
1457. VENKATESWARLU, K. and K. RAMASWAMY, *Ind. J. Pure Appl. Phys.* **2**, 209 (1964).
1458. VENKATESWARLU, K. and K. SATHIANANDAN, *Optika i Spektroskopiya* **11**, 46 (1961).
1459. VENKATESWARLU, K. and S. SUNDARAM, *J. Chem. Phys.* **23**, 2365 (1955).
1460. VENKATESWARLU, K. and R. THANALAKSHMI, *Proc. Ind. Acad. Sci.* **57A**, 181 (1963).
1461. VENKATESWARLU, K. and P. THIRUGNANASAMBANDAM, *Proc. Ind. Acad. Sci.* **48A**, 344 (1958).
1462. VENKATESWARLU, K. and P. THIRUGNANASAMBANDAM, *Trans. Far. Soc.* **55**, 1993 (1959).
1463. VERE, G. J. and C. C. PATEL, *Can. J. Chem.* **40**, 1576 (1962).
1464. VILTANGE, M., *Compt. rend.* **256**, 2816 (1963).
1465. VILTANGE, M., *Mikrochim. Ichnoanal. Acta* **3**, 461 (1964).
1466. VILTANGE, M., *Mikrochim. Ichnoanal. Acta* **4**, 609 (1965).
1467. VILTANGE, M., *Mikrochim. Ichnoanal. Acta* **4**, 809 (1965).
1468. VITEK, A., *Coll. Czech. Chem. Comm.* **33**, 1601 (1968).
1469. VRATNY, F., *Appl. Spectroscopy* **13**, 59 (1959).
1470. VRATNY, F., C. N. R. RAO and M. DILLING, *Anal. Chem.* **33**, 1455 (1961).
1471. WADDINGTON, T. C., *J. Chem. Soc.* 1708 (1958).
1472. WADDINGTON, T. C., *J. Chem. Soc.* 4340 (1958).
1473. WADDINGTON, T. C., *J. Chem. Soc.* 2499 (1959).
1474. WADDINGTON, T. C. and F. KLANBERG, *J. Chem. Soc.* 2339 (1960).
1475. WAGNER, E. L. and E. L. BULGOZDY, *J. Chem. Phys.* **19**, 1210 (1951).
1476. WAGNER, E. L. and D. F. HORNIG, *J. Chem. Phys.* **18**, 296 (1950).
1477. WAGNER, E. L. and D. F. HORNIG, *J. Chem. Phys.* **18**, 305 (1950).
1478. WALDRON, R. D., *Phys. Rev.* **99**, 1727 (1955).
1479. WALKER, A. and J. R. FERRARO, *J. Chem. Phys.* **43**, 2689 (1965).
1480. WALRAFEN, G. E., *J. Chem. Phys.* **36**, 90 (1962).
1481. WALRAFEN, G. E., *J. Chem. Phys.* **37**, 1468 (1962).
1482. WALRAFEN, G. E., *J. Chem. Phys.* **40**, 3749 (1964).
1483. WALRAFEN, G. E. and D. M. DODD, *Trans. Far. Soc.* **57**, 1286 (1961).
1484. WALRAFEN, G. E., D. E. IRISH and T. F. YOUNG, *J. Chem. Phys.* **37**, 662 (1962).
1485. WALTON, R. A., *Spectrochim. Acta* **24A**, 1527 (1968).
1486. WALTON, R. A. and B. J. BRISDON, *Spectrochim. Acta* **23A**, 2489 (1967).
1487. WARRIER, A. V. R., P. RAMAMURTHY, C. C. PATEL and O. S. NARAYANAN, *Indian J. Chem.* **2**, 131 (1964).
1488. WATT, G. W. and D. S. KLETT, *Inorg. Chem.* **3**, 782 (1964).
1488a. WEAVER, E. E., B. WEINSTOCK and C. P. KNOP, *J. Amer. Chem. Soc.* **85**, 111 (1963).
1489. WEBB, A. N., J. T. NEU and K. S. PITZER, *J. Chem. Phys.* **17**, 1007 (1949)
1490. WEBER, A. and S. M. FERIGLE, *J. Chem. Phys.* **20**, 1497 (1952).

1491. WEGDAM, G. H., R. BONN and J. VAN DER ELSKEN, *Chem. Phys. Letters*, 182 (1968).
1492. WEIDLEIN, J. and K. DEHNICKE, *Z. anorg. allgem. chem.* **337**, 113 (1965).
1492a. WEIDLEIN, J. and K. DEHNICKE, *Angew. Chem. (Int. Ed.)* **5**, 1041 (1966).
1493. WEIDLEIN, J. and K. DEHNICKE, *Z. anorg. allgem. chem.* **348**, 278 (1966).
1494. WEIDLEIN, J., U. MUELLER and K. DEHNICKE, *Spectrochim. Acta* **24A**, 253 (1968).
1494a. WEIL, A. and J.-P. MATHIEU, *Compt. rend.* **238**, 2510 (1954).
1495. WEINSTOCK, B., H. H. CLAASSEN and C. L. CHERNICK, *J. Chem. Phys.* **38**, 1470 (1963).
1496. WEINSTOCK, B., H. H. CLAASSEN and J. G. MALM, *J. Chem. Phys.* **32**, 181 (1960).
1497. WEINSTOCK, B. and G. L. GOODMAN, *Advan. Chem. Phys.* **9**, 169 (1965).
1498. WEINSTOCK, B. and J. G. MALM, *J. Inorg. Nucl. Chem.* **2**, 380 (1956).
1499. WEIR, C. E. and E. R. LIPPINCOTT, *J. Res. Nat. Bur. Stds.* **65A**, 173 (1961).
1499a. WEIR, C. E., E. R. LIPPINCOTT, A. V. VALKENBURG and E. N. BUNTING, *J. Res. Nat. Bur. Stds.* **63A**, 55 (1959).
1500. WEIR, C. E. and R. A. SCHROEDER, *J. Res. Nat. Bur. Stds.* **68A**, 465 (1964).
1501. WENDLING, E., *Bull. Soc. Chim. France* 16–23 (1967).
1502. WENDLING, E. and R. ROHMER, *Bull. Soc. Chim. France* 8–16 (1967).
1503. WENDLING, E., R. ROHMER and R. WEISS, *Compt. rend.* **256**, 1117 (1963).
1504. WENTINK, T., *J. Chem. Phys.* **29**, 188 (1958).
1504a. WENTINK, T., *J. Chem. Phys.* **30**, 105 (1959).
1505. WENTINK, T. and V. H. TIENSUU, *J. Chem. Phys.* **28**, 826 (1958).
1506. WERDER, R. D., R. A. FREY and HS. H. GUNTHARD, *J. Chem. Phys.* **47**, 4159 (1967).
1507. WESTON, R. E. and T. F. BRODASKY, *J. Chem. Phys.* **27**, 682 (1957).
1507a. WHITE, D., D. E. MANN, P. N. WALSH and A. SOMMER, *J. Chem. Phys.* **32**, 481 (1960).
1508. WHITE, W. B. and B. A. DE ANGELIS, *Spectrochim. Acta* **23A**, 1069 (1967).
1509. WIENER, R. N. and E. R. NIXON, *J. Chem. Phys.* **26**, 906 (1957).
1510. WILKINSON, G. R., *Molecular Spectroscopy (Conference Proceedings 1968)*. (Institute of Petroleum, London, 1968) 43–81.
1511. WILMSHURST, J. K., *J. Chem. Phys.* **23**, 2463 (1955).
1512. WILMSHURST, J. K., *J. Mol. Spectroscopy* **5**, 343 (1960).
1513. WILMSHURST, J. K. and H. J. BERNSTEIN, *Can. J. Chem.* **35**, 191 (1957).
1514. WILMSHURST, J. K. and H. J. BERNSTEIN, *J. Chem. Phys.* **27**, 661 (1957).
1515. WILMSHURST, J. K. and S. SENDEROFF, *J. Chem. Phys.* **35**, 1078 (1961).
1516. WILSON, E. B., *J. Chem. Phys.* **7**, 1047 (1939).
1517. WILSON, E. B., *J. Chem. Phys.* **9**, 76 (1941).
1518. WILSON, M. K. and R. M. BADGER, *J. Chem. Phys.* **17**, 1232 (1949).
1519. WILSON, M. K. and S. R. POLO, *J. Chem. Phys.* **20**, 1716 (1952).
1520. WINNEWISSER, B. P. and M. WINNEWISSER, *Z. Naturforsch* **23A**, 832 (1968).
1521. WINSTON, H. and R. S. HALFORD, *J. Chem. Phys.* **17**, 607 (1949).
1522. WISE, J. H. and J. T. ELMER, *J. Chem. Phys.* **18**, 411 (1950).
1523. WISE, J. H. and H. H. HANNAN, *J. Inorg. Nucl. Chem.* **23**, 31 (1961).
1524. WOLTZ, P. J. H., E. A. JONES and A. H. NIELSEN, *J. Chem. Phys.* **20**, 378 (1952).
1525. WOLTZ, P. J. H. and A. H. NIELSEN, *J. Chem. Phys.* **20**, 302 (1952).
1526. WOODHEAD, J. L. and J. M. FLETCHER, *J. Chem. Soc.* 5039 (1961).
1527. WOODWARD, L. A., *Trans. Far. Soc.* **54**, 1271 (1958).

1528. Woodward, L. A. and L. E. Anderson, *J. Chem. Soc.* 1284 (1957).
1529. Woodward, L. A. and P. T. Bill, *J. Chem. Soc.* 1699 (1955).
1530. Woodward, L. A. and J. A. Creighton, *Spectrochim. Acta* **17**, 594 (1961).
1531. Woodward, L. A., J. A. Creighton and K. A. Taylor, *Trans. Far. Soc.* **56**, 1267 (1960).
1532. Woodward, L. A., G. Garton and H. L. Roberts, *J. Chem. Soc.* 3723 (1956).
1533. Woodward, L. A., N. J. Greenwood, J. R. Hall and I. J. Worrall, *J. Chem. Soc.* 1505 (1958).
1534. Woodward, L. A. and A. A. Nord, *J. Chem. Soc.* 2655 (1955).
1535. Woodward, L. A. and A. A. Nord, *J. Chem. Soc.* 3721 (1956).
1536. Woodward, L. A. and H. L. Roberts, *J. Chem. Soc.* 1170 (1956).
1537. Woodward, L. A. and H. L. Roberts, *Trans. Far. Soc.* **52**, 615 (1956).
1538. Woodward, L. A. and H. L. Roberts, *Trans. Far. Soc.* **52**, 1458 (1956).
1539. Woodward, L. A. and H. L. Roberts, *Trans. Far. Soc.* **53**, 1557 (1957).
1540. Woodward, L. A. and G. H. Singer, *J. Chem. Soc.* 716 (1958).
1541. Woodward, L. A. and M. J. Taylor, *J. Chem. Soc.* 4473 (1960).
1542. Woodward, L. A. and M. J. Taylor, *J. Chem. Soc.* 407 (1962).
1543. Woodward, L. A. and M. J. Ware, *Spectrochim. Acta* **20**, 711 (1964).
1544. Woodward, L. A. and M. J. Ware, *Spectrochim. Acta* **24A**, 921 (1968).
1545. Yamaguchi, A., *J. Chem. Soc. Japan* **80**, 1109 (1959).
1546. Yellin, W. and R. A. Plane, *J. Amer. Chem. Soc.* **83**, 2448 (1961).
1547. Yeranos, W. A., *Bull. Soc. chim. belges* **74**, 407 (1965).
1548. Yeranos, W. A., *Bull. Soc. chim. belges* **74**, 414 (1965).
1549. Yeranos, W. A., *Mol. Phys.* **9**, 449 (1965).
1550. Yeranos, W. A., *Mol. Phys.* **9**, 455 (1965).
1551. Yeranos, W. A., *Z. Physik. Chem.* (Frankfurt) **45**, 72 (1965).
1552. Yeranos, W. A., *Z. Physik. Chem.* (Frankfurt) **45**, 77 (1965).
1553. Yeranos, W. A. and F. D. Foss, *Mol. Phys.* **9**, 87 (1965).
1554. Yeranos, W. A. and M. J. Joncich, *Mol. Phys.* **13**, 263 (1967).
1554a. Yost, D. M. and J. E. Sherborne, *J. Chem. Phys.* **2**, 125 (1934).
1555. Zengin, N. and P. A. Giguere, *Can. J. Chem.* **37**, 632 (1959).
1556. Ziomek, J. S., *J. Chem. Phys.* **22**, 1001 (1954).
1557. Ziomek, J. S., J. R. Ferraro and D. F. Peppard, *J. Mol. Spec.* **8**, 212 (1962).
1558. Ziomek, J. S. and C. B. Mast, *J. Chem. Phys.* **21**, 862 (1953).
1559. Ziomek, J. S. and E. A. Piotrowski, *J. Chem. Phys.* **34**, 1087 (1961).
1560. Ziomek, J. S. and M. D. Zeidler, *J. Mol. Spec.* **11**, 163 (1963).
1561. Zumwalt, L. R. and P. A. Giguere, *J. Chem. Phys.* **9**, 458 (1941).
1562. Abramowitz, S. and I. W. Levin, *J. Chem. Phys.* **51**, 463 (1969).
1563. Adams, D. M. and R. G. Churchill, *J. Chem. Soc.* **A**, 697 (1970).
1564. Adams, D. M. and D. C. Newton, *J. Chem. Soc.* **A**, 2822 (1970).
1565. Addison, C. C., R. Davis and N. Logan, *J. Chem. Soc.* **A** 3333 (1970).
1566. Agiorgitis, G., *Z. Naturforsch* **24B**, 932 (1969).
1567 Ahlijah, G. Y. and M. Goldstein, *J. Chem. Soc.* **A**, 326 (1970).
1568. Ahlijah, G. Y. and M. Goldstein, *J. Chem. Soc.* **A** 2590 (1970).
1569. Al-Baldawi, S. A., M. H. Brooker, T. E. Gough and D. E. Irish, *Canad. J. Chem.* **48**, 1202 (1970).
1570. Allan, A., J. L. Duncan, J. H. Holloway and D. C. McKean, *J. Mol. Spectrosc.* **31**, 368 (1969).
1571. Allavena, M., R. Rysnik, D. White, V. Calder and D. E. Mann, *J. Chem. Phys.* **50**, 3399 (1969).

1572. ALLEN, G. C. and G. A. M. EL-SHARKAWY, *Inorg. Nucl. Chem. Letters* **6**, 281 (1970).
1573. AMOS, D. W., *Chem. Comm.* 19 (1970).
1574. ANDERSON, A. and L. Y. WONG, *Canad. J. Chem.* **47**, 2713 (1969).
1575. ANDREWS, L. and D. L. FREDERICK, *J. Amer. Chem. Soc.* **92**, 775 (1970).
1576. APPELMAN, E. H., *Inorg Chem.* **8**, 223 (1969).
1577. ASHBY, E. C. and B. D. JAMES, *Inorg. Chem.* **8**, 2468 (1969).
1578. AYMONINO, P. J., H. SCHULZE and A. MUELLER, *Z. Naturforsch.* **24b**, 960 (1969).
1579. AZEEM, M., M. BROWNSTEIN and R. J. GILLESPIE, *Canad. J. Chem.* **47**, 4158 (1969).
1580. BANDY, A. R., H. B. FRIEDRICH and W. B. PERSON, *J. Chem. Phys.* **53**, 674 (1970).
1581. BARAN, E. J. and P. J. AYMONINO, *Anales. Assoc. quim. Argentina* **56**, 91 (1968).
1582. BARAN, E. G., P. J. AYMONINO and A. MUELLER, *Z. Naturforsch* **24B**, 271 (1969).
1583. BARANIAK, E., H. C. FREEMAN, J. M. JAMES and C. E. NOCKOLDS, *J. Chem. Soc.* **A**, 2558 (1970).
1584. BARRACLOUGH, C. G. and D. J. KEW, *Austral. J. Chem.* **23**, 2387 (1970).
1585. BARRACLOUGH, J. M., D. H. BROWN, A. L. LANE, D. W. A. SHARP and D. A. PATERSON, *Chem. Comm.* 608 (1969).
1586. BAYERSDORFER, L., U. ENGELHARDT, J. FISCHER, K. HOEHNE and J. SANDER, *Z. anorg. chem.* **366**, 169 (1969).
1587. BEAR, I. J. and W. G. MUMME, *Spectrochim. Acta* **26A**, 755 (1970).
1588. BEATTIE, I. R., M. J. GALL and G. A. OZIN, *J. Chem. Soc.* **A**, 1001 (1969).
1589. BEATTIE, I. R., T. R. GILSON and G. A. OZIN, *J. Chem. Soc.* **A**, 534 (1969).
1590. BEATTIE, I. R. and J. R. HORDER, *J. Chem. Soc.* **A**, 2655 (1969).
1591. BEATTIE, I. R. and J. R. HORDER, *J. Chem. Soc.* **A**, 2433 (1970).
1592. BEATTIE, I. R., J. R. HORDER and P. J. JONES, *J. Chem. Soc.* **A**, 329 (1970).
1593. BEATTIE, I. R., K. M. S. LIVINGSTON, G. A. OZIN and D. J. REYNOLDS, *J. Chem. Soc.* **A**, 958 (1969).
1594. BEATTIE, I. R., K. M. S. LIVINGSTON, G. A. OZIN and D. J. REYNOLDS, *J. Chem. Soc.* **A**. 449 (1970).
1595. BEATTIE, I. R., M. MILNE, M. WEBSTER, H. E. BLAYDEN, P. J. JONES, R. C. G. KILLEAN and J. L. LAWRENCE, *J. Chem. Soc.* **A**, 482 (1969).
1596. BEATTIE, I. R. and G. A. OZIN, *J. Chem. Soc.* **A**, 2615 (1969).
1597. BEATTIE, I. R. and R. O. PENY, *J. Chem. Soc.* **A**, 2429 (1970).
1958. BEAUMONT, M., B. CLAUDEL and B. MENTZEN, *J. Inorg. Nucl. Chem.* **32**, 1165 (1970).
1599. BENAZETH, S., A. LOUTELLIER and M. BIGORGNE, *J. Organometallic Chem.* **24**, 479 (1970).
1600. BEN-DOR, L. and I. FELNER, *Inorg. Chim. Acta* **4**, 49 (1970).
1601. BESSETTE, F. and A. CABANA, *Spectrochim. Acta* **25A**, 157 (1969).
1602. BESSETTE, F., A. CABANA, R. P. FOURNIER and R. SAVOIE, *Canad. J. Chem.* **48**, 410 (1970).
1603. BINNIE, W. P., M. J. REDMAN and W. J. MALLIO, *Inorg. Chem.* **9**, 1449 (1970).
1604. BONADEO, H. A. and E. SILBERMAN, *J. Mol. Spectrosc*, **32**, 214 (1969).
1605. BONADEO, H. A. and E. SILBERMAN, *Spectrochim. Acta* **26A**, 2337 (1970).
1606. BOORMAN, P. M., P. J. CRAIG and T. W. SWADDLE, *J. Chem. Soc.* **A**, 2970 (1969).

1607. Bragin, J. and M. V. Evans, *J. Chem. Phys.* **51**, 268 (1969).
1608. Brinkmann, F. J., *Inorg. Nucl. Chem. Letters* **6**, 453 (1970).
1609. Brinkmann, F. J., H. Gerding and K. Olie, *Rec. Trav. Chim.* **88**, 1358 (1969).
1610. Brisdon, B. J., G. A. Ozin and R. A. Walton, *J. Chem. Soc.* **A**, 342 (1969).
1611. Britnell, D., G. W. A. Fowles and R. Mandyczewsky, *Chem. Comm.* 608 (1970).
1612. Brooker, M. H. and D. E. Irish, *Inorg. Chem.* **8**, 219 (1969).
1613. Brooker, M. H. and D. E. Irish, *Canad. J. Chem.* **48**, 1183 (1970).
1614. Brooker, M. H. and D. E. Irish, *Canad. J. Chem.* **48**, 1198 (1970).
1615. Brooker, M. H., D. E. Irish and G. E. Boyd, *J. Chem. Phys.* **53**, 1083 (1970).
1616. Brown, C. W. and J. Overend, *Spectrochim. Acta* **25A**, 1535 (1969).
1617. Brown, D. H., K. D. Crosbie, J. I. Darragh, D. S. Ross and D. W. A. Sharp, *J. Chem. Soc.* **A**, 914 (1970).
1618. Brown, D. H. and D. T. Stewart, *Spectrochim. Acta* **26A**, 1344 (1970).
1619. Brown, L. C., G. M. Begun and G. E. Boyd, *J. Amer. Chem. Soc.* **91**, 2250 (1969).
1620. Brown, R. D. and G. P. Pez, *Spectrochim. Acta* **26A**, 1375 (1970).
1620a. Brown, R. G. and S. D. Ross, *Spectrochim. Acta* **26A**, 945 (1970).
1620b. Brown, R. G. and S. D. Ross, *Spectrochim. Acta* **26A**, 955 (1970).
1621. Brown, R. G. and S. D. Ross, *Spectrochim. Acta* **26A**, 1149 (1970).
1622. Brown, T. L., W. G. McDugle and L. G. Kent, *J. Amer. Chem. Soc.* **92**, 3645 (1970).
1623. Brownstein, M., P. A. W. Dean and R. J. Gillespie, *Chem. Comm.* 9 (1970).
1624. Bruna, P. J., J. E. Sicre and H. J. Schumacher, *Chem. Comm.* 1542 (1970).
1625. Bullock, J. I., *J. Chem. Soc.* **A**, 781 (1969).
1626. Bullock, J. I. and F. W. Parrett, *Canad. J. Chem.* **48**, 3095 (1970).
1627. Bullock, J. I. and F. W. Parrett, *Chem. Comm.* 157 (1968).
1628. Burgard, M. and J. McCordick, *Inorg. Nucl. Chem. Letters* **6**, 599 (1970).
1629. Cahill, J. E. and G. E. Leroi, *J. Chem. Phys.* **51**, 1324 (1969).
1630. Caillet, P. and P. Saumagne, *J. Mol. Structure* **4**, 191 (1969).
1631. Campbell, J. A., D. P. Ryan and L. M. Simpson, *Spectrochim. Acta* **26A**, 2351 (1970).
1632. Capwell, R. J. and G. M. Rosenblatt, *J. Mol. Spectrosc.* **33**, 525 (1970).
1633. Carter, H. A. and F. Aubke, *Canad. J. Chem.* **48**, 3456 (1970).
1634. Carter, H. A., A. M. Qureshi, J. R. Sams and F. Aubke, *Canad. J. Chem.* **48**, 2853 (1970).
1635. Cavanagh, K. and D. F. Evans, *J. Chem. Soc.* **A**, 2890 (1969).
1636. Charlton, T. L. and R. G. Cavell, *Inorg. Chem.* **9**, 379 (1970).
1637. Chopin, F. and G. Turrell, *J. Mol. Structure* **3**, 57 (1969).
1638. Christe, K. O., *Inorg. Chem.* **9**, 2801 (1970).
1639. Christe, K. O. and W. Maya, *Inorg. Chem.* **8**, 1253 (1969).
1640. Christe, K. O. and W. Sawodny, *Inorg. Chem.* **8**, 212 (1969).
1641. Christe, K. O. and W. Sawodny, *Z. anorg. Chem.* **374**, 306 (1970).
1642. Christe, K. O. and C. J. Schack, *Inorg. Chem.* **9**, 1852 (1970).
1643. Christe, K. O. and C. J. Schack, *Inorg. Chem.* **9**, 2296 (1970).
1644. Christe, K. O., C. J. Schack, D. Pilipovich and W. Sawodny, *Inorg. Chem.* **8**, 2489 (1969).
1645. Chugtai, A. R. and R. N. Keller, *J. Inorg. Nucl. Chem.* **31**, 633 (1969).
1646. Claassen, H. H. and E. H. Appelman, *Inorg. Chem.* **9**, 622 (1970).

1647. Claassen, H. H., G. L. Goodman, J. H. Holloway and H. Selig, *J. Chem. Phys.* **53**, 341 (1970).
1648. Clark, R. J. H., L. Maresca and P. J. Smith, *J. Chem. Soc.* **A**, 2687 (1970).
1649. Clarke, J. H. R. and R. E. Hester, *Inorg. Chem.* **8**, 1113 (1969).
1650. Clase, H. J., A. M. Noble and J. M. Winfield, *Spectrochim. Acta* **25A**, 293 (1969).
1651. Cleveland, J. M., *Inorg. Nucl. Chem. Letters* **6**, 535 (1970).
1652. Coe, J. S. and J. R. Lyons, *Inorg. Chem.* **9**, 1775 (1970).
1653. Coker, E. H. and D. E. Hofer, *J. Chem. Phys.* **53**, 1652 (1970).
1654. Commeyras, A. and G. A. Olah, *J. Amer. Chem. Soc.* **91**, 2929 (1969).
1655. Converse, J. G., J. B. Hamilton and R. E. McCarley, *Inorg. Chem.* **9**, 1366 (1970).
1656. Cooney, R. P. J. and J. R. Hall, *Austral. J. Chem.* **22**, 331 (1969).
1657. Cooney, R. P. J. and J. R. Hall, *Austral. J. Chem.* **22**, 337 (1969).
1658. Cooney, R. P. J. and J. R. Hall, *Austral. J. Chem.* **22**, 2117 (1969).
1659. Cox, L. E. and D. G. Peters, *Inorg. Chem.* **9**, 1927 (1970).
1660. Creighton, J. A., *Chem. Comm.* 163 (1969).
1661. Crouch, P. C., G. W. A. Fowles and R. A. Walton, *J. Chem. Soc.* **A**, 972 (1969).
1662. Cubois, M. L., M. B. Delhaye and F. Wallart, *Compt. rend.* **269B**, 260 (1969).
1663. Cyvin, S. J., B. Vizi, A. Mueller and B. Krebs, *J. Mol. Structure* **3**, 173 (1969).
1664. Davies, N., D. Saunders and M. G. H. Wallbridge, *J. Chem. Soc.* **A**, 2915 (1970).
1665. Dean, P. A. W., D. F. Evans and R. F. Phillips, *J. Chem. Soc.* **A**, 363 (1969).
1666. Debeau, M., *Spectrochim. Acta* **25A**, 1311 (1969).
1667. Debeau, M. and H. Poulet, *Spectrochim. Acta* **25A**, 1553 (1969).
1668. Dehnicke, K., G. Pausewang and W. Ruedorff, *Z. anorg. Chem.* **366**, 64 (1969).
1669. Devlin, J. P., P. C. Li and G. Pollard, *J. Chem. Phys.* **52**, 2267 (1970).
1670. Dickson, F. E., *J. Inorg. Nuclear Chem.* **31**, 2636 (1969).
1671. Dillon, K. B. and T. C. Waddington, *Chem. Comm.* 1317 (1969).
1672. Dilts, J. A. and E. C. Ashby, *Inorg. Chem.* **9**, 855 (1970).
1673. Dinerman, C. E. and G. E. Ewing, *J. Chem. Phys.* **53**, 626 (1970).
1674. Doorenbos, H. E., J. C. Evans and R. O. Kagel, *J. Phys. Chem.* **74**, 3385 (1970).
1675. Dostal, K., V. Kocman and V. Ehrenbergrová, *Z. anorg. Chem.* **367**, 80 (1969).
1676. Drake, J. E. and C. Riddle, *J. Chem. Soc.* **A**, 2114 (1969).
1677. Drake, J. E., C. Riddle and D. E. Rogers, *J. Chem. Soc.* **A**, 910 (169).
1678. Drew, M. G. B., D. F. Lewis and R. A. Walton, *Inorg. Chem. Nucl. Letters* **6**, 136 (1970).
1679. Durig, J. R. and D. J. Antion, *Appl. Spectrosc.* **24**, 16 (1970).
1680. Durig, J. R., D. J. Antion and C. B. Pate, *J. Chem. Phys.* **51**, 4449 (1969).
1681. Durig, J. R. and J. M. Casper, *J. Mol. Structure* **5**, 351 (1970).
1682. Durig, J. R. and K. L. Hellams, *Inorg. Chem.* **8**, 944 (1969).
1683. Durig, J. R., K. K. Lau, G. Nagarajan, M. Walker and J. Bragin, *J. Chem. Phys.* **50**, 2130 (1969).
1684. Eckhardt, R., D. Eggers and L. J. Slutsky, *Spectrochim. Acta* **26A**, 2033 (1970).

1685. EDWARDS, D. A. and R. T. WARD, *J. Chem. Soc.* **A**, 1617 (1970).
1686. EDWARDS, D. A. and R. T. WARD, *J. Mol. Structure* **6**, 421 (1970).
1687. EDWARDS, H. G. M. and L. A. WOODWARD, *Spectrochim. Acta* **26A**, 897 (1970).
1688. EDWARDS, H. G. M. and L. A. WOODWARD, *Spectrochim. Acta* **26A**, 1077 (1970).
1689. EDWARDS, H. G. M., L. A. WOODWARD, M. J. GALL and M. J. WARE, *Spectrochim. Acta* **26A**, 287 (1970).
1690. ERICKSON, N. E. and A. G. MADDOCK, *J. Chem. Soc.* **A**, 1665 (1970).
1691. ETCHEPARE, J., *Spectrochim. Acta* **26A**, 2147 (1970).
1692. EVANS, C. A. and M. J. TAYLOR, *Chem. Comm.* 1201 (1969).
1693. EVANS, J. C. and G. Y-S. LO, *J. Phys. Chem.* **73**, 448 (1969).
1694. EWING, G., R. BLICKSENDERFER, C. DINERMAN and D. T. SHENG, *Chem. Eng. News* 42 (1969).
1695. EYSEL, H. H., *J. Mol. Structure* **5**, 275 (1970).
1696. FEREDAY, R. J., N. LOGAN and D. SUTTON, *J. Chem. Soc.* **A**, 2699 (1969).
1697. FERRARO, J. R., *J. Chem. Phys.* **53**, 117 (1970).
1698. FERRARO, J. R. and M. BECKER, *J. Inorg. Nucl. Chem.* **32**, 1495 (1970).
1699. FERRARO, J. R., D. POTTS and A. WALKER, *Canad. J. Chem.* **48**, 711 (1970).
1700. FERRARO, J. R., A. QUATTROCHI, K. C. PATIL and C. N. R. RAO, *J. Inorg. Nucl. Chem.* **31**, 3667 (1969).
1701. FLINT, C. D. and M. GOODGAME, *J. Chem. Soc.* **A**, 442 (1970).
1702. FORNERIS, R. E. and C. E. HENNIES, *J. Mol. Structure* **5**, 449 (1970).
1703. FORNERIS, R., J. HIRAISHI, F. A. MILLER and M. UEHARA, *Spectrochim. Acta* **26A**, 581 (1970).
1704. FOWLES, G. W. A., D. A. RICE and R. A. WALTON, *Spectrochim. Acta* **25A**, 1035 (1969).
1705. FRAIS, P. W., A. GUEST and C. J. L. LOCK, *Canad. J. Chem.* **47**, 1069 (1969).
1706. FRENZEL, C. A., K. E. BLICK, C. R. BENNETT and K. NIEDENZU, *J. Chem. Phys.* **53**, 198 (1970).
1707. GARG, C. L. and K. V. NARASIMHAN, *Spectrochim. Acta* **26A**, 627 (1970).
1708. GEDDES, A. L. and G. L. BOTTGER, *Inorg. Chem.* **8**, 802 (1969).
1709. GERDING, H., H. GIJBEN and D. J. STUFKENS, *Rec. Trav. Chim.* **88**, 667 (1969).
1710. GERDING, H., D. J. STUFKENS and H. GIJBEN, *Rec. Trav. Chim.* **89**, 619 (1970).
1711. GILLESPIE, R. J. and M. J. MORTON, *Inorg. Chem.* **9**, 616 (1970).
1712. GILLESPIE, R. J. and M. J. MORTON, *Inorg. Chem.* **9**, 811 (1970).
1713. GILLESPIE, R. J. and A. WHITLA, *Canad. J. Chem.* **48**, 657 (1970).
1714. GIORGIO, G., B. MENTZEN, M. BREYSSE and B. CLAUDEL, *J. Inorg. Nucl. Chem.* **32**, 1517 (1970).
1715. GOEDKEN, V. L., L. M. VALLARINO and J. V. QUAGLIANO, *J. Amer. Chem. Soc.* **92**, 303 (1970).
1716. GOODGAME, D. M. L., M. A. HITCHMAN and D. F. MARSHAM, *J. Chem. Soc.* **A**, 1933 (1970).
1717. GOUBEAU, J., F. HAENSCHKE and A. RUOFF, *Z. anorg. Chem.* **366**, 113 (1969).
1718. GREENWOOD, N. N. and R. L. TRANTER, *J. Chem. Soc.* **A**, 2878 (1969).
1719. GREY, I. E. and P. W. SMITH, *Austral. J. Chem.* **22**, 1627 (1969).
1720. GRIFFITH, W. P., *J. Chem. Soc.* **A**, 1372 (1969).
1721. GRIFFITH, W. P., *J. Chem. Soc.* **A**, 286 (1970).
1722. GRIFFITH, W. P. and P. J. B. LESNIAK, *J. Chem. Soc.* **A**, 1066 (1969).

1723. GRIFFITHS, J. E., *Spectrochim. Acta* **25A**, 965 (1969).
1724. GRIFFITHS, J. E. and D. F. STURMAN, *Spectrochim. Acta* **25A**, 1415 (1969).
1725. GRIMM, F. A. and R. F. PORTER, *Inorg. Chem.* **8**, 731 (1969).
1726. GUAY, M. and R. SAVOIE, *Canad. J. Chem.* **47**, 201 (1969).
1727. GUILLORY, W. A. and C. E. HUNTER, *J. Chem. Phys.* **50**, 3516 (1969).
1728. GUILLORY, W. A. and C. E. SMITH, *J. Chem. Phys.* **53**, 1661 (1970).
1729. GUTTERMAN, D. F. and H. B. GRAY, *J. Amer. Chem. Soc.* **91**, 3105 (1969).
1730. HAGEL, R. B. and L. F. DRUDING, *Inorg. Chem.* **9**, 1496 (1970).
1731. HARMON, K. M., L. HESSE, L. P. KLEMANN, C. W. KOCHER, S. V. MCKINLEY and A. E. YOUNG, *Inorg. Chem.* **8**, 1054 (1969).
1732. HASTIE, J. W., R. HAUGE and J. L. MARGRAVE, *Chem. Comm.* 1452 (1969) and *J. Chem. Phys.* **51**, 2648 (1969).
1733. HATHAWAY, B. J., R. J. DUDLEY and P. NICHOLLS, *J. Chem. Soc.* **A**, 1845 (1969).
1734. HATHAWAY, B. J. and F. STEPHENS, *J. Chem. Soc.* **A**, 884 (1970).
1735. HAUCK, J., *Z. Naturforsch* **25b**, 224 (1970).
1736. HAUCK, J., *Z. Naturforsch* **25b**, 468 (1970).
1737. HAUCK, J., *Z. Naturforsch* **25b**, 647 (1970).
1738. HAUCK, J. and A. FADINI, *Z. Naturforsch* **25b**, 422 (1970).
1739. HERLINGER, A. W. and T. V. LONG, *Inorg. Chem.* **8**, 2661 (1969).
1740. HOEFLER, F., W. SAWODNY and E. HENGGE, *Spectrochim. Acta* **26A**, 819 (1970).
1741. HOLLOWAY, J. H. and J. G. KNOWLES, *J. Chem. Soc.* **A**, 756 (1969).
1742. HOPPER, M. A. and D. W. JAMES, *Spectrochim. Acta* **25A**, 569 (1969).
1743. HORN, R. W., E. WEISSBERGER and J. P. COLLMAN, *Inorg. Chem.* **9**, 2367 (1970).
1744. HOSKINS, L. C., *J. Chem. Phys.* **50**, 1130 (1969).
1745. HUGGINS, K. G., F. W. PARRETT and H. A. PATEL, *J. Inorg. Nucl. Chem.* **31**, 1209 (1969).
1746. IQBAL, Z., *J. Chem. Phys.* **53**, 3763 (1970).
1747. IRISH, D. E., G. CHANG and D. L. NELSON, *Inorg. Chem.* **9**, 425 (1970).
1748. IYER, V. J. and M. A. NABAR, *Z. Naturforsch.* **25b**, 1077 (1970).
1749. JACOX, M. E. and D. E. MILLIGAN, *J. Chem. Phys.* **51**, 4143 (1969).
1750. JAMES, D. W., R. D. CARLISLE and W. LEONG, *Austral. J. Chem.* **23**, 1779 (1970).
1751. JAMES, D. W. and G. M. KIMBER, *Austral. J. Chem.* **22**, 2287 (1969).
1752. JAMES, D. W. and G. M. KIMBER, *Inorg. Nucl. Chem. Letters* **5**, 609 (1969).
1753. JAMES, D. W. and G. M. KIMBER, *Austral. J. Chem.* **23**, 829 (1970).
1754. JAMES, D. W. and W. H. LEONG, *Austral. J. Chem.* **23**, 1087 (1970).
1755. KATSAROS, N. and V. W. GEORGE, *Inorg. Chim. Acta* **3**, 165 (1969).
1756. KEITH, J. N. and I. J. SOLOMON, *Inorg. Chem.* **9**, 1560 (1970).
1757. KHOO, K. H. and J. D. MURRAY, *J. Inorg. Nuclear Chem.* **31**, 2437 (1969).
1758. KIEFER, W., *Z. Naturforsch* **25a**, 1101 (1970).
1759. KING, G. W. and K. SRIKAMESWARAN, *J. Mol. Spectrosc.* **29**, 491 (1969).
1760. KLEBOTH, K., *Monatsh.* **101**, 357 (1970).
1761. KLEE, W. E., *Spectrochim. Acta* **26A**, 1165 (1970).
1762. KLEE, W. E. and G. ENGEL, *J. Inorg. Nucl. Chem.* **32**, 1837 (1970).
1763. KLEE, W. E. and G. WEITZ, *J. Inorg. Nucl. Chem.* **31**, 2367 (1969).
1764. KNOECK, J., *Anal. Chem.* **41**, 2069 (1969).
1765. KNOP, O., F. BRISSE and L. CASTELLIZ, *Canad. J. Chem.* **47**, 971 (1969).
1766. KNOX, G. F. and T. M. BROWN, *Inorg. Chem.* **8**, 1401 (1969).

1767. KNUTH, K., J. JANDER and U. ENGELHARDT, *Z. Naturforsch.* **24b**, 1476 (1969).
1768. KOENIG, E., *Inorg. Chem.* **8**, 1278 (1969).
1769. KOMISSAROVA, L. N., E. G. TETERIN, N. P. ANOSHINA and V. I. SPITSYN, *Dokl. Chem.* **188**, 746 (1969).
1770. KRASSER, W. and H. W. NUERNBERG, *Spectrochim. Acta* **26A**, 1059 (1970).
1771. KRUCK, T. and B. HERBER, *Angew. Chem. Int. Ed.* **8**, 679 (1969).
1772. LAANE, J., JONES, L. H., R. R. RYAN and L. B. ASPREY, *J. Mol. Spectrosc.* **30**, 489 (1969).
1773. LADWIG, G., *Z. Anorg. Chem.* **364**, 225 (1969).
1774. LANE, A. P. and D. W. A. SHARP, *J. Chem. Soc.* **A**, 2942 (1969).
1775. LARSSON, R. and A. MIEZIS, *Acta Chem. Scand.* **23**, 27 (1969).
1776. LEVIN, I. W., *J. Chem. Phys.* **50**, 1031 (1969).
1777. LEVIN, I. W., *J. Mol. Spectrosc.* **33**, 61 (1970).
1778. LEVITT, S. R. and R. A. CONDRATE, *Appl. Spectrosc.* **24**, 288 (1970).
1779. LINDSAY, J. W., H. N. ROBINSON, H. L. BRAMLET and A. J. JOHNSON, *J. Inorg. Nucl. Chem.* **32**, 1559 (1970).
1780. LIVINGSTON, K. and G. A. OZIN, *J. Chem. Soc.* **A**, 2840 (1969).
1781. LOEHR, T. M. and R. A. PLANE, *Inorg. Chem.* **8**, 73 (1969).
1782. LOEWENSCHUSS, A., A. RON and O. SCHNEPP, *J. Chem. Phys.* **50**, 2502 (1969).
1783. LONG, T. V., A. W. HERLINGER, E. F. EPSTEIN and I. BERNAL, *Inorg. Chem.* **9**, 459 (1970).
1784. LOUB, J. and J. ROSICKY, *Z. Anorg. Chem.* **365**, 308 (1969).
1785. LUU, D. V. and M. ABENZOA, *Compt. rend.* **269B**, 208 (1969).
1786. MACCOLL, C. R. P., *Coord. Chem. Rev.* **4**, 147 (1969).
1787. MACCORDICK, J., G. KAUFMANN and R. ROHMER, *J. Inorg. Nucl. Chem.* **31**, 3059 (1969).
1788. MAGUIRE, P. M. and H. E. RUBALCAVA, *Inorg. Chem.* **8**, 246 (1969).
1789. MAMANTOV, G., D. G. VICKROY, E. J. VASINI, T. MAEKAWA and M. C. MOULTON, *Inorg. Nucl. Chem. Letters* **6**, 701 (1970).
1790. MARONI, V. A. and E. J. CAIRNS, *J. Chem. Phys.* **52**, 4915 (1970).
1791. MARONI, V. A., D. M. GRUEN, R. L. MCBETH and E. J. CAIRNS, *Spectrochim. Acta* **26A**, 418 (1970).
1792. MATHUR, M. S., C. A. FRENZEL, K. E. BLICK and E. B. BRADLEY, *J. Mol. Structure* **5**, 412 (1970).
1793. MATTES, R., *Z. Naturforsch* **24b**, 772 (1969).
1794. MCALPINE, I. and H. SUTCLIFFE, *Spectrochim. Acta* **25A**, 1723 (1969).
1795. MEINERT, H. and G. KAUSCHKA, *Z. Chem.* **9**, 114 (1969).
1796. MELVEGER, A. J., R. K. KHANNA and E. R. LIPPINCOTT, *J. Chem. Phys.* **52**, 2747 (1970).
1797. MICHALSKI, J., M. MIKOLAJCZYK and B. MLOTKOWSKA, *Chem. Ber.* **102**, 90 (1969).
1798. MILLIGAN, D. E. and M. E. JACOX, *J. Chem. Phys.* **53**, 2034 (1970).
1799. MILLIGAN, D. E., M. E. JACOX and W. A. GUILLORY, *J. Chem. Phys.* **52**, 3864 (1970).
1800. MITCHELL, P. C. H. and F. TRIFIRO, *J. Chem. Soc.* **A**, 3183 (1970).
1801. MOULE, D. C. and C. R. SUBRAMANIAM, *Canad. J. Chem.* **47**, 1011 (1969).
1802. MUELLER, A., E. J. BARAN, F. BOLLMAN and P. J. AYMONINO, *Z. Naturforsch.* **24b**, 960 (1969).
1803. MUELLER, A., E. J. BARAN and P. J. HENDRA, *Spectrochim. Acta* **25A**, 1654 (1969).

1804. MUELLER, A., F. BOLLMANN and E. J. BARAN, *Z. anorg. Chem.* **370**, 238 (1969).
1805. MUELLER, A., F. BOLLMANN and B. KREBS, *Z. anorg. Chem.* **368**, 155 (1969).
1806. MUELLER, A. and E. DIEMANN, *Z. anorg. Chem.* **373**, 57 (1970).
1807. MUELLER, A., E. DIEMANN and E. J. BARAN, *Z. anorg. Chem.* **375**, 87 (1970).
1808. MUELLER, A., E. DIEMANN and U. HEIDBORN, *Z. anorg. Chem.* **376**, 125 (1970).
1809. MUELLER, A., E. DIEMANN and V. V. K. RAO, *Chem. Ber.* **103**, 2961 (1970).
1810. MUELLER, A., E. DIEMANN and H. SCHULTZE, *Z. anorg. Chem.* **376**, 120 (1970).
1811. MUELLER, A., A. RUOFF, B. KREBS, O. GLEMSER and W. KOCH, *Spectrochim. Acta* **25A**, 199 (1969).
1812. MULLER, O., W. B. WHITE and R. ROY, *Spectrochim. Acta* **25A**, 1491 (1969).
1813. MURRELL, J. N., *J. Chem. Soc.* **A**, 297 (1969).
1814. MURRELL, J. N. and A. B. NIKOLSKII, *J. Chem. Soc.* **A**, 1363 (1970).
1815. NEWBERRY, J. E., *Spectrochim. Acta* **25A**, 1699 (1969).
1816. NEWKIRK, A. E. and V. B. HUGHES, *Inorg. Chem.* **9**, 401 (1970).
1817. NOLAN, M. J. and D. W. JAMES, *Austral. J. Chem.* **23**, 1043 (1970).
1818. OEHLER, O. and H. H. GUNTHARD, *J. Chem. Phys.* **51**, 4719 (1969).
1819. OERTEL, R. P. and R. A. PLANE, *Inorg. Chem.* **8**, 1188 (1969).
1820. OHWADA, K., *J. Inorg. Nucl. Chem.* **32**, 1209 (1970).
1821. OHWADA, K., *Spectrochim. Acta* **26A**, 1035 (1970).
1822. OHWADA, K., *Spectrochim. Acta* **26A**, 1723 (1970).
1823. OMLOO, W. P. F. A. M. and F. JELLINEK, *Rec. Trav. Chim.* **88**, 1205 (1969).
1824. OSKAM, A., R. ELST and J. C. DUINKER, *Spectrochim. Acta* **26A**, 2021 (1970).
1825. OUELETTE, T. J., C. T. RATCLIFFE and D. W. A. SHARP, *J. Chem. Soc.* **A**, 2351 (1969).
1826. OZIN, G. A., *Chem. Comm.* 1325 (1969).
1827. OZIN, G. A., *J. Chem. Soc.* **A**, 116 (1969).
1828. OZIN, G. A., *J. Chem. Soc.* **A**, 2952 (1969).
1829. OZIN, G. A., *Canad. J. Chem.* **48**, 2931 (1970).
1830. OZIN, G. A., *J. Chem. Soc.* **A**, 2307 (1970).
1831. OZIN, G. A., G. W. A. FOWLES, D. J. TIDMARSH and R. A. WALTON, *J. Chem. Soc.* **A**, 642 (1969).
1832. OZIN, G. A. and D. J. REYNOLDS, *Chem. Comm.* 884 (1969).
1833. OZIN, G. A. and A. V. VOET, *Chem. Comm.* 896 (1970).
1834. OZIN, G. A. and A. V. VOET, *Chem. Comm.* 1489 (1970).
1835. PACANSKY, J. and V. CALDER, *J. Chem. Phys.* **53**, 4519 (1970).
1836. PÂQUES-LEDENT, M. TH. and P. TARTE, *Spectrochim. Acta* **25A**, 1115 (1969).
1837. PARRETT, F. W. and N. J. TAYLOR, *J. Inorg. Nucl. Chem.* **32**, 2458 (1970).
1838. PATEL, S. J., *J. Inorg. Nucl. Chem.* **32**, 3708 (1970).
1839. PATEL, S. J. and D. G. TUCK, *Canad. J. Chem.* **47**, 229 (1969).
1840. PAUL, R. C., V. P. KAPILA and K. C. MALHOTRA, *J. Chem. Soc.* **A**, 2267 (1970).
1841. PAUL, R. C., K. K. PAUL and K. C. MALHOTRA, *J. Inorg. Nucl. Chem.* **31**, 2614 (1969).
1842. PAUSEWANG, G. and W. RUEDORFF, *Z. anorg. Chem.* **364**, 69 (1969).
1843. PERCHARD, C. and A. NOVAK, *Spectrochim. Acta* **26A**, 871 (1970).
1844. PERRIN-BILLOT, C., A. PERRIN and J. PRIGENT, *Chem. Comm.* 676 (1970).
1845. PERRY, R. O., *Chem. Comm.* 886 (1969).
1846. PÉZOLET, M. and R. SAVOIE, *Canad. J. Chem.* **47**, 3041 (1969).

1847. PUGLISI, C., *J. Inorg. Nucl. Chem.* **32**, 692 (1970).
1848. QURESHI, A. M. and F. AUBKE, *Canad. J. Chem.* **48**, 3117 (1970).
1849. REICHMAN, S. and F. SCHREINER, *J. Chem. Phys.* **51**, 2355 (1969).
1850. REISFELD, M. J., *J. Mol. Spectrosc.* **29**, 120 (1969).
1851. ROBINSON, E. A., D. S. LAVERY and S. WELLER, *Spectrochim. Acta* **25A**, 151 (1969).
1852. ROESKY, H. W., F. N. TEBBE and E. L. MUETTERTIES, *Inorg. Chem.* **9**, 831 (1970).
1853. ROSE, D. and G. WILKINSON, *J. Chem. Soc.* **A**, 1791 (1970).
1854. ROSS, S. D. and N. A. THOMAS, *Spectrochim. Acta* **26A**, 971 (1970).
1855. RUOFF, A., J. B. MILNE, G. KAUFMANN and M. LEROY, *Z. anorg. Chem.* **372**, 119 (1970).
1856. RUOFF, A. and J. WEIDLEIN, *Z. anorg. Chem.* **370**, 113 (1969).
1857. RUSH, J. J., A. J. MCIVIGER and E. R. LIPPINCOTT, *J. Chem. Phys.* **51**, 2947 (1969).
1858. SAKAKIBARA, M., Y. TAKAHASHI, S. SAKAI and Y. ISHII, *Inorg. Nucl. Chem. Letters* **5**, 427 (1969).
1859. SCHACK, C. J. and D. PILIPOVICH, *Inorg. Chem.* **9**, 387 (1970).
1860. SCHMIDT, A., *Chem. Ber.* **102**, 380 (1969).
1861. SCHMIDT, M. and H. H. J. SCHROEDER, *Angew. Chem. Int. Ed.* **9**, 808 (1970).
1862. SCHMIDT, M. and H. H. J. SCHROEDER, *Z. anorg. Chem.* **378**, 192 (1970).
1863. SCHMIDT, M. and W. SIEBERT, *Chem. Ber.* **102**, 2752 (1969).
1864. SCHETTINO, V. and I. C. HISATSUNE, *J. Chem. Phys.* **52**, 9 (1970).
1865. SCHEUERMANN, W. and G. J. RITTER, *J. Mol. Structure* **6**, 240 (1970).
1866. SCHEUERMANN, W. and G. J. RITTER, *Z. Naturforsch.* **24a**, 408 (1969).
1867. SCHEUERMANN, W., G. J. RITTER and C. J. H. SCHUTTE, *Z. Naturforsch.* **25a**, 1856 (1970).
1868. SCHUTTE, C. J. H. and M. A. HEYNS, *J. Mol. Structure* **5**, 37 (1970).
1869. SCOTT, J. F., *J. Chem. Phys.* **53**, 852 (1970).
1870. SEETL, F., R. BUDENZ and K-P WANCZEK, *Chem. Ber.* **103**, 3946 (1970).
1871. SEIDL, V., O. KNOP and M. FALK, *Canad. J. Chem.* **47**, 1361 (1969).
1872. SELIG, H., H. H. CLAASSEN and J. H. HOLLOWAY, *J. Chem. Phys.* **52**, 3517 (1970).
1873. SELIG, H., J. H. HOLLOWAY, J. TYSON and H. H. CLAASSEN, *J. Chem. Phys.* **53**, 2559 (1970).
1874. SEPPELT, K. and W. SUNDERMAYER, *Naturwiss.* **56**, 281 (1969).
1875. SESHADRI, K. S., L. A. NIMON and D. WHITE, *J. Mol. Spectrosc.* **30**, 128 (1969).
1876. SHAMIR, J. and J. BINENBOYM, *J. Mol. Structure* **4**, 100 (1969).
1877. SHERWOOD, P. M. A. and J. J. TURNER, *Spectrochim. Acta* **26A**, 1975 (1970).
1878. SHRIVER, D. F. and I. WHARF, *Inorg. Chem.* **8**, 2167 (1969).
1879. SHURVELL, H. F., *Spectrochim. Acta* **25A**, 973 (1969).
1880. SHURVELL, H. F. and H. J. BERNSTEN, *J. Mol. Spectrosc.* **30**, 153 (1969).
1881. SIEBERT, H. and H. H. EYSEL, *J. Mol. Structure* **4**, 29 (1969).
1882. SIMO, C., E. BANKS and S. L. HOLT, *Inorg. Chem.* **9**, 183 (1970).
1883. SLADKY, F. O., P. A. BULLINER and N. BARTLETT, *J. Chem. Soc.* **A**, 2179 (1969).
1884. SLADKY, F. O. and N. BARTLETT, *J. Chem. Soc.* **A**, 2188 (1969).
1885. SPRAGG, R. A., H. STAMMREICH and Y. KAWANO, *J. Mol. Structure* **3**, 305 (1969).
1886. STEGER, E., I. C. CIUREA and A. FADINI, *Spectrochim. Acta* **25A**, 1649 (1969).

1887. STERZEL, W., *Z. anorg. Chem.* **368**, 308 (1969).
1888. STERZEL, W. and W-D. SCHNEE, *Z. anorg. Chem.* **376**, 134 (1970).
1889. STEUDEL, R. and D. LAUTENBACH, *Z. Naturforsch.* **24b**, 350 (1969).
1890. STIRLING, G. C., C. J. LUDMAN and T. C. WADDINGTON, *J. Chem. Phys.* **52**, 2730 (1970).
1891. STUFKENS, D. J., *Rec. Trav. Chim.* **89**, 755 (1970).
1892. STUFKENS, D. J. and H. GERDING, *Rec. Trav. Chim.* **89**, 417 (1970).
1893. STUFKENS, D. J., A. M. WILMANS and H. GERDING, *Rec. Trav. Chim.* **89**, 1297 (1970).
1894. SUBBA RAO, G. V., C. N. R. RAO and J. R. FERRARO, *Appl. Spectrosc.* **24**, 436 (1970).
1895. SWADDLE, T. W., P. J. CRAIG and P. M. BOORMAN, *Spectrochim. Acta* **26A**, 1559 (1970).
1896. TARTE, P. and J. PREUDHOMME, *Spectrochim. Acta* **26A**, 747 (1970).
1897. TARTE, P. and J. PREUDHOMME, *Spectrochim. Acta* **26A**, 2207 (1970).
1898. TAKEMOTO, J. and K. NAKAMOTO, *Chem. Comm.* 1017 (1970).
1899. THIRUGNANASAMBANDAM, P. and G. J. SRINIVASAN, *J. Chem. Phys.* **50**, 2467 (1969).
1900. TOMAR, V. S., H. D. BIST and D. P. KHANDELWAL, *Appl. Spectrosc.* **24**, 598 (1970).
1901. TURCOTE, R. P., J. O. SAWYER and L. EYRING, *Inorg. Chem.* **8**, 238 (1969).
1902. TURNBULL, A. G., *Austral. J. Chem.* **22**, 2237 (1969).
1903. VERBEEK, W. and W. SUNDERMEYER, *Angew. Chem. Int. Ed.* **8**, 376 (1969).
1904. BRONSWYK, W. VON, R. J. H. CLARK and L. MARESCA, *Inorg. Chem.* **8**, 1395 (1969).
1905. LAMPE, F. VON, *Z. anorg. Chem.* **367**, 170 (1969).
1906. LAMPE, F. VON, *Z. anorg. Chem.* **368**, 93 (1969).
1907. WAMSER, C. A., W. B. FOX, B. SUKORNICK, J. R. HOLMES, B. B. STEWART, R. JUURIK, N. VANDERKOOI and D. GOULD, *Inorg. Chem.* **8**, 1249 (1969).
1908. WEBSTER, M. and H. E. BLAYDEN, *J. Chem. Soc.* **A**, 2443 (1969).
1909. WHARF, I. and M. ONYSZCHUK, *Canad. J. Chem.* **48**, 2250 (1970).
1910. WHARF, I. and D. F. SHRIVER, *Inorg. Chem.* **8**, 914 (1969).
1911. WIEGHARDT, K. and H. H. EYSEL, *Z. Naturforsch.* **25b**, 105 (1970).
1912. WIESER, H., P. J. KRUEGER, E. MULLER and J. B. HYNE, *Canad. J. Chem.* **47**, 1633 (1969).
1913. WING, R. M. and K. P. CALLAHAN, *Inorg. Chem.* **8**, 871 (1969).
1914. WITT, J. D. and R. M. HAMMAKER, *Chem. Comm.* 667 (1970).
1915. WOLFE, D. F. and G. L. HUMPHREY, *J. Mol. Structure* **3**, 293 (1969).
1916. YELLIN, W. and W. A. CILLEY, *Spectrochim. Acta* **25A**, 879 (1969).

Part II

References prefixed with W are taken directly from Wyckoff's 'Crystal Structures'; the Roman numeral refers to the chapter and the rest of the designation to the paragraph. Roman numerals V, VI, VII refer to chapters in Volume 2 and VIII, IX, X to chapters in Volume 3.

References prefixed with WN relate to Winchell and Winchell; the number is the page number.

All other references in this section are prefixed by X.

A. N. WINCHELL and H. WINCHELL, *The Microscopical Characters of Artificial Inorganic Solid Substances*, Academic Press (New York), 3rd edition 1964.

R. W. G. WYCKOFF, *Crystal Structures*, Volumes 2 and 3, Interscience (New York), 2nd edition 1965.

X1. ALBRIGHT, J. G., *Z. Krist.* **84**, 150 (1932).
X2. ALEXAKOS, L. G., C. D. CORNWELL and S. B. PIERCE, *Proc. Chem. Soc.* 341 (1963).
X3. ATOJI, M. and R. E. RUNDLE, *J. Chem. Phys.* **29**, 1306 (1958).
X4. AYNSLEY, E. and S. SAMPATH, *J. Chem. Soc.* 3099 (1959).
X5. BACON, G. E. and N. A. CURRY, *Proc. Roy. Soc.* **266A**, 95 (1962).
X6. BARCLAY, G. A. and E. G. COX, *Z. Krist.* **113**, 23 (1960).
X7. BARCLAY, G. A., E. G. COX and H. LYNTON, *Chem. and Ind.* **11**, 178 (1956).
X8. BARNES, W. H. and R. G. HUNTER, *Nature* **130**, 96 (1932).
X9. BARTELL, L. S., R. M. GAVIN, H. B. THOMPSON and C. L. CHERNICK, *J. Chem. Phys.* **43**, 2547 (1965).
X10. BARTRAM, S. M. and E. J. FELTEN, *Rare Earth Research* (Ed. Nachman and Lundin), pub: Gordon & Breach. Proc. 2nd Conf. Denver 1961, p. 329.
X11. BAUR, W. H., *Naturwiss.* **49**, 464 (1962).
X12. BEEVERS, C. A. and H. LIPSON, *Z. Krist.* **83**, 123 (1932).
X13. BEEVERS, C. A. and C. M. SCHWARTZ, *Z. Krist.* **91A**, 157 (1935).
X14. BERGER, S. V., *Acta Chem. Scand.* **3**, 660 (1949).
X15. BIRNSTOCK, R., *Z. Krist.* **124**, 310 (1967).
X16. BOEHM, G., *Z. Krist.* **63**, 619 (1926).
X17. BOONSTRA, E. G., *Acta Cryst.* **24B**, 1053 (1968).
X18. BOOTH, H. S. and J. M. CARTER, *J. Phys. Chem.* **36**, 1359 (1932).
X19. BOWATER, I. C., R. D. BROWN and F. R. BURDEN, *J. Mol. Spec.* **28**, 454 (1968).
X20. BRADLEY, W. F., D. L. GRAF and R. S. ROTH, *Acta Cryst.* **20**, 283 (1966).
X21. BRANDON, J. K. and I. D. BROWN, *Can. J. Chem.* **46**, 933 (1968).
X22. BREDIG, M. A., *J. Amer. Chem. Soc.* **63**, 2353 (1941).
X23. BREDIG, M. A., *J. Phys. Chem.* **46**, 747 (1942).
X24. BRIXNER, L. M., P. A. FLOURNOY and K. BABCOCK, *J. Electrochem. Soc.* **111**, 873 (1964).
X25. BROWN, D. H., K. R. DIXON, R. D. W. KEMMITT and D. W. A. SHARP, *J. Chem. Soc.* 1559 (1965).
X26. BURNS, J. H., *Acta Cryst.* **15**, 89 (1962).
X27. BURNS, J. H., *Acta Cryst.* **15**, 1098 (1962).
X28. BUSING, W. R., M. ZOCCHI and H. A. LEVY, *Ann. Meeting Am. Cryst. Assocn.* Paper N-3, August 1961.
X29. CALVO, C., *Can. J. Chem.* **43**, 1147 (1965).
X30. CALVO, C., *Inorg. Chem.* **7**, 1345 (1968).
X31. CAMPBELL, J. A., *Acta Cryst.* **9**, 192 (1956).
X32. CIPRIANI, C., *Rend. Soc. Min. Ital.* **11**, 58 (1955).
X33. CLARK, D., *J. Chem. Soc.* 1615 (1952).
X34. COING-BOYAT, J., *Acta Cryst.* **12**, 939 (1959).
X35. COING-BOYAT, J., *Compt. rend.* **248**, 2109 (1959).
X36. COLBY, M. Y., *Z. Krist.* **77**, 168 (1931).
X37. COLLIN, R. L. and W. N. LIPSCOMB, *Acta Cryst.* **4**, 10 (1951).
X38. COX, B., *J. Chem. Soc.* 3521 (1954).
X39. COX, B., *J. Chem. Soc.* 876 (1956).
X40. CRUICKSHANK, D. W. J., *Acta. Cryst.* **17**, 672 (1964).
X41. CRUICKSHANK, D. W. J., H. LYNTON and G. A. BARCLAY, *Acta Cryst.* **15**, 491 (1962).

X42. DESHPANE, V. T., D. B. SIRDESMUKH and W. M. MUDHOLKER, *Acta Cryst.* **12**, 257 (1959).
X43. FERGUSSON, J. E., C. J. WILKINS and J. F. YOUNG, *J. Chem. Soc.* 2136 (1962).
X44. FERRARI, A. A., BRAIBANTI, G. BIGLIARDI and A. M. M. LANFREDI, *Acta Cryst.* **21**, 681 (1966).
X45. FERRARI, A. and L. CAVALCA, *Gazz. Chim. Ital.* **80**, 151 (1950).
X46. FERRONI, E., A. SABATINI and P. ORIOLI, *Gazz. Chim. Ital.* **87**, 630 (1957).
X47. FERRONI, E., A. SABATINI and P. ORIOLI, *Ricerca sci.* **27**, 1557 (1957).
X48. FINBAK, C. and O. HASSEL, *Z. Physik* **5**, 460 (1937).
X49. FINBAK, C. and O. HASSEL, *Z. Phys. Chem.* **35B**, 25 (1937).
X50. FLEMING, J. E. and H. LYNTON, *Chem. and Ind.* 1416 (1960).
X51. FORNASERI, M., *Periodico Mineral.* (Rome) **18**, 103 (1949).
X52. GAETZ, W., *Naturwiss* **50**, 567 (1963).
X53. GALLEZOT, P. and D. WEIGEL, *Compt. rend.* **256**, 2349 (1963).
X54. GALLEZOT, P., D. WEIGEL and M. PRETTRE, *Acta Cryst.* **22**, 699 (1967).
X54a. GARNER, C. D. and S. C. WALLWORK, *J. Chem. Soc.* **A**, 1496 (1966).
X55. GAULTIER, M. and G. PANNETIER, *Bull. Soc. chim. France* 105 (1968).
X55a. GIGUERE, P. A., I. D. LIU, J. S. DUGDALE and J. A. MORRISON, *Can. J. Chem.* **32**, 117 (1954).
X56. GOODWIN, T. H. and J. WHETSTONE, *J. Chem. Soc.* 1455 (1947).
X57. GOSSNER, B. and K. R. BRUECKL, *Z. Krist.* **69**, 422 (1929).
X58. GRAN, S. and C. RØMMING, *Acta Chem. Scand.* **22**, 1686 (1968).
X58a. GRIFFITH, R. L., *J. Chem. Phys.* **11**, 499 (1943).
X59. GUTOWSKY, H. S., G. E. PAKE and R. BERSOHN, *J. Chem. Phys.* **44**, 1722 (1966).
X60. HAMILTON, W. C., *Acta Cryst.* **10**, 103 (1957).
X61. HARDY, A., *Ann. Chim.* (Paris) **7**, 281 (1962).
X62. HARDY, A., C. PIEKARSKI and P. HAGEN-MULLER, *Compt. rend.* **249**, 2579 (1959).
X63. HARRIS, P. M. and E. P. MEIBOHM, *J. Amer. Chem. Soc.* **69**, 1231 (1947).
X64. HARSHBARGER, W., R. K. BOHN and S. H. BAUER, *J. Amer. Chem. Soc.* **89**, 6466 (1967).
X65. HASSEL, O. and H. M. VIERVOLL, *Tidsskr. Kjemi. Bergvesen Met.* **3**, 7 (1943).
X66. HELMHOLTZ, L., *J. Chem. Phys.* **4**, 316 (1936).
X67. HENDRICKS, S. B., E. POSNJAK and F. C. KRACEK, *J. Amer. Chem. Soc.* **54**, 2780 (1932).
X68. HERRMAN, W. and W. ILGE, *Z. Krist.* **80**, 402 (1931).
X69. HIGGINS, T. H. S. and C. J. G. RAW, *J. Chem. Phys.* **23**, 606 (1955).
X70. HIROTA, E., *J. Chem. Phys.* **28**, 839 (1959).
X71. HOARD, J. L. and V. BLAIR, *J. Amer. Chem. Soc.* **57**, 1983 (1935).
X72. HOFFMANN, W., *Z. Krist.* **75**, 158 (1930).
X73. HOFFMANN, W., *Z. Krist.* **78**, 279 (1931).
X74. HOFFMANN, W., *Z. Krist.* **82**, 323 (1932).
X75. ITO, T., N. MORIMOTO and R. SADANAGA, *Acta Cryst.* **7**, 53 (1964).
X76. ITO, T. and J. WEST, *Z. Krist.* **83**, 1 (1932).
X77. IVERONOVA, V. I., V. P. TARASOVA and M. M. UMANSKII, *Izv. Akad. Nauk SSSR Ser. Fiz.* **15**, 164 (1951).
X78. IVERONOVA, V. I., V. P. TARASOVA, Z. K. ZOLINA, G. V. MARKHASIN and I. M. SUKHODREVA, *Z. Fiz. Khim. SSSR* **29**, 314 (1955).
X78a. JEFFREY, G. A. and G. S. PARRY, *J. Chem. Soc.* 4864 (1952).

X78b. JEFFREY, G. A. and G. S. PARRY, *J. Amer. Chem. Soc.* **76**, 5283 (1954).
X79. KAMB, B., *Acta Cryst.* **17**, 1437 (1964).
X80. KAMB, B., *Science*, **150**, 205 (1965).
X81. KAMB, W. B. and S. K. DATTA, *Nature* **187**, 140 (1960).
X82. KANNAN, K. K. and M. A. VISWAMITRA, *Acta Cryst.* **19**, 151 (1965).
X83. KARTHA, G., *Current Sci.* (India) **20**, 151 (1951).
X84. KARTHA, G., *Acta Cryst.* **5**, 845 (1952).
X85. KARTHA, G., *Proc. Ind. Acad. Sci.* **36A**, 501 (1952).
X86. KARTHA, G., *Proc. Ind. Acad. Sci.* **38A**, 1 (1953).
X87. KETELAAR, J. A. A., *Z. Krist* **92A**, 155 (1935).
X88. KIMURA, K. and S. H. BAUER, *J. Chem. Phys.* **39**, 3172 (1963).
X89. KLINKENBERG, L. J., *Rec. trav. chim.* **56**, 36 (1937).
X90. KOKKOROS, P. A. and P. J. RENTZEPERIS, *Acta Cryst.* **11**, 361 (1958).
X91. KRACEK, F. C., E. POSNJAK and S. B. HENDRICKS, *J. Amer. Chem. Soc.* **53**, 3339 (1931).
X92. KRACEK, F. C., E. POSNJAK and S. B. HENDRICKS, *Nature* **128**, 410 (1931).
X93. KRACEK, F. C., E. POSNJAK and S. B. HENDRICKS, *J. Amer. Chem. Soc.* **54**, 2766 (1932).
X94. KREBS, B., G. GATTOW and A. MUELLER, *Z. anorg. allgem. chem.* **337**, 279 (1965).
X95. LEVI, G. R. and G. PEYRONEL, *Z. Krist* **92A**, 190 (1935).
X96. LINDQUIST, I., *Acta Chem. Scand.* **4**, 1066 (1950).
X96a. LINDQUIST, I. and M. MORTSELL, *Acta Cryst.* **10**, 406 (1957).
X97. LIPSON, H., *Proc. Roy. Soc.* **A156**, 462 (1936).
X98. LU, C. and J. DONOHUE, *J. Amer. Chem. Soc.* **66**, 818 (1944).
X99. LUKASZEWICZ, K., *Bull. Acad. polon. Sci., Ser. Sci. chim.* **15**, 47 (1967).
X100. LUKASZEWICZ, K., *Bull. Acad. polon. Sci., Ser. Sci. chim.* **15**, 53 (1967).
X101. LYTLE, F. W., *J. Appl. Phys.* **35**, 2212 (1964).
X102. MATKOVIC, B. and B. RIBAR, *Croat. chim. acta* **35**, 147 (1963).
X103. MACGILLAVRY, C. H. and C. L. VAN PANTHALEON VAN ECK, *Rec. Trav. Chim.* **62**, 729 (1943).
X104. MANI, N. V. and S. RAMASESHAN, *Z. Krist.* **114**, 200 (1960).
X104a. MASSEY, J. T. and D. R. BIANCO, *J. Chem. Phys.* **22**, 442 (1954).
X104b. MASSEY, J. T. and R. W. HART, *J. Chem. Phys.* **23**, 942 (1955).
X104c. MCDONALD, R. R., A. C. LARSON and D. T. CROMER, *Acta Cryst.* **17**, 1104 (1964).
X105. MCDOUGALL, L. A. and J. E. KILPATRICK, *J. Chem. Phys.* **42**, 2311 (1965).
X106. MOONEY, R., *J. Chem. Phys.* **16**, 1002 (1948).
X107. MOREAU, R., *Bull. Soc. Roy. Sci. Liège* **32**, 252 (1963).
X108. MURCHISON, C., S. REICHMAN, D. ANDERSON, J. OVEREND and F. SCHREINER, *J. Amer. Chem. Soc.* **90**, 5690 (1968).
X109. NARAY-SZABO, I. and J. NEUGEBAUER, *J. Amer. Chem. Soc.* **69**, 1280 (1947).
X110. NARAY-SZABO, I. and J. POCZA, *Z. Krist.* **104A**, 28 (1942).
X111. NIGGLI, A., *Acta Cryst.* **7**, 776 (1954).
X112. NITTA, I., Y. TOMIIE and C. H. KOO, *Acta Cryst.* **5**, 292 (1952).
X112a. OGG, R. A. and J. D. RAY, *Diss. Far. Soc.* **19**, 246 (1955).
X113. OSWALD, H. R., *Helv. Chem. Acta* **48**, 590 (1965).
X114. PAULING, L. and R. G. DICKINSON, *J. Amer. Chem. Soc.* **46**, 1615 (1924).
X115. PENNEY, W. G. and G. B. B. M. SUTHERLAND, *J. Chem. Phys.* **2**, 492 (1934).
X116. PIERCE, S. B. and C. D. CORNWELL, *J. Chem. Phys.* **47**, 1731 (1967).

X117. PISTORIUS, C. W. F. T. and W. E. SHARP, *Acta Cryst.* **14**, 316 (1961).
X118. PITZER, K. S. and S. J. STRICKLER, *J. Chem. Phys.* **41**, 730 (1964).
X119. PLUMB, R. C. and J. W. SWAINE, jr., *J. Phys. Chem.* **68**, 2057 (1964).
X120. PLYUSCHEV, V. E. and M. B. VARFOLOMEEV, *Dokl. Akad. Nauk SSSR* **158**, 664 (1964).
X121. PUSHKINA, G. YA. and L. N. KOMISSAROVA, *Zhur. neorg. Khim.* **12**, 2852 (1967).
X122. RENTZEPERIS, P. J., *Z. Krist.* **117**, 431 (1962).
X123. RIBAR, B. and B. MATKOVIC, *Croat. Chem. Acta* **37**, 117 (1965).
X124. RIVOIR, L. and M. ABBAD, *Anal. Fis. Quim.* **39**, 306 (1943).
X125. ROBERTSON, B. E. and C. CALVO, *Can. J. Chem.* **46**, 605 (1968).
X126. RUBEN, H. W., D. H. TEMPLETON, R. D. ROSENSTEIN and L. OLOVSSON, *J. Amer. Chem. Soc.* **83**, 220 (1961).
X127. RUMANOVA, I. M., *Dokl. Akad. Nauk SSSR* **123**, 78 (1958).
X128. RUMANOVA, I. M. and G. I. MALITSKAYA, *Kristallografiya* **4**, 510 (1959).
X129. RUMANOVA, I. M., G. F. VOLODINA and N. V. BELOV, *Kristallografiya* **9**, 655 (1964).
X130. RUSH, J. J., *J. Chem. Phys.* **44**, 1722 (1966).
X130a. SCHMEISSER, M. and K. LANG, *Angew. Chem.* **67**, 156 (1955).
X131. SCHMITT, A. and W. ZEIL, *Z. Naturforsch* **18A**, 428 (1963).
X132. SCHOLDER, R. and W. KLEMM, *Angew. Chem.* **66**, 461 (1954).
X133. SCHWARZ, H., *Z. anorg. allgem. chem.* **322**, 1 (1963).
X134. SCHWARZ, H., *Z. anorg. allgem. chem.* **322**, 129 (1963).
X135. SCHWARZ, H., *Z. anorg. allgem. chem.* **322**, 137 (1963).
X136. SCHWARZ, H., *Z. anorg. allgem. chem.* **323**, 44 (1963).
X137. SCHWARZ, H., *Z. anorg. allgem. chem.* **323**, 275 (1963).
X138. SIMONOV, V. I. and N. V. BELOV, *Kristallografiya*, 4, 163 (1959).
X139. SINGER, J. T. and D. T. CROMER, *Acta Cryst.* **12**, 719 (1959).
X140. SMITH, A. J., *Acta Cryst.* **13**, 749 (1960).
X141. SMITH, D. F., *Phys. Rev.* **86**, 608 (1952).
X142. STANLEY, E., *Acta Cryst.* **9**, 897 (1956).
X143. STARITZKY, E. and D. T. CROMER, *Anal. Chem.* **28**, 555 (1956).
X144. STARITZKY, E. and D. I. WALKER, *Anal. Chem.* **28**, 914 (1956).
X145. SUCHOW, L., I. FANKUCHEN and R. WARD, *J. Amer. Chem. Soc.* **74**, 1678 (1952).
X146. SUTOR, D. J., *Acta Cryst.* **12**, 72 (1959).
X147. TAKEUCHI, Y., *Acta Cryst.* **5**, 574 (1952).
X148. TANG, Y. C. and L. J. KUE, *Acta Chim. Sinica* **22**, 572 (1956).
X149. TASMAN, H. A. and K. H. BOSWIJK, *Acta Cryst.* **8**, 59 (1955).
X150. TASMAN, H. A. and K. H. BOSWIJK, *Acta Cryst.* **8**, 857 (1955).
X151. TEMPLETON, D. H., *Acta Cryst.* **13**, 684 (1960).
X152. TEMPLETON, D. H., A. ZALKIN, J. D. FORRESTER and S. M. WILLIAMSON, *J. Amer. Chem. Soc.* **85**, 816 (1953).
X153. TOLLES, W. M. and W. D. GWINN, *J. Chem. Phys.* **36**, 1119 (1962).
X154. TROTTER, J., *Z. Kristallogr. Kristallgeom* **121**, 81 (1965).
X155. UEDA, T., *Mem. Coll. Sci. Univ. Kyoto* **20B**, 227 (1954).
X156. VAN NIEKERK, J. N. and F. J. SCHOENING, *Acta Cryst.* **6**, 227 (1953).
X157. VDOVENKO, V. M., E. V. STROGANOV and A. P. SOKOLOV, *Radiokhimiya* **3**, 19 (1961).
X158. VDOVENKO, W. M., E. V. STROGANOV, A. P. SOKOLOV and G. LUNGU, *Radiokhimiya* **4**, 59 (1962).

X159. Vdovenko, W. M., E. V. Stroganov, A. P. Sokolov and V. N. Zandin, *Radiokhimiya* **2**, 24 (1960).
X160. Vegard, L., *Z. Physik* **9**, 395 (1922).
X161. Voigt, A. and W. Biltz, *Z. anorg. allgem. chem.* **133**, 298 (1924).
X162. Volodina, G. F., I. M. Rumanova and N. V. Belov, *Kristallografiya* **6**, 919 (1961).
X163. Waldbauer, L. and D. C. McCann, *J. Chem. Phys.* **2**, 615 (1934).
X164. Wallwork, S. C. and W. E. Addison, *J. Chem. Soc.* 2925 (1965).
X165. Webb, N. C., *Acta Cryst.* **21**, 942 (1966).
X166. Weigel, D., *Bull. Soc. Chim. France* 2087 (1963).
X167. Weigel, D., B. Imelik and P. Lafitte, *Bull. Soc. Chim. France* 544 (1962).
X168. Weigel, F., V. Scherer and H. Henschel, *Radiochim. Acta* **4**, 18 (1965).
X169. Weigel, F., V. Scherer and H. Henschel, *J. Amer. Ceram. Soc.* **48**, 383 (1965).
X170. West, C. D., *J. Amer. Chem. Soc.* **54**, 2256 (1932).
X171. West, C. D., *Z. Krist.* **88**, 19 (1934).
X172. West, C. D., *Z. Krist.* **88**, 198 (1934).
X173. West, C. D., *Z. Krist.* **91**, 480 (1935).
X174. Westenbrink, H. G. K., *Proc. Acad. Sci. Amsterdam* **29**, 1374 (1926).
X175. Wilhelmi, K. A., *Naturwiss* **44**, 580 (1957).
X176. Wilhelmi, K. A., *Acta Chem. Scand.* **12**, 1965 (1958).
X177. Wilhelmi, K. A. and O. Jonsson, *Acta Chem. Scand.* **15**, 1415 (1961).
X178. Wilson, I. A., *Arkiv. Kemi. (Min. Geol.)* **15B** (1941).
X179. Wooster, W. A., *Z. Krist.* **94**, 375 (1936).
X180. Yu, S. H. and C. A. Beevers, *Z. Krist.* **95A**, 426 (1936).
X181. Zachariasen, W. H. and H. A. Plettinger, *Acta Cryst.* **14**, 229 (1961).
X182. Zachariasen, W. H., *Z. Krist.* **71**, 501 (1929).
X183. Zachariasen, W. H., *Z. Krist.* **71**, 517 (1929).
X184. Zachariasen, W. H. and F. A. Barta, *Phys. Rev.* **37**, 1626 (1931).
X185. Zachariasen, W. H. and H. E. Buckley, *Phys. Rev.* **37**, 1295 (1931).
X186. Zemann, J., *Naturwiss* **45**, 182 (1958).
X187. Zemann, J., *Acta Cryst.* **13**, 863 (1960).

Formula index

THE ARRANGEMENT OF THE FORMULA INDEX

Page references are in italics, bibliography references are in ordinary type. The index is divided into sections, each corresponding to one of the chapters 4–10, and arranged in that order.

For the references to chapter 4, which deals with systems having no discrete molecular vibrating unit, the materials of formula types AB, AB_2, AB_3, A_2B_3 are listed in alphabetical order of A (the cation). For any A, B is listed in the order F, Cl, Br, I, OH, O, S, Se, Te. For the perovskites, ABX_3, and spinels, AB_2X_4, where A, B are both cations and X an anion, the arrangement is first by alphabetical order of A and then by alphabetical order of B. Chapters 5–10 deal with the vibrations of discrete molecules and ions. A molecule or ion of formula type XY_n is always so listed; this leads to the formulation of a few common systems in an unusual manner. For instance, H_2O, H_2S are listed as OH_2, SH_2. If the system is one in which all bonds radiate from a central atom, this will usually be the atom X. Exceptions to this are the mixed-halide ions such as BrI_2^-, where the atomic arrangement is Br—I—I, and triatomic species XYZ, in which Y is the central atom. Acyclic systems X_2Y_2 are listed such that the X atoms are the central atoms; e.g. O_2H_2, O_2D_2, S_2F_2, N_2F_2.

For a vibrating unit of any formula type, molecules are listed before ions. Molecules are arranged with X alphabetically, and, for any X, Y is listed in the order D, H, T, F, Cl, Br, OH, O, S, Se, Te and alphabetically thereafter. Further substituents Z follow the same rules as Y; for example, BF_2H precedes BF_2Cl, which precedes BF_2Br. Where there are additional substituents W, V, etc., X is the central atom and the substituents are given in order of increasing atomic mass. Thus, in POFClBr, of type XYZWV, P=X, O=Y, F=Z, Cl=W, Br=V.

Ions are listed in the first place according to the total formal charge on the ion; the ion with the greatest formal positive charge is listed first. Within any such group, the arrangement is first by alphabetical order of X, then in the order of Y specified for molecules (see above) and finally by alphabetical order of the counter ion (if given) according to valency. Thus for ions of formula type XY_2, the order of groups is XY_2^{2+}, XY_2^+, XY_2^-, XY_2^{2-}. Therefore ClF_2^+ and NO_2^+ (in that order) precede ClF_2^- and NO_2^-. Within the group of ions of formula type XY_2^-, the order is BO_2^-, $BrCl_2^-$, BrI_2^-, BrO_2^-, ClF_2^-, $ClBr_2^-$, ClO_2^-, DF_2^-, HF_2^-, HCl_2^-, HBr_2^-, HI_2^-, ICl_2^-, IBr_2^-, NO_2^-. This illustrates the ordering with respect to X and Y. As an example of the ordering with respect to the counter ion, consider the compounds containing the group BO_2^-.

The first references listed are those in which no counter ion is specified, e.g., studies in alkali halide matrices. Where the cation is given, the order is Cs^+, K^+, Li^+, Na^+, Rb^+, Ca^{2+}, Mg^{2+}, Pb^{2+}, Sr^{2+}, Eu^{3+}, La^{3+}, Nd^{3+}, Sm^{3+}. At the end of the list is a reference to a compound in which the ion forms part of a complex-$Zn_4O(BO_2)_6$.

SYSTEMS WITH NO DISCRETE MOLECULAR VIBRATING UNIT

TRIATOMIC MOLECULES AND IONS

Formula type X_3

Formula type XY_2

Formula type XYZ

TETRATOMIC MOLECULES AND IONS

Formula type XY_3 and X_4

Formula index

Formula type X_2Y_2

Formula index

AQUO- AND HYDROXO-COMPLEXES: MO_6 COORDINATION POLYHEDRA: MO_6 ANIONS

AMMINE AND NITRO COMPLEXES

Formula index

Subject index

Subject index

Subject index

Subject index

MADE AND PRINTED IN GREAT BRITAIN BY
WILLIAM CLOWES & SONS LIMITED, LONDON, BECCLES AND COLCHESTER